5th International Conference on New Advances in Civil Engineering (ICNACE 2019)

IOP Conference Series: Materials Science and Engineering
Volume 800

Kyrenia, Cyprus
8 - 10 November 2019

ISBN: 978-1-7138-2799-3
ISSN: 1757-8981

Printed from e-media with permission by:

Curran Associates, Inc.
57 Morehouse Lane
Red Hook, NY 12571

CURRAN ASSOCIATES INC. proceedings .com

Some format issues inherent in the e-media version may also appear in this print version.

Institute of Physics
Dirac House, Temple Back
Bristol BS1 6BE UK

Phone: 44 1 17 929 7481
Fax: 44 1 17 920 0979

techtracking@iop.org

Additional copies of this publication are available from:

Curran Associates, Inc.
57 Morehouse Lane
Red Hook, NY 12571 USA
Phone: 845-758-0400
Fax: 845-758-2633
Email: curran@proceedings.com
Web: www.proceedings.com

TABLE OF CONTENTS

THE EFFECT OF SILICA FUME INCLUSION ON THE PROPERTIES OF RECYCLED CONCRETE AGGREGATE- CONTAINING CONCRETE 1
P Akpinar, H M E Al Attar

CHARACTERIZATION OF ASPHALT BINDERS BLENDED WITH NANOMATERIAL AND POLYMER 8
S I A Ali, A Abubaker, A B Z Alshetwi

ANALYSIS OF INCREMENT OF ROAD TRAFFIC ACCIDENTS IN LIBYA: CASE STUDY CITY OF TRIPOLI 18
S I A Ali, F A A Elturki, S N Jibrel

THE IMPORTANCE OF DESIGNER-CONTRACTOR COMMUNICATION IN GEOTECHNICAL DESIGN AS EXEMPLIFIED BY A CUT SLOPE FAILURE 28
D Anafarta, E Turk̇, S O Akbas

MEASUREMENT OF VIBRATIONS GENERATED BY HIGH SPEED RAILWAY TRAFFIC AND EVALUATION ACCORDING TO INTERNATIONAL NORMS 36
S Ates, E Celebi, O Kirtel, A C Zulfikar, F Goktepe

PHYSICO-MECHANICAL CHARACTERIZATION OF HIGH-VOLUME FLY ASH CEMENT PASTE COMPOSITES 47
E Aydin

UTILIZATION OF LARGE AMOUNTS OF INDUSTRIAL WASTES TO IMPROVE SUSTAINABILITY OF THE CONSTRUCTION SECTOR 59
E Aydin

PERFORMANCE INDICATORS IN THE CONSTRUCTION INDUSTRY: A STUDY WITH PORTUGUESE COMPANIES 69
A Barros, S Sousa, E Nunes

APPLICATION OF GENETIC ALGORITHM METHOD FOR SOIL NAILING PARAMETERS OPTIMIZATION 79
F Benayoun, D Boumezerane, S Rehab Bekkouche, L Bendada

COAL ASH USE AS A CEMENT REPLACEMENT IN CONCRETE PRODUCTION 89
S C Bostanci

MODIFICATION OF HYDRAULIC CONDUCTIVITY OF SANDY SOIL USING SEAWATER AND ALKALINE SOLUTIONS 98
Osama Dawoud

SEISMIC DESIGN OF MID-RISE REINFORCED CONCRETE STRUCTURES ACCORDING TO TEC 2007 AND TBEC 2018 106
S Derogar, I Safkan, B Odabas

INFLUENCE OF FILLERS ON STRUCTURE AND BEHAVIOUR OF GYPSUM MORTARS 117
M Doleželová, J Krejsová, L Scheinherrová, A Vimmrová

USE OF SLAB COMPOSITE WITH GALVANIZED STEEL DECK IN REINFORCED CONCRETE FRAME 125
M Eissa, M Celikag

PERFORMANCE OF ENCASED SINGLE STONE COLUMN IN LAYERED SOIL 131
R E Farah, Z Nalbantoglu

OXIDATION OF REACTIVE BLACK 5 SOLUTIONS WITH SULFATE RADICALS ACTIVATED BY IRON 138
N Gamadze, S Pirgalioglu

AN EXPERIMENTAL STUDY ON THE BEHAVIOR OF FIBER-REINFORCED CONCRETE FLEXURAL MEMBERS UNDER CYCLIC LOADING 143
H R Gatabi, M Celikag, H A Bengar

PROPERTIES OF HIGH-STRENGTH CONCRETE CONTAINING WELL GRADED RUBBER PARTICLES 153
A Habib, U Yildirim, O Eren

INTEGRATING AHP AND GIS AS A DECISION-MAKING TOOL FOR THE OPTIMAL ALLOCATION OF WIND FARM: A CASE STUDY OF SYRIA 160
M Habib, A Matouk

A FRAMEWORK FOR PAVEMENT CRACK DETECTION AND CLASSIFICATION 170
A Habib, M M Kunt

VISCOELASTIC PROPERTIES OF EVA INTERLAYER USED IN LAMINATED GLASS STRUCTURES 177
T Hána, M Eliášová, M Vokác, K V. Machalická

DESIGN AND VERIFICATION OF PROPERTIES OF HEAVY-WEIGHT CONCRETE FOR THE PRODUCTION OF WEIGHTS 186
R Hela, L Bodnarova

EVALUATION OF CARBONATION DEPTH EVOLUTION TENDENCIES OF REINFORCED CONCRETE BUILDINGS LOCATED IN COASTAL AND INLAND AREAS OF NORTH CYPRUS 192
Q Al Haj Houseen, P Akpinar

ESTIMATION OF COMPACTION CHARACTERISTICS OF SOILS USING ATTERBERG LIMITS 202
A Hussain, C Atalar

TWO DIMENSIONAL DETERMINATION OF PEAK ACCELERATION DISTRIBUTIONS DURING EARTHQUAKES 208
Y Ipek, M Akgün, Ö C Ozdag, T Gönenç, E Pamuk, H Dindar, C Atalar

A COMPARATIVE STUDY OF CRITICAL FAILURE SURFACE DETERMINATION IN SLOPE STABILITY ASSESSMENT 215
A Iravanian, A Shlash

SWELL AND COMPRESSIBILITY CHARACTERISTICS OF EXPANSIVE CLAYS IN NORTH CYPRUS 226
Anoosheh Iravanian, Amr Abdeh

SURFACE WATER RESOURCE AND EFFECT OF WEATHER PARAMETERS IN ESTIMATING THE ANNUAL RAINFALL: A CASE STUDY IN LEBANON 234
Y Kassem, H Gökçekus, J Aljamal

ACOUSTIC AND STRUCTURAL DESIGN OF A HIGHWAY NOISE BARRIER 248
S Kesten, Ö Umut, B Ayva

ADAPTION OF CONSTRUCTION MANAGEMENT ASPECTS AND BUILDING ECONOMIC PRINCIPLES TO PREVENT COST OVERRUNS IN MODERN TIMBER CONSTRUCTION 261
J Koppelhuber, M W Bok

BUILDING SERVICES IN TIMBER CONSTRUCTION – INVESTIGATION ON DIFFICULTIES AND RELIEFS DURING THE INSTALLATION PROCEDURES IN MULTI-STOREY TIMBER RESIDENTIAL BUILDINGS 271
D Koppelhuber, A Magg

MORINGA OLEIFERA SEEDS AND CHITOSAN AS ALTERNATIVES TO CONVENTIONAL COAGULATION AGENTS 278
T Kucera, L Hofmanová

MODULAR TIMBER STRUCTURES 290
D Kuda, M Petrícková

AN EVALUATION OF LEVEL OF KNOWLEDGE, BEHAVIOUR AND ATTITUDE TOWARDS ENVIRONMENTAL FOOTPRINT IN SECONDARY SCHOOL STUDENTS IN LIBYA 297
I F G Laama, D Ozdal, T Onay

SERVICE QUALITY FACTORS IN THE CONSTRUCTION SECTOR: A LITERATURE REVIEW 310
M F B Landy, S Sousa, F Romero

THE INFORMATION EXCHANGE IN THE CONSTRUCTION SECTOR IN THE AGE OF BIM: A MULTIPLE CASE STUDY FROM ITALY 319
A D Leto

BIO-PRECIPITATION OF CACO3 FOR SOIL IMPROVEMENT: A REVIEW 324
A Miftah, H Khodadadi Tirkolaei, H Bilsel

ASSESSMENT OF BEDESTAN (ST. NICHOLAS CHURCH) ON THE FIRST DECADE OF CONSERVATION WORKS 331
D Misirlisoy, N Bilsel, I Safkan

EFFECT OF FLY ASH-LIMESTONE DUST STABILIZATION ON VOLUME CHANGE AND STRENGTH OF EXPANSIVE SOIL 341
J Naji, H Bilsel, A P Balkis

SOME REFURBISHMENT PARTICULARITIES: AN OLDER BUILDING CASE STUDY 347
R A F Oliveira, J Lopes, M I Abreu

THE EFFECT OF STONE COLUMN AREA REPLACEMENT RATIO ON SEISMIC BEHAVIOUR OF FOUNDATION 360
I Safkan, S Derogar, J Anywar

PIPE TYPE AND SEISMIC PERFORMANCE IN CHRISTCHURCH, NEW ZEALAND 366
S Toprak, E Nacaroglu, M Ceylan, G Dundar, O Y Cirmiktili

Author Index

IOP Conf. Series: Materials Science and Engineering **800** (2020) 011001 doi:10.1088/1757-899X/800/1/011001

Preface

The International Conference on New Advances in Civil Engineering (ICNACE) is a conference series, organized with the aim of forming a venue for researchers and professionals to discuss the current issues, recent progresses and future directions on a wide range of cutting-edge topics in the areas of Construction Technology and Management, Geotechnical Engineering, Materials and Structural Engineering, and Water and Environmental Engineering. The first four conferences were held in Dubai/UAE (2015), Zagreb/Croatia (2016), Helsinki/Finland (2017), and in St. Petersburg/Russia (2018). The 5th one was held in Kyrenia, North Cyprus on 8-10 November 2019.

The three keynote speakers contributed to ICNACE2019 were Prof. Dr. Feyza Çinicioğlu (Turkey) on "Soft soil mechanics through teachings of the behaviour beneath embankments", Prof. Dr. Murat Saatçioğlu (Canada) on "Protection of built infrastructure against extreme loads: Bomb blasts and earthquakes" and Assoc. Prof. Dr. Rosario Montuori (Italy) on "Advances in the design of steel structure for seismic protection". The invited speakers who have delivered presentations regarding field-implementation and case studies were Asst. Prof. Dr. Eriş Uygar (N. Cyprus) and Paul A. Cowin (UK) on "Design and field experiences on the use of PVD for ground modification", and Assoc. Prof. Dr. Bertuğ Akıntuğ (N. Cyprus) on "Analysis of floods and droughts in North Cyprus".

The proceedings of ICNACE2019 are comprised of 42 papers, which include fundamental and applied research, the relationship between research and practice, and the development of well-instrumented case studies that define the quality of design procedures. Our intention was to put together a scientific programme to encourage open and constructive discussions. It is the hope of the organizers that this aim was fulfilled and the participants shared experiences and ideas on the conference themes and the related subjects during the two days of the conference.

We would like to extend our appreciations to the keynote and invited speakers, the sponsors and all who contributed to the smooth flow and friendly environment of the conference during the two days in a lovely seaside resort hotel in the eastern part of Kyrenia.

Huriye Bilsel
İsmail Safkan
Ayşe P. Balkıs

February 2020

IOP Conf. Series: Materials Science and Engineering **800** (2020) 011001 doi:10.1088/1757-899X/800/1/011001

International Advisory Committee
Rifat Bulut (USA)
David Cameron (Australia)
Feyza Çinicioğlu (Turkey)
Turan Durgunoğlu (Turkey)
Tuncer Edil (USA)
David Frost (USA)
Antonio Gens (Spain)
Emre Gürcanlı (Turkey)
Erol Güler (Turkey)
Suzanne Lacasse (Norway)
Tim Länsivaara (Finland)
Herbert Mang (Austria)
Charles Ng (Hong Kong)
Harianto Rahardjo (Singapore)
Leena Korkiala-Tanttu (Finland)
Devendra Narain Singh (India)
Alessandro Tarantino (UK)
Hywel Thomas (UK)
David Toll (UK)
Snehasis Tripathy (UK)
Erol Tutumluer (USA)
Mehmet T. Tümay (USA)

Local Advisory & Scientific Committee
Bertuğ Akıntuğ (Middle East Technical University Cyprus Campus)
Pınar Akpınar (Near East University)
Ertan Akün (Cyprus International University)
Roozbeh Amoly (Girne American University)
Cavit Atalar (Near East University)
Ertuğ Aydın (European University of Lefke)
İbrahim Bay (European University of Lefke)
Şevket Can Bostancı (European University of Lefke)
Tahir Çelik (Cyprus International University)
Tolga Çelik (Eastern Mediterranean University)
Ece Çelik (Eastern Mediterranean University)
Mürüde Çelikağ (Eastern Mediterranean University)
Ömer Damdelen (Cyprus International University)
Shahram Derogar (European University of Lefke)
Ceren İnce Derogar (Middle East University Cyprus Campus)
Özgür Eren (Eastern Mediterranean University)
Hüseyin Gökçekuş (Near East University)
Anoosheh Iravanian (Near East University)
Mehmet Metin Kunt (Eastern Mediterranean University)
Khaled Marar (Eastern Mediterranean University)
Mohammed Ali Mosaberpanah (Cyprus International University)
Şerife Öncü (Eastern Mediterranean University)
Ali Erhan Öztemir (Cyprus International University)
Rifat Reşatoğlu (Near East University)
Amin Riazi (Cyprus International University)
Ismail Safkan (European University of Lefke)
Serhan Şensoy (Eastern Mediterranean University)
Zalihe Nalbantoğlu Sezai (Eastern Mediterranean University)
Göktuğ Tenekeci (Cyprus International University)
Kemal D. Tözer (Cyprus International University)
Umut Türker (Eastern Mediterranean University)
Eriş Uygar (Eastern Mediterranean University)

Sponsor or funding acknowledgements

We would like to thank our sponsors, who has supported the 5th International Conference on New Advances in Civil Engineering (ICNACE 2019).

TUNALI İNŞAAT group of companies Since 1976

TOSUNOĞLU İNŞAAT

Onay F. Demirciler Eğitim ve Bilim Vakfı OFD

ÖZYALÇIN CONSTRUCTION

KIBRIS MÜHENDİSLİK VE MİMARLIK PROJE OFİSİ

ERTEK GROUP

IOP Conf. Series: Materials Science and Engineering **800** (2020) 011003 doi:10.1088/1757-899X/800/1/011003

Peer review statement

All papers published in this volume of *IOP Conference Series: Materials Science and Engineering* have been peer reviewed through processes administered by the proceedings Editors. Reviews were conducted by expert referees to the professional and scientific standards expected of a proceedings journal published by IOP Publishing.

IOP Conf. Series: Materials Science and Engineering **800** (2020) 012001 doi:10.1088/1757-899X/800/1/012001

The effect of silica fume inclusion on the properties of recycled concrete aggregate- containing concrete

P Akpinar[1*], H M E Al Attar[2]

[1]Near East University, Civil Engineering Department, Lefkosa, N. Cyprus, Mersin 10, Turkey
[2]ACTED N.G.O. Ankawa, Erbil, KR-Iraq

*Email:pinar.akpinar@neu.edu.tr

Abstract. Protecting the environment against the hazards of construction and demolishing (C&D) wastes is crucial. Elimination of C&D wastes by using them as recycled concrete aggregates (RCA) within the manufacture of new concretes is a promising solution for this problem. Since RCA coming from demolished structures have poor qualities, overall performance of new concrete should be monitored carefully. Without achieving a satisfactory strength performance, manufacturing concretes with RCA could not be widely accepted in construction sector. This study investigates the effects of silica fume, which is an effective pozzolanic material, on the properties of concrete mixtures made with 50%RCA as a replacement to quarried aggregates. RCA used was obtained from a 25-years old public building in Nicosia that was structurally failed and demolished. Results obtained indicated a 12.3% water demand increase for obtaining the targeted slump when SF was added in fresh concrete. The increase in compressive strength of concrete with SF in 28 days was 1.5%, where this increase was recorded as 8.1% at 90 days, yielding more than 65MPa strength. Also, addition of SF was observed to improve flexural strength of concrete samples up to a 4.3% at the age of 28 days, enabling the manufacture concrete mixtures apt for structural use.

1. Introduction

The increase in the world population and in the demand for construction activities lead to two major environmental problems. This increased demand in construction activities may lead to the depletion of natural resources for quarried aggregates, which are known to fill approximately the 75% of concrete volume. Secondly, the resultant construction and demolishing wastes that are majorly disposed as landfills, pose a threat for the nature. Construction and demolishing wastes disposed as landfills end up occupying agricultural lands with a considerable amount of non-biodegradable wastes, affecting the soil quality and making the land unsuitable for agriculture purposes. Moreover, leachate from construction and demolishing wastes are known to negatively ground water quality as well refs here.

Re-using demolished concrete wastes in the form of recycled concrete aggregates (RCA) in the manufacture of new concretes is considered as a potentially effective solution for both of the depletion of natural aggregates and landfill disposal-related problems. A recycled concrete aggregate particle is obtained by crushing the old demolished concrete elements to different sizes. Such a RCA particle consists of an original old natural aggregate with old mortar (from the demolished concrete) surrounding it. However, it is known that the porous and permeable nature of the old and damaged mortar in RCA may yield a lower quality for the new concrete that it is added to refs here. The use of RCA as a replacement of natural aggregates in concrete manufacture for sustainable development of construction activities could only become widespread if the strength and the quality of new concrete produced with RCA is within acceptable ranges. The results of the previous studies showed that by carrying out an attentive mix design and by correct selection of concrete constituent materials and their quantities, the compressive strength of the concrete mixes could be improved greatly [1, 2]. Addition of silica fume into concrete has been reported to enhance different quality aspects of concrete as well [3, 4].

The aim of this study is to evaluate the effects of silica fume addition on the fresh and hardened concrete which is made with inclusion of recycled concrete aggregates obtained from a 25 years old reinforced concrete structure that had performance failure and hence demolished. The details on the experimental procedures and the results obtained are presented in below sections.

2. Materials and Methods

CEM II/B-S 42.5 N type of cement containing approximately 25% blast-furnace slag that conforms EN 197-1 (2011) [5] has been used in this study for the manufacture of concrete mixes. A constant cement content of 310 kg/m3 was used in the mixes, in which the cement: fine aggregates: coarse aggregates contents' ratio employed was equivalent to 1: 3.01: 2.85. In the case of silica fume (SF) addition, the content of SF was set to 10 % of the cement by weight. The chemical composition of the used SF is as shown in Table 1 below.

Table 1. Chemical composition of silica fume used in this study

Chemical Composition								
Compound	SO_3	SiO_2	CaO	Na_2O	MgO	Al_2O_3	Fe_2O_3	K_2O
Content (%)	-	94.3	0.30	0.27	0.43	0.09	0.01	0.830

Slump class of S4 as described in related standards [6] was targeted for the concrete mixes and a constant (1.29% of cement weight) superplasticizer content was employed in the manufacture of all concrete mixes. The required water/cement ratio for the studied concrete mixes to have the targeted S4 slump was determined through trial batches.

The sizes of coarse aggregates used in this study were graded in two size groups as 5-12mm and 13-19mm for both natural origin and recycled aggregates, where only natural (i.e. quarried)sand having a size range of 0-4mm was used in all mixes. The use of fine recycled concrete aggregates as a replacement of sand was defined to be out of the scope of this study. The natural (quarried, virgin) aggregates used in this study were obtained from the active quarries of Besparmak (Pentadaktylos) Mountains, where the recycled concrete aggregates were acquired from the wastes of a 25-years old public school building located in Nicosia, North Cyprus. This building was demolished by local authorities after a series of investigations that yielded the result that the building was experiencing structural failure, since its compressive strength was observed to be reduced to 12MPa. Figure 1 a and b shown the source building of the recycled concrete aggregates used in this study before and after being demolished. Once the building was demolished, the concrete parts freed from reinforcing steels and other contaminations such as broken bricks and glasses, were brought to laboratory and crushed down to desired sizes in order to replace natural aggregates in concrete manufacture. Prior to their use in the manufacture of new concrete mixes, detailed experimental studies were carried out on the specific gravity, density, absorption capacity and abrasion resistance of RCA used in this study. As a result of these tests, it was observed that the density, specific gravity and the abrasion resistance of RCA was lower than the natural aggregates, where the absorption capacity of RCA was found to be higher than natural aggregates used in this stud. Information on the followed test procedures and the analysis of obtained results have been reported and discussed in detail in a previous work [7].

Figure 1. Failed public school building in N. Cyprus used as the source of RCA. (**a**)before being demolished**;** (**b**)after being demolished.

Three concrete mixes were designed in order to verify the influence of SF addition to the fresh and hardened properties of RCA-containing concrete. Mix 1 shown in Table 2 was casted for control purposes and it contained only natural aggregates (NA) and no RCA or no SF contents. Mix 2 was prepared by replacing 50 % of the coarse natural aggregates with RCA but with no SF addition; where, mix 3 was prepared similar to mix 2 with the only difference of 10% (by mass of cement) silica fume addition

Table 2. Content information for the concrete mixes used in this study

Concrete Mix No:	**Replacement of coarse-NA with coarse-RCA**	**Silica fume addition**
Mix 1	0	0
Mix 2	50%	0
Mix 3	50%	10%*

** by mass of cement.*

Besides the workability (slump) tests that are done to determine the water demand of these three mixtures in order to yield S4 class slump, a total of 27 cube samples, having 15x15x15cm^3 dimensions, were casted in order to verify the compressive strength evolution of these mixes in 7, 28 and 90 days, following the related European standard [8]. In addition to the compressive strength behavior determination, 18 unreinforced beam samples, having 15x15x60cm^3 dimensions, were also casted to study the flexural strength behavior of these mixes at the ages of 7 and 28 days, following the procedure of the related European standard [9]. The results obtained on the studies carried out on the fresh and hardened states of the employed three concrete mixes are presented in below sections.

3. Results and discussions

3.1. Effect of silica fume on workability characteristics of RCA-containing concrete

The workability characteristics of concrete mixes having a constant superplasticizer content that contain no RCA and no SF (mix 1), 50%RCA replacement (mix 2) and 50% RCA replacement together with SF addition (mix3) have been studied in order to verify their water demand to maintain an S4 slump. After testing some trial batches, the final results regarding the water demands of these three mixes that yielded 18mm slump (i.e. within S4 class) have been reported in Table 3.

Table 3. Water/cement ratio for each mix needed for acquiring targeted slump

	Mix Content Summary*	Water/Cement Ratio	Slump (cm)
Mix 1	0%RCA (100%NA)	0.53	18
Mix 2	50% RCA	0.57	18
Mix 3	50% RCA & SF	0.64	18

**Where; NA: natural aggregates; RCA: recycled concrete aggregates; SF: Silica fume*

Results obtained show that replacement of 50 % of natural aggregates with RCA yielded a 7.5% increase in w/c needed for maintaining the defined slump value. This finding is expected to be due to the water absorption of adhered old mortar in RCA with a porous nature. Similar results have also been reported in the previous studies in the related literature (ref) . When 10% (by mass of cement) silica fume was added to RCA-containing concrete mixture, an additional w/c increase of 12.3% was observed to be required in order to maintain the defined slump. Due to its fine particle character SF is known to yield a water demand increase in concrete (ref); however, addition of increased quantities of water is generally tried to be avoided in concrete mixes due to its expected decreasing effects on final strength of hardened concrete [10].

3.2. Effect of silica fume on compressive strength evolution of RCA-containing concrete

The average compressive strength values obtained for the ages of 7, 28 and 90 days for all mixes and their corresponding standard deviation values are presented in Table 4. Figure 2 illustrates the evolution of the compressive strength of studied mixes throughout 90 days. When the results shown in Table 4 and Figure 2 are considered, it can be observed that mix 1 which has no recycled concrete addition yielded the highest compressive strength values at every testing age. Replacing 50% of natural aggregates in mix1 with recycled concrete aggregates as in the case of mix 2, has been observed to yield a drop in the compressive strength performance of concrete at each age. This drop in strength (compared to mix1) is observed to be 12.9%, 23.7% and 17.2% at the ages of 7, 28 and 90 days, respectively. Poor quality of adhered old mortar in the RCA has been reported in the previous studies in the literature to cause a lower strength performance for the new concrete that it is added to [11-13].

Table 4. Average compressive strength and standard deviation values for all concrete mixes

		7 days		28 days		90 days	
	Mix Content Summary*	Strength (MPa)	Std. Dev.	Strength (MPa)	Std. Dev.	Strength (MPa)	Std. Dev.
Mix 1	0%RCA (100%NA)	52.20	1.68	64.77	1.50	73.10	4.53
Mix 2	50% RCA	45.47	1.72	49.40	4.73	60.53	2.15
Mix 3	50% RCA &SF	48.97	0.64	50.13	0.31	65.43	0.85

** NA: natural aggregates; RCA: recycled concrete aggregates; SF: Silica fume*

The results shown in Table 4 and Figure 2 also indicate that the addition of silica fume to RCA-containing concrete, as in the case of mix3, leads to higher strength values of compressive strength compared to the strength values yielded by mix 2 at every testing age. The increase in strength of RCA-containing concrete with the addition of SF compared to mix 2 is recorded to be 7.7%, 1.5% and 8.1% for the ages of 7, 28 and 90 days respectively.

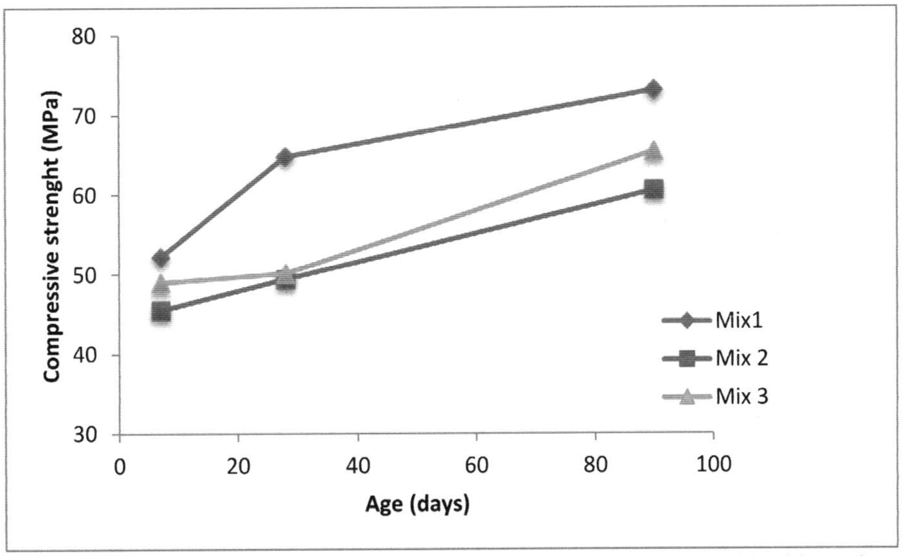

Figure 2. Compressive strength evolution of concrete mixtures used in this study

It was observed that the increase in compressive strength acquired with 10% silica fume addition was not sufficient to compensate the decrease in strength yielded by replacement of natural aggregates by recycled concrete aggregates. Further experimental studies should be carried out on concrete mixes with increased silica fume contents in order to provide more insights on the effects of increased percentages of SF on the RCA-containing mixes.

3.3. Effect of silica fume on flexural strength evolution of RCA-containing concrete

Flexural strength evolution of employed three mixes was determined by testing three unreinforced beam samples of each mix at 7 and 28 days. The average flexural strength values, as well as the standard deviation values for each mix at every age are reported in Table 5. Figure 3 illustrates the evolution of flexural strength of the employed mixes within 28 days.

Table 5. Average flexural strength and standard deviation values for concrete mixes used

		7-days		28-days	
	Mix Content Summary*	**Strength (MPa)**	**Std. Dev.**	**Strength (MPa)**	**Std. Dev.**
Mix 1	0%RCA (100%NA)	7.82	0.47	9.12	0.59
Mix 3	50%RCA	4.91	0.16	6.34	0.19
Mix 7	50%RCA &SF	5.10	0.38	6.61	0.44

* *NA: natural aggregates; RCA: recycled concrete aggregates; SF: Silica fume*

The results presented in Table 5 and Figure 3 show that flexural strength decrease of 37.2% and 30.5% were observed at the ages of 7 and 28 days respectively, when 50% of the natural aggregates were replaced with recycled concrete aggregates (i.e. the case of mix 2). Other researchers also report parallel findings in their works regarding the reduced flexural strength behavior of concretes made with RCA inclusion [14].

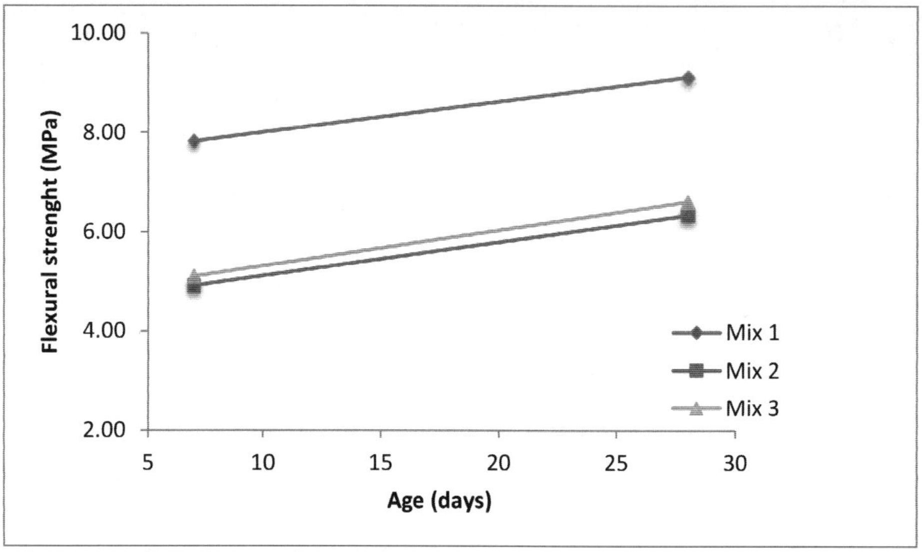

Figure 3. Flexural strength evolution of concrete mixtures used in this study

In the case of mix 3, the addition of SF was observed only a slight increase in the flexural strength; that was recorded as 3.9% and 4.3% at the ages of 7 and 28 days, with the comparison of values obtained by mix 2 and mix 3. The increase in the flexural strength of RCA-containing concrete with the addition of SF was observed to be of lower percentages, compared to the increase in compressive strength with SF addition. These obtained results indicate that the effect of SF addition to the flexural strength of RCA-containing concrete is less than its effects on compressive strength performance.

The cases of increased SF contents addition with the use of varying % RCA replacements in concrete mixtures should be further investigated with detailed experimental studies, in order to provide a broader view on the influence of silica fume on the performance of RCA-containing concretes in hardened state.

4. Conclusive remarks

Recycled concrete aggregates obtained from a 25-years old structurally failed reinforced concrete building were used to replace 50% of the natural aggregates in the manufacture of new concrete mixes. As it is known that the use RCA reduces performance of new concrete manufactures, 10% silica fume was added to the concrete mix and the effect of silica fume on the fresh and hardened characteristics of the RCA-containing concrete was studied systematically. The conclusive remarks acquired from this experimental study are as the following:

- The addition of silica fume to RCA-containing concrete caused a water demand increase of 12.3% in order to maintain the defined concrete slump value. This additional water included in concrete is expected to cause potentially adverse effects on the ultimate strength of concrete to an extent that should be further investigated.
- The addition of silica fume to RCA-containing concrete was observed to positively affect the compressive strength. An increase of 8.1% in the compressive strength of RCA-containing concrete was observed at the age of 90 days when 10% SF was added to the mix. However, this quantity was observed to be not sufficient to compensate the decrease in strength experienced due to 50% RCA replacement in concrete.
- Including 10% SF to RCA-containing concrete has also observed to be also yielding a slight improvement in the flexural strength of concrete. The percentage increase recorded for flexural strength with the addition of SF is lower than the percentage increase recorded for compressive strength of RCA-containing concrete.

- Further experimental studies should be carried out with varying SF quantities added to concrete having different RCA contents, in order to provide more insight on the resultant influence of SF addition in a more detailed way for both fresh and hardened concrete.

Acknowledgments
Authors would like to acknowledge Tüfekçi Ready-mix Concrete Company in N. Cyprus and to Samir Jabal for their technical support and sponsorship throughout the study.

References

[1] Akpinar P and Khashman A 2017 Intelligent classification system for concrete compressive strength *Procedia Computer Science* **120** 712-8 https://doi.org/10.1016/j.procs.2017.11.300
[2] Khashman A and Akpinar P 2017 Non-destructive prediction of concrete compressive strength using neural networks *Procedia Computer Science* **108** 2358-62
[3] Pedro D, de Brito J and Evangelista L 2017 Evaluation of high-performance concrete with recycled aggregates: Use of densified silica fume as cement replacement *Construction and Building Materials* **147** 803–14
[4] Pedro D, de Brito J and Evangelista L 2017 Mechanical characterization of high-performance concrete prepared with recycled aggregates and silica fume from precast industry. *Journal of Cleaner Production* **164** 939–49.
[5] EN 197-1 2011 Cement - Part 1: Composition, specifications and conformity criteria for common cements. *European Standards*.
[6] EN 12350-2 2009 Testing fresh concrete - Part 2: Slump-test *European Standards*
[7] Al Attar H and Akpinar P Performance of Environmentally-friendly Concrete Made Using Increased Quantities of Construction and Demolishing Wastes, *2nd International Conference on Environment: Survival and Sustainability- October 2019 (in press)*
[8] EN 12390-3 2000 Testing Hardened Concrete- Part 3: Compressive strength of test specimens *European Standards*.
[9] EN 12390-5 2009 Testing hardened concrete - Part 5: Flexural strength of test specimens. *European Standards*
[10] Neville AM 2005 *Properties of Concrete* (New York: John Wiley & Sons)
[11] Abdel-Hay A S 2017 Properties of recycled concrete aggregate under different curing conditions. *HBRC Journal* **13(3)** 271–6 https://doi.org/10.1016/j.hbrcj.2015.07.001
[12] Bulatović V, Melešev M, Radeka M, Radonjanin V and Lukić I 2017 Evaluation of sulfate resistance of concrete with recycled and natural aggregates *Construction and Building Materials* **152** 614–31 https://doi.org/10.1016/j.conbuildmat.2017.06.161
[13] González-Fonteboa B, Seara-Paz S, De Brito J, González-Taboada I, Martínez-Abella F and Vasco-Silva R 2018 Recycled concrete with coarse recycled aggregate. An overview and analysis. *Materiales de Construcción* **68(330)** 151 https://doi.org/10.3989/mc.2018.13317
[14] Lau T L, Elleithy W, Choong W K, Tze T Y, Lee C M and Modhwadia A L 2014 Effects of recycled aggregates on concrete strengths *Materials Research Innovations* **18(6)** 372- 4 https://doi.org/10.1179/1432891714Z.000000000982

IOP Conf. Series: Materials Science and Engineering **800** (2020) 012002 doi:10.1088/1757-899X/800/1/012002

Characterization of asphalt binders blended with nanomaterial and polymer

S I A Ali [1*], A Abubaker[2], A B Z Alshetwi[3]

[1]Civil Engineering Department, Near East University, Near East Boulevard, 99138 Nicosia, North Cyprus, Mersin 10, Turkey.
[2]Faculty of Graduate Studies, Cyprus International University, Nicosia, 99138, North Cyprus [3]Faculty Higher Institute of Science and Technology, Al-Zahra, Libya

*E-mail: shabarofking10@gmail.com

Abstract. Modification of bitumen binders with polymers and nanomaterials has successes in mitigating various major causes of pavement failures. The aluminum oxide nanoparticles (Al_2O_3) and Acrylate-Styrene-Acrylonhrtilrei polymer (ASA) were poured on the base bitumen binder 60/70 penetration grade with percentages of 3, 5 and 7% by the weight of bitumen binder. The outcome shows that the inclusion of polymer modified binders (ASMBs) and nanoparticles modified binders (ALMBs) matter has a great deal of impact on the rheological behavior of the bitumen binder at intense temperatures. However, the complex shear modulus (G*) improves the mix of the two modifiers. The increment at 75 °C was 63.70% for five percent ASBMs polymer, while for ALMBs, 71.12% for five percent. Although, the modified bitumen binders show high resistivity to rutting in intense temperatures, upon increment to 80% and 59% for five percent concentration of ASMBs as well as ALMBs chronologically. The refinement of bitumen binders gives an increase in the property of the binder which is the creep recovery for approximately about 69.23% and 62.53%. The use of ASMBs and ALMBs as discovered in this study shows that both modifiers above can mitigate the issues of the bitumen binder at intense temperatures, and five percent is regarded as the maximum constituent of the two modifiers.

1. Introduction

Bitumen is a naturally occurring dark brown to black cementation material, in other cases, it can be obtained by processing petroleum and petroleum products [1]. One of the significant uses of bitumen binder is its vast application in the construction of highways and roads network [2]. A typical highway pavement usually contains aggregates coated with a blend of bitumen binder, and the performance of pavements is associated with the rheological properties of bitumen binder [3]. Pavement might show distress when exposed to low and elevated temperatures. At low temperatures, fatigue cracks occur and lead to weakness in bitumen layers, while at elevated temperatures, permanent deformation (rutting) occurs and create channels on the roads because of vehicles [4, 5]. Therefore, to mitigate those distress, the modification of bitumen binder and mixture is essential to enhance the performance and reduce pavement distress [6, 7]. Various kinds of bitumen binder modifiers were used to improve its performance including; polymers, rubbers, fibers, and nanomaterials [8-11]. It has been identified that polymers show better bitumen performance when the used as modifiers, the behavioral pattern and content of the polymer have an essential part to play in improving the characteristics of the bitumen binder. The most common polymers used worldwide as modifiers of bitumen binder are 75% elastomers and 15% plastomers, the remaining percentage which is 10, and that constitutes rubber or other refinements like; steel slags and waste glass [12, 13]. One of the major weaknesses of polymers modified bitumen binder is the segregation if stored at increased temperatures. The probability of separation under inert conditions is a major challenge for the actual usage blends [14]. The nanomaterials were used as additives to the mixture and to significantly enhance the stability of modified binders at high temperatures to decrease the phase segregation and improve the stability of bitumen binder and polymers [15]. Presently nanotechnology is rapidly integrated into the application of bitumen binder, with several kinds of nanomaterials being used as modifiers of bitumen binder and mixture. The usage of nanomaterials has been observed to be able to enhance the performance of bitumen binder and mixture. Adding a small amount of nanoclay based bitumen binder has a significant effect lead to increase its stiffens, reducing the strain failure rate of base bitumen binder and decrease the moisture susceptibility of the mixture [16]. Furthermore, using nanomaterials as a modifier of the bitumen binder results in an increase in the performance grade and enhancement in the elastic properties compared with the base bitumen binder [17]. Also, the viscosity rate was slightly decreased with the additional increase of nanomaterials in bitumen binder, as well as it works to reduce and delay the aging process. This study characterizes and investigates the influences of nano-aluminum oxide particles (Al_2O_3) and the Acrylate Styrene Acrylonhrtilrei (ASA) on the rheological properties of bitumen binder. Moreover, to establish a comparative study regarding the best performance in bitumen.

2. Literature review

Several highway engineers have recognized the failure of asphalt layers, which result in a decrease in the service life of the road surface. Because of this, most of the highway engineers used the additive to enhance the performance of asphalt and reduce the maintenance cost. Many researchers have been studying the influences of polymers on modified bitumen binder and bitumen mixture [11]. Bernier has evaluated the rutting susceptibility of polymer modified bitumen mixtures containing recycled pavements, to compare the rutting resistance of bitumen binders and bitumen mixtures containing different polymer modifiers and different recycle bitumen pavement sources. The results show a significantly changed with the increment of the polymer. The additional increase of polymer significantly leads to reduce rutting susceptibility [18]. Glaoui determined the development of complex shear modulus, creep recovery and phase angle of polymer modified binder. The results showed that the complex modulus was considerably improved, and the phase angle was reduced at low temperatures, this behavior is not preferable for fatigue cracking. Moreover, the creep recovery results indicated that the polymer-modified binder after the aging condition has negative effects on the resistance of permanent deformation [19]. The engineering characteristic of polymer modified binders containing wax additives was evaluated by Mazumder et al. through Superpave binder tests. The investigation shows that polymers can improve rutting resistance and reduce the effects of fatigue and low-temperature cracking [20]. On the other hand, several types of research were conducted on nanomaterials modified bitumen binder. Shiman investigated the impact nanocomposites have on the properties of bitumen binders under extreme temperatures. A base bitumen was PG 58, and the nanomaterial was Single-Wall Nanotube (SWNT). The results showed that the addition of 0.3% wt. of the nanocomposite was able to increase the PG of the modified bitumen. Moreover, it was found that the elastic component to the complex modulus (G*) of modified bitumen binders is higher than the base bitumen binder and also the modified binders have lower thermal sensitivity and aging effects compared to base bitumen binder (Shiman et al., 2011). Golestani evaluates the performance of modified bitumen binders with nanoclay and the results showed that nanoclay is able to enhance the physical and rheological properties of the modified bitumen binders [21]. Khodary used Calcium hydroxide ($Ca\ (OH)_2$) nanoparticles as a bitumen modifier to evaluate the physical and mechanical properties of the modified bitumen binder and mixtures. The results show that the influence of $Ca\ (OH)_2$ nanoparticles on the mechanical properties such as bitumen mixtures was significant. Also, the addition of 5% $Ca\ (OH)_2$ nanoparticles decrease the penetration of modified bitumen up to 30% and 7% respectively while Softening increased by nearly 45%. Also, it was found that the $Ca\ (OH)_2$ nanoparticles modified bitumen mixtures specimens have the highest value of indirect tensile strength, while base bitumen has the lowest value. Finally, it was recommended that the use of Ca $(OH)_2$ as a modifier of bitumen binder and mixture is preferable to in hot climate and for heavy traffic load area [22]. Estimating the moistre damage of the bitumen mixture modified with nano zinc oxide (ZnO) was conducted by Hamedi et al. [23]. The results showed that the modification of bitumen using nano-ZnO leads to reduce bitumen binder's acidity and can improve adhesion between bitumen and aggregate compared to base bitumen binder. It was found that by adding nano-ZnO to bitumen binder, the unconditioned and conditioned indirect tensile strength (ITS) values were significantly enhanced [23].

3. Experimental design

3.1Materials

Three primary materials used in the study includes local Malaysian bitumen binder 60/70 penetration grade, Acrylate Styrene Acrylonhrtilrei polymer as well as the oxide of nano-aluminum oxide were obtained from China. The physical properties of the base bitumen binder, polymer, and nano-aluminum particles are shown in Table 1.

Table 1. The physical properties of the asphalt cement and nanoparticles

Material	Properties	Test Method	Value
	Specific Gravity	ASTM D70	1.03
Bitumen 60/70	Penetration @ 25 ˚C	ASTM D5	70
	Softening point (˚C)	ASTM D36	46.0
	Viscosity @ 135 ˚C (Pa.s)	ASTM D4402	0.5
	Ductility (cm) @ 25 ˚C,	ASTM D113	≥125
ASA	Specific Gravity	-	0.30
	Size mm	-	2
Nano Al_2O_3	Size nm	-	13
	Density	g/cm3	3.5-3.9

3.2Preparation of modified bitumen binders

The melt blending process was used to produce modified bitumen binders, this is carried out by putting these percentages chronologically, i.e., 3%, 5% and 7 % the two modifiers which are ASA polymer and Al_2O_3 nanoparticles respectively by weight of the base bitumen binder. Heat is applied to the bitumen until there is a transformation in the state of the particle, i.e., from its solid state to fluid. After which it was mixed using a Silverson high shear mixer at a temperature of 170˚C (±1 ˚C) within the speed limit of a 5000 rpm for an hour and thirty minutes to create homogenous mixes. The mixing temperature was consistency 170 (±1 ˚C), while the speed was reduced to 1500 rpm and the Al_2O_3 and ASA were added gradually to the blend, to make sure that the particles uniformly dispersed then the speed was increased to 5000 rpm. Dispersion is essential because nanoparticles have solid surface-active forces and it is difficult for them to disperse at high concentration. The samples were coded in the Figures and Tables to be for base bitumen binder 0% BB, for ASA polymer modified bitumen binders N% ASMBS, while for Al_2O_3 nanoparticles is N% ALMBs, where N is the concentration of the modifier.

3.3 Viscosity Test procedure

The refined bitumen binders, as well as the base viscosity, are to be to determine using a Brookfield rotational viscometer. The number 27 Spindle was employed at a given speed of 20 rpm as given by the Superpave test parameters. The average of three readings for each test temperature was recorded and identified to be the result. The experimental study is carried out under these temperatures, 135°C up to 200°C for all the samples in use.

3.4 Storage Stability test procedure
The storage stability of the refined bitumen binders alongside the base was measured according to the following procedure. A 3 cm width aluminum foil tube that is 16cm high is prepared, base and modified samples were poured into it. The aluminum foil tubes were placed in an oven at a temperature of 163 ± 5°C for exactly two days. This is later cooled at room temperature and shared into three equal parts in a horizontal form. The specimens removed from the up and downward parts were used to examine the storage stability of both elements (modifiers) by ascertaining the softening points of the sections. In a scenario where there is a difference of < 2.5°C on both the up and bottom areas, these samples were seen as those with great high-temperature storage stability. And on the other hand, where the softening points have a difference of more than 2.5 °C, samples of those refined bitumen binders were noted as the unstable ones.

4. Experimental design

4.1 High-temperature viscosity
Further increase in concentration and temperature on the viscosity of ASMBs and ASMBs refined bitumen binders have certain effects which are presented in Figure 1. It is visibly clear as shown in this figure that the viscosity of modified bitumen binders, irrespective of concentration, reduces as the temperature increases, the same pattern was observed in the base bitumen binder. A great deal of the bitumen binder viscosity can be enhanced by increasing the concentration level of both polymer and nanoparticles, thus implying that the bitumen thickness and better coating of aggregates in the bitumen mixture will be increased as well. Also, the bitumen refined with ALMBs is of much greater value regarding viscosity when compared to those refined with ASMBs. Moreover, from Figure 2 the mixing and compaction temperatures range were (157-145) for 0% BB, the ranges for ALSMBs (3, 5 and 7%) were around (148 and 160, 153 and 168, 165 and 167) respectively. While the mixing and compaction temperatures for ALMBs (3, 5 and 7%) were nearly (151 and 163, 158 and 173, 163 and 179) respectively. This indicates that modifiers type and amount have a significant role in temperature and viscosity.

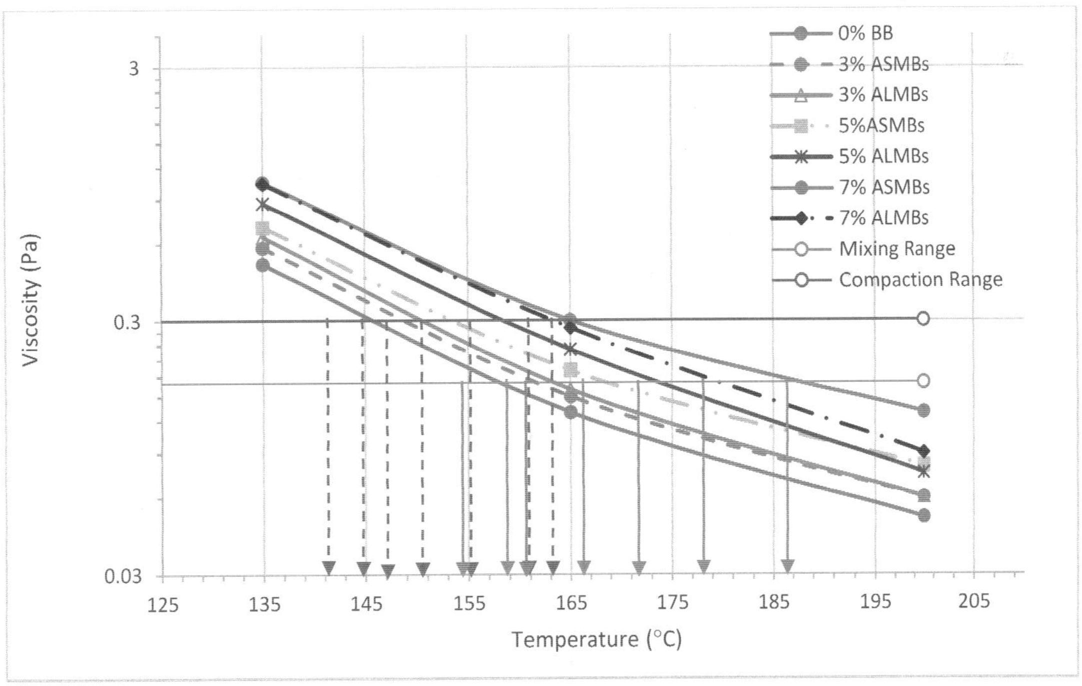

Figure 1. The viscosity of Base Bitumen and ASMBs and ALMBs.

4.2 High-temperature storage stability

Regarding the difference in density between the bitumen binder as well as the polymer, it is not able to state that phase separation is possible in the refined bitumen binder at intense temperatures during the storage process. Bearing in mind that the minute size of the nanoparticles offers the ability to possess high workability and also compatibility. Although it is possible for an agglomeration to occur in ALMBs refined bitumen binders. The base bitumen binder ASMBs and ALMBs modified bitumen binder results for their storage stability are shown in Figure 2. As illustrated in this figure, the difference in the softening points within the base bitumen binder and the modifiers is at about 12 °C. Note that the enhancement of the softening point of refined bitumen binders implies that the bitumen is getting harder than the base bitumen binder. Also, the difference between the top and down parts of the ASA polymer modified bitumen binders for 3, and 5% were are less than 2.5 °C, which means the modifier (ASA polymer) had good stability at elevated temperatures. Meanwhile, 7% ASMBs the variances surpassed the acceptable range, which shows the presence of phase separation between the base bitumen binders, there is an increment in the ASMBs when polymer percentage is increased. When bitumen is modified using ALMBs, it is an indication that all refined or modified bitumen samples are of high storage stability at really high temperatures, the dissimilarity between the upper and lower parts were < 2.5 °C. It is seen that even though 7% ALMBs are considered to be stable, the variation in the upper and lower parts became higher when compared to 5% ALMBs, this implies that agglomeration occurs among the nanoparticles which thus increases their content in bitumen binder. In a nutshell, modified bitumen binders with ALMBs possess better storage stability than that of the samples modified using ALMBs.

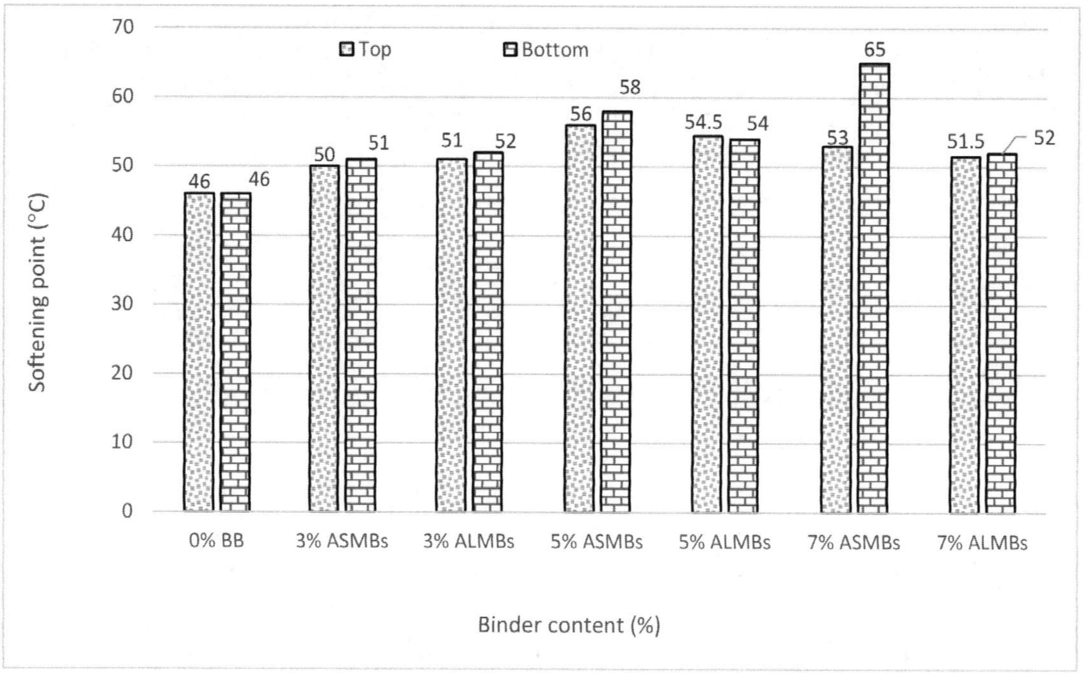

Figure 2. Storage stability of base, ASMBs and ALMBs modified bitumen binders.

4.3 Isochronal plot

This can be described as a plot of certain visco-elastic variables like complex shear modulus or phase angle, which contrasted with the temperature at a constant frequency or loading time. The isochronal scheme is used to present viscoelastic data over a range of temperatures at a specified frequency. Thus, the isochronal plots of the temperature (°C) versus complex modulus (G*) and phase angle (δ) at the frequency of 0.159 Hz (1 rad/s) are shown in Figure 3. It indicates that the increase in G* was consistently detected for modified bitumen binders when contrasted to that of the base bitumen binder and also the modified bitumen is recorded to have the greatest performance when 5% ASMBs sample for polymer and 5% ALMBs was added. Also, the phase angle was a decline for all modified binder responding with an increase in G* value. Higher G* and lower δ present the best viscoelastic properties of bitumen binder to resist the rutting and fatigue distress. Moreover, 7% concentration for both ASMBs and ALMBs indicates a small reduction in the value of G* due to phase segregation and agglomeration as detected by FE-SEM, and interfacial interactions respectively.

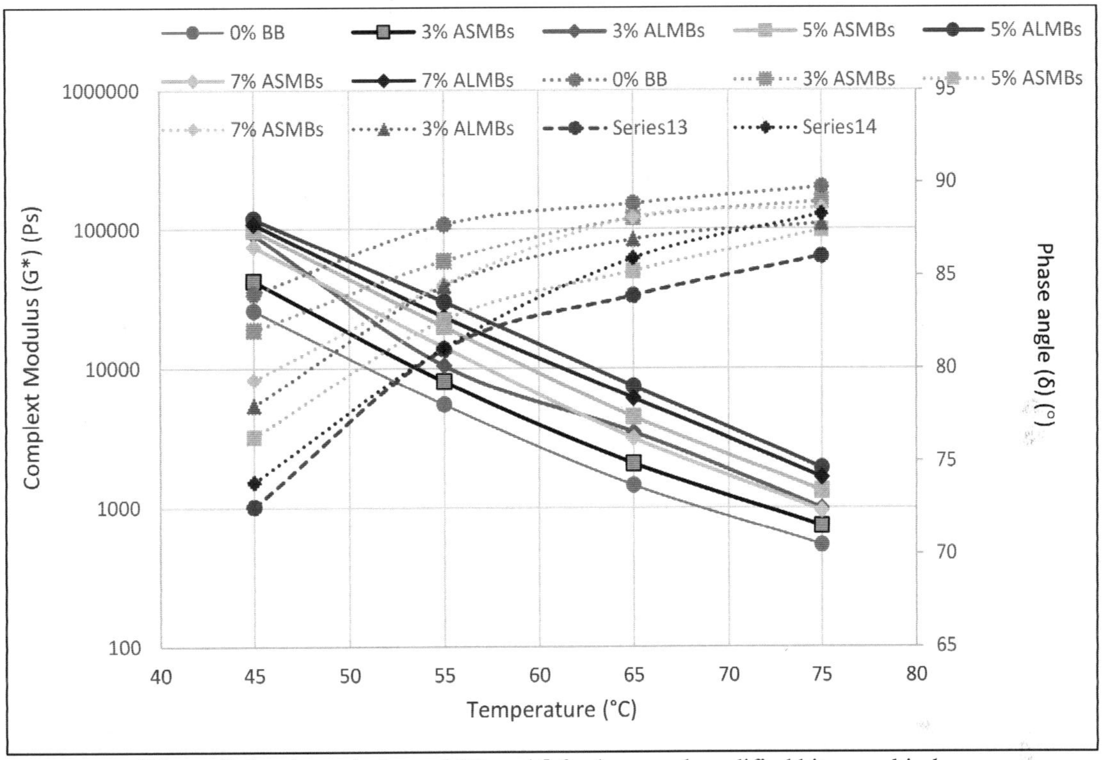

Figure 3. Isochronal plots of G* and δ for base and modified bitumen binders.

4.4 Influence of temperature on rutting

Irreversible or permanent deformation (rutting) of the bitumen binder. The description of Superpave recognizes the least number of 1.0 kPa for the G*/sin δ of base bitumen binder at performance grade temperature. Shenoy suggested the term |G*|/(1-(1/tanδ sinδ)) as an improvement and an alternative to the Superpave specification parameter (G*/sinδ) [24]. Figure 4 shows the comparison of the rutting parameter of base bitumen binder, polymer, and nanoparticles modified bitumen binders at 65 and 75 °C using Shenoy and Superpave specification parameters. It can be realized that the addition of both modifiers can greatly improve the resilience of bitumen to permanent deformation. The base bitumen binder takes the lowest of G*/sin δ value and 5% ALMBs is seen to have the greatest value of G*/sin δ in all the samples that were tested. It is seen from the careful examination that the resistance of bitumen binder to rutting parameter increased up to 5% with increment in the concentration of both modifiers. Various behavior is displayed at 7% when the G*/sin δ value declined slightly as the compatibility of polymer modified bitumen binders reduces due to phase segregation and agglomeration which causes softness in modified bitumen binders. The evaluation of both rutting parameters; Superpave specification (SPS) and Aroon Shenoy parameter (ASP) shows that there are non-huge differences in the values as noted in Fig. 6, which indicates that both methods are valid to evaluate the rutting parameter.

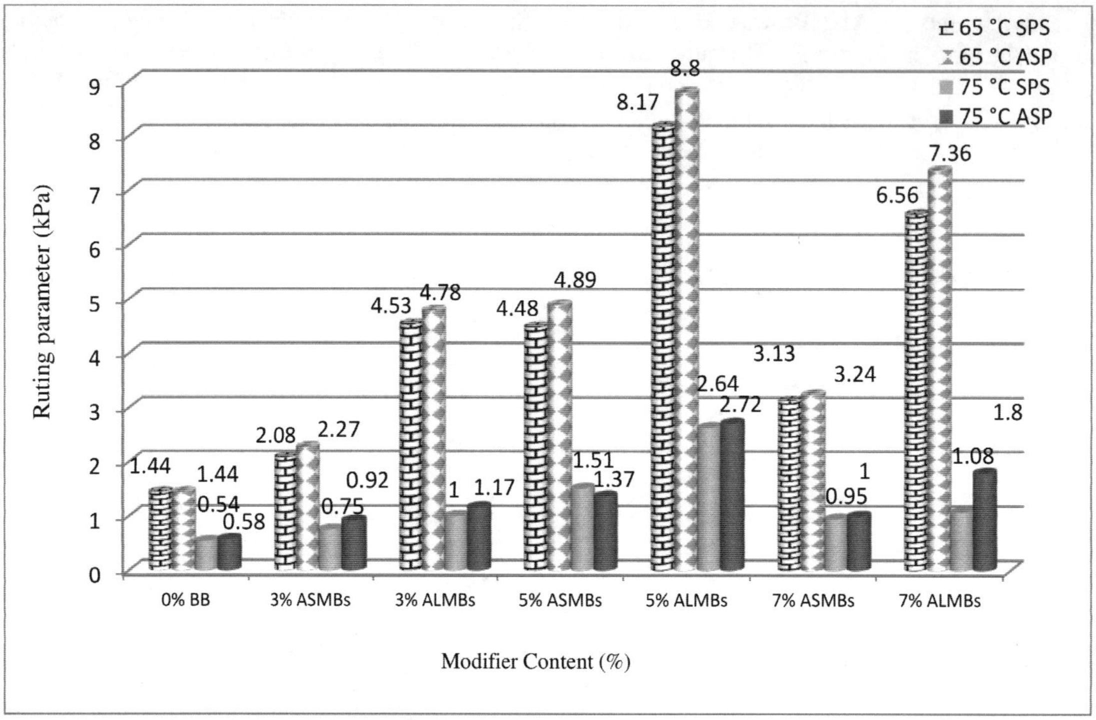

Figure 4. Comparison between SPS and ASP rutting parameters of base and modified bitumen binders at 65°C.

4.5 Loss and storage modulus of base and modified bitumen binders

The dissipated energy during the loading cycle is defined as the loss modulus (G”), like the energy loosed in the heating of bitumen. It is an amount of vibrational energy that has been transformed during vibration, which cannot be recovered, whereas the rigidity of viscoelastic materials and it proportional to the energy stored during a loading cycle represents the storage modulus (G′). A low tan δ value indicates one that is more elastic. Since G' measures the binder elasticity, while a high value is indicative of a material that has a high non-elastic strain component, the binders at high temperatures should have a high value of storage modulus (G’) to have good resistance to deformation. To avoid the flow of the binder at elevated temperatures high elasticity is required. The elastic (storage) and viscous (loss) modulus values of base and modified binders tested at temperatures of 55, 65 and 75°C are shown in Figures 5 and 6. Similar to previous results, the addition of both modifiers resulted in an increase in elastic modulus and viscous modulus compared to base bitumen binder. The highest improvement for both visco and elastic properties was observed with a concentration of 5% ALMBs followed by 5% ASMBs, wears as expected the base bitumen binder has the lowest viscoelastic values among all the blends.

5. Conclusions

This study was carried out to assess the characteristics of bitumen binder refined with ASMBs and ALMBs at elevated temperatures, conducting storage stability, a viscosity the MSRC test and frequency sweep. The following conclusions deduced based on the results:

- In respect to viscosity, the values of modified bitumen binders increased with an increment in the concentration of the modifier, each of the viscosity values was within the limit of the Superpave specifications (less than 3 Pa at 135 °C).
- It is possible to store the modified bitumen binders at high temperatures, particularly for the samples refined with ALMBs, while the ASMBs can store up to 5% concentration of polymer.

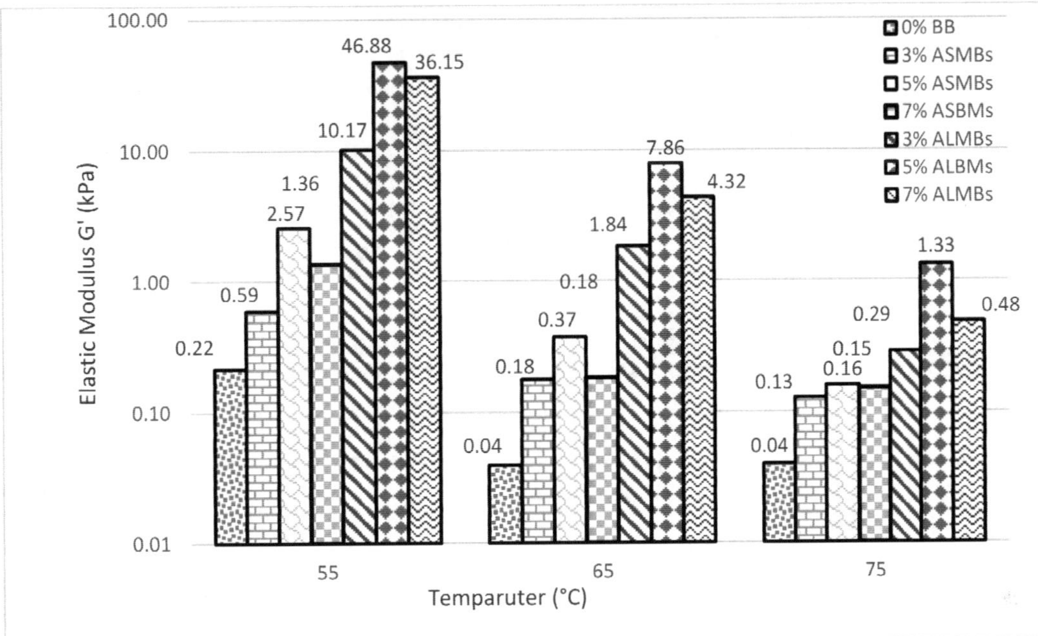

Figure 5. Elastic Modulus of Bas and Modified Bitumen Binders

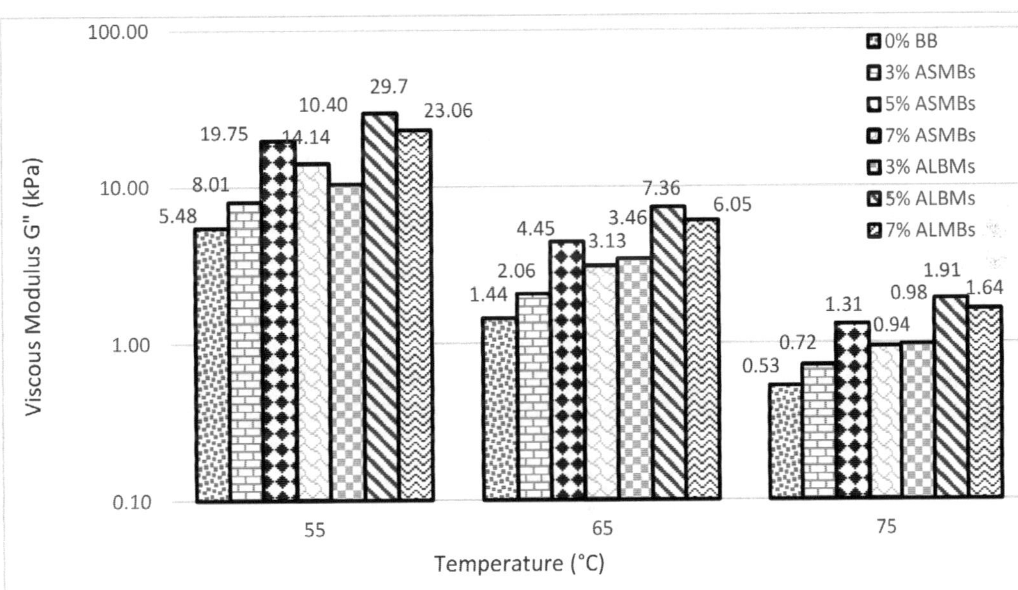

Figure 6. Viscous Modulus for the base and Modified Bitumen Binders

- Complex modulus (G*) and rutting parameters (G*/sin δ) values of base bitumen binders were the lowest compared with modified bitumen binders. Furthermore, the 5% ALMBs have the greatest value within all the modified bitumen binders which imply the better performance of modified bitumen binder concerning rutting resistance. It is also important to note that the investigation of rutting parameters shows that there are no many differences between Aroon Shenoy parameter (ASP) and Superpave specification (SPS) parameters and both of them can represent the rutting behavior of the base and modified bitumen binder.

- In a nutshell general, the refinement or modification of bitumen binder using ASMBs and ALMBs has a great effect on the characteristics and properties of bitumen binder. Furthermore, the samples modified with nanoparticles indicate better performance than those modified with the polymer. Lastly, 5% can be regarded as the ideal content for both modifiers.

Acknowledgments

The authors would like to acknowledge the Civil Engineering department at the national university of Malaysia, for providing the necessary equipment to finish this study.

References

[1] Asphalt-Institute, Aspahlt Handbook 7th edition. 7th ed. 2007, USA: Asphalt Institute. p.788.
[2] Chen, J.S., M.C. Liao, and H.-H. Tsai, Evaluation, and optimization of the engineering properties of polymer-modified asphalt. Practical Failure Analysis, 2002. 2(3): p. 75-83.
[3] Paliukaitė, M., A. Vaitkus, and A. Zofka, Influence Of Bitumen Chemical Composition And Ageing On Pavement Performance. Baltic Journal of Road & Bridge Engineering, 2015. 10(1).
[4] Gajewski, M., D. Sybilski, and W. Bańkowski, The influence of binder rheological properties on asphalt mixture permanent deformation. The Baltic Journal of Road and Bridge Engineering, zaakceptowany do druku, 2015.
[5] Zheng, C., Li, G., Xu, Y., Wang, D. and Lv, D., 2018. Analysis of the Effects of Ageing on the Cohesive Strength of Polymer-Modified Bitumen at Low Temperatures. The Baltic Journal of Road and Bridge Engineering, 13(2), pp.156-164.
[6] Iwański, M. and G. Mazurek, Rheological Properties Of The Bituminous Binder Extracted From Sma Pavement With Hydrated Lime. Baltic Journal of Road & Bridge Engineering, 2016. 11(2).
[7] Fang, C., Yu, R., Liu, S. and Li, Y., 2013. Nanomaterials applied in asphalt modification: a review. Journal of Materials Science & Technology, 29(7), pp.589-594.
[8] Yao, H., You, Z., Li, L., Shi, X., Goh, S.W., Mills-Beale, J. and Wingard, D., 2012. Performance of asphalt binder blended with non-modified and polymer-modified nanoclay. Construction and Building Materials, 35, pp.159-170.
[9] Zare-Shahabadi, A., A. Shokuhfar, and S. Ebrahimi-Nejad, Preparation and rheological characterization of asphalt binders reinforced with layered silicate nanoparticles. Construction and Building Materials, 2010. 24(7): p. 1239-1244.
[10] Ghuzlan, K., G. Al-Khateeb, and A.A. Damrah, Using oil shale ash waste as a modifier for asphalt binders. Journal of Material Cycles and Waste Management, 2013. 15(4): p. 522-529.
[11] Al-Khateeb, G.G. and K.Z. Ramadan, Investigation of the effect of rubber on rheological properties of asphalt binders using Superpave DSR. KSCE Journal of Civil Engineering, 2015. 19(1): p. 127-135.
[12] Gonzalez, O., Munoz, M.E., Santamarıa, A., Garcıa-Morales, M., Navarro, F.J. and Partal, P., 2004. Rheology and stability of bitumen/EVA blends. European Polymer Journal, 40(10), pp.2365-2372.
[13] Habib, N.Z., Kamaruddin, I., Napiah, M. and Isa, M.T., 2011. Effect of thermoplastic copolymers on microstructure and viscoelastic behaviour of bitumen. In Proceeding of Malaysian Universities Transportation Research Forum and Conferences 2010.
[14] Masson, J.F., Collins, P., Robertson, G., Woods, J.R. and Margeson, J., 2003. Thermodynamics, phase diagrams, and stability of bitumen− polymer blends. Energy & Fuels, 17(3), pp.714-724.
[15] Alhamali, D.I., Wu, J., Liu, Q., Hassan, N.A., Yusoff, N.I.M. and Ali, S.I.A., 2016. Physical and rheological characteristics of polymer modified bitumen with nanosilica particles. Arabian Journal for Science and Engineering, 41(4), pp.1521-1530.
[16] Yao, H. and Z. You, Effectiveness of micro-and nanomaterials in asphalt mixtures through dynamic modulus and rutting tests. Journal of Nanomaterials, 2016. 2016: p. 10.

[17] Shiman, L., Shiman, A., Spitsyna, N. and Lobach, A., 2011. Effects of nanocomposites on the high temperature rheological properties of a PG58 asphalt-binder. In Road Materials and New Innovations in Pavement Engineering, pp. 229-237.
[18] Bernier, A., A. Zofka, and I. Yut, Laboratory evaluation of rutting susceptibility of polymer-modified asphalt mixtures containing recycled pavements. Construction and Building Materials, 2012. 31: p. 58-66.
[19] Glaoui, B., Merbouh, M.H., van de Ven, M., Chailleux, E. and Youcefi, A., 2013. How Thermal Fatigue Cycles Change the Rheological Behavior of Polymer Modified Bitumen?. Energy Procedia, 36, pp.844-851.
[20] Mazumder, M., H. Kim, and S.-J. Lee, Performance properties of polymer modified asphalt binders containing wax additives. International Journal of Pavement Research and Technology, 2016. 9(2): p. 128-139.
[21] Golestani, B., F.M. Nejad, and S.S. Galooyak, Performance evaluation of linear and nonlinear nanocomposite modified asphalts. Construction and Building Materials, 2012. 35: p. 197-203.
[22] Khodary, F., M. Abdel-Sadek, and H. El-Sheshtawy, Mechanical properties of modified asphalt concrete mixtures using Ca (OH) 2 nanoparticles. Int. J. Civ. Eng. Technol, 2014. 5(5): p. 61-68.
[23] Hamedi, G.H., F.M. Nejad, and K. Oveisi, Estimating the moisture damage of asphalt mixture modified with nano zinc oxide. Materials and Structures, 2016. 49(4): p. 1165-1174.
[24] Shenoy, A. (2001). Refinement of the Superpave specification parameter for performance grading of asphalt. Journal of transportation engineering, 127(5), 357-362.

IOP Conf. Series: Materials Science and Engineering **800** (2020) 012003 doi:10.1088/1757-899X/800/1/012003

Analysis of increment of road traffic accidents in Libya: Case study city of Tripoli

S I A Ali [1*], F A A Elturki [1], S N Jibrel [2]

[1]Civil Engineering Department, Near East University, Near East Boulevard, 99138 Nicosia, North Cyprus, Mersin 10, Turkey.
[2]Higher Institute of Science and Technology, Albrket, Ghat, Libya

*E-mail: shabarofking10@gmail.com

Abstract. Traffic and highway engineers working with law enforcement officials are constantly seeking for better methods to ensure safety for motorists and pedestrians. The planning process requires that engineers collect and maintain traffic safety data, identify the hazards location, conduct studies and establish project priorities. Unfortunately, in Libya, the increase in demand for private transportation in recent years, due to poor or lack of public transportation led to some traffic problems, especially in the capital (Tripoli). Also, the growth of private transportation has significant influences on the society regarding road traffic accidents (RTAs). This study investigates the most critical factors affecting RTAs in Tripoli, the capital city of Libya. Four main classifications were chosen to build the questionnaire, namely; human factors, road factors, vehicle factors, and environmental factors. Moreover, the data was collected from 400 respondents in the study area, and the relative importance index (RII) was used to rank the factors. The results show that human factors have the most significant impacts compared with other factors. Moreover, 84% of respondents considered the over speeding as the most affect factor causes of RTAs while 81% considered the disobedience to driving regulations as the second most influential factor in human factors. Besides, 74% of respondents considered poor brakes or brake failure factors has a high impact on the RTAs among the vehicle factors, while 79% classified poor or non-street lighting factors as one of the most effective factors on RTAs in road factors. The environmental factors have slight influences on RTAs.

1. Introduction

Road traffic accidents (RTAs are considered as one of the main problems in the world. It was found that the RTAs were ranked as the 11th leading cause of death globally, where about 1.24 million people die each year as a result of road traffic crashes, and they were the primary cause of physical disability for drivers, passengers, and pedestrians in developing countries [1]. Growth in urbanization and an increase in the number of vehicles has led to traffic congestion in urban centres and subsequently to RTAs. The increasing number of vehicles and growth in urbanization consequent lead to increased RATEs, so it is essential to improve road traffic safety procedures. Countries such as Libya faced many precarious road safety problems by several factors including human factors, road factors, vehicle factors, and environmental factors. Without taking action, road traffic crashes are predicted to result in the deaths of around 1.9 million people annually by 2020. Approximately, only 28 countries, representing 416 million people (7% of the world's population) have adequate laws that address all five risk factors (speed, drink-

IOP Conf. Series: Materials Science and Engineering **800** (2020) 012003 doi:10.1088/1757-899X/800/1/012003

driving, helmets, seat-belts, and child restraints) [2]. This study aims to investigate and classify the factors influencing the RTAs in Tripoli.

2. Literature review

The increasing number of private vehicles used in the city area has caused problems, such as traffic congestion, environmental pollution, and traffic accidents. Previous studies have shown some factors affecting accident in Tripoli, such as the increase in the number of private vehicles due to poor public transport in Tripoli [3]-[7]. Over 90% of victims of these accidents are in low-and-middle-income countries (approximately one million people). This difference continues when accounting for population distribution; the traffic casualty rate in countries of low-income was 17.4 per 100,000 people, and now they have highest annual road traffic fatality rates, at 24.1 per 100,000, while the rate in high-income countries was 12.6 and decreased to 9.2 per 100,000 [8]. Libya as a developing country is faced with the issue of excessive road crash rate. Figure 1 shows traffic fatality for several countries, and it is obvious that Libyan RTAs increased.

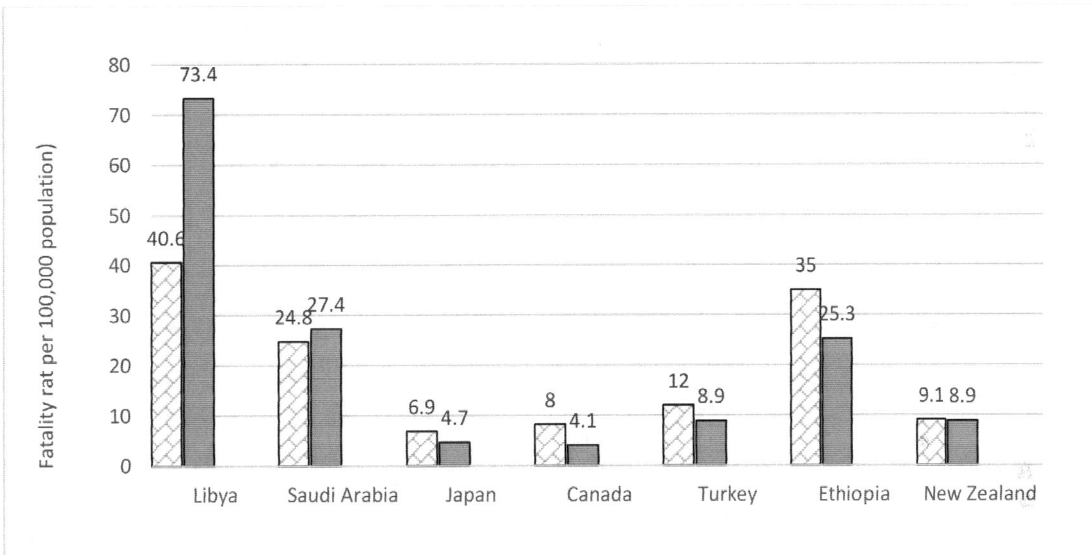

Figure 1. Traffic fatality for selected countries [8].

RTAs remain a worldwide public health issue. Presently placed ninth, RTAs are projected to become the fifth leading cause of death in 2030. Each year, over a million people perish worldwide as a consequence of RTAs, over 2500 casualties daily. Low and middle-income countries account for 90% of fatalities linked to RTA [8]. By 2030, road accidents are expected to become the fifth leading cause of accidents around the world [8]. Also, in high-income countries, traffic accidents are expected to reach 2% of the gross national product (GNP), while in middle-income and low-income countries it will reach 1.5% and 1% respectively [9].In a study that has been in Tripoli-Libya, age and gender differences in traffic law knowledge and behavior of road users in Libya were investigated. The results indicated a distinction in (younger) males from females in the understanding of traffic laws. Males (18-25 years of age) are more likely to break traffic regulations than females in Tripoli- Libya. Road traffic crashes cannot be eliminated entirely. However, some casualties can be kept to the minimum or removed by simple protective measures like driver behavior, car conditions, road conditions, use of seatbelts and speed control [11].

3. Methodology

The quantitative technique was used in this study to collect data from the field of study, and the distribution of questionnaires was conducted to targeted drivers, pedestrians, passengers and also ordinary people. The questionnaire was designed with the objective of obtaining information from drivers, passengers, pedestrians to get more respondents about the factors affecting the RTAs in Libya. The questionnaire was grouped into four separate sections. The first part was regarding respondents' personal information, while the second part was asking the respondents about their daily trip characteristics. Also, in the third section, the respondents were asked to answer the questions to assess the factors that are affecting the RTAs. In particular, this section was divided into four parts which include: human factors, vehicle factors, road factors, and environment Factors. However, each part contains questions regarding each factor and its impact on the RTAs. Finally, subsequent remedial measures that may be taken to avoid more injuries and harmless likely to occur by car accidents by giving their own opinion on closed-ended questions. Moreover, two ways were used to distribute the surveys; the first one is to create google forum on the internet to reach a large number of people, and this method was used to reach to the significant number of people in the short period compared to the second one which convenient way to distribute printed questionnaires. The questionnaire collected the data, and the number of respondents targeted was 400. The distribution of the questionnaire was in two ways, namely; online questionnaires by google form and printed questionnaires. Google form was constructive as it reached as many participants in a short time about 350 respondents which represent more than 85% of the total. The printed questionnaire needs more time to distribute to the respondents, about 50 which represents less than 15% of the total respondents as shown in Figure 2.

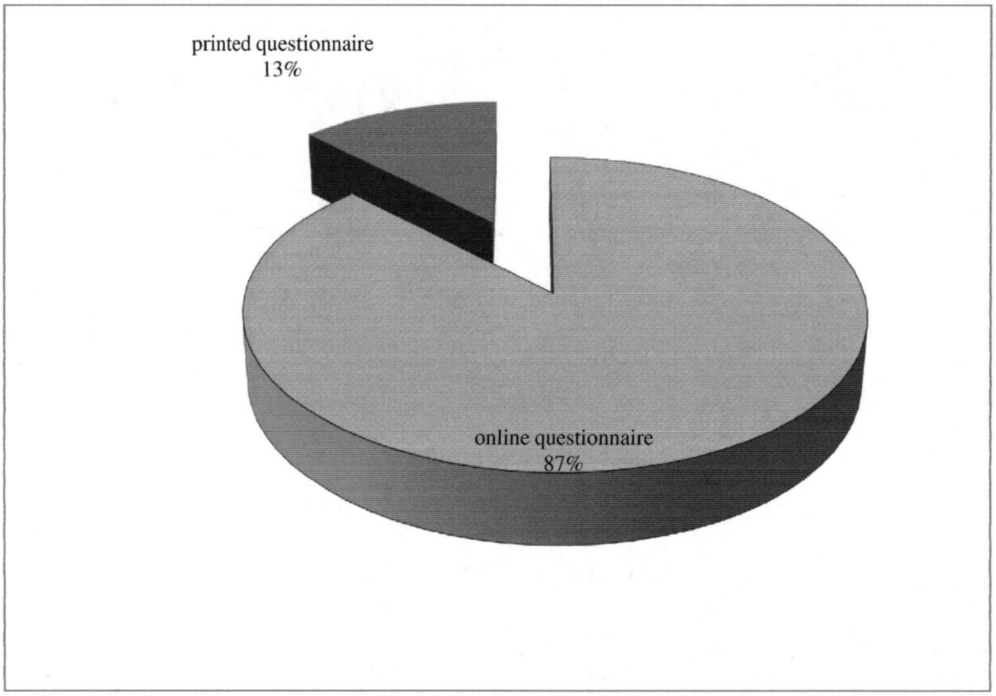

Figure 2. Distribution of questionnaire on the respondents.

4. Results and discussions

4.1 Demographic characteristics

For instance part of the demographic information, the research requested the respondents to indicate their gender, age, marital status, occupation, nationality, education level, income per month and also the place of work or study. This information required to have more information about the respondents who represent a random sample of the study. The respondents of the questionnaire were 400, where the target group was road users involved in RTAs such as; drivers, passengers, and pedestrians, 78.5% of respondents were males while 21.5% were females as listed in Figure 3.

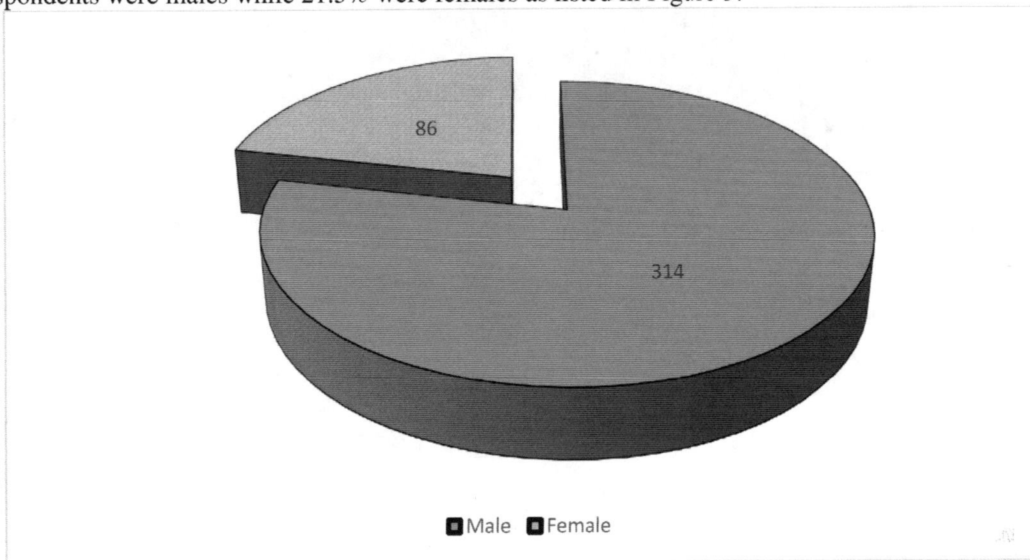

Figure 3. Gender frequency and percentage.

The age of respondents as shown in Table 1 it was divided into four sections, which starts from 18 years until more than 45 years. It was noted that the majority of the respondents were between 26 and 35 years old with 41.5%. The second highest percentage was between the ages of 18 - 25 years with 32.8%, while 7% of their age was more than 45 years old. The results indicate that there was a significantly good proportion of young respondents, which might mean the probability to obtain significant results as the young drivers are more involved in the RTAs.

Table 1. The age of respondents.

Ages (Years)	Frequency	Percent	Cumulative %
18 - 25	131	32.8	32.8
26 - 35	166	41.5	74.3
36 – 45	75	18.8	93.0
> 45	28	7.0	100
Total	400	100	

4.2 Education level

The education level is critical in this study as educated people have awareness about any phenomenon and issues more than poorly educated. Figure 3 shows the education profile of the respondents was 88.3% of respondents were educated to undergraduate or post-graduate level and 9.3% were educated to the high school level. Moreover, 2% were educated to less than high school, and a slight number of respondents were with no qualification 0.5%. Therefore, most of the applicants have been educated to a high level, so they must entirely understand and critically review the issues of traffic and get benefits from its awareness programs;

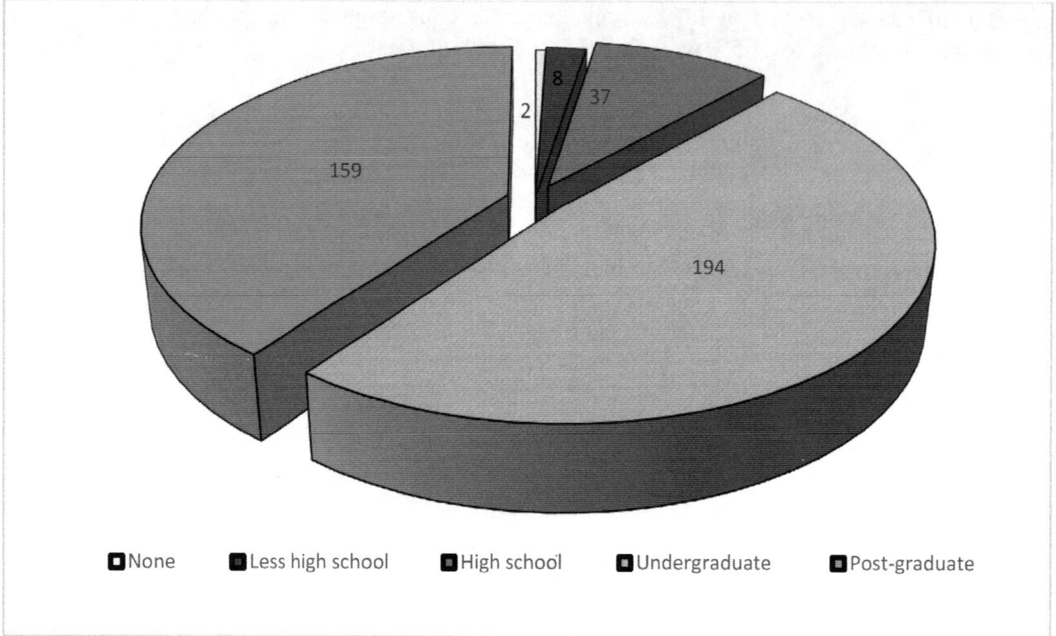

Figure 4. Educational profile of the respondents.

4.3 Driving experience

The experience in driving including obtaining a license can be seen that a majority of respondents were had to experience more than 12 years with a rate of 29.8%. Also, 22.8% of respondents were had experience in driving between 8 and 12 years, while 47.5 % of respondents were less than 8 years' experience driving as demonstrated in Table 2.

Table 2. Experience in driving including obtaining a license

Years	Frequency	Percent	Valid	Cumulative%
1-3	79	19.8	19.8	19.8
4-7	111	27.8	27.8	47.5
8-12	91	22.8	22.8	70.3
> 12	119	29.8	29.8	100
Total	400	100	100	

4.4 Human factors group

Human factors are one of the most critical factors affecting RTAs in Tripoli. The results show that many accidents occur as a result of these factors. Therefore these factors should be considered in particular to reduce these accidents, which leads to minimizing the casualties resulting from those accidents. Table 3 shows the human factors which the first group of elements in the questionnaire; this group contains nine factors, where the impact and importance of these factors vary from one factor to another. Hence, these factors were classified according to the significance of each of these factors and their impact on road accidents in Tripoli.

Table 3. Ranking of human factors group.

Factors No.	Factor description	RII	Group	Overall
1	Decision-making ability	0.755	4	9
2	Disobedience to driving codes	0.844	2	2
3	Drinking /taking drugs and driving	0.737	6	13
4	Following the vehicle in front too closely	0.752	5	10
5	Inexperienced drivers	0.723	8	17
6	Over-speeding	0.8555	1	1
7	Passenger's attitude	0.7065	9	22
8	Sleepiness and fatigue	0.729	7	15
9	Using a mobile phone	0.812	3	4

It was observed from Table 3 that over-speeding was the most effective factor among the human factors. Also, it has the most effect in the overall groups, which means this factor has a significant impact on the RTAs in Tripoli with a value of RII 0.8555. The second factor which has a considerable effect on RTAs was the disobedience to driving code such as children who are playing with the car on the roads which also ranked as the second overall group with the RII value of 0.844. In Libya, it a typical behavior that children are playing with cars on intercity road networks due to the less awareness of kids. Therefore, it is essential to pay attention to these issues. Also, ignoring the road control devised such as signs and traffic lights which only way to control driver's behavior on road networks results in many accidents and that causes an increase in injuries and fatalities. Using a mobile phone during driving or crossing streets was as accepted has a significant influence on RTAs, it was ranked the third factor affecting the human factors group, while it ranks as fourth in overall categories with a value of RII was 0.812. Using mobile phones considered one of the most critical factors that lead to the occurrence of severe accidents worldwide due to a lack of attention to the geometry of roads and objects on the streets during driving. The Sleepiness and fatigue, inexperienced drivers and passenger attitude were classified as the last factors in this group, where the ranked in seventh, eighth and ninth with RII of 0.729, 0.723 and 0.706 respectively. The distances inner the city is not that far to make the drivers be under stress that is why this factor has a low impact on the other factors. Also, inexperienced drivers are not that significant matter since all of the Libyan start driving at a young age like 12-13 years old without any driving license, so most of the drives have the initial experience. Regarding passenger's attitude, in recent years the altitude of Libyan drivers enhanced because of the blockage of low and the proliferation of weapons inside the city.

4.5 Vehicle factors

The vehicle factors contain several factors that might increase the possibility of accidents occurring on the roads in the city of Tripoli, where the importance of these factors varies among each other regarding their impact on RTAs and Table 4 demonstrations the ranking of those factors regarding their importance. From Table 4, it was found that poor brake or brake failure has great implications on RTAs as ranked the first factor in the group and the fifth factor in the overall categories with RII of 0.808. Furthermore, the faulty light factor is classified as the second factor affecting the RTAs in the group, and fourteenth in overall the groups with RII of 0.73 which means it has a medium effect on the RTAs in Tripoli. Also, the bad tires were ranked as the third factor affecting RTAs in this group, while it was classified as sixteenth in the overall of the groups with RII of 0.726. From field observation it's well known that the Libyans usually do not care about changing tires unless it explodes, this indicates that they are careless and put their lives at risk due to the possibility of increased risk of accidents due to the tire explosion.

Table 4. Ranking of vehicle factors.

Factors No.	Factor Description	RII	Group	Overall
1	The absence of rear view	0.705	4	23
2	Bad tires	0.726	3	16
3	Breakdown of engine	0.576	7	30
4	Defective horns	0.52	10	33
5	Defective steering	0.562	8	31
6	Faulty gearbox	0.5415	9	32
7	Faulty light	0.73	2	14
8	Poor brakes/brake failure	0.808	1	5
9	Windshield (break, not clean)	0.614	5	27
10	Other mechanical failures	0.587	6	29

4.6 Road factors group

The road humps factor ranked as the 4th causing factor in the group while it is listed as the 8th in overall groups with RII of 0.7625. The absence of illuminated roads due to the shortage of electric nowadays with over-speed and illegal humps which constructed by residents without following humps standards then the increase in accidents will be an inevitable consequence of all those factors. Further, the roadway geometrics and Animals out of control (e.g., cows, dogs) factors have the same rank 5th in the road factors group resulted in the same listed in all groups18th with 0.717 RII. The roadway geometrics is not the main problem, but the lacked sings in the road networks is the primary matter, as the drives will not recognize for example the T-junction of deep left or right curves and allowable speed on the curve which all depends on the experience of the drivers. On the other hand the low in Libya stand with the owners of animals in any collusion, the driver must pay the offense in case of a collision with an animal which leads to careless of the owners regarding controlling their animals. The road site details (e.g., steep hill, narrow road, bend/winding road, slippery road) factor has a slight effect on the RTAs, which is located at the 7th effecting factor in the group and it is ranked as the 20th in overall with RII 0.7155 as shown in Table 5. Meanwhile, the factors such as regularly roadworks and the road surroundings (e.g., buildings, fences, and vegetation) were found to have the values of RII 0.634 and 0.6115 respectively, the value of RII less than 0.7 which means they have no adequate effects on the RTAs in Tripoli. In General, illumination is essential to ensure the safety of road users, and emphasis on street lighting is necessary to reduce accidents. On the other hand, the existence of roads signings one of the sufficient keys to guide the drivers and in the lack of knowledge of the roads networks, while the poor road surface factor also is so crucial in term of efficient on the RTAs, as there are many holes and cracks in the roads, that may lead to the punctured tires and damaged of vehicles as well as increase fuel consumption, therefore it is essential to perform a regular maintenance to road networks to prevent accident-related to poor road service.

Table 5. Ranking of road factors.

Factors No.	Factor Description	RII	Group	Overall
1	Poor road surface	0.785	3	7
2	Poor/no street lighting	0.827	1	3
3	Insufficient signing	0.783	2	6
4	Road humps	0.763	4	8
5	Regularly roadworks	0.634	8	26
6	Road surroundings	0.612	9	28
7	Animals out of control	0.717	5	18
8	Road site details	0.716	7	20
9	Roadway geometrics	0.717	5	18

The road humps factor ranked as the 4th causing factor in the group while it is listed as the 8th in overall groups with RII of 0.7625. The absence of illuminated roads due to the shortage of electric nowadays with over-speed and illegal humps which constructed by residents without following humps standards then the increase in accidents will be an inevitable consequence of all those factors. Further, the roadway geometrics and Animals out of control (e.g., cows, dogs) factors have the same rank 5th in the road factors group resulted in the same listed in all groups18th with 0.717 RII. The roadway geometrics is not the main problem, but the lacked sings in the road networks is the primary matter, as the drives will not recognize for example the T-junction of deep left or right curves and allowable speed on the curve which all depends on the experience of the drivers. On the other hand the low in Libya stand with the owners of animals in any collusion, the driver must pay the offense in case of a collision with an animal which leads to careless of the owners regarding controlling their animals. The road site details (e.g., steep hill, narrow road, bend/winding road, slippery road) factor has a slight effect on the RTAs, which is located at the 7th effecting factor in the group and it is ranked as the 20th in overall with RII 0.7155 as shown in Table 7. Meanwhile, the factors such as regularly roadworks and the road surroundings (e.g., buildings, fences, and vegetation) were found to have the values of RII 0.634 and 0.6115 respectively, the value of RII less than 0.7 which means they have no adequate effects on the RTAs in Tripoli. In General, illumination is essential to ensure the safety of road users, and emphasis on street lighting is necessary to reduce accidents. On the other hand, the existence of roads signings one of the sufficient keys to guide the drivers and in the lack of knowledge of the roads networks, while the poor road surface factor also is so crucial in term of efficient on the RTAs, as there are many holes and cracks in the roads, that may lead to the punctured tires and damaged of vehicles as well as increase fuel consumption, therefore it is essential to perform a regular maintenance to road networks to prevent accident-related to poor road service.

4.7 Environmental factors group

The environmental factors group considered one of the groups that may affect RTAs. This category contains a small number of factors when compared with other groups. However, it provides essential factors that have a significant impact on RTAs in Tripoli. The results displayed in Table 6 show that the glare which the difficulty in seeing in the presence of bright light factor is the first factor affecting the environmental factors group and it is ranked as tenth in overall groups with RII 0.752, which indicates a moderate effect on the RTAs.

Table 6. Ranking of the environmental factors.

Factors NO.	Factor Description	RII	Group	Overall
1	Slick roads	0.664	4	24
2	Glare	0.752	1	10
3	View obstructions	0.748	2	12
4	Other highway-related conditions	0.641	5	25
5	Weather conditions	0.715	3	20

5. Conclusions

The results show that the factors differ in their impact on RTAs from one factor to another. Each factor has a specific effect on RTAs, some of them have high effects and lead to increase RTAs, and others have moderate effects while there are some factors with slight impact or without impact on RTAs. Also, the major contributing cause of many crashes as observed in the results is driver error. The human errors leading to highway crashes are related to the complex interaction for the driver's psychological and physiological condition, the system design, and the existing environmental condition. The driver's actions that lead to errors include driving at an inappropriate speed or over speeding which ranked as first among all factors 84% for existing physical and/or environmental conditions, driver inattentiveness, failure to yield the right-of-way, wrongly negotiating curves, the use of mobile phones. Moreover, driver's disobedience to driving code with 81% has significant impacts on the RTAs. The main aim of the traffic engineer regarding human factors in Tripoli networks to ensure safety is to understand how those factors influence the drivers so that the probability of occurrence of these associated safety consequences of these factors is reduced. For example, in the city of Tripoli rumple strips are placed at side and medians of high-speed roads in rural areas to alert inattentive or sleepy drivers when they are leaving the travel lane. The mechanical situation of an automobile might be involved in highway crashes which known as vehicle factors group. Faulty breaks in vehicles were the 1st-factor affecting RTAs with more than 74% compared to other factors; this indicates a higher influence on the RTAs. The 2nd influencing was Faulty light approximately 61% of respondents seeing this factor significantly affecting the RTAs followed by other factors such as worn tiers. However, many vehicle manufacturers are now installing in-vehicle equipment that helps to reduce the potential of the vehicle being involved in crashes and/or the severity of a collision when the vehicle is involved in one. These include seat belt reminders, antilock braking systems (ABS), collision avoidance sensor and speed limiter/intelligent speed adaption. In Libyan, due to the political issues and lacking in control of the importing of vehicles the condition was not in good condition, which was prohibited before the current situation. In this case, the government should take action to reduce these issues and importing the vehicles which installing all safety equipment. Furthermore, the quality and the condition of the roads is one of the factors involve the RTAs. As a result, the highway needs to be designed to afford adequate stopping sight distance (SSD) at the design speed. Otherwise, the drivers will be incompetent to take remedial actions to evade the accidents. One of the solutions, traffic signs should afford sufficient SSD when the sings go from green to yellow and then to red. In Libya, the pavement surface is in a critical situation and most of the roads need to be reconstructed at the time of maintenance is over. Moreover, the signs should be placed on the importance of the information contained in them and avoid placing it at the location where they are not necessary or essential. Also, in situations where the required information cannot be placed on one sign or several signs at a single location, the information should be given in portions along the road to reduce the information load. The physical and climatic environment surrounding an automobile are factors occurrence of highway accidents; the most common environmental factor is the weather. Weather conditions are the primary road accidents contribution to the highway, but as shown from the results in the city of Tripoli, it has a slight impact on RTAs. The fog has been caused many severe accidents because vehicles traveling at high speeds are not capable of realizing other automobiles have been ahead that may have slowed down or stopped. Recommendations to Stakeholders' that should be taken into consideration for the government, to reduce RTAs in Tripoli, to achieve the objective of this research.

- Over Speed is the most critical factor affecting RTAs in Tripoli, where the majority of respondents considered that speeding has a massive impact on the increase in accidents. Recently, it has also been noted that the rate of accidents has been increased as a result of extreme speed and this has resulted in increased casualties. Hence, the government should increase awareness of the dangers of speed by giving specialists to lectures in high schools and universities to increase awareness of accidents as a result of excessive speed. Moreover, the government should also install speed-monitoring cameras in all places, as well as increase traffic campaigns and punish those who exceed the speed limit with high fines that may reach to high penalties and withdrawal of license.
- The handling of mobile phones during driving is one of the reasons for the increase in RTAs. Hence, awareness campaigns should be conducted regarding the dangers of using phones while driving, also not being lenient in performing traffic violations. Additionally, the use of a mobile phone for pedestrians while crossing the roads leads to increased accidents; pedestrians must be alerted and warned not to use the phones while crossing the roads.
- Obtaining a driver's license in Libya has become very easy so that it reached the stage where the driver's license can be obtained without a driving test. Either paying a bribe to a traffic man or knowing someone in the traffic department, which leads to many crimes regarding increasing accident rates thus increasing the proportion of casualties. Hence, the emphasis should be on the issue of obtaining driving licenses. So the traffic man participatory in giving the license without an exam should be expelled.
- Most of the participants considered that the roads in Libya are terrible and need many improvements. The government should pay attention to these roads regarding providing

functional lighting. The government should also ensure a good quality of the road surfaces and make the periodic maintenance of these roads. Also, providing road signs and especially roads linking cities.

- The public transport system in Libya is inadequate and almost is non-exist, which leads to an increase in the percentage of vehicles, resulting in increased traffic congestion and increased accidents. The government should concern to public transport, establish trains and an excellent public transport system to reduce RTAs

References

[1] S. Ameratunga, M. Hijar, and R. Norton, Road-traffic injuries: confronting disparities to address a global-health problem. Lancet, 2006. 367(9521): p. 1533-1540.

[2] World Health Organization. (2013). Global status report on road safety 2013: supporting a decade of action: summary.

[3] A. Ishtiaque. "Road infrastructure and road safety." Transport and Communications Bulletin forAsia and the Pacific 83 (2013): 19-25.

[4] S. I. Albrka, A. Ismail, H. A. Yahia, M. A. Ladin. Application of Transyt-7F on signalised road junction networks in Shah Alam and Petaling Jaya. Jurnal Teknologi. 2014 Feb 15;69 (2).

[5] A. A. Hammoudi, "Causes and strategies to reduce road traffic accidents in Abu Dhabi." PhD diss., Cardiff Metropolitan University, 2014.

[6] A. Ismail and H. A. Yahia, H. A. M. (2011). Causes and effects of road traffic accidents in Tripoli–Libya. ISBN 978-602-8605-08-3.

[7] K. B. Lankarani, S. T. Heydari, M. R. Aghabeigi, G. Moafian, A. Hoseinzadeh, and M. Vossoughi. (2014). The impact of environmental factors on traffic accidents in Iran. Journal of injury and violence research, 6(2), 64.

[8] World Health Organization. (2009). European status report on road safety. WHO Regional Office for Europe.

[9] S. M. Sabbour, and J. M Ibrahim. (2010). Driving behavior, driver style and road traffic accidents among the young medical group. Injury Prevention, 16(1), A33.

[10] World Health Organization. (2015). Global status report on road safety 2015. World Health Organization.

[11] H. A. M. Yahia, A. Ismail, S. I. Albrka, A. S. Almselati, and M.A. Ladin. (2014). Attitudes and awareness of traffic safety among drivers in Tripoli-Libya. Research Journal of Applied Sciences, Engineering and Technology, 7(24), 5297–5303.

IOP Conf. Series: Materials Science and Engineering **800** (2020) 012004 doi:10.1088/1757-899X/800/1/012004

The importance of designer-contractor communication in geotechnical design as exemplified by a cut slope failure

D Anafarta[1], E Turk[1] and S O Akbas[2,3*]

[1]Akademi Etut Proje Muh. Mus. Ltd. Sti., Ankara, Turkey
[2]Gazi University, Dept. of Civil Engineering, Ankara, Turkey

*Email: soakbas@gazi.edu.tr

Abstract. A shallow slope failure has occurred on the cut slopes of the Bigadic-Simav-Abide highway at the section located between Km:125+530-125+870, during construction. A detailed investigation of the landslide mechanism, which includes site specific surveys, laboratory studies, and stability analyses indicates that the main reason of the failure lies within the construction procedure details. This study focuses on the forensic geotechnical engineering procedures applied for determining the cause of the slope failure. An emphasis was placed on the importance of strictly following the construction sequence as illustrated in the design documents in detail, as well as on the indispensable role of continuous communication between the designer and the contractor for successful performance of geotechnical works.

1. Introduction

Observation of deformations and local slumps at the slopes located between Km:125+530-125+870 along the Bigadic-Simav-Abide highway, which is under construction, has necessitated a geotechnical investigation to be undertaken to explore the possible causes of the failure and the remedial measures that can be taken to ensure the stability. This investigation included an assessment of the engineering geology of the site and the slopes, as well as geotechnical modeling studies, which led to an improved design of the roadway cut.

However, during the construction of the roadway cut slopes according to design, shallow slope failures have occurred unexpectedly. Thus, a study was planned to understand the main reason of this failure, which is necessary to ensure the future roadway safety. This study presents the geotechnical methodology applied to understand the root cause of the failures, inevitably focusing on the effects of construction sequence and procedures on the design assumptions.

2. The original design

The site under investigation where signs of slope instability were observed in terms of localized slumps and deformations at various portions of the cut slopes, is located between Km: 125+530-125+870 along the Bigadic-Simav-Abide highway, which is still under construction (Figures 1 and 2).

[3] Corresponding author

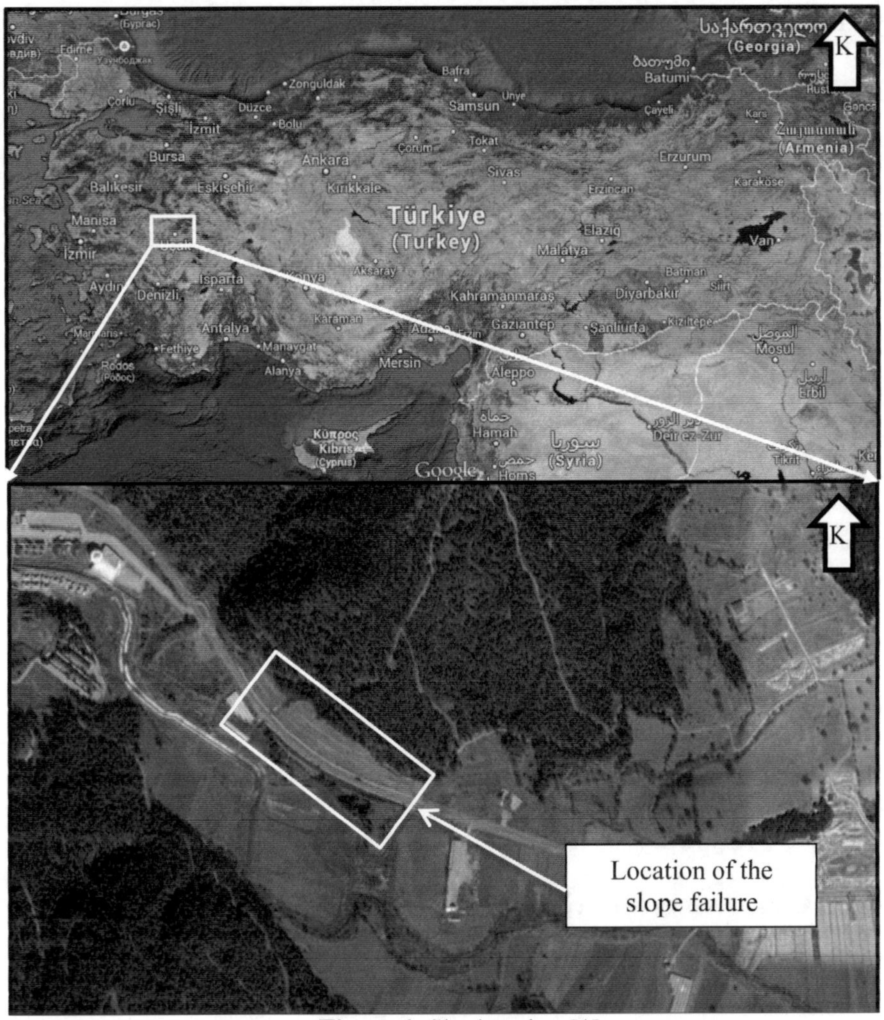

Figure 1. Site location [1].

Figure 2. A view of the cut slope deformations between Km: 125+530-125+870 [1].

In order to determine the reasons of the signs of slope instability and to estimate the geotechnical properties for remediation studies, a site investigation program that included drilling of exploration boreholes at two locations with a total depth of 50 m was planned. According to the results of this study, the site is underlain by lower- middle Miocene aged sediments that consists of claystone-siltstone and mudstone units of varying degree of weathering. A close look at the boring logs indicate that highly weathered claystone and mudstone is overlain by from 2.0 m to to as much as 18.0 m of clay and clayey sand/gravel especially on the upper portions along the cut slope. Ground water was observed mainly at the interface between the soil and rock layers. A cross-section of the slope under investigation is shown in Figure 3.

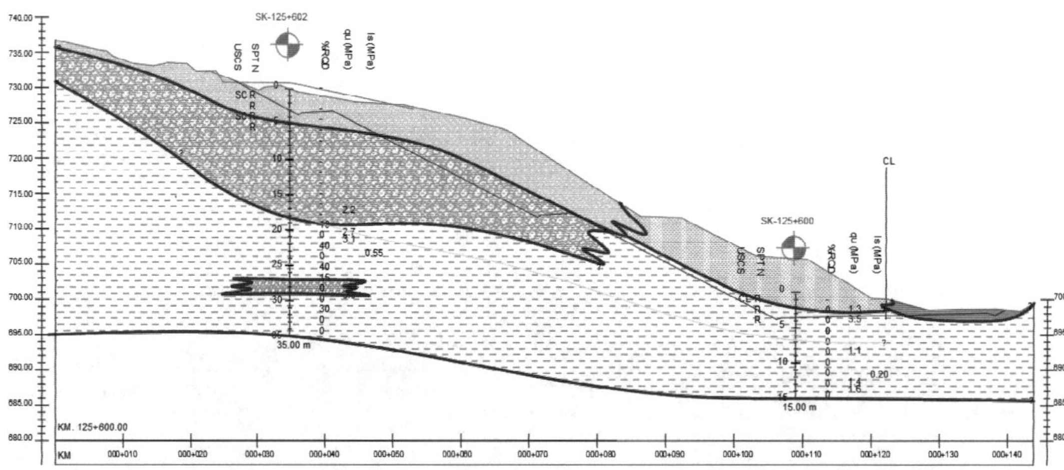

Figure 3. A cross-section of the slope under investigation [1].

The geotechnical parameters required for slope stability analyses were estimated based on in-situ and laboratory test results, which were quite limited both in terms of quality and quantity, especially for the soil layers mainly due to the urgent nature of the project. Note that Mohr-Coulomb failure model has been used for the rock unit modeling also, but the parameters were derived as a fit to Hoek-Brown criterion. The employed parameters for idealized strata are summarized in Table 1.

Table 1. Geotechnical parameters employed in the slope stability analyses [1]

	Clayey sand	Silty-clayey gravel	Clay	Weathered claystone/mudstone
Unit weight, γ (kN/m^3)	18	19	18	21
Effective stress friction angle, ϕ' (^{o})	30	32	30	17
Effective cohesion, c′(kPa)	3	5	5	68
Undrained shear strength, s_u (kPa)	-	-	110	

For all slope stability analyses, a limit equilibrium-based computer program named Slide 2018 [2] was employed. For seismic conditions, the horizontal earthquake acceleration applicable to earth structures was determined to be 0.23g for the site, using the interactive web application available at AFAD website.

As an initial step, the stability of the slope was assessed for the existing conditions using the long-term, i.e., effective parameters given in Table 1. As shown in Figure 4, the results indicate a factor of safety (FS) of about 1.2 under static conditions, which is lower than the required value of 1.5 as specified by the General Directorate of Highways. Note that the corresponding FS is less than 1.0 under seismic loads, and it is expected that even lower FS values may be attainable locally at different locations within the investigated section of the highway.

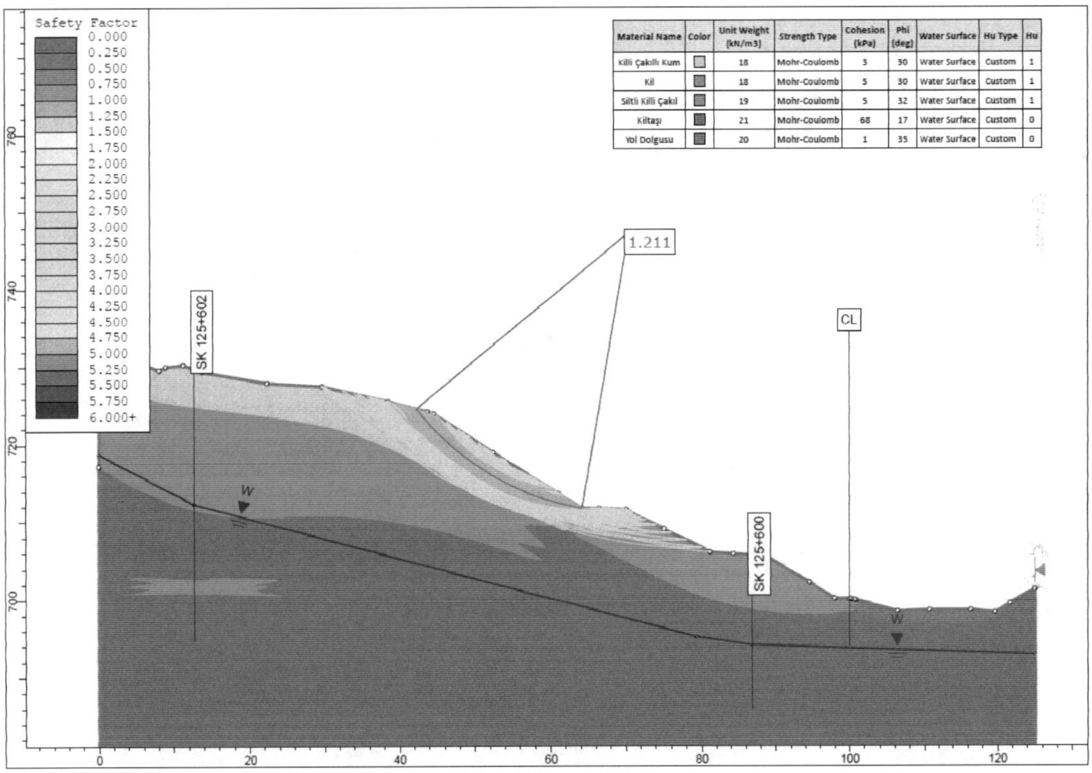

Figure 4. The results of stability analyses for the slope in its existing geometry under static loading conditions [1]

Considering the low FS values obtained from the limit equilibrium analyses along with the signs of instability observed at the site, remedial measures was decided to be undertaken, which include construction of an eight-meter high masonry wall at the toe of the slope and backfilling it using crushed rock, effectively forming a 5 m wide berm behind it (Figures 5 and 6).

As can be seen from Figure 5, a temporary small-scale excavation with a 45° slope has to be made in order to construct the foundation of the masonry wall. Therefore, care was taken to investigate the stability of the slope during the excavation stage as this might be a relatively critical condition due to the removal of material from the toe of the slope. Assuming that the excavation will not stay open for a long time until the masonry wall will be completed, short-term strength parameters were used for the analyses at this stage. Based on this assumption, and considering the observed facts that slopes with 45° angle remain only temporarily stable at the site, the design documents clearly stated to perform the excavation and wall construction at maximum ten-meter long sections, preferably during the dry season to minimize the time during which the excavation stays open.

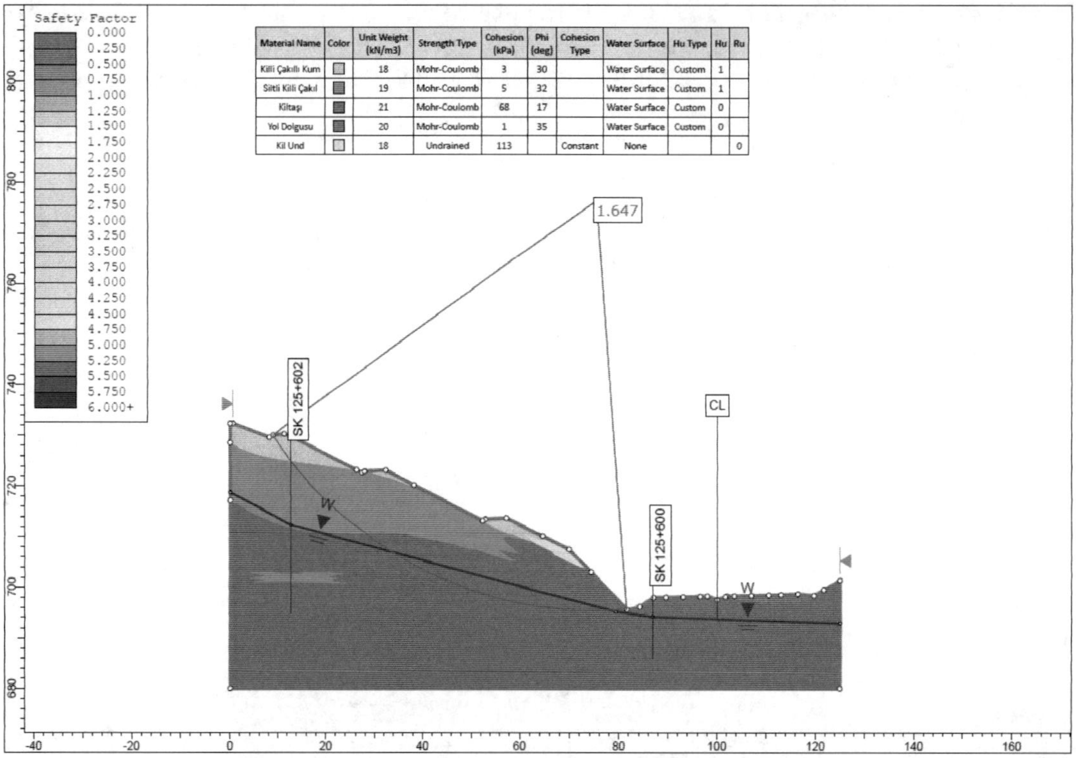

Material Name	Color	Unit Weight (kN/m3)	Strength Type	Cohesion (kPa)	Phi (deg)	Cohesion Type	Water Surface	Hu Type	Hu	Ru
Killi Çakıllı Kum		18	Mohr-Coulomb	3	30		Water Surface	Custom	1	
Siltli Killi Çakıl		19	Mohr-Coulomb	5	32		Water Surface	Custom	1	
Kiltaşı		21	Mohr-Coulomb	68	17		Water Surface	Custom	0	
Yol Dolgusu		20	Mohr-Coulomb	1	35		Water Surface	Custom	0	
Kil Und		18	Undrained	113		Constant	None			0

Figure 5. The results of stability analyses during remediation works [1]

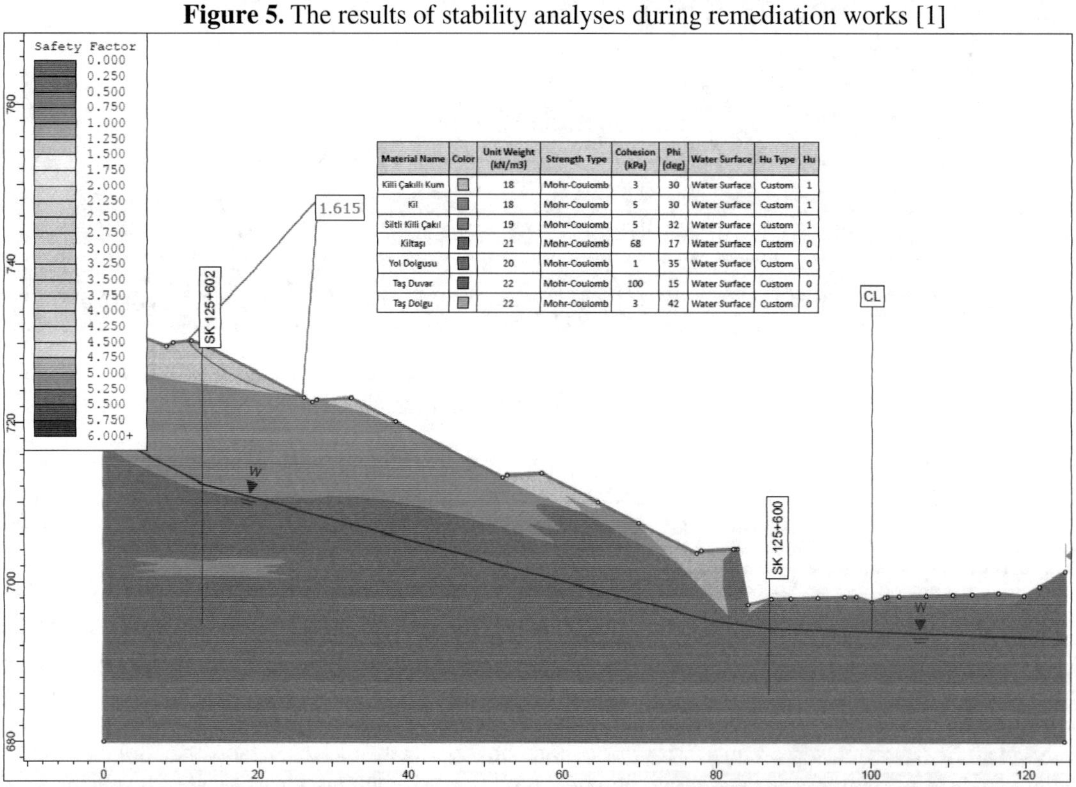

Material Name	Color	Unit Weight (kN/m3)	Strength Type	Cohesion (kPa)	Phi (deg)	Water Surface	Hu Type	Hu
Killi Çakıllı Kum		18	Mohr-Coulomb	3	30	Water Surface	Custom	1
Kil		18	Mohr-Coulomb	5	30	Water Surface	Custom	1
Siltli Killi Çakıl		19	Mohr-Coulomb	5	32	Water Surface	Custom	1
Kiltaşı		21	Mohr-Coulomb	68	17	Water Surface	Custom	0
Yol Dolgusu		20	Mohr-Coulomb	1	35	Water Surface	Custom	0
Taş Duvar		22	Mohr-Coulomb	100	15	Water Surface	Custom	0
Taş Dolgu		22	Mohr-Coulomb	3	42	Water Surface	Custom	0

Figure 6. The results of stability analyses for the remediated slope after construction [1]

3. The second failure

During remediation works of the slope, before the masonry wall was constructed, shallow slope failures were observed at the site at several locations (Figure 7).

Figure 7. A photo of the shallow slope failure during construction

As soon as the designer became aware of the problem, a forensic geotechnical engineering study was initiated for determining the cause of the slope failures. Unfortunately, time and budget constraints prevented performing additional site or laboratory investigations. However, a detailed look at the slope stability problem based on the field records was enough to indicate at least two procedures that are not in compliance with those stated in the design documents.

First of all, as can clearly be seen from the field records and also from some of the photographs, the very important instruction in the design documents that stated to perform the excavation and wall construction at maximum ten-meter long sections to minimize open excavation time was not followed. Instead, first the whole excavation was made along the length of the cut slope, later to be followed by the wall construction. This clearly nullified the undrained behavior assumption made during the slope stability analyses for the temporary excavation case. In addition, it is understood that, probably after the first signs of slope instability, which might be cracks or some deformations, the contractor laid down a layer of crushed rock on the slope face to prevent further damage without the consent of the designer. The crushed rock particles can indeed be observed on the failed slope in Figure 7. This had a further adverse effect on the slope by increasing the destabilizing weight.

Limit equilibrium analyses simulating the above-mentioned conditions have been performed to investigate the slope stability (Figure 8). Note that, due to the long elapsed time of open excavation conditions, drained strength parameters were employed this time. Also, based on the fact that the clay layer was remolded to some extent, a minimal 10% reduction was applied to its strength parameters. The results of the limit equilibrium analyses clearly indicate that, even without the effect of crushed rock layer above the slope face, the slope is on the verge of sliding with a lower FS compared to that during the initial conditions before any remediation work took place.

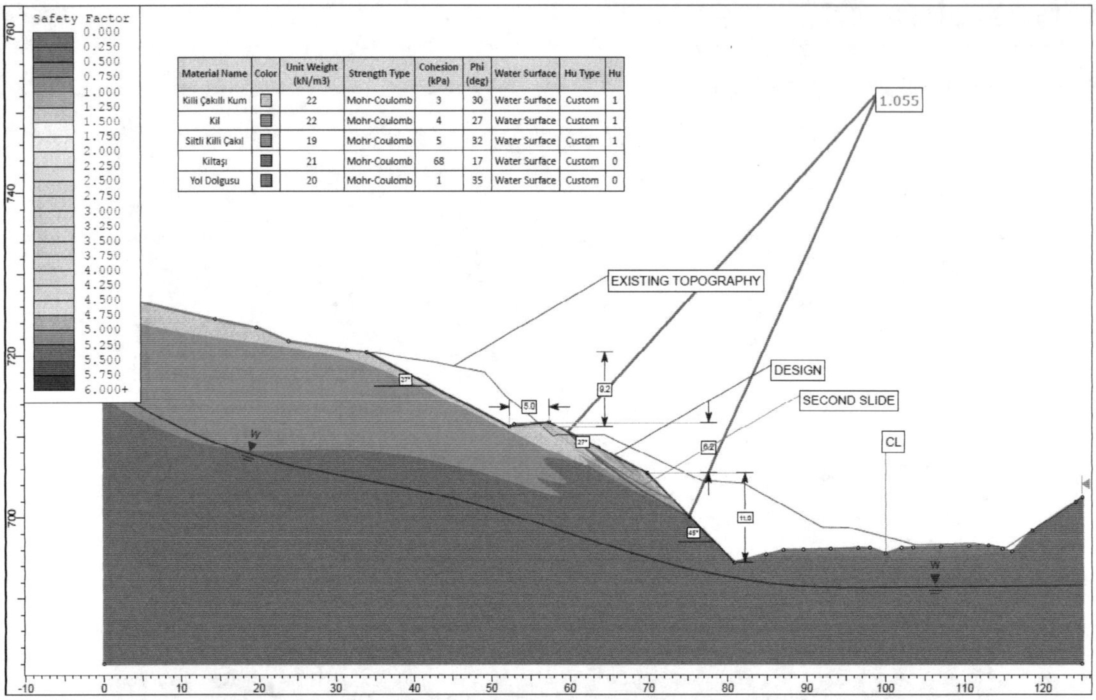

Figure 8. Analysis of the second slide using existing conditions

Based on the slope stability problems summarized above, a new remediation project, which is an extended and surely more expensive version of the previous one was undertaken. The details are illustrated in Figure 9.

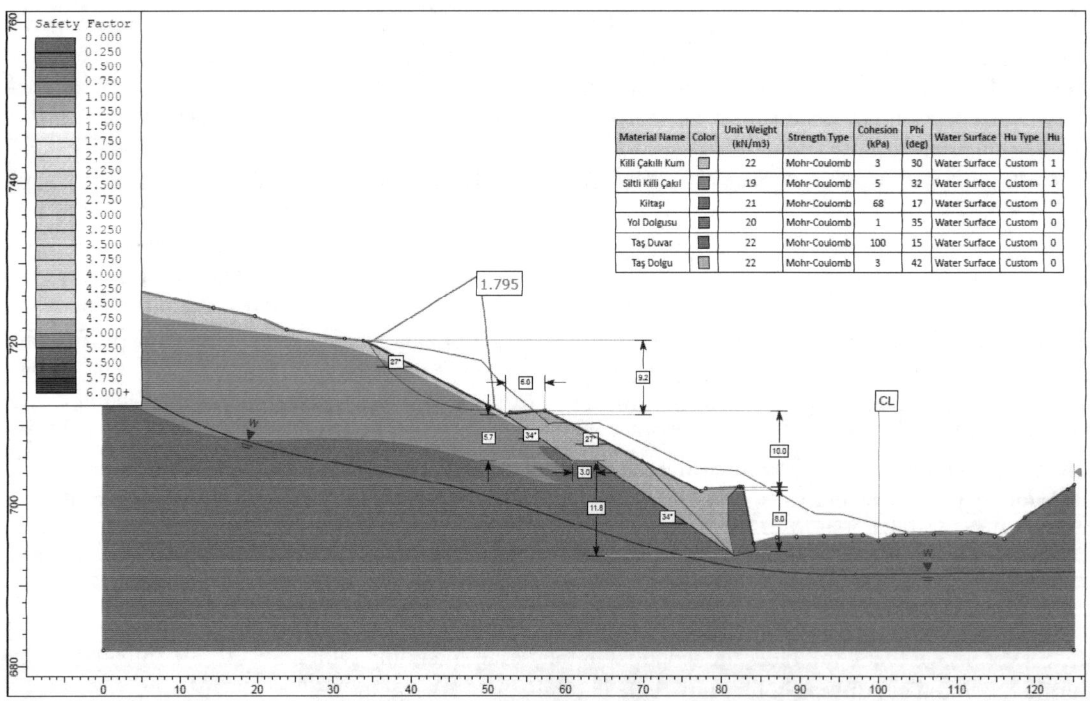

Figure 9. Final remediation project for the slope

IOP Conf. Series: Materials Science and Engineering **800** (2020) 012004 doi:10.1088/1757-899X/800/1/012004

4. Summary and Conclusions

This study exemplified the importance of designer-contractor communication and the significance of attention to design details by a relatively small-scale slope failure. It was clearly demonstrated that not following the recommended procedure of performing the excavation at a maximum of ten-meter long sections, thus elongating the open excavation time, decreased the FS against slope stability failure to a value even less than the originally existing conditions. In addition, a well-meant slope stabilization measure that was taken without the consent of the designer, which is clearly a lack of communication issue, aggravated the conditions even more. Design details and communication clearly become more important when there is limited amount of geotechnical data, which is the usual case for small-scale projects like the one described herein.

References

[1] Aral B, Turk E and Tufenkci O O 2019 *Bigadiç-Simav-Abide Yolu Demirci Ayr.- (Uşak-Kütahya) Ayr. Arasi Km: 114+700-129+685 Toprak Sanat Yapilari, Kopru Ust Yapi (bsk) Isleri Insaat Işi Km:125+530-125 Heyelan Projesi Jeolojik Jeoteknik Etut ve Geoteknik Değerlendirme Raporu* (Ankara: Akademi Etut Proje Muh. Ltd. Sti.)

[2] Rocscience Inc. 2018 Slide 2018 (Toronto, ON, Canada)

Measurement of vibrations generated by high speed railway traffic and evaluation according to international norms

S Ates [1*], E Celebi [2], O Kirtel [3], A C Zulfikar [3], and F Goktepe [5]

[1, 2] Department of Civil Engineering, Sakarya University 54187 Esentepe Campus/Sakarya/TURKEY
[3] Department of Civil Engineering, Sakarya University of Applied Sciences 54187 Esentepe Campus/Sakarya/TURKEY
[4] Department of Civil Engineering, Gebze Technical University, 41400 Cayirova Campus/Kocaeli/TURKEY
[5] Department of Civil Engineering, Bartin University, 74100 Bartin/TURKEY

*Email: sami.ates@ogr.sakarya.edu.tr

Abstract. High-speed trains are one of the most preferred ways of transportation. The fact that it is fast and safe, cost effective and not being effected by weather conditions provides many advantages for passengers but there are also many disadvantages for the buildings close to the train track and the people who reside in these buildings. In this study, a series of site tests were performed to obtain the acceleration recordings of the vibrations existing in free field of the soil and building foundation. The average shear wave velocity of soil was considered as 200m/s and the velocity of the high-speed train was taken as 250km/h. In this study, it is aimed to reveal the kinematic interaction between the free field motion and foundation vibration of the building while high-speed train passed. Furthermore, vibrations occurring under the current soil conditions of the study area were evaluated according to the key parameters defined in regulations of the Federal Transit Administration Report (FTA) and Human Exposure to Vibration in Buildings/Effects of Vibration on Structures (DIN 4150-2/3). It is seen that the compliance with the limit values specified in the regulations changes according to the purpose of use. The results showed that rigid foundations significantly modify the wave propagation and vibration level due to kinematic interaction effects.

1. Inyroduction

High-speed railways in Turkey are designed with a maximum speed of 250-300km/h and a maximum axle load of 22.5 tons. High-frequency vibrations produced by high-speed trains during railway crossings are spread to the nearby surroundings via the underlying soil. The train induced ground-born vibration energy is dissipated in the soil medium and affects the structures close to the railway track by vibrating. It is well known that the considerable portion of the vibration energy is transmitted by Rayleigh surface waves with undesirable effects [1-2]. These high amplitude environmental oscillations not only damage nearby buildings and their foundations, but also affect human comfort level with its unfavorable shaking. It is seen that the vibrations spread from the train lines passing through the densely residential district exceed the acceptable limits given for nearby structures in regulations in terms of human comfort and building health. Therefore, the problem of train induced high level vibrations associated with the dynamic soil-structure interaction and wave propagation is remarkable issue to be handled from to point of reducing structural vibrations, determining how far the structure should be built from the railway line, eliminating the adverse effects related with strong ground motion in near environment, and knowing what precautions should be taken to provide for building safety, human health and comfort. For this purpose, several experimental studies on in-situ have been carried out to measure the environmental and structural vibrations [1-3]. The importance of this type of geo-dynamic problem increases especially in weak soil deposits for constructions.

The goal of this study is to investigate the acceptability level of high-speed train traffic induced structural vibrations in densely populated area and to assess the recorded vibration data according to the threshold values of the international norms given [4-6]. Furthermore, the effect of the kinematic interaction on free field motion caused by the rigidity of the building foundation is examined experimentally by comparing the obtained response spectrum curves. It is intended to determine the characteristics of the soil-structure interaction problem with field tests and to provide the necessary information infrastructure for the measures that can be taken to ensure that the vibration level limit values specified in international regulations are not exceeded. The transfer functions are derived from the measurements to better understanding the effect of soil-structure interaction, especially in soft ground conditions. With the help of the data obtained from the free field vibration and basement motion of the considered building, it is thought that this function will allow us to estimate the vibration level that may occur in the structure and location where a new building will be constructed near the train line can be determined more appropriately.

2. Research method

In this study, field tests were performed by using high performance seismometers which have 32-bit high resolution and have 120 decibel (dB) dynamic range with built in (Global Positioning System) GPS antenna to measure the vibrations radiated from the railway to the environment. The accelerations are recorded by data loggers inside seismometers for assessing the ground motion and structural vibration generated by repeated train crossings, taking into account the local soil conditions where the high-speed train line passes. This location was chosen because of the soft soil conditions as well as the structures close to the train line. During the site measurement, the accelerations of the high-speed trains were recorded at different times and the vibrations were monitored repeatedly. The test location was selected near the Arifiye station of the Istanbul-Ankara high-speed train line with a total length of 533 km. The average shear wave velocity of the test area was measured as 200m/s. The speed of the passing trains during the field measurements was determined to be 250 km/h. The test site is located in the social facility of the Sakarya University of Applied Sciences, just 8-10 m from the railway line, as shown on the Figure 1.

2.1. Location of the investigated region

As illustrated in Figure 1a, the high-speed railway line focused on this study passes through the village of Kırkpınar at the western end of the place in the Sapanca District. The Figure 1b shows the measurement site where the accelerometers are located and the building considered.

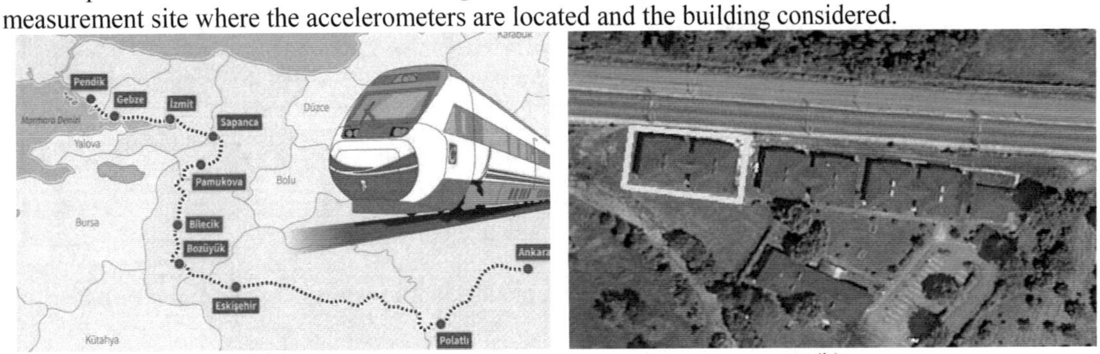

(a) (b)

IOP Conf. Series: Materials Science and Engineering **800** (2020) 012005 doi:10.1088/1757-899X/800/1/012005

Figure 1. Location of the railway line (a) and testing site (b)

(a) (b) (c)

Figure 2. Field view and installation of accelerometers.

2.2. Layout of the experimental site

The position of the accelerometers in the working area is shown in Figure 3 along with their distance. There are four different railway lines in experimental site. Two of them denoted as the line 1 and the line 2 belong to high-speed trains. While line 1 connects the direction between Istanbul and Ankara, line 2 serves the route of Ankara-Istanbul. Railway lines defined as 3 and 4 is used for freight transportation and suburban passenger traffic. A total of 12 accelerometers were installed in the test field, 8 of which were placed to observe ground vibrations and 4 to monitor structural vibrations at the basement and first floor of the building considered. In order to reveal the effects of kinematic interaction, which is an important component of the dynamic soil-structure interaction, on free field motion, devices with the designations SZ173 and BH153 were placed in the research area (Figure 3).

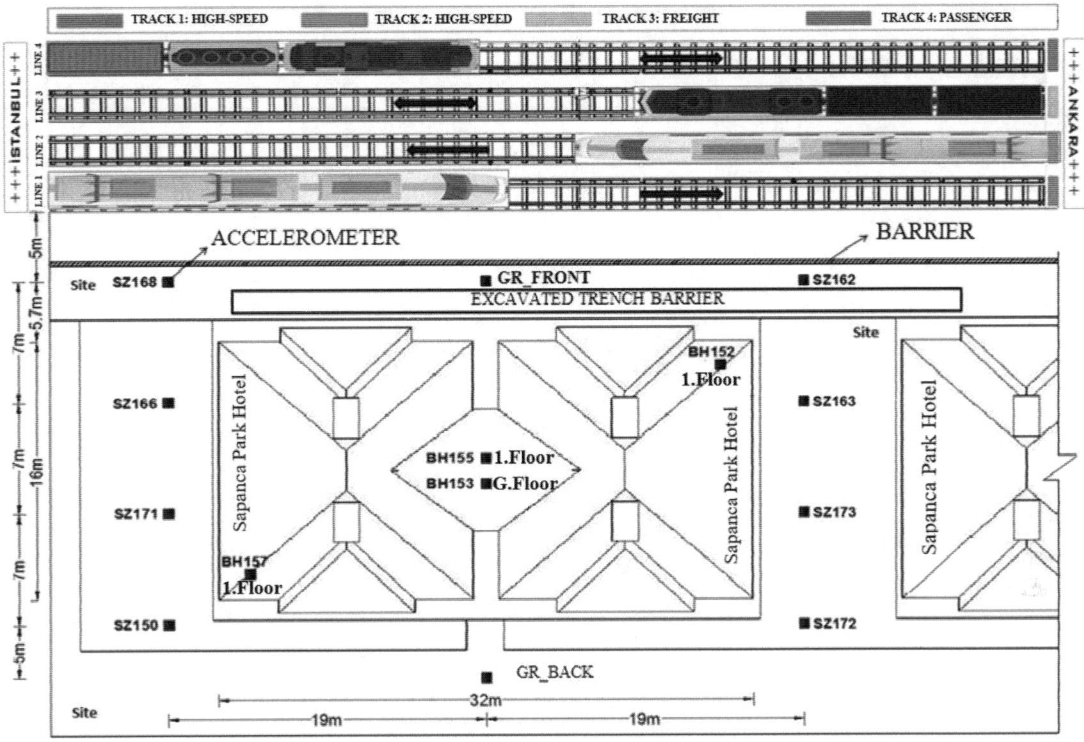

Figure 3. The layout of testing site and instrumentation

2.3. General information of train records

The timetable for repeated train crossings is given in the Table 1 below with the device numbers on which the vibrations are recorded. Structural vibrations recordings in the building are realized by accelerometers identified as BH153 and BH155, while free ground excitations are measured by instruments coded as SZ168 and SZH173.

Table 1. Timetable for repeated train passes.

Device No	2. Track	1. Track
SZ168	10:11	11:04
BH153	10:11	11:04
BH155	10:11	11:04
SZH173	10:11	11:04

2.4. Analysis of in-situ measurements

2.4.1 Kinematic Interaction: The data obtained from SZ173 and BH153 instruments were compared to investigate the kinematic interaction problem. The raw data obtained from the devices were transformed into interpretable data with the help of computer-aided software program such as SeismoSignal. Of all the data recorded by the device on observation point, only the 15-second train transit time was considered. This part of data recorded was opened in SeismoSignal, first aligned (calibrated) to the zero line which is called as baseline correction, then a Butterworth-type filter was applied to the vibration in the frequency range of 10 to 100 Hz to remove the ambient noise vibration. Finally, acceleration spectrum curves were obtained using processed time histories on free ground surface and structure. All analyses were carried out separately for vibrations propagating parallel and perpendicular to the train line during the passage of trains. Response spectra and transfer functions obtained from acceleration time histories are given in Figures 4-6 for two different times of train transitions.

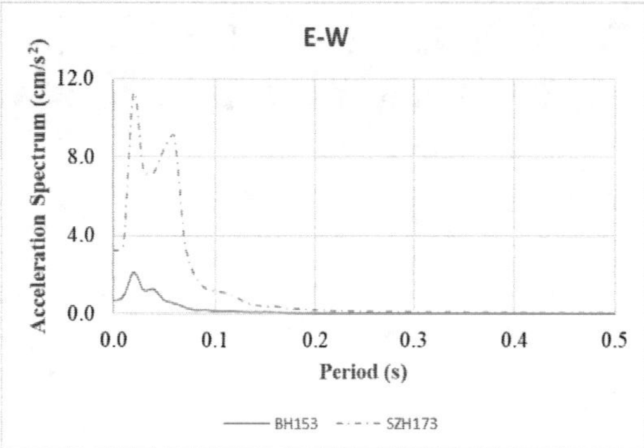

(a) Train at 10.11

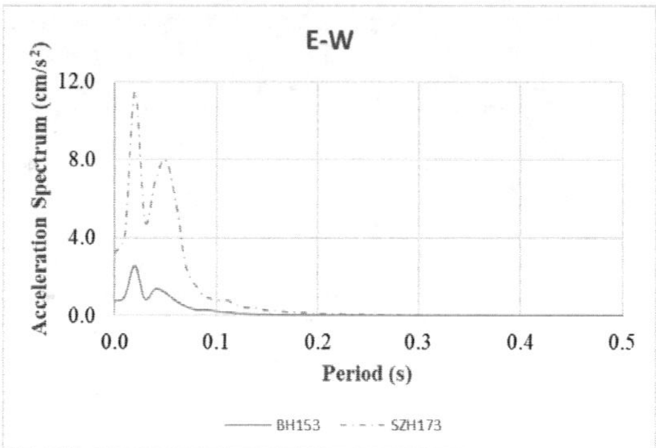

(b) Train at 11.04

Figure 4. Acceleration spectra for parallel to the railway.

When the measurement results obtained at different times are compared, it is clear seen that kinematic interaction in the parallel direction reduces the vibration level of the free field motion. As shown in Figure 4, the peak acceleration values for both recording times decreased for buildings with natural periods of 0.25 seconds from 11cm/s^2 to 3cm/s^2.

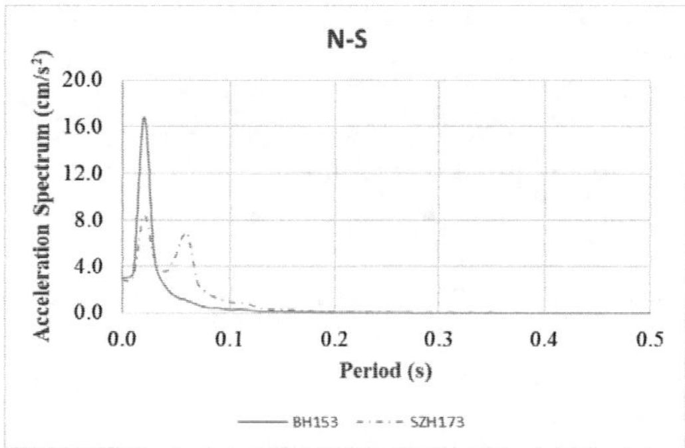

(a) Train at 10.11

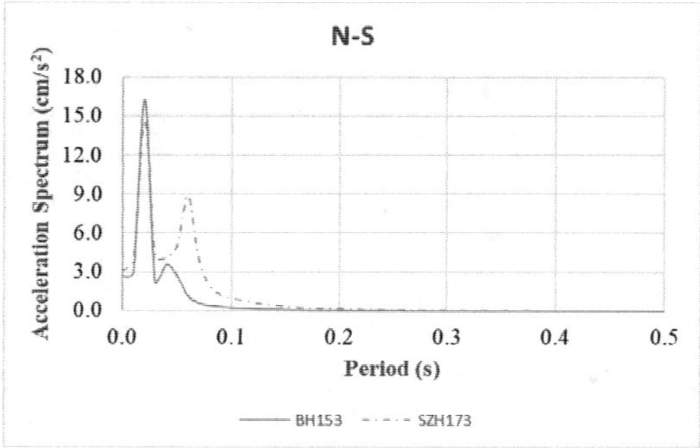

(b) Train at 11.04

Figure 5. Acceleration spectra for perpendicular to the railway.

It is even seen in Figure 5 for both recording times that the kinematic interaction in the perpendicular direction increases the vibration level of free field motion.

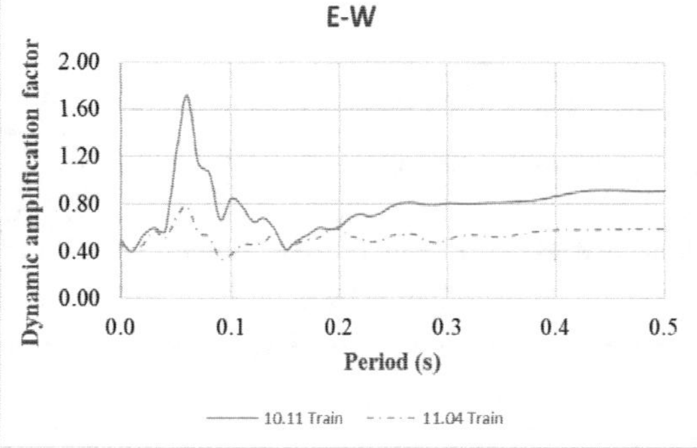

(a) Parallel to the Line

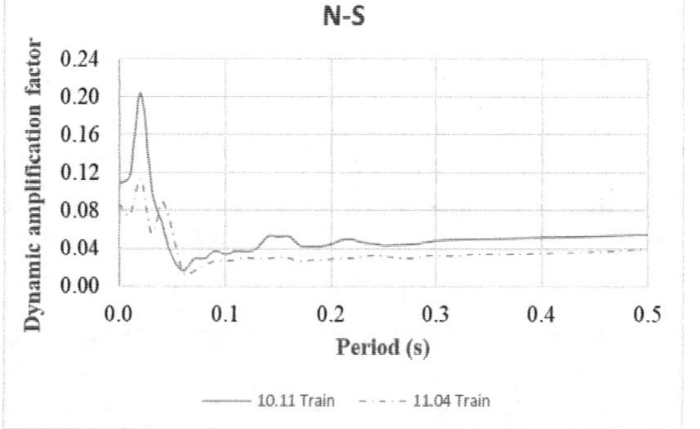

(b) Perpendicular to the line

Figure 6. Transfer functions.

The discrepancy of the transfer functions for two different transit times of trains that directly indicate the kinematic interaction effect in the considered direction to the train line is shown in Figure 6a-b.

2.4.2 Evaluation of the vibration data according to regulations

2.4.2.1 *Federal Transit Administration (FTA)*: Decibel notations are employed for assessing the vibration level according to the FTA criteria. Root Mean Square (RMS) values were achieved by taking the square root of the mean squares of the amplitude of the velocity values after the raw data were processed and the acceleration data were transferred to the velocity data. The maximum RMS value can be calculated as follow,

$$L_v = 20\log\left(\frac{v}{v_{ref}}\right)$$

where υ is the maximum RMS amplitude in m/s and the υ_{ref} is the reference velocity amplitude taken as $2.54\text{x}10^{-8}$ m/s. The obtained velocity level is evaluated in terms of threshold values in the regulations for human response and structural damage are given in Table 2 and 3, respectively.

Table 2. Threshold vibration speed levels in the regulation for human perception.

Vibration Velocity Level	Human Response
65 VdB	Approximate threshold of perception for many humans. Low-frequency sound usually inaudible. Mid-frequency sound: excessive for quiet sleeping areas.
75 VdB	Approximate dividing line between barely perceptible and distinctly perceptible. Many people find transit vibration at this level annoying. Low-frequency noise: tolerable for sleeping areas. Mid-frequency noise: excessive in most quiet occupied areas.
85 VdB	Vibration tolerable only if there are an infrequent number of events per day. Low-frequency noise: excessive for sleeping areas. Mid-frequency noise: excessive even for infrequent events for some activities.

Table 3. Regulation Limit Values for Building Damage

Ground-Borne Vibration and Noise Impact Criteria				
Land Use Category	Ground-Borne Vibration Impact Levels (VdB re 1 micro inch/sec)		Ground-Borne Vibration Impact Levels (dB re 20 micro Pascals)	
	Frequent[1] Events	Infrequent[2] Events	Frequent[1] Events	Infrequent[2] Events
Category 1: Buildings where vibration would interfere with interior operations.	65 VdB[3]	65 VdB[3]	N/A[4]	N/A[4]
Category 2: Residences and buildings where people normally sleep.	72 VdB	80 VdB	35 dBA	43 dBA
Category 3: Institutional land uses with primally daytime use.	75 VdB	83 VdB	40 dBA	48 dBA

Notes:
1. *Frequent Events* is defined as more than 70 vibration events per day.
2. *Infrequent Events* is defined as fewer than 70 vibration events per day.
3. This criterion limit is based on levels that are acceptable for most moderately sensitive equipment such as optical microscopes. Vibration-sensitive manifacturing or research will require detailed evaluation to define the acceptable vibration levels. Ensuring lower vibration levels in a building often requires special design of the HVAC system and stiffened floors.
4. Vibration-sensitive equipment is not sensitive to ground-borne noise.

The transformed measurement data of train passage at 10.11 for the evaluation criteria defined in the FTA are summarized in Table 4.

Table 4. Transformed measurement data for evaluation criteria defined in FTA of train passage. at 10.11

Device No	Max RMS		Max Velocity (mm/s)		Max Lv (VdB)	
	E-W	N-S	E-W	N-S	E-W	N-S
SZ168	0.175	0.200	0.248	0.283	76.782	77.935
BH153	0.014	0.069	0.020	0.098	54.870	68.682
BH155	0.020	0.071	0.029	0.100	57.989	68.901

The data obtained from the SZ168 device at 10.11 for train passage shows that the vibration level in parallel and perpendicular direction to the train line exceeds the limit values of the regulation for human perception as indicated in Table 4. When the structural vibration measured with BH153 and BH155 accelerometers are compared with the criteria values specified in the regulation, it is seen that the vibration velocity level exceeds the threshold values only in the perpendicular direction to the railway line.

Regarding the regulation for structural damage (Table 3), the data obtained from SZ168 device indicate that the vibration level exceeds the acceptable values in parallel and perpendicular direction to the train line for the land use category 1 in the case of infrequent events. Likewise, when considering the records monitored from BH153 and BH155 devices, the vibration level does not exceed the regulation limit values parallel to the train line in terms of structural damage for land use category 1. However, the same situation cannot be mentioned for other vibration direction as shown in Table 4.

Table 5. Transformed measurement data for evaluation criteria defined in FTA of train passage at 11.04.

Device No	Max RMS		Max Velocity (mm/s)		Max Lv (VdB)	
	E-W	N-S	E-W	N-S	E-W	N-S
SZ168	0.189	0.218	0.268	0.308	77.440	78.667
BH153	0.020	0.052	0.028	0.074	57.774	66.289
BH155	0.032	0.049	0.045	0.070	62.015	65.745

The converted measurement results for the evaluation criteria defined in the FTA for the train passage at 11.04 are shown in Table 5. The same results concerning human perception and structural damage were obtained when the vibration recordings were also evaluated for this transition time of high-speed trains.

2.4.2.2 Human Exposure to Vibration in Buildings/Effects of Vibration on Structures (DIN 4150-2/3) In order to make a comparison with the limit values defined in this German standard, it is necessary to obtain the frequency content of the vibration records, especially the dominant frequency of the dynamic source which is obtained by Fast Fourier Transform depicted in Figures 7-10. The dominant frequency and maximum speed values of the measured records were compared with the limit values in the DIN4150 as given in Tables 6-9. Evaluation criteria for human health and comfort according to DIN 4150-2 are given in Table 6.

Table 6. Threshold vibration speed levels in the regulation for human perception.

Approximate Vibration Level	Degree of Perception
0.1 mm/s	Not felt
0.15 mm/s	Threshold of perception
0.35 mm/s	Barely noticeable
1.0 mm/s	Noticeable
2.2 mm/s	Easily noticeable
6 mm/s	Strongly noticeable
14 mm/s	Very strongly noticeable

Evaluation criteria for structural damage according to DIN 4150-3 are given in table 7.

Table 7. Regulation limit values for building damage.

Type of Structure	Vibration Threshold for Structural Damage, PPV (mm/s)				
	Short-Term				Long-Term
	At Foundation			Uppermost Floor	Uppermost Floor
	0 to 10 Hz	10 to 50 Hz	50 to 100 Hz	All Frequencies	All Frequencies
Commercial /indusrial	20	20 to 40	40 to 15	40	10
Residential	5	5 to 15	15 to 20	15	5
Sensitive/Historic	3	3 to 8	8 to 10	8	2.5

The measured peak velocity values parallel to and perpendicular to the train line are given in Table 8 for transition time at 10.11 of high-speed train.

Table 8. Speed values obtained from acceleration data at 10.11 of train.

Device No	Max Velocity (mm/s)	
	E-W	N-S
SZ168	0.248	0.283
BH153	0.020	0.098
BH155	0.029	0.100

Table 9 summarizes the peak velocity values parallel to and perpendicular to the train line obtained from the acceleration time history for passage time at 11.04 of high-speed train.

Table 9. Speed values obtained from acceleration data at 11.04 of train.

Device No	Max Velocity (mm/s)	
	E-W	N-S
SZ168	0.268	0.308
BH153	0.028	0.074
BH155	0.045	0.070

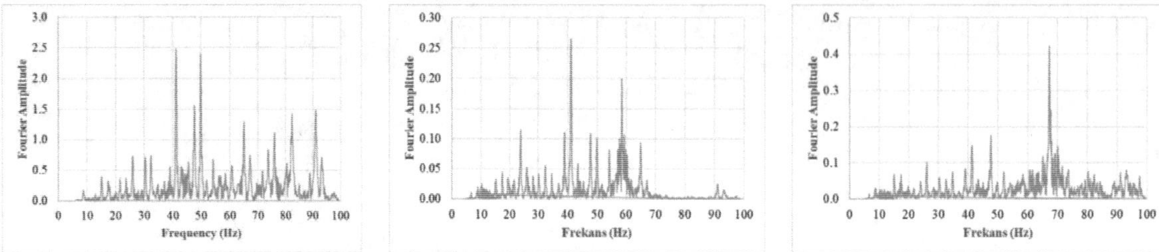

(a) SZ168 (b) BH153 (c) BH155

Figure 7. Fourier amplitudes parallel to the railway at 10.11 of train.

Evaluating the frequency content of the vibration recordings in the parallel direction to the railway shown in Figure 7, the dominant frequencies are determined as 41 Hz, 42 Hz and 68 Hz, for accelerometers defined as SZ168, BH153 and BH155, respectively. When the vibration record of SZ168 is examined in terms of structural damage, it is seen that the measured maximum velocity does not exceed the acceptable limit value given in the regulation. Furthermore, the data obtained from BH153 and BH155 devices do not exceed the regulation limit values in parallel direction to the train line according to the dominant frequency indicated in Figure 7b-c.

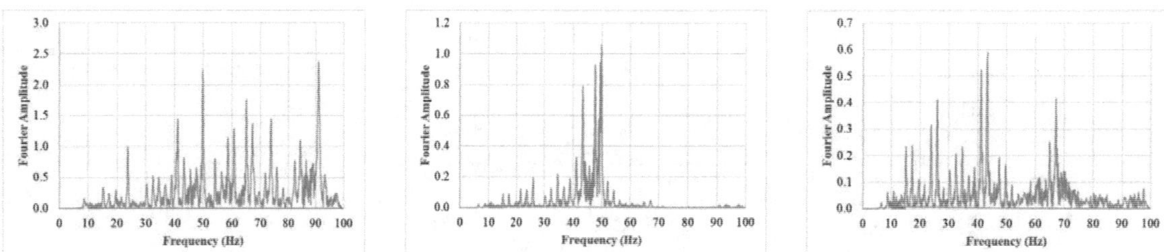

(a) SZ168 (b) BH153 (c) BH155

Figure 8. Fourier amplitudes perpendicular to the railway at 10.11 of train.

Reviewing the frequency content of the vibration recordings in the perpendicular direction to the railway illustrated in Figure 8, the dominant frequencies are observed as 91 Hz, 50 Hz and 43 Hz for accelerometers denoted as SZ168, BH153 and BH155, respectively. Herein, it is seen that the limit values are not exceeded for the vibration data received at all measuring points in terms of structural damage.

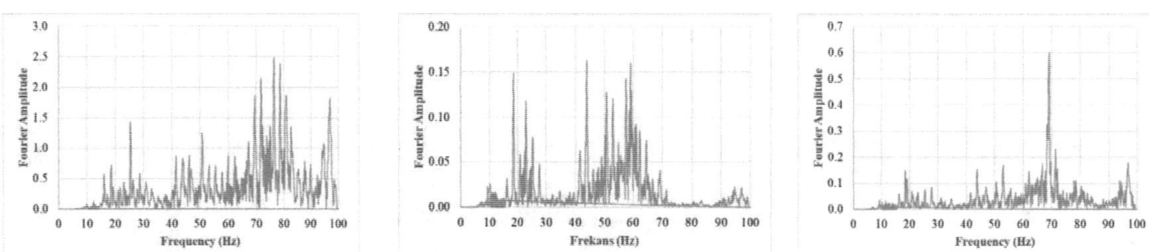

(a) SZ168 (b) BH153 (c) BH155

Figure 9. Fourier amplitudes parallel to the train line at 11.04 of train.

As shown in Figure 9, the Fourier amplitudes are obtained by using Fast Fourier Transform as 77 Hz, 42 Hz and 69 Hz for all measurement devices. In terms of structural damage, the acceptable velocity values of the regulation is not exceeded in the parallel direction to the railway of the high-speed train for all considered observation points.

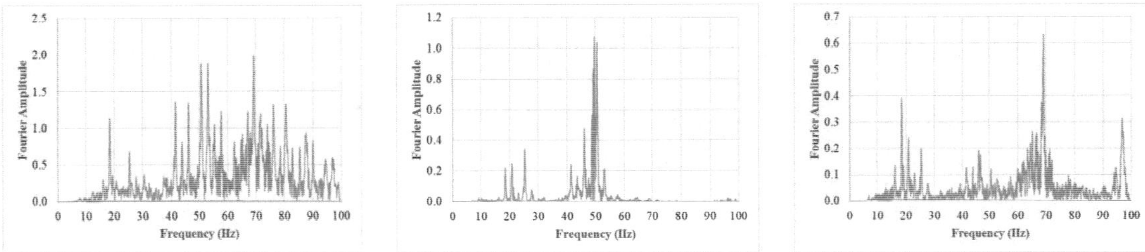

(a) SZ168 (b) BH153 (c) BH155

Figure 10. Fourier amplitudes perpendicular to the train line at 11.04 of train.

Examining the Fourier response of the vibration records for assessing the threshold values given in regulation, measured vibration level in both direction of the test field doesn't make any favourable effect on human comfort and building safety.

3. Conclusions

Turkey has launched a high-speed railway transport to be completed until 2023 and ten thousand km high-speed railway line is planned. The increase in the high-speed rail network has made the passage of urban area inevitable. With the increase of train traffic in urban environments, the vibration effects propagating to the densely populated residential district during the train passing have come out as a problem. In this study, various field studies have been conducted to observe the vibration effects of high-speed trains. Structural and non-structural vibration recordings were analysed. The following results are inferred from the analysis results:

- It was observed that the effect of kinematic interaction on free field vibration caused by the rigidity of the building foundation was different in both directions.
- Analysing the obtained vibration data, it was concluded that the vibration effects that propagate in parallel and perpendicular directions during the high-speed train passing are different and therefore vibration measurements should be performed in all directions.
- When vibration measurement results are compared according to the criteria given in terms of human comfort in international standards;
  - when FTA is considered, it is seen that vibration on the free field can be perceived at a disturbing level, but the vibration in the structure is below the threshold value in terms of human response.
  - when the same measured data are evaluated according to DIN4150-2, it is observed that the free ground motion is above the threshold of perception but it cannot be felt inside the structure.
- When vibration records are evaluated according to the criteria given in terms of structural response in international standards;
  - according to FTA, the vibration effects propagated in the parallel direction to the train line remained under the acceptable limit for moderately sensitive equipment, but the vibration effects propagated in the direction perpendicular to the train line exceeded this level,
  - Considering the DIN4150-3, all vibration effects were observed to be under the defined threshold levels.

Acknowledgement

This research is funded by TUBITAK (The Scientific and Technological Research Council of Turkey) under the grant No: 217M427. The support of TUBITAK is greatly appreciated.

IOP Conf. Series: Materials Science and Engineering **800** (2020) 012005 doi:10.1088/1757-899X/800/1/012005

References

[1] Krylov, V.V. (1996), “Vibration impact of high-speed trains effects of track dynamics”, J. Acoust. Soc. Am., 100(5), 3121-3133.

[2] Massarsch, K.R. (2000), “Settlements and damage caused by construction-induced vibration”, Proceedings of international workshop wave 2000, Chouw N, Schmid G, editors, pp. 299-315, Bochum, Germany, Balkema, Rotterdam, 13-15 December 2000.

[3] Xia, H., Zhang, N. and Cao, Y.M. (2005), “Experimental study of train-induced vibrations of environments and buildings”, J. Sound Vib., 280(3-5), 1017-1029.

[4] Goktepe F., Kuyuk H.S., Celebi E., In-situ measurement of railway-traffic induced vibrations nearby the liquid-storage tank, Earthquakes and Structures, Sakarya University, Institute of Science, Department of Civil Engineering, 12 (5), 583-589, (2017).

[5] Celebi E., Zulfikar A.C., Kırtel O., Goktepe F. In-situ measurement of environmental vibrations induced by high-speed trains and mitigation of structural vibrations, TUBITAK 1001 Project no: 217M427, (2018-present).

[6] FTA (2005), Federal Transit Administration Report, High-speed Ground Transportation Noise and Vibration Impact Assessment. U.S.A.

[7] DIN 4150-2 (1999), Structural vibration - Human exposure to vibration in buildings. Germany.

[8] DIN 4150-3 (1999), Structural vibration - Effects of vibration on structures. Germany.

Physico-mechanical characterization of high-volume fly ash cement paste composites

E Aydın
Department of Civil Engineering, European University of Lefke, Lefke, Northern Cyprus, TR-10

Email: eraydin@eul.edu.tr

Abstract. The current annual worldwide production of coal ash is estimated at one billion tons of which at least 70-75 % is fly ash. A huge amount of such wastes can seriously damage the environment and ecology of the planet. In this study, a high volume of fly ash was used to produce environmentally friendly building products. The composites were designed as pure cement, and as a result, no aggregates were used. Composites are classified as non-load-bearing elements in the construction sector. Physico-mechanical characterisation of laboratory-produced composites was evaluated based on dry density, water absorption, apparent specific gravity, porosity, and compressive and flexural strength tests. Based on the test results, fly ash can be used as an alternative binder for building products. Water absorption test results of composites produced by 40% fly ash showed satisfactory results. Dry density measurements proved the lightweight nature of the composites. Compressive and flexural strength test results were within the limit specified by the international authorities in the building sector. The composites comprising 60% fly ash attained 30 Mpa compressive strength at 28 days. The produced samples can be used effectively for various civil engineering works. The utilisation of fly ash in construction could reduce the carbon dioxide emission and produce a better sustainable solution.

1. Introduction

Many research studies have been conducted on high volume fly ash (HVFA) optimization in cementitious composites since the 1970s [1]. The utilization of fly ash as a binder in cement paste and mortar applications has increased rapidly in the 2000s [2], and this trend is assumed to continue in the upcoming periods due to the industrial and technical feasibility [3]. To universally recognize fly ash in cement and concrete as an alternative binder material depends on a detailed understanding of its performance both in the fresh and hardened state [4]. A significant achievement has been reached in recent years based on various evaluating techniques that have been used to characterize fly ash to fulfil its possible applications in bulk utilization. These evaluations assessed appropriate physico-chemical properties, such as chemical stability against the aggressive environment, carbon content, particle size distribution, water absorption, pozzolanicity, and porosity [5]. However, the aforementioned properties have not fully satisfied the criteria adopted by international authorities for evaluating the overall quality of the pure cement composites. This might be because of unsatisfactorily analyzing the properties of fly ash in real and standard testing applications [6].

Many studies have indicated that adding fly ash to cement paste and mortar mixes results in decreased frictional forces among the particles and thus improved flow characteristics [7], excellent pumpability, and reduced tendency of water gain [8]. Many previous studies have mentioned that the decrease in water demand is because of "ball-bearing effects of spherical fly ash particles" [9]. Helmuth suggested that "flocculation and dispersion effects" are more important than the "lubricating effects" in water reduction of spherical particles [10]. Water requirement and carbon content are directly proportional unlike the presence of coarse fly ash particles. Helmuth stated that appropriate dispersion of fly ash particles on cement grains decreases water amount. This effect is the same as that of water reducers [11].

Tokyay proved that Class C fly ashes may not be as effective as Class F fly ashes in reducing heat rise during hydration due to their relative reactivity [12, 13]. Low lime fly ashes generally reduce the degree of heat rise when used as a cement replacement. High lime fly ashes are usually more reactive than are Class F fly ashes, so they do not essentially cause a decrease in heat evolution. Based on field applications in Canada, 30 % of fly ash replacement have been reported to decrease the heat of hydration [13].

Previous studies showed that for low strength and high strength application, 30-80 % fly ash replacement level can be used [14]. In load-bearing applications, the level of cement replacement ranges from 15 % up to 30 %, the maximum acceptable limit for strength development [15].

Carette and Malhotra investigated the performance of eleven different types and various origins of Canadian fly ashes. Their study consisted of two lignitic, three subbituminous, and six bituminous fly ashes. They stated that, besides the fly ash type, a 20 % fly ash replacement with cement caused lower compressive strength in the early period of hardening [7]. Several studies have been performed to optimize the fly ash level for mortars and concretes in load-bearing elements [16]. Swamy et al. indicated that concrete mixes composed of 30 % Class F fly ash by weight of binder amount could be proportioned to have satisfactorily workable mixes. Additionally, the percentage of water reducer was adjusted to attain the desired workability with a slump value of more than 100 mm for easy placing in structural grade reinforced concretes [17].

For roller-compacted concrete or mass concrete applications, high replacement level of fly ash is usually restricted. However, fly ash level of 40 to 60 % can be used adequately in normal structural concrete [18]. By contrast, it is believed that higher replacement percentages are feasible in low strength and non-structural grade concrete applications, such as in highways and road-bases [16] as well as in low w/c with 50 % ASTM Class F structural grade concrete [7].

For HVFA concrete's site applications, compressive strength at 28 and 90 days was reported as 35-50 MPa and 50-70 MPa, respectively. Samples obtained from large blocks composed of HVFA concrete attained 110 MPa compressive strength beyond 10 years [19]. Wilbert et al. mentioned that optimum fly ash replacement level should be between 55 % and 60 % for structural grade concrete [7]. Gopalan determined the optimum quantity of fly ash for structural grade concrete applications. Strengths of the mixtures were evaluated at 7 and 28 days. He mentioned that the ideal fly ash replacement level was directly related to the method of mixture proportioning. Additionally, the author stated that evaluation of composite behaviour using "cementing efficiency of fly ash" was unpredictable and the rate of strength gain in concrete was directly related to the "gel/space ratio." The study proved that a "gel/space ratio" of 0.70 is the optimum for fly ash concrete considering the strength properties [20].

Taniguchi et al. evaluated the performance of HVFA concrete for use in the construction of marine structures. They reported that the strength characteristics of HVFA concrete depended on amount and type of chemical admixture and curing regime. In their study, NaCl was used as a chemical activator, and test results showed higher early strength and excellent rate of strength gain with curing. Additionally, HVFA concrete exhibited good resistance against seawater with respect to volume changes [21].

Cuijuan et al. conducted experiments to examine the effect of the optimum high lime fly ash replacement on concrete strength. The experimental data showed that concrete composed of 25 % fly ash or less was an optimal level to achieve higher strength. When the replacement level increased to 40 %, concrete either increased or attained the equal strength without fly ash concrete for both curing ages of 90 days and 180 days [22].

An experimental study conducted at the University of North Dakota on high calcium fly ash concretes selected ash replacement level of 30 %. As a result, high lime fly ash resulted in higher early strength compared to Class F fly ash [23].

Hooton examined the effect of incorporating high alkaline fly ash in concrete applications. The results showed that the availability of the high amount of alkalis in fly ash does not produce expansive gel and does not contribute to an alkali silicate reaction. The fly ash replacement level of 35 % showed lower compressive strength of concrete at 1-day and identical strength beyond 7 days. However, beyond 28-days, higher strengths were recorded [24].

Sivasundaram et al. proportioned several concrete mixtures composed of HVFA (low lime), superplasticizer, and AEA. Blocks with dimensions of 1.52x1.52x1.52 m were manufactured to evaluate the rate of heat rise on the hydration of cementitious materials. The HVFA concrete composed of superplasticizer showed superior performance. The concrete attained higher compressive strength at all testing ages. Based on this research, the samples attained 40 MPa compressive strength at 91-days (w/c ratio of 0.33) and 53 MPa (w/c ratio of 0.28). Based on the field performance evaluation of this concrete, a concrete block with 8.84x7.32x 2.74 m dimensions was also produced under controlled conditions. Thermal cracks were not observed in this study, and strengths were similar to those produced in the laboratory [25]. In another study, Sivasundaram et al. showed that the long-term performance of HVFA concrete was excellent with respect to compressive strength [26].

Naik et al. examined the engineering properties of concrete composed of a high amount of low lime fly ashes obtained from two different sources. Fly ash replacement level was between 0-60 % to attain 41 MPa at 28-days. The superplasticizer was used to preserve the consistency of the design mixture groups and adjust the w/c ratio. The compressive and tensile strength of the produced concrete increased with testing age and decreased with fly ash amount. For compressive strength, 30 MPa was measured at 28 days at a 60 % fly ash replacement. The study revealed that the concrete containing up to 60 % fly ash could be used to meet fresh and hardened requirements for load-bearing applications [27].

Naik et al. prepared three mixture groups for concrete which can be used in paving work. The mixtures comprised 20 % and 50 % Class C fly ash and the third mix contained 40 % Class F fly ash. Total binder amount for three mixture groups was about 270 kg and contained an AEA and superplasticizer. The compressive strength of mixtures containing either 40 % low lime fly ash or 50 % high lime fly ash had lower compressive and tensile strengths compared to the reference mixture containing 20 % high lime fly ash. The HVFA concrete mixes showed similar results with respect to compressive and tensile strengths. The authors stated that for paving applications, concrete could be produced with 40 % of the low lime fly ash at a w/c below 0.36. Additionally, concrete containing 50 % high lime fly ash showed good performance in paving applications [27].

Langley evaluated the flexural strength of concrete composed of 55 % low lime fly ash. The flexural strength varied from 7.2-7.5 MPa (w/c ratio: 0.27) and from 5.6-6.3 MPa (w/c ratio: 0.49) considering the 91-365 days of hardening [28]. Sivasundaram et al. reported that 14 days and 91-days flexural strength values of concretes comprising 58 % of total cementitious materials ranged between 4-5 MPa and 5-6 MPa, respectively [26]. Naik et al. reported that the flexural strength of concrete containing HVFA varied from 4-4.7 MPa at 28-days and varied from 4.4-4.9 MPa at 56-days. Additionally, the splitting tensile strength was recorded as 3.5 MPa at 28-days for the medium strength grade HVFA concrete produced at Canada Center for Mineral and Energy Technology (CANMET) [27].

Mather found that cement composed of high C3A content replaced with 30 % level can decrease sulfate resistance. Conversely, Mehta, Hooton, and Manz et al. showed that high lime fly ashes may satisfactorily increase the concrete's sulfate resistance [29].

Wesche and Schubert produced concrete containing 50 % fly ash and recorded an increase in sulfate resistance when low sulfate resistance cement was used. The positive contribution of fly ash in sulfates is that it reduced the overall pH level in the pore solution due to reduction of the amount of calcium hydroxide [30-32].

High lime fly ash has been proven its performance, and it is commercially available as a building material used mostly in cement production, various concrete elements, embankments, structural fill, and road sub-bases [33].

Freezing and thawing tests results of masonry units composed of fly ash showed that these units could be expected to perform well in vertical wall construction. For severe lateral exposure, a minimum compressive strength of 21 MPa is recommended [31].

Based on the quality and safety issues, international codes had revised the minimum strength requirements for concrete masonry units and structural grade concrete construction. Block manufacturers suggested that for both light and normal weight units, minimum compressive strength should be 24.5 MPa and 34.5 MPa, respectively [7].

Based on the previous attempts stated above, the fly ash utilization level in the construction sector is still low. This study focused on the HVFA utilization as a partial replacement to cement. Pure cement paste composites were produced. The research also considered the sustainable development and manufacturing of ecological construction materials composed of HVFA.

2. Materials and methodology

OPC, conforming to ASTM C150M-12 standard, was used. The fineness and specific gravity of the cement were 293 m2/kg and 3.09, respectively. High lime fly ash was brought from Soma thermal power plant in Turkey. Water reducer was used in 70FA30C mixture groups. Tap water was used in all stages of the study. The chemical compositions of cement and fly ash are presented in Table 1.

Table 1. Chemical composition of cement and fly ash.

Chemical (%)	Soma fly ash	Cement
SiO_2	41.28	18.91
SiO_2 (insoluble)	–	0.68
Al_2O_3	22.51	4.96
Fe_2O_3	5.33	3.71
CaO	20.96	62.61
MgO	2.05	1.72
SO_3	1.78	2.47
LOI*	2.46	3.73

2.1. Methodology

Four mixture groups composed of a high amount of fly ash with mass ranging from 60-100% were chosen to produce cement paste composites. Casting samples were demoulded 1 day after mixing. The samples were preserved at a controlled condition (T: 24°C and RH: 95%). Physical properties of the composites were evaluated based on a flow table, fresh unit mass, porosity, apparent specific gravity, dry unit mass and water absorption tests. Mechanical characterisation was evaluated by compressive and flexural strength tests. Tests were conducted at 7 and 28 days. Average of the six samples were used for evaluation. The studied mixture groups are presented in Table 2. In Table 2, FA denotes fly ash, C denotes cement by % mass and Adm denotes admixture addition by mass.

Table 2. Mixture Groups of the study.

Group	Fly ash (FA) (%)	Cement (%)	Admixture (%)	Water/binder (w/b)
100FA	100	0	0	0.41
80FA20C	80	20	0	0.41
60FA40C	60	40	0	0.41
70FA30C0.6 Adm	80	20	0.6	0.41

3. Results and discussions

Flow table values of the composites are shown in Figure 1. Fresh unit weights (FUW) of composites varied from 17.1 kN/m3 to 18.0 kN/m3. Generally, reduction in the frictional force can improve the fresh fly ash cement paste properties. Opposing charges on the adjacent particles of cement can exert electrostatic attraction. Increasing solid volume by pore-filling mechanisms causes also an increase in the fresh unit weight [34]. Additionally, flowability of the composites improved with increased fly ash replacement level in the paste shown in Figure 1.

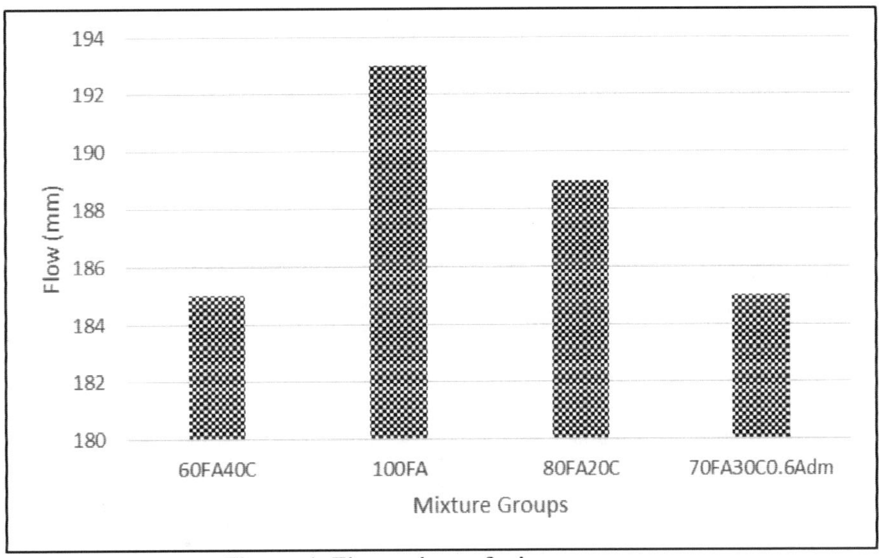

Figure 1. Flow values of mixture groups.

The reaction products of fly ash are very effective in pore refinement of large capillary voids, improving the many intrinsic properties of the matrix. Admixture addition most probably adsorbs on the water-cement interface and directly induces the surface tension of water. They begin to disintegrate into their ionic components. This affects the workability and friction among the particles [2].

Water to binder ratio (w/b) had a greater effect on the overall performance of the final composites. The reduction in w/b ratio had a positive effect on water reducer admixture in the mix. When w/b ratios increased, the WRA became ineffective. Degree of the cohesion and the size of the liquid phase caused a significant change in the fresh properties. In Figure 2, the apparent specific gravity of the mixture groups is shown. Apparent specific gravity (ASG) is an important property to evaluate the pore structure of the composite materials. As seen in Figure 2, increased replacement level of fly ash increased the ASG values.

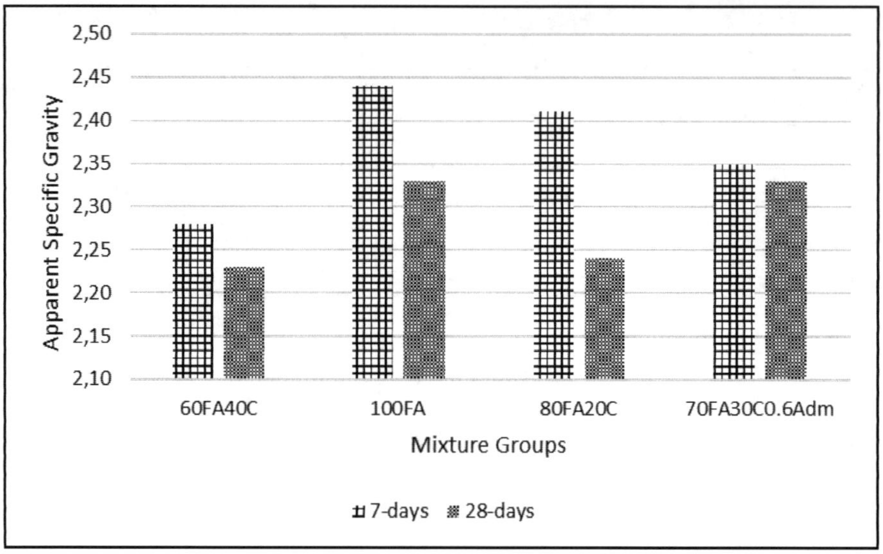

Figure 2. ASG values of mixture groups (7 and 28-days).

Figure 3 shows dry unit mass (DUM) of mixture groups is shown. Based on the DUM values of Soma fly ash mix groups, the laboratory produced composites can adequately be used in the production of aerated composites. DUM value of aerated composites varies from 300- 2100 kg/m3, and because of their air trapping capacity, composites are an accepted sound permeable and thermal comfort material commonly used in the production of lightweight wall composites. This can be considered as a perfect condition for the production of marketable lightweight panels and factories as well as residential building materials. Because of their partial reaction, the hydration products become mature. PC pastes result from the advanced crystallization of the hydration products. Thus, soluble chemicals can affect either the "degree of ionization of cement compounds" or the "degree of crystallization of the hydration products." This influences the setting characteristics of the pure paste composites [35].

A DUM value of the final composites indicates that final composites are in the range of lightweight concrete. The material is a lightweight composite with dry unit weight ranging between 11.9 and 16.5 kN/m3, which is quite lower compared to that of conventional construction materials promising beneficial outcomes in terms of construction costs and earthquake safety [4]. The final product can be used adequately in the production of lightweight aggregates and semi-isolating elements. The mixture proportion of lower mass is particularly beneficial under poor soil conditions.

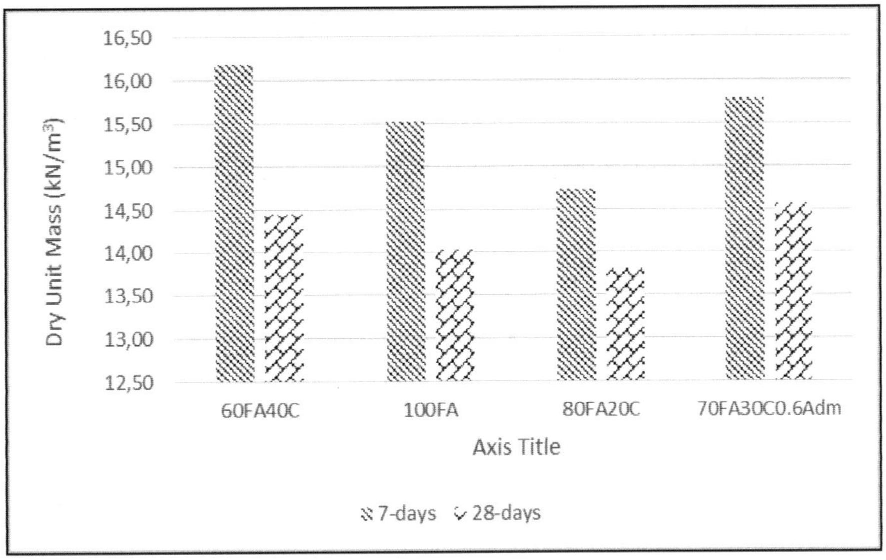

Figure 3. DUM values of mixture groups (7 and 28-days).

The water-reducing effect increases with dosages of fly ash. At a higher fly ash replacement level, pore solution occupies the voids between the fly ash particles. The behaviour is controlled by particle packing phenomena. Large pores can only be occupied by hydration products which are also required for the self-healing of cement hydrates or fly ash itself [36]. Figure 4 shows the water absorption (WA) of mixture groups. Water absorption values indicate that mixtures composed of 60% fly ash are adequate for the production of construction material, such as bricks, tiles and paving stone manufacturing. The standard limit for water absorption is nearly 16% for building materials. Although the increases of the amount of fly ash above 60% showed higher standard limits, as shown in Figure 4, it could be possible to decrease the specified limits after longer curing periods (i.e., after 28-days).

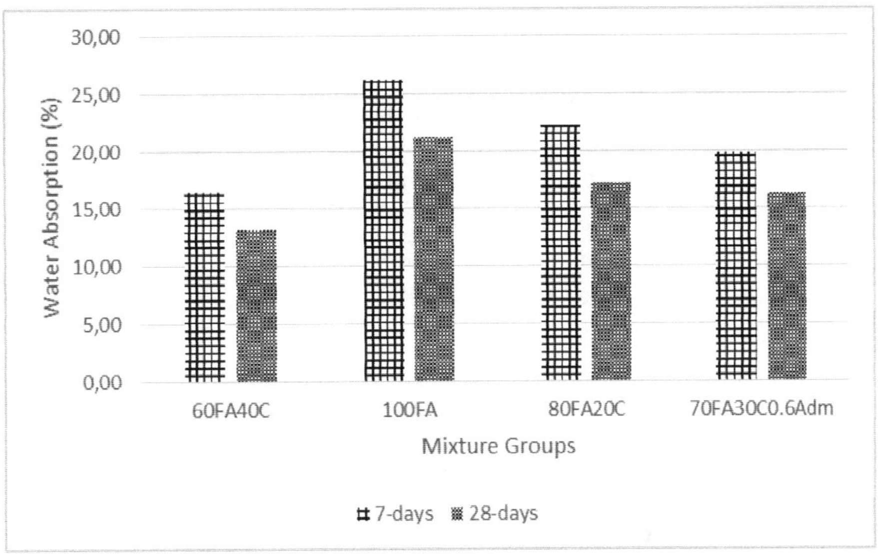

Figure 4. Water Absorption of mixture groups (7 and 28-days).

The decrease in WA values revealed that increased density resulted in the pore refinement. Figure 5 shows the porosity values of mixture groups. Fly ash replacement increases the porosity values.

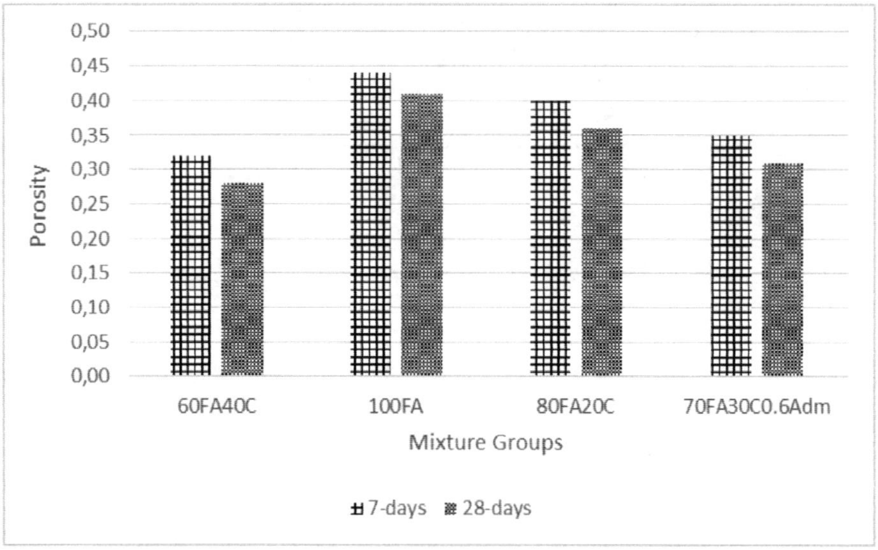

Figure 5. Porosity values of mixture groups (7 and 28-days).

As illustrated in Figure 5, the WRA shows less efficiency with fly ash, although the reactivity of available glassy phase is higher to absorb water. "Fly ash particles are spherical in shape" and Class C particles are not easily dispersed through the solution. However, the mixtures composed of WRA decreased the water demand. This might be due to fly ash particles absorbing the molecules of WRA to form "double electrical layer" and retaining the particles away from each other. The effect of surface area, porosity and water content varied little before and after the replacement level. However, the effects reduced gradually with the increase of replacement [37]. The weak phase between the fly ash particles seemed to have decreased considerably, though certain parts had higher porosity. The bond among those two phases may be weak during the early stages of hydration. Formation of stable hydrate products due to modification of unstable regions indicated that bond develops as hydration continues [38].

Figure 6 shows the unconfined compressive strength (UCS) values. The UCS results of mix design groups of Class C fly ash at 28 days showed that the final composites can be classified as C25/C30 concrete strength class for 60FA40C groups. The increase in fly ash amount caused a slight reduction in UCS values. The optimum level for fly ash replacement seems to be 60%.

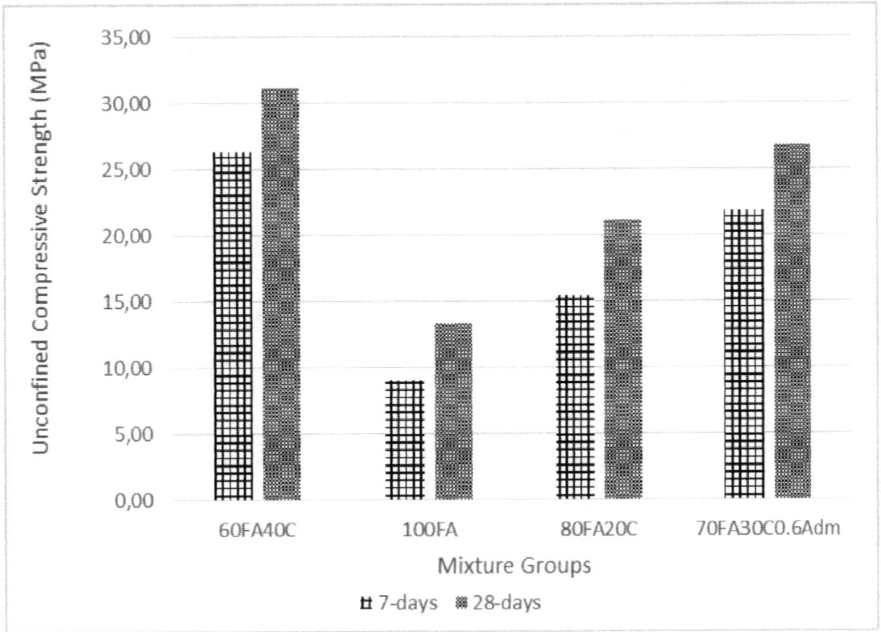

Figure 6. UCS values of mixture groups (7 and 28-days).

The rate of strength increase based on the cement hydration causes the development of bonds, and continuous hydration can decrease in overall porosity of the composites. The volume of hydration products in cement paste remains nearly constant during the early period. However, the volume of solids within the bulk increases with age and thus causing a reduction in the porosity of the paste. The UCS increases with decreasing porosity of the composites. This improvement in UCS may be due to the development of hydration products that are gradually replaced by liquid between cement particles. The space between the particles decreases and the overall porosity of the paste reduces with ongoing hydration [39]. Considering the pozzolanic activity and hydration, the packing effect is more important for the arrangement of small particles that fill the small pores and contribute to the increase of UCS [30]. Generally, finer particles produce higher compressive strength value compared to the larger ones [40]. It can be concluded that fly ash and cement combinations significantly enhance the composite's UCS value. In general, the fly ash replacement level (above 50%) reduced the strength of the hardened cement paste at the 7-days. Based on the UCS tests results, composites have adequate strength for civil engineering applications and building construction elements, such as road bases, bricks, tiles, and ceramic.

Figure 7 shows the flexural strength (FS) values of the produced composites. FS and UCS are highly correlated, and the same trend with UCS was observed. The lowest FS values in pure fly ash mix groups are due to higher water requirement of the fly ashes that considerably reduces early and late FS values. For pavements, UCS value is less critical compared with FS values. Adequate curing is essential for fly ash-cement mixes. The rate of hydration is slow and aggressive environments, such as freeze-thaw, sulfate and chloride attack, are very risky for composite's strength. The fly ash mixes are vulnerable to those attacks, and any expansion and contraction of gel during ion-exchange of pore solution by adsorption of water in pores may reduce FS of the composites. This decrease in FS can be overcome by using chemical admixtures to stabilize the fly ash particles.

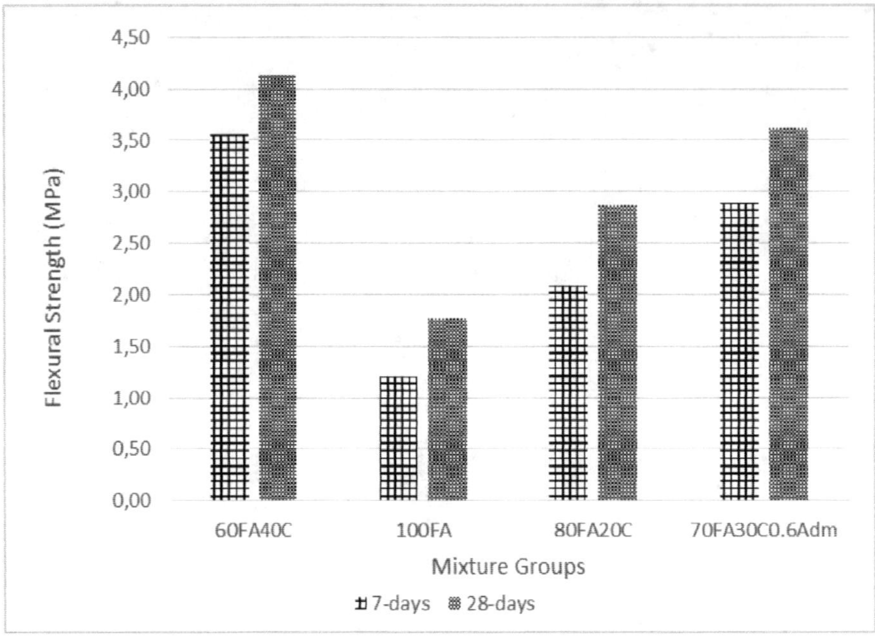

Figure 7. FS values of mixture groups (7 and 28-days).

4. Conclusions and recommendations

1- The final composites can be used satisfactorily in the manufacturing of aerated composites. DUM values of the laboratory-produced composites vary from 300- 2100 kg/m3.

2- The material is a lightweight and sound composite; the unit weight ranges between 11.9– 16.5 kN/m3, which is quite lower than those of conventional material of construction.

3- The efficiency of the water reduction in high lime fly ash is low in the absence of water reducers.

4- Most water absorption (WA) values of the HCP composites are higher than that of WA values specified in the related standards. However, the absorption value of the mixture groups composed of 60% fly ash satisfies the standard limits (lower than 16%).

5- The 28 days UCS results of the 60FA40C mix groups show that the final composites can be classified as C25/C30 concrete strength class.

6- FS value of 6.9 to 8.8 is required for vehicular paving applications. The design groups had failed for load-bearing applications.

7- It is recommended to study fly ashes with different fineness values to investigate in more detail the microstructure, freezing-thawing resistance, and abrasion resistance of HVFA cement paste composites.

References

[1] EPA-452/R-03-006 Report 2003 *Economic Impact Analysis for the Brick and Structural Clay Products Manufacturing NESHAP: Final Rule*, 25 pp.

[2] Aydin E and Arel H Ş 2017 *Constr. Build. Mat.* **157** 96-107

[3] Aydin E and Arel H Ş 2018 *Data in Brief* **16** 321-326.

[4] Aydin E 2009 *Fifth International Conference on Construction in the 21st Century (CITC-V) Collaboration and Integration in Engineering, Management and Technology* May 20-22, Istanbul, Turkey p.1526-1535.

[5] Bouzoubaa N Zhang M H and Malhotra V M 2000 *Cement Concrete Res.* **30** 1037-1046.
[6] Bouzoubaa N and Fournier B 2003 *Cement Concrete Res.* **33** 1029-1037.
[7] Naik T R 1992 *Report on CBU-1992-15* 169 pp.
[8] Erdoğan T Y 1997 *Admixtures for Concrete* (Middle East Technical University Press) 188 pp.
[9] Joshi R C 1979 *Proceedings of the Fifth International Fly Ash Utilization Symposium, Atlanta* p. 610-623.
[10] Helmuth R A 1987 (*Portland Cement Association*) 203-205.
[11] Helmuth R A 1986 *Proceedings of the Second International Conference on the Use of Fly Ash, Silica Fume, Slag and Natural Pozzolans in Concrete, Madrid, Spain* (Malhotra V M Ed) **ACI SP-91** 723-740.
[12] Tokyay M 1988 *Cement Concrete Res.* **18** 957-960.
[13] Siddique R and Khan M I 2011 **Supplementary Cementing Materials** (Springer-Verlag Berlin Heidelberg) 288 p.
[14] Aydin E 2017 *Constr. Build. Mater.* **153** p. 393-401.
[15] Ravina D and Mehta P K 1988 *Cement Concrete Res.* **18** 571-583.
[16] Aydin E and Balkis A P 2017 *ASTM J Testing Eval* **45 (6)** 2029-2038
[17] Swamy R N Sami A R A and Theodorakopoulos D D 1983 *ACI Mat Journal* **1** 414-423.
[18] ACI 318-19 2019 (*ACI Publications*) 391 pp.
[19] Zhang M H Bilodeau A and Malhotra V M 1998 *Cement Concrete Res.* **28 (11)** 1555-1569.
[20] Gopalan M K 1991 *Proceedings: Shanghai Ash Utilization Conference*, **2 (EPRI Report No. GS-7388**) 59-1 to 59-10.
[21] Taniguchi K Suzuki T Shimomura Y Ohga H and Nagataki S 1989, *Proceedings of the Third CANMET/ACI International Conference on Fly Ash, Silica Fume, Slag, and Natural Pozzolans in Concrete, Supplementary Papers, Trondheim, Norway* pp. 66-91.
[22] Cuijuan S Loushu G and Haimin W 1986 *Proceeding of the Second International Conference on Fly Ash, Silica Fume, Slag and Natural Pozzolans in Concrete, Madrid, Spain* (Malhotra V M Ed., **1 ACI Publication SP-91**) 387-412.
[23] Manz O E 1970 *Electric Power Symposium on the Use of Ash, in Particular in Production of Concrete and Prefabricated Construction Elements, Ankara.*
[24] Hooton R D 1986 *Proceedings of the Second International Conference on Fly Ash, Silica Fume, Slag and Natural Pozzolans in Concrete, Madrid, Spain (*Malhotra V M Ed., **1 ACI Publication SP-91**) 333-345.
[25] Sivasundaram V Carette G G and Malhotra V M 1987 *Proceedings of the Eight International Coal Ash Utilization Symposium* **2** (ACAA Washington, D C) 34-1 to 34-13.
[26] Sivasundaram V Carette G G and Malhotra V M 1989 *Third International Conference on the Use of Fly Ash, Silica Fume, Slag and Natural Pozzolans in Concrete, Trondheim, Norway.*
[27] Naik T R Ramme B W and Tews J H 1991 *Proceedings: Shanghai Ash Utilization Conference* **2 (EPRI Report No. GS-7388**) 60-1 to 60-20.
[28] Langley W S 1991 *70th Annual meeting of transportation research board, Washington, D.C.*
[29] Ramyar K 1993 Effects of Turkish Fly Ashes on the Portland Cement Fly Ash Systems *PhD Thesis*, 208 pp.
[30] Wesche K 1991 *Fly Ash in Concrete: Properties and Performance* (Chapman and Hall, Newyork) 256 pp.
[31] Mather K 1982 *George Verbeck Symposium on Sulfate Resistance of Concrete* (**SP-77, ACI, Detroit**) 63-74.
[32] Tütünlü F 2000 The Utilization of Fly Ash in Manufacturing of Building Bricks. *M.Sc. Thesis.*
[33] Aydin E 2009 *International Conference on Concrete Repair, Rehabilitation and Retrofitting, Cape Town, South Africa (*London: CRC Press) 109-115.
[34] Alshamsi A M Alhosani K I and Yousri K M 1997 *Magazine of Concrete Research* **49 (179)** 111-115.
[35] Mehta P K and Monteiro P J M 2001 **Microstructure,Properties and Materials**, 2nded. 239 p.

IOP Conf. Series: Materials Science and Engineering **800** (2020) 012006 doi:10.1088/1757-899X/800/1/012006

[36] Xu A and Sarkar S L 1994 *J. Mater. Civil Eng.* **6 (1)** 117-136.
[37] Sun W Handong Y Binggen Z 2003 *Cement Concrete Res.* **2301** 1-7.
[38] Zhang M H 1995 *Cement Concrete Res.* **25 (6)** 1165-1178.
[39] Hewlett P C 1988 *Lea's Chemistry of Cement and Concrete* (John Wiley and Sons Inc, Newyork) 1053 p.
[40] Aydin E 2019 *10th international concrete conference, May 2-4, Bursa, Turkey* 419-429.

IOP Conf. Series: Materials Science and Engineering **800** (2020) 012007 doi:10.1088/1757-899X/800/1/012007

Utilization of large amounts of industrial wastes to improve sustainability of the construction sector

E Aydın
Department of Civil Engineering, European University of Lefke, Lefke, Northern Cyprus, TR-10

Email: eraydin@eul.edu.tr

Abstract. The concept of "sustainability" in the construction sector has increasingly led to the manufacturing of thermal comfort products composed of natural or recycled materials. Several such products are readily available in the market while others are still in the early stages of production or under research. The assessment of the sustainability potential of building materials as well as of entire buildings or structures is a significantly complex issue as it involves the quantification of three interacting and interdependent parameters: the environmental impact, the technical performance, and the lifetime. From this it is decidedly apparent that simply protecting the environmental impact (e.g., during the construction process) will not benefit the environment as long as an equivalent technical performance and lifetime cannot be ensured. This situation can be analysed using the definition of the so-called "sustainability potential". To this end, bottom ash and marble dust were used as a replacement for cement (up to 70%). In this study, technical performance of the laboratory-produced composites was evaluated using mechanical and durability tests. The sustainability potential of the composites was also analysed using the reduction in cost and the reduction in carbon dioxide emissions of the produced composites. The test results showed that the pure cement paste composites are adequate for targeting controlled low strength and non-load-bearing elements in civil engineering works. Durability performance also proved that the composites are moderately resistant to sulphates.

1. Introduction

Understanding the effects of global warming is more important than ever before [1]. A systematic sustainability approach for better construction has continued [2-3], and research on industrial wastes in concrete construction has become popular, providing a sustainable future [4-7]. Various types of industrial waste, such as bottom ash [8-10], fly ash [11-13], marble dust [14-16], and agricultural wastes like rice husk ash and coconut husk ash [17-18], have been satisfactorily used in building construction.

In recent years, climate change and its effect on our planet have led world leaders to take action. In a framework of growing international awareness focused on sustainability, numerous policy makers have recognized the need to decrease the amount of industrial wastes and enhance the utilization rate of those wastes. To date, the commonly accepted strategies have been either adopting a waste management plan on site or increasing taxes for landfills [19-20].

From the government's perspective, local authorities should engage in a precise evaluation of industrial waste in order to launch appropriate policies and adopt new regulations, guidelines, and practical application for sustainable construction. In addition, the correct level of waste charge, suitable encouragements for construction companies to consider practical methods, and the development of ideal waste treatment facilities have been considered in recent years for green building applications [21].

The movement towards sustainable strategies has put pressure on the acceptance of appropriate approaches to protect the planet throughout all engineering activities. The majority of pollution currently comes from construction activities and accounts for 70% of the total waste production. Furthermore, approximately 25% of the world's entire energy is depleted during the production of building materials such as cement and steel. The construction sector consumes raw materials during production, including nearly two billion tons of cement, ten billion tons of sand and gravel, and one billion tons of water. Several efforts to improve concrete sustainability and to alter concrete to a "low impact construction material" have been investigated in recent years [1, 22, 23]. An estimated 25 billion tons of concrete are produced globally every year [1]. In sustainable building strategies, especially in the case of green buildings, structures have been designed in such a way to reduce carbon emissions and optimize resources from the beginning of construction to the end of service life [1, 19]. When considering the social and environmental concerns, green buildings have the highest rank in the construction industry and have become popular in recent years [2, 19].

Evaluations of industrial by-products in pure paste and mortar are increasing. The use of such wastes as a cement replacement in paste, mortar, and/or concrete shows excellent performance. In addition, applications can help with the safe disposal of industrial wastes and the protection of the dumping site or nearby surroundings. Such efforts have a positive impact on social, environmental, and economic sustainability [24]. The European Union Directive No. 2008/98/CE encourages the reuse and recycling of waste materials. It is projected that by 2020 new buildings will be composed of at least 5% of recycled materials. [25].

Bottom ash is composed of granular particles that collect at the "bottom of coal-fueled boilers"; it has to be sieved through smaller fractions and grinded to increase its "pozzolanic activity" [10]. Bottom ash is replaced as either cement or an aggregate (fine aggregate or coarse aggregate) to produce concrete used for manufacturing building blocks and low-strength applications [26].

Singh and Siddique [25] evaluated the workability and strength of concrete containing bottom ash as a fine aggregate replacement. Their experiments revealed that the compressive strength beyond 28 days of hardening the concrete incorporated with bottom ash was identical to that of the samples produced without bottom ash. Siddique et al. [26] cast self-compacting concrete specimens containing fly ash and bottom ash with different water-to-cement ratios and evaluated the strength properties. The study revealed that the ideal percentage for replacing bottom ash with cement was less than 20%.

Maschio et al. [27] evaluated the rheological behaviour of paste and mortar samples containing fly ash and bottom ash. The composite's rheological properties were improved with bottom ash and fly ash. However, when considering the same samples, a substantial decrease in strength and an increase in the level of replacement beyond 180 days of hardening were observed.

Marble wastes have long been widely used in concrete construction to achieve the sustainability requirements. In Turkey, annual production approaches 7 million tons, accounting for 40% of worldwide production. During manufacturing processes, approximately 25% of the bulk product is transformed into large pieces or dust. Thus, dumping the products has caused serious environmental problems. However, these wastes can be effectively managed in the construction sector to produce various concrete elements, such as sewage pipes, drainage utility units, and paving for walk-ways [22, 28, 29, 30].

Mixture proportioning of self-compacting concrete composed of marble powders was proposed in one study [29]. Gesoğlu et al. [30] prepared a highly workable concrete mixture containing marble powder and fly ash and evaluated the performance of the produced samples in fresh and hardened states. Their study revealed that strength and durability properties were improved with the addition of marble powder. Topçu et al. [31] examined the self-compacting concrete properties to propose the ideal level of marble necessary to ensure a high performance. The study showed that blending no more than 200 kg/m3 of marble wastes promises excellent engineering properties. Ergun [32] studied the effect of marble wastes on the mechanical characterization of concrete. The results demonstrated that the maximum replacement level was 5% of marble wastes in high strength applications.

Li et al. [5] and Li et al. [6] prepared mortar samples composed of marble wastes and proposed a new technique for mixture proportioning: the paste replacement method. In this technique, they replaced the marble powder with a portion of paste, instead of replacing cement or fine aggregate, and considered the fixed water-to-cementitious ratio. The model developed in these studies offers numerous advantages. The model consists of large waste utilization and necessitates using less cement content in mixture proportioning. In addition, the mortar samples prepared using this technique have a smaller carbon footprint and are superior in strength and durability.

The current study focuses on the performance of pure cement paste composites. To date, few studies have considered the physical, mechanical, and durability properties in one research program. In this research, a high amount of bottom ash and marble powder were used as replacements in cement compositions to manufacture sustainable construction materials.

2. Materials and Methodology

2.1Materials

In this study, Ordinary Portland cement (Type I) was used. The Blaine fineness and specific gravity of the cement are 302 m^2/kg and 3.10, respectively. Bottom ash was obtained from the local brick factory plant. The specific gravity was 1.40. Bottom ash particles were sieved through a 200 micrometre sieve and were then used to prepare the pure paste composites. The marble wastes were collected from a dumping site near Stonite Marble Ltd. Tap water was used in all stages of the study. The chemical composition of the materials used in this study is presented in table 1

Table 1. Chemical composition of cement, bottom ash and marble dust.

Oxides (%)	Cement	Bottom ash	Marble dust
SiO_2	20.8	53.3	9.2
Al_2O_3	5.7	27.2	1.1
Fe_2O_3	2.3	9.6	0.6
CaO	65.3	0.9	44.1
MgO	0.7	0.6	5.8
K_2O	0.4	1.2	0.04
SO_3	1.2	0.4	0.07
Loss on ignition	2.9	4.1	38.7

3.2.2. Methodology

Four mixture groups contained a high volume of bottom ash and marble dust. The bottom ash and marble powder replacement level varied from 70%–100% by mass. Samples were demoulded after 1 day. The composites were immersed in a curing tank containing lime-saturated water at a temperature of 25°C and a relative humidity of 90%. Fresh and hardened tests were used to evaluate the overall performance of the laboratory-produced composites. Flow table, dry unit mass, water absorption, and porosity tests were used to characterize the quality of the cement paste composites. The compressive and flexural strength tests were used to evaluate the mechanical performance. Sulphate tests were used to analyse the behaviour of the samples under aggressive environment. All tests were performed after 7 and 28 days. Sulphate tests were conducted at 28 days. The averages of the eight samples were used for analysis. The study's mixture groups are presented in table 2, where BA denotes bottom ash, M denotes marble dust, and C denotes cement by % mass. Figure 1 shows the flow table and mini slump test measurement of pure cement paste composites. The water-to-binder ratio was chosen based on the previous laboratory works of the author [1, 4, 10]. Figure 2 shows the flow measurement results for the tested composites.

Table 2. Mixture groups of the study.

Group	Bottom Ash (BA) (%)	Marble dust (%)	cement (%)	Water/binder (w/b)
M100C0	0	100	0	0.42
BA100C0	100	0	0	0.42
M70C30	0	70	30	0.42
BA70C30	70	0	30	0.42

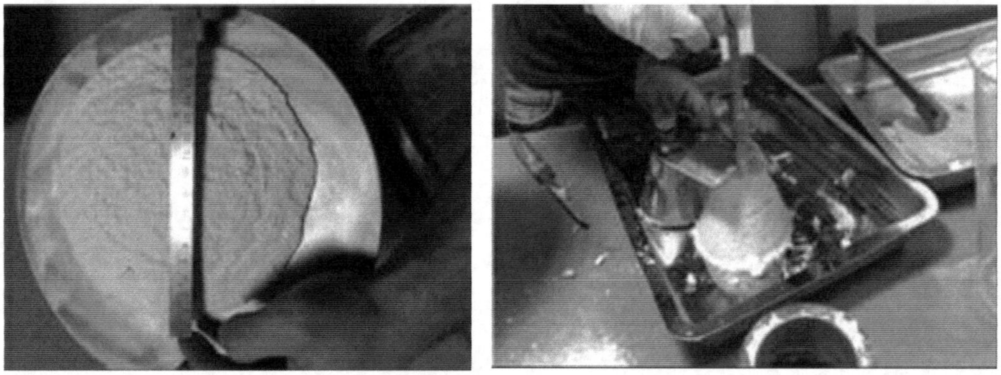

Figure 1. Flow table and mini slump test measurement for marble-bottom ash cement paste composites.

3. Results and discussions

Figure 2 demonstrates that pure bottom ash decreased the flow values more than pure marble composites. This can be attributed to the coarser character of the bottom ash particles. For the same flow values, bottom ash particles need more water than marble particles. Comparable results were obtained in previous researches [1, 4, 18, 20, 33]. Interparticle friction loss and cohesion of the mixture by the addition of marble wastes contribute to better dispersion in the matrix, thereby improving the workability of the mix.

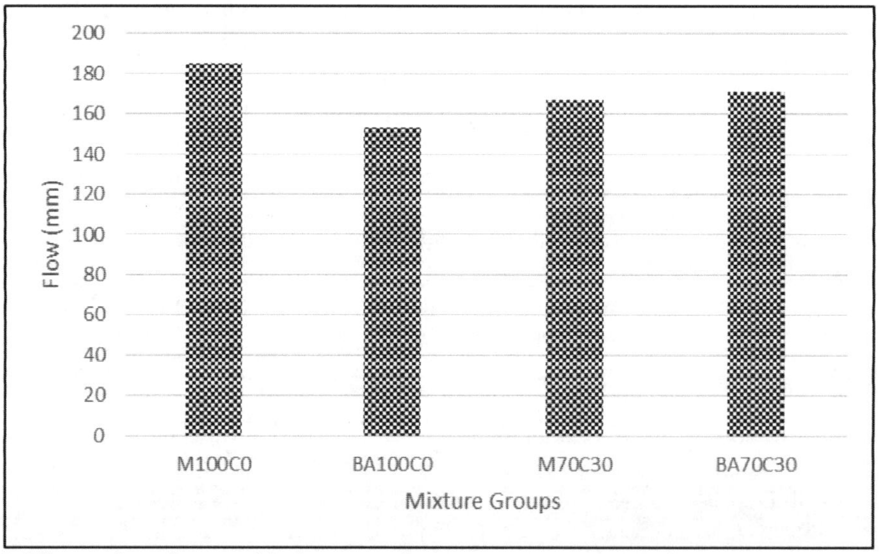

Figure 2. Flow values for marble-bottom ash cement paste composites.

Figure 3 shows the dry unit mass (DUM) for the tested composites. The DUM values for all mixture groups were in the range of the lightweight material category. Incorporating marble wastes and bottom ash into DUM had the same effect for all ages. Due to the ongoing hydration process, water was used by cementitious material and DUM values tended to decrease after 7 days.

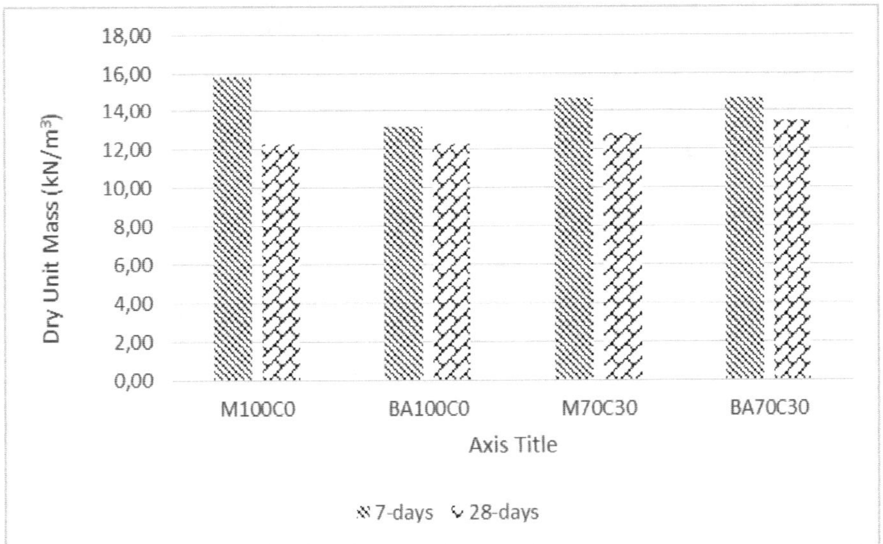

Figure 3. Dry unit mass values for marble-bottom ash cement paste composites.

DUM values for pure marble wastes are higher than pure bottom ash mixture groups. This can be due to the available water amount in pure marble composites as also recorded in flow table values. Since more water available in pure marble composites, more gel formation is expected (i.e. amount of calcium silicate hydrates more in M100C0 mix groups).

Figure 4 shows the water absorption values for laboratory produced composites. All water absorption values are higher than the specified international standard limit (WA<16%) for manufacturing of building materials [34, 35]. Pure bottom ash mixture groups due to the coarser nature of the bottom ash particles had higher water absorption values compared to marble wastes. A decrease in water absorption values were observed for both bottom ash and marble waste groups. When the amount of cement increased in the mixture water absorption values tends to decrease.

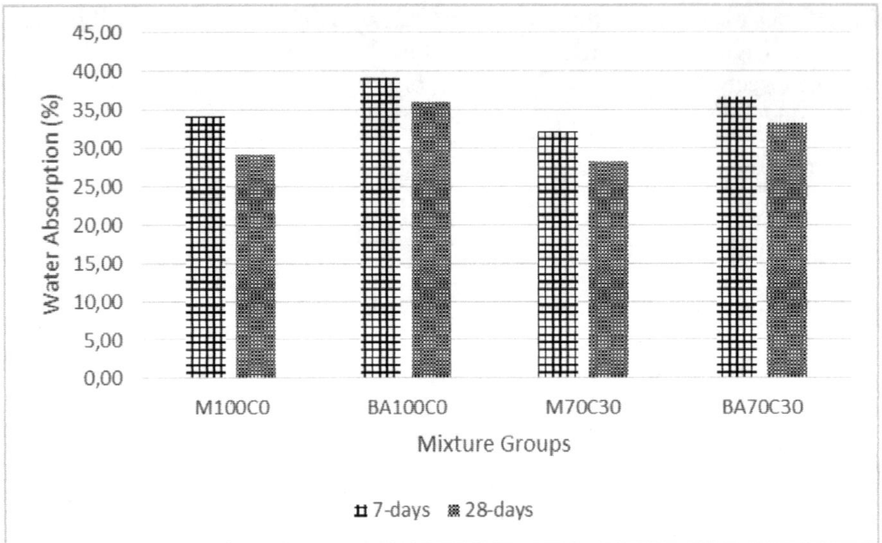

Figure 4. Water absorption values for marble-bottom ash cement paste composites.

Figure 5 shows the apparent specific gravity (ASG) results of the tested composites. ASG values of the marble wastes are higher than those of the bottom ash mixture groups. Due to ongoing hydration, ASG values have a tendency to decrease, which can be attributed to the pore-refinement of the matrix. The densely packed arrangement of particles and the reduction of the overall pore of the system causes ASG values to decrease. Interestingly, the bottom ash replacement was more efficient with cement than the marble replacement. Figure 6 shows the porosity values of the tested composites.

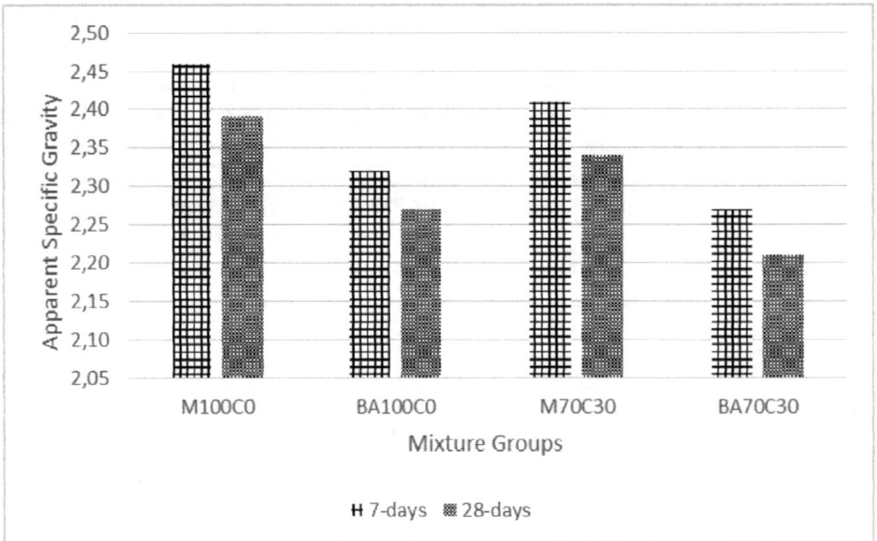

Figure 5. Apparent specific gravity values for marble-bottom ash cement paste composites.

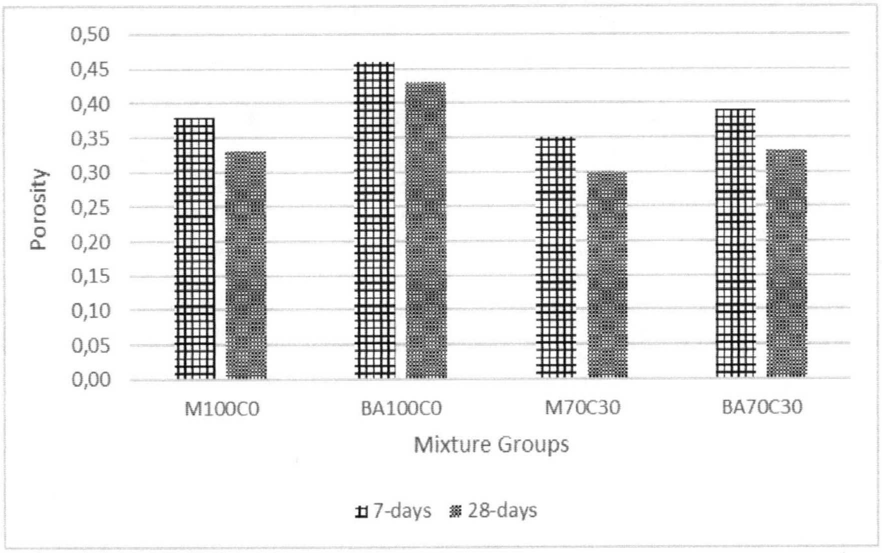

Figure 6. Porosity values for marble-bottom ash cement paste composites.

As expected, the porosity values of the bottom ash groups were higher than the marble waste groups. Bottom ash particles are coarser than marble wastes and, thus, produce a more porous structure. A decrease in porosity was recorded based on the lower replacement level. The increase in the amount of cement also led to the formation of more gel products, which can reduce the porosity of the composites.

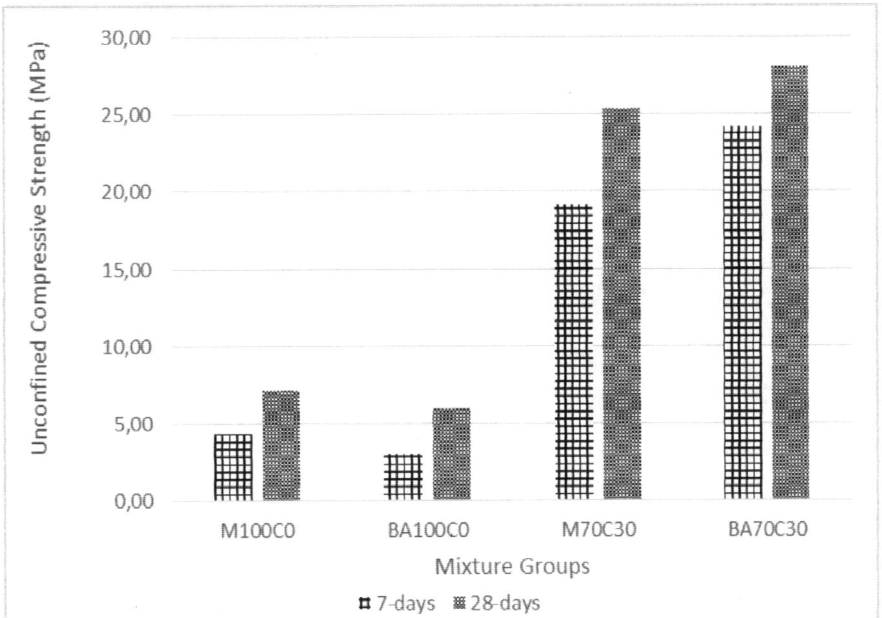

Figure 7. Unconfined compressive strength for marble-bottom ash cement paste composites.

Figure 7 shows the unconfined compressive strength (UCS) results of the composites. As expected, the pure bottom ash and pure marble waste groups had lower UCS values. Cement showed a greater contribution in the bottom ash groups, which could be due to the amount of SiO_2, which is higher in bottom ash. In addition, the total amount of Al_2O_3+Fe_2O_3+SiO_2 in bottom ash is 90.1%, which helps generate more gel formation and higher strength. When compared with the porosity values, which are higher in the bottom ash group, the strength was expected to be lower in the bottom ash groups. However, as previously mentioned, the chemical reaction—not the filler effect or closer packing of the particles—governed the overall behaviour. The M70C30 mixture groups attained 25 MPa compressive strength at 28 days. This value was approximately 30 MPa for the BA70C30 mixture groups at the same testing age. Figure 8 illustrates the flexural strength (FS) of the tested composites. The same trend as in the UCS was observed here. However, when comparing the UCS values of the M70C30 and BA70C30 mix groups and the FS values, the increase in FS in B70C30 is higher. The experimental results showed that the behaviour of composites composed of bottom ash are more sensitive to bending. A significant reduction in flexural strength was recorded in the bottom ash mix groups, and the contribution of cement in bottom ash was more pronounced based on the test results shown in Figure 8.

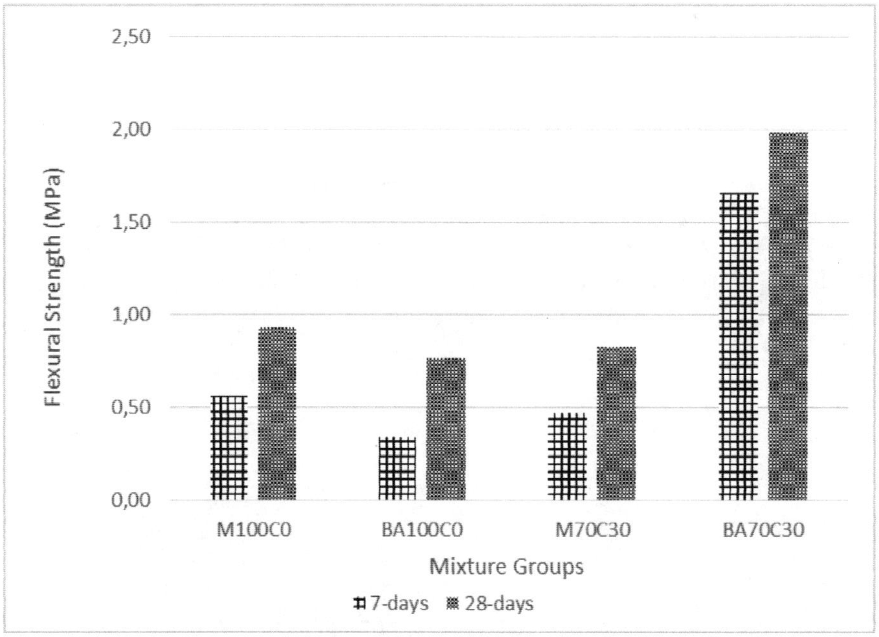

Figure 8. Flexural strength for marble-bottom ash cement paste composites.

Figure 9 shows the mass loss from the sulphate solution. Mass loss was greater in the bottom ash groups. However, as the amount of cement increased in the system, this reduction almost diminished and values quite similar to those at 28 days were measured in the bottom ash mixture groups. These results are compatible with the porosity values. The mass loss values were higher in the pure bottom ash and pure marble waste groups, but remained in the range of the building materials specified by the standards. The mass loss values were less than 16% for all mixture groups, indicating that the composites have moderate resistance to sulphate attack.

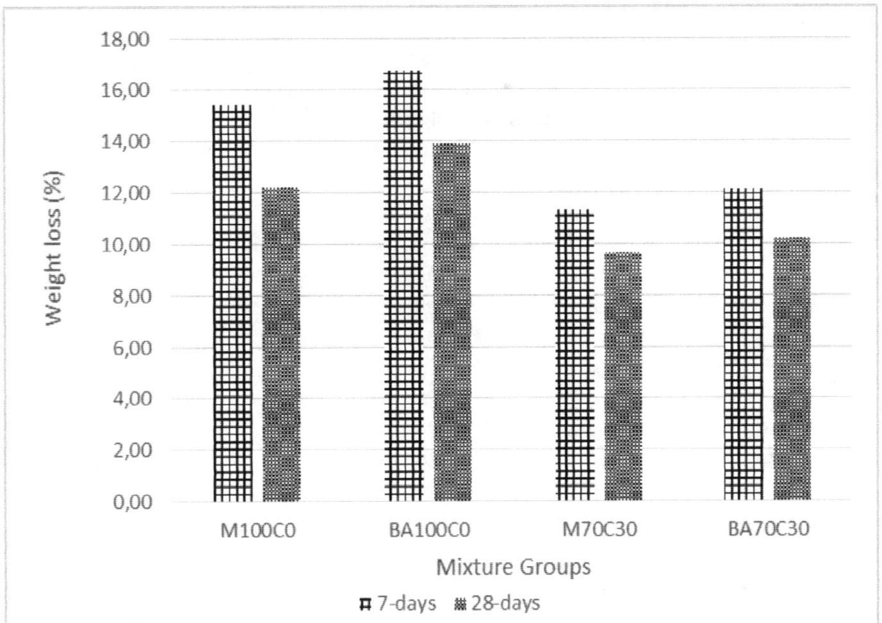

Figure 9. Mass loss by sulphate solution for marble-bottom ash cement paste composites.

The results showed that the bottom ash and marble wastes can be satisfactorily utilized at higher rates (i.e., up to 70%). The flexural strength of marble wastes proved that these composites are weak in tension. The bottom ash mixture groups showed better improvement when the replacement level did not exceed 70%. The chemical stability was more pronounced in the bottom ash mixture groups. The results demonstrated the superior characteristics of both the bottom ash and marble waste mixture groups. Thus, it is believed that the utilization of a large amount of these wastes in the concrete sector could reduce carbon dioxide emissions. The amount of the reduction would be based on the replacement level. In a previously published study [1], the author reported that "replacing cement with 10% marble powder would reduce CO_2 by 2.62 million tons for Turkey and 3.55 billion tons worldwide". Furthermore, when considering the cost of C20 concrete in Turkey (approximately US$35) and the reduction in cost for the 30% replacement level (i.e., around 35%), the reduction in cost also depends on the replacement level. Thus, it would be more economical and environmentally feasible to utilize large amounts of industrial wastes to become more sustainable. Such wastes could also be an alternative binder for the concrete industry.

4. Conclusions and recommendations

Based on the test results, the following conclusions can be drawn:

1. Bottom ash decreased the flow values more than marble wastes due to its granular nature.
2. The composites were classified as lightweight according to their dry unit mass values. The highest and lowest dry unit mass values were measured as 15.82 kN/m^3 and 12.32 kN/m^3, respectively.
3. Water absorption values were higher than the specified limits for the manufacturing building materials.
4. Marble wastes had higher apparent specific gravity values than bottom ash groups.
5. Bottom ash composites had higher porosity values than marble waste groups.
6. Unconfined compressive strength values proved that up to 70% of bottom ash and marble waste can be satisfactorily used to produce a strength of 25 MPa at 28 days.
7. The flexural strength of the BA70C30 mixture groups showed better performance than the M70C30 mixture groups.

8. The mass loss values for both bottom ash and marble wastes showed that the composites are classified as having moderate resistance to sulphate attack.
9. The behaviour of bottom ash cement paste composites under bending should be investigated in detail to understand the overall mechanisms encountered for reductions in strength values

References

[1] Aydin E and Arel H Ş 2019 *J. Clean. Prod.* **237**, 117801
[2] Aydin E 2019 *10th international concrete conference*, May 2-4, Bursa, Turkey 419-429.
[3] Yüksel İ 2017 *Environ. Dev. Sustain.* **19** 369-384.
[4] Aydin E and Arel H Ş 2019 *Data in Brief* **26** 104528.
[5] Li L G Huang Z H Tan Y P Kwan A K H and Liu F 2018 *Constr. Build. Mater.* **166** 423-432.
[6] Li L G Huang Z H Tan Y P Kwan A K H and Chen H Y 2019 *J. Clean. Prod.* **210** 55-65.
[7] Mashaly A O Shalaby B N Rashwan M A 2018 *Constr. Build. Mater.* **169** 800–818.
[8] Yüksel İ and Bilir T 2007 *Constr. Build. Mater.* **21** 686-694.
[9] Aydin E 2016 *Constr. Build. Mater.* **124** 582-588
[10] Aydin E and Arel H Ş 2017 *Constr. Build. Mat.* **157** 96-107
[11] Aydin E and Arel H Ş 2018 *Data in Brief* **16** 321-326.
[12] Aydin E 2009 Fifth International Conference on Construction in the 21st Century (CITC-V) Collaboration and Integration in Engineering, Management and Technology May 20-22, Istanbul, Turkey p.1526-1535.
[13] Arel H Ş 2016 *J. Clean. Prod.* **131** 179-188.
[14] Mashaly A O El-Kalioubу B A Shalaby B N El-Gohary A M 2016 *J. Clean. Prod.* **112** 731-741.
[15] Aydin E 2019 *J. Polytechnic* **22 (2)** 259-267 DOI: 10.2339/politeknik.441707
[16] Arel H Ş and Aydin E 2018 *ACI Mater. J.* **115 (1)** 55-64
[17] Lothenbach B Scrivener K and Hooton R D 2011 *Cement Concrete Res.* **41** 217-229.
[18] Supino S Malandrino O Testa M and Sica D 2016 *J. Clean. Prod.* **112** 430-442.
[19] Siddique R and Khan M I 2011 **Supplementary Cementing Materials** (Springer-Verlag Berlin Heidelberg) 288 p.
[20] Silvestre B S and Ţîrcă D M 2019 *J. Clean. Prod.* **208** 325-332.
[21] Ulubeyli G C and Artir R 2015 *Procedia Soc. Behav. Sci.* **195** 2191-2198.
[22] Arel H Ş and Aydin E 2017 *Constr. Build. Mater.* **157** 1084-1091.
[23] Aydin E and Balkis A P 2017 *ASTM J. Testing Eval.* **45 (6)** 2029-2038
[24] European Union Directive *2008/98/EC on waste* (**Waste Framework Directive**)
[25] Singh M and Siddique R 2016 *J. Clean. Prod.* **112** 620-630.
[26] Siddique R Aggarwal P and Aggarwal Y 2012 *Constr. Build. Mater.* **29** 73-81.
[27] Maschio S Tonello G Piani L and Furlani E 2011 *Chemosphere* **85** 666-671.
[28] Alyamaç K E and Ince R 2009 *Constr. Build. Mater.* **23** 1201-1210.
[29] Gesoğlu M Guneyisi E Kocabağ M E Bayram V and Mermerdaş K 2012 *Constr. Build. Mater.* **37** 160-170.
[30] Topçu İ B Bilir T and Uygunoğlu T 2009 *Constr. Build. Mater.* **23** 1947-1953.
[31] Ergun A 2011 *Constr. Build. Mater.* **25** 806-812.
[32] Sathonsaowaphak A Chindaprasirt P and Pimraksa K 2009 *J. Hazard. Mater.* **168** 44-50.
[33] Aydin E 2009 *International Conference on Concrete Repair, Rehabilitation and Retrofitting, Cape Town, South Africa* (London: CRC Press) 109-115.
Aydin E 2017 *Constr. Build. Mater.* **153** 393-401.

IOP Conf. Series: Materials Science and Engineering **800** (2020) 012008 doi:10.1088/1757-899X/800/1/012008

Performance indicators in the construction industry: a study with Portuguese companies

A Barros[1], S Sousa[1,2,*], and E Nunes[1,2]

[1]Department of Production and Systems, School of Engineering, University of Minho, Campus de Gualtar, 4710-057 Braga, Portugal

[2]ALGORITMI Research Centre, School of Engineering, University of Minho, Portugal

*Email: sds@dps.uminho.pt

Abstract. This paper aims to study the use of performance indicators (PIs) by business managers in the civil construction sector in a region that encompasses eight districts located in the north of Portugal. Through the literature review, it was possible to gather a list of twenty potential PIs used in this sector. Subsequently, a questionnaire was applied to a sample of construction companies from this region. A statistical analysis of the data collected allowed to identify the twelve most used and most important PIs for the companies surveyed. The results indicate that the companies involved in this study mainly use the traditional financial measures, however, recognize that non-financial measures, namely the customer satisfaction index, employee satisfaction, internal customer satisfaction index, and the training of employees, are increasingly important for the success of companies. The study also analysis the criteria to select PIs, its main benefits, and difficulties faced by companies on their usage.

1. Introduction

The growth of business competitiveness in the construction industry (CI) sector has led companies to seek performance improvements to become more competitive and to meet their stakeholders' needs [1]. However, for many companies, defining and using the most relevant performance indicators (PIs), which can support improvement initiatives, can still be a challenge [2]. Neely [3] emphasizes that companies have difficulties in determining what to measure and how to measure. Tripathi and Jha [4] add that these difficulties are aggravated when it comes to construction companies, depending on the particularities and complexity of the sector.

The main objective of this work is to identify gaps and improvement opportunities for the management processes of CI companies. To that end, this work aims to investigate: (i) the PIs used by the administrators of the CI companies; (ii) the PIs they consider more important; (iii) the criteria used to select the PIs; and (iv) the obstacles and benefits arisen from its usage. The work will focus on the sub-sector of residential construction, in a region that encompasses eight districts located in the north of Portugal.

The remaining of this paper is organised as follows. Section 2 presents a literature review on performance measurement (PM), covering PIs and PM systems in the CI sector, highlighting the obstacles and advantages of using such indicators and ending with specific research questions. Section 3 presents the research methodology, including questionnaire development. Section 4 presents the questionnaire results and analysis. The paper ends with conclusions.

2. Performance Measurement

2.1. Measurement Process

Performance measurement has been gaining increasing recognition of its important role as an element for the effective and efficient management of organizations [5]. Neely et al. [6] define PM as the process of quantifying the efficiency and effectiveness of an action. Performance measurement provides information that helps in the planning, process control, and also enables the management of strategic objectives and targets [7]. The standard ISO 9001:2015 [8] under "Performance Evaluation" addresses the need for the organization to evaluate the performance and effectiveness of the Quality Management System (QMS) and through the measurement process it is possible to identify the organization's capabilities and the expected performance, both at processes and organizational level [9]. Neely [3] adds that PIs should be considered an integral part of the planning and control process, providing a means of capturing data. According to [10], the measurement process, includes design, data collection and analysis, synthesis, and presentation (Figure 1). Thus, there is a need to design or select and use PIs, however, this task is not trivial [11].

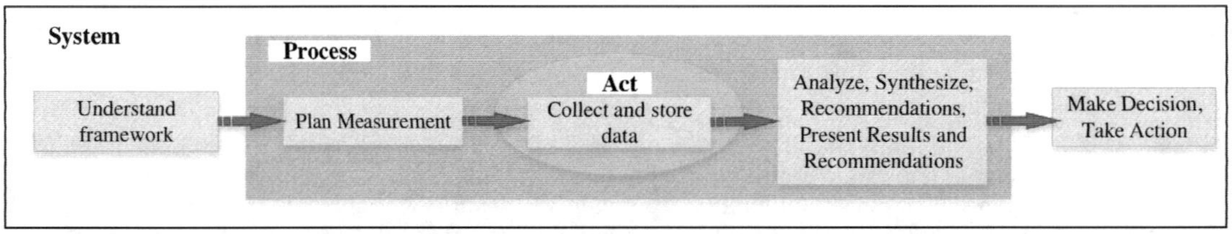

Figure 1. Measurement process (adapted from [10])

2.2. Performance Indicators

Traditionally, PM in the CI relied on financial indicators and, during the 1990s, companywide approaches to measure performance became common practice, using financial and non-financial PIs [1]. In the literature, there are several classifications of PIs, according to the company information needs and its organization and decision structure. For [12], the PIs are classified as Results Indicators, Process Indicators, and Variable Indicators. However, regardless of the type of classification adopted, the PIs must follow the changes that occur in the strategic objectives of the organization, not only be improved but also replaced if necessary.

Several authors [13, 14], have been defining criteria for the design and selection of PIs that are efficient and effective from the point of view of strategic management. These authors suggest that PIs should be transparent, useful, easy to implement, simple to understand, with visual impact, focus on improvement, low cost, and related to the organization's strategy and objectives.

Considering that PM can adversely affect performance [15], to improve overall efficiency of CI, PIs must be designed or selected, implemented and used assuming that its benefits outweigh its costs [4, 16].

2.3. Performance Measurement in the CI

2.3.1 Benefits. The CI is a very fragmented industry, in a competitive environment, in which performance improvement is a necessity and can be supported by PIs [1]. According to [17], there are two main drivers that influence successful implementation of a PM System: (1) top management commitment and (2) the perceived benefits arising from designing, implementing and using PIs. One key advantage of PM is to provide key information to support decision-making, driving company performance towards higher levels of efficiency and effectiveness [2, 18].

Overall, in the literature the following advantages of using PIs can be recognized: Increased productivity [19]; Costs reduction [18]; Support decision-making [2]; Help to implement strategies [20]; Establish targets [1]; Provide feedback [4]. Nevertheless, the existence of obstacles or barriers to the adoption of PIs may influence generalised use, hampering the development of the CI.

2.3.2 Obstacles. There are barriers to the implementation of the PM system, both structural and in the form of implementation. It can be concluded that there is a set of potential obstacles/difficulties on the adoption of PIs:

- Lack of time to do data collection [17];
- Lack of knowledge and experience of employees [15, 17, 19];
- Difficulty in selecting and choosing PIs [16];
- Time and difficulty doing data analysis [11];
- Lack of top management interest [21];
- Short-term emphasis on objectives by top management [16];
- Lack of human resources attached to this task [22];
- Some resistance from the "shop floor" personnel [17].

2.4. Performance Measurement Systems in the CI

In spite of the above-mentioned difficulties, models of PM specific for the CI were proposed, based mainly on Benchmarking systems using KPIs [1, 18].

Table 1 presents the PI proposed by some of these models used by construction companies in the world. Each PI definition is not presented and can be different from model to model. It can be concluded that by adopting different reference models and depending on specific context and objectives, PIs can be tailored to each company.

Table 1. Performance Indicators proposed by models used in the construction sector.

Initiative	Performance Indicator (category)
Key Performance Indicators (KPI) United Kingdom [23]	Cost deviation (F); Productivity (F); Cost of construction (F); Profitability (F); Deadline deviation (P); Equality and diversity (P); Defects (P); Construction time (P); Vehicle movements (P); Absenteeism (due to illness) (P); Lifetime performance (P); Weekly hours worked (P); Customer satisfaction (C); Safety (E&S); Water and energy consumption (E&S); Waste generation (E&S); Impact on the environment (E&S); Impact on biodiversity (E&S); Habitat area created/conserved (E&S); Workers Satisfaction (L&G); Staff turnover (L&G).
Danish Benchmarking System – Denmark [24]	Cost of m^2 (F); Productivity (F); Work intensity (P); Real-time construction versus planned (P); Actual construction time (P); Index of good practices in construction sites (P); Defects during the first year after delivery (C); Customer satisfaction during construction (C); Defects on delivery (C); Accidents frequency (E&S).
Performance and Productivity Indicators (ICBENCH) Portugal [2, 25]	Cost deviation (F); Profitability (F); Investment in technology (F); Sales growth (F); Pending Billing (F); Profitability (F); Impacts of defects in delivery (F); Defects (P); Deadline deviation (P); Successful proposals (P); Subcontracting (P); Customer satisfaction (C); Business Repetition (C); Accidents frequency (E&S); Water consumption (E&S); Permanent staff (L&G); Training (L&G); Employees satisfaction (L&G).
CI Institute Benchmarking & Metrics Programme USA [26]	Deviation from construction cost (F); Deviation from the cost per phase of the work (F); Cost of design changes during construction (F); Cost of rework (F); Cost per phase growth (F); Deviation of term by phase of the work (P); Duration of work/phase (P); Total duration of the Construction (P); Hiring rate (P); Accidents rate (E&S); Workdays lost due to accidents (E&S).

Overall, PIs can be classified in different categories, for example, according to the Balanced Scorecard [7] in internal processes (P), customer (C), financial (F) and learning and growth (L&G). However, in this specific activity sector, PIs related to environment and safety can be considered relevant. Thus, PIs in Table 1 were classified using the BSC categories (P, C, F and L&G plus Environment and Safety (E&S). As this classification depends on the specific definition of a PI its classification can vary. For example, productivity can be considered an internal process PI, or as a financial PI. Table 1 suggests a classification, mainly to ascertain if the above categories are present in most models.

Despite the existence of several PM frameworks specifically designed to the CI [2, 4, 27, 28, 29, 30], its utilization resulted essentially in the development of benchmarking activities allowing companies to identify best practices, compare their performance with best practices identifying gaps and defining projects to reduce those gaps. To identify such gaps, the following research questions were posed:

- What are the criteria that the administrators of the CI companies use when choosing a PI?
- What are the PIs that companies can use to bridge the gap between their current status and industry best practices?
- What are the main difficulties associated with the use of PIs, and the benefits that the implementation of the PIs bring to these companies?

3. Research Methodology

3.1. Introduction

The research methodology adopted can be classified as exploratory, by using a research to organize knowledge about PI and performance evaluation. The technique used for data collection was the questionnaire, which is a systematic process of collecting observable and quantifiable data.

Regarding the problem approach, the research is characterized as quantitative, considering the use of statistical techniques in data analysis. The quantitative approach is characterized by the use of quantification, both in the modalities of collecting the information, and in the treatment of these through statistical techniques, to guarantee the accuracy of the work done. Thus, the various stages of the methodology used to carry out this study are presented in Figure 2.

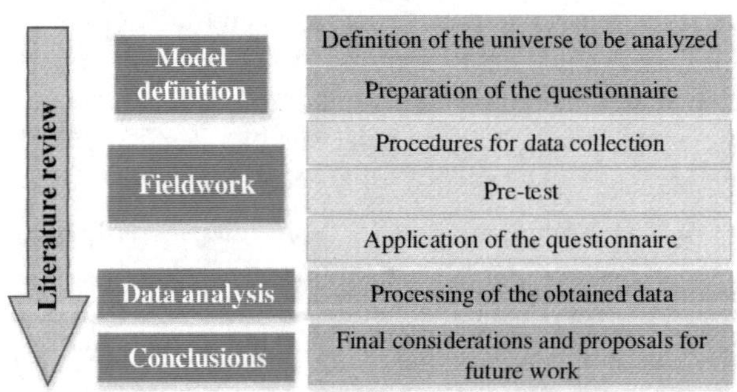

Figure 2. Stages of the research methodology.

3.2. Definition of the Study Area

This paper is an exploratory research since it seeks to organize knowledge about PM and PIs in companies in the CI sector in the northern region of Portugal. To this end, a list was obtained from the Institute of Public Markets of Real Estate and Construction with 23188 companies from the CI sector registered in this Portuguese Institute throughout the eight districts located to the north of the country, from the subsector of residential and non-residential building construction. Of these companies, 250 companies were randomly selected to constitute the sample for this study, since the means available for carrying out the work were limited. The sample was stratified by district to have the same proportion of companies as in the defined universe (see Table 2).

Table 2. Universe of study and sample size.

District	Companies		Sample	
	number	%	%	number
Aveiro	3593	15,6	15,6	39
Braga	4706	20,3	20,4	51
Bragança	692	3,0	2,8	7
Porto	7521	32,4	32,4	81
Guarda	1177	5,1	5,2	13
V. Castelo	1999	8,6	8,8	22
Vila Real	1064	4,6	4,4	11
Viseu	2436	10,5	10,4	26
Total	**23188**	**100**	**100**	**250**

3.3. Questionnaire Development

Based on the assumptions of the research developed on PIs and in the CI sector, a questionnaire was elaborated with the objective of answering the three research questions, to identify (i) the PIs used by the companies of the subsector of building construction and the criteria used to choose PIs; (ii) the degree of importance of each PI for companies, and (iii) the main difficulties and benefits related with its use. The questionnaire is structured into four sets of questions (Table 3).

Based on the literature review, 20 PIs (see Table 4) were selected for the questionnaire. These PIs span over the five groups/categories identified in the literature regarding PMS and used in this work.

Table 3. Groups of questions presented in the questionnaire.

Subject	Description
Respondents' profile	Information about the position, experience and literacy of respondents.
Company characterization	Main characteristics of the company, such as time in the market and the presence of certificates, and the use of PM systems and Benchmarking.
Use and importance of PIs	PI used and their level of importance for companies (assessed using a Likert scale (with 1 = very low, 2 = low, 3 = medium, 4 = high, and 5 = very high).
Aspects related to PIs	It addresses issues such as the criteria for choosing an indicator, the benefits and difficulties associated with PI.

Table 4. Performance indicators used in the questionnaire.

Group	Performance indicator
Financial (F)	Cost of project changes made during construction; Productivity; Cost of reworking; Cost of customer complaints; Cost of construction; Deviation from the cost of construction.
Client (C)	Customer Satisfaction Index; Internal customer satisfaction index.
Internal processes (P)	Index of good practices in construction sites; Defects detected in the delivery of the property; Evaluation of suppliers' materials; Term deviation; Number of non-conformities in audits.
Learning and Growth (L&G)	Employees' satisfaction; Training of employees; Subcontracting rate; Absenteeism.
Environment and Safety (E&S)	Water and energy consumption; Accident frequency rate; Impact on the environment.

3.4. Sample Size and Data Collection

In the first phase, a pre-test was carried out, which consisted of the face-to-face completion of the survey by five companies included in the sample, in order to observe and analyse the doubts raised by the survey, thus preventing possible misinterpretations or even failures of the questionnaire that lead to their possible incomprehension. Thus, after the application of the pre-test, it was possible to analyse some factors, such as inconsistency or complexity of the issues, the presence of inaccessible language in some sentences, and if the application of the questionnaire was done in a timely manner.

According to the identified factors, corrections were made in the questionnaire. It was then sent by email to the companies that make up the sample. In the email sent, the purpose of the study and its methodology was explained, requesting the completion of the questionnaire by the manager or the person considered most suitable for this in the company.

After the questionnaires were sent out to the companies, it was expected that they would be completed within three weeks. In cases where it was not possible to get a response within the stipulated deadline, the company was contacted again via email in an attempt to understand, the reason for such delay or lack of collaboration. To increase the response rate, companies were contacted by telephone, requesting the completion of the questionnaire.

With this, a total of 85 questionnaires were completed and returned, resulting in a response rate of 34%. According to Dillon, Madden and Firtle [31], the response rate in scientific studies ranges from 10 to 20%, so the results obtained can be accepted as representative of the universe studied, or at least with a response rate higher than similar studies.

4. Questionnaire Results and Analysis

4.1. Respondents Profile

The majority of respondents (80%) who participated in this study are Managers. Regarding the experience of the respondents, two situations were analysed (i) the time in the current position and; (ii) the time in the construction industry sector. The results show that 24% of the respondents are in the actual position over than 10 years, 27% between 5 and 10 years, 28% between 2 and 5 years, and 21% less than 2 years. On the other hand, 60% of the respondents work in the sector for over 10 years, 19% between 5 and 10 years, 14% between 2 and 5 years, and only 7% of the respondents work in the sector less than 2 years. Thus, it can be concluded that the majority of respondents have experience in this sector. It is also noted that the majority of respondents have good academic qualifications, since 62% have higher education, being 42% undergraduate and 20% graduate.

4.2. Companies' Characterization

This work addressed the main characteristics of the companies surveyed, such as (i) time in the market, (ii) held certifications, (iii) use of PM, and (iv) use of Benchmarking.

According to this survey, more than 70% of companies have been in the market for more than five years, which allows us to conclude that the companies analysed in the research show a good consolidation in the market. As far as certification is concerned, only 28 companies (about 1/3) held some type of certification, especially the ISO 9001 certification (12 companies).

Most companies (74%) use financial reports, but this figure drops to about 54% when it comes to the use of non-financial reports. These reports are obtained, in their vast majority, without the support of a PM system since 79% of the companies do not have any PM system and only 21% have one.

Regarding the use of Benchmarking, the survey results indicate that only 25% of the companies surveyed perform Benchmarking. The main sources of information that companies use to do Benchmarking are data comparison with similar companies (12.1%) and consultants' information (9.6%). Regardless of the use of PMS companies may use PIs. Thus, the following section presents results related to PIs.

4.3. Criteria Used to Select Performance Indicators

Table 5 shows the percentage of respondents that used the presented criteria to select PIs. The main criteria in choosing a PI by the surveyed companies are "Usefulness to the company" and "Focus on improvement". The majority of respondents referred these two factors. The remaining criteria were referred by almost half of respondents.

IOP Conf. Series: Materials Science and Engineering **800** (2020) 012008 doi:10.1088/1757-899X/800/1/012008

Table 5. Criteria used to select performance indicators.

Criteria	Percentage of respondents
Usefulness to the company	76.9
Focus on improvement	59.0
Easiness to implement	43.6
Related with strategy/objectives	43.6
Low cost	41.0

4.4. Importance and Use of Indicators

This shows the PI most widely used among the surveyed companies, as well as PI that the same companies regard as most important, regardless of whether they use them or not.

Table 6 shows the 20 PIs considered in this study ordered in descending order of the frequency of use. Overall, the majority of respondents said that they use 18 of the 20 PIs. This suggests that even without a formal PM system, companies use non-financial PIs.

Table 6. Most frequent and most important indicators for the surveyed companies.

Performance Indicator	Group	Percentage of use	Importance	
			Average	SD
Cost of construction	F	85.9	4.14	0.79
Productivity	F	83.5	4.39	0.71
Employees satisfaction	L&G	82.4	4.22	0.70
Training of employees	L&G	82.4	4.06	0.82
Customer Satisfaction Index	C	78.8	4.44	0.81
Cost of project changes made during construction	F	76.5	3.73	0.99
Term deviation	P	74.1	4.01	1.05
Deviation from the cost of construction	F	72.9	4.05	0.96
Evaluation of material suppliers	P	68.2	3.72	0.84
Defects detected in the delivery of the property	C	67.1	3.8	1.12
Cost of reworking	F	65.9	3.95	0.98
Internal customer satisfaction index	C	65.9	4.14	0.95
Index of good practices in construction sites	P	62.4	3.84	0.86
Impact on the environment	E&S	58.8	3.71	1.05
Cost of customer complaints	F	57.6	3.85	1.04
Accident frequency rate	E&S	57.6	3.62	1.30
Water and energy consumption	E&S	55.3	3.55	0.92
Absenteeism	P	51.8	3.22	1.21
Subcontracting rate	P	42.4	3.24	1.05
Number of non-conformities in audits	P	38.8	3.26	1.21

In the top ten of the most used PIs are four financial indicators (*cost of construction, productivity, cost of project changes made during construction*, and *deviation from the cost of construction*), and six non-financial indicators (*employees satisfaction, training of employees, customer satisfaction index, term deviation, evaluation of material suppliers*, and *defects detected in the delivery of the property*). Amongst these, *cost of construction* is the one most used (85.9% of surveyed companies).

Table 6 also shows the importance attributed to each of the 20 PI considered (Average column) by the companies surveyed. The shaded lines correspond to the 10 PIs considered with the highest degree of importance, distributed as follows: five are financial PIs (*productivity, cost of construction, deviation from the cost of construction, cost of reworking*, and *cost of customer complaints*) and five are non-financial PIs (*customer satisfaction index, employees satisfaction, internal customer satisfaction index, training of employees*, and *term deviation*). The *customer satisfaction index* is the most important PI for companies surveyed with an average of 4.44.

Seven of the ten most important PIs are in the top ten list of the most used by the companies. Thus, there are three PIs in the top ten list of importance that are not included in the top ten list of use: the *cost of rework, the index of internal customer satisfaction in the works,* and *the cost of customer complaints*. These data suggest that companies should be using these three PI because they consider them important, although more than 1/3 of the companies surveyed do not use them.

The assignment of high values of importance to certain PI reflects the perception that the respondents have about the advantages of their use. It turns out that the most important PI are not always the most used by companies, as this study shows. In this respect the literature refers to the advantages of adopting PIs and the obstacles to their use.

4.5. Benefits and difficulties associated with performance indicators

There are several benefits and difficulties that many companies face in making full use of a given PI. Table 7 shows the percentage of respondents that reported benefits and difficulties of using PIs.

Table 7. Benefits and difficulties associated with performance indicators

Benefits and difficulties		Percentage of respondents
Benefits associated with PIs	Increase productivity	62.2
	Reduce costs	60.8
	Increase revenues	44.6
	Support decision-making	41.9
	Help to implement strategies	37.8
	Establish targets	36.5
	Provide feedback	27.0
Difficulties associated with PIs	Time to do data collection	64.4
	Lack of knowledge and experience of employees	32.9
	Difficulty in selecting and choosing indicators	30.1
	Time and difficulty doing data analysis	26.0
	Lack of top management interest	9.6
	Short term emphasis on objectives by top management	8.2
	Other difficulties (the lack of human resources attached to this task, some resistance from the "shop floor" personnel)	8.2

To *increase productivity* and to *reduce costs* are the main benefits reported by these companies regarding PIs use. The other listed benefits were also reported by many respondents (27%-45%).

In terms of implementation of PIs, *time to do data collection* is the main difficulty faced by companies. There are a significant number of difficulties referred in the literature that respondents did not consider relevant (<10%) such as: *lack of top management interest; short-term emphasis on objectives by top management; the lack of human resources attached to this task;* and *some resistance from the "shop floor" personnel*.

5. Conclusions

Based on a sample of 250 companies and a response rate of 34%, the conclusions apply to this sample and generalisation should be made with caution since the companies that did not answer the questionnaire could provide data that might influence these results.

The use of PMS by the reporting companies is reduced since 79% of the companies surveyed do not have one in place. However, 18 out of the 20 PIs referred to in the questionnaire were used by the majority of respondents.

In general, the participating companies have a correct understanding of the importance of PIs in the implementation of good management practices, highlighting a balanced set of PIs with a broad scope. These PIs are distributed over the groups: financial, internal processes, learning and growth and environment and safety. With this, it can be concluded that companies in the Northern Region of Portugal use PIs.

Regarding the degree of importance of PIs, it was noticed that in the list of the top ten indicators, half of them are financial indicators. The most important PI for the companies studied is the *customer satisfaction index*. This result shows that although construction companies give more importance to financial indicators they realize that these alone are insufficient to measure organizational performance.

This study allowed to identify the most important criteria that managers of the surveyed companies use to select PIs as *the usefulness to the company* and *the focus on improvement*. The most important PIs in each of the five groups considered are: *customer satisfaction index* (C); *productivity* (F); *term deviation* (P); *employees satisfaction* (L&G) and *Impact on the environment* (E&S). The list of the most PIs is not the same as the most used. Thus, it can be suggested that the use of these most important PIs may allow companies to shorten the gap between their current status and industry best practices.

This study highlights *increasing productivity* and *reducing costs* as the main benefits resulting from the use of IPs. These advantages have been acknowledged by [18], and [20]. As the main obstacle is identified *the time to the data collection*, which may denote the inadequacy of the companies' information system to support the data collection. This obstacle has been referred to in the literature over the past two decades [17].

Acknowledgments
This work has been supported by FCT – Fundação para a Ciência e Tecnologia within the Project Scope: UID/CEC/00319/2019.

References

[1] Horta I and Camanho A 2014 Competitive positioning and performance assessment in the construction industry *J. Expert Systems with Applications: An Int. Journal* **41** 974-983.
[2] Costa J M, Horta I, Guimarães N, Nóvoa M H and Cunha J F 2007. icBench – A benchmarking tool for Portuguese construction industry companies *Int. J. for Housing Science* **31** 33-41.
[3] Neely A 1999 The performance measurement revolution: why now and what next? *Int. J. of Operations & Production Management* **19** 205–228.
[4] Tripathi K K and Jha K N 2018 An Empirical Study on Performance Measurement Factors for Construction Organizations *J Civ Eng* **22** 1052.
[5] Kennerley M and Neely A 2002 A framework of the factors affecting the evolution of performance measurement system *Int. J. of Operations and Production Mgmt.* **22** 1222–1245.
[6] Neely A, Gregory M and Platts K 1995 Performance measurement system design - a literature review and research agenda *Int. J. of Operations and Production Mgmt.* **15** 80–116.
[7] Kaplan R and Norton D 1992 *The balanced scorecard - measures that drive performance* Harvard Business Review, January-February 71-79.
[8] NP EN ISO 9001:2015 Sistemas de Gestão da qualidade, Requisito. Instituto Português da Qualidade. Portugal.
[9] Neely A, Adams C and Kennerley M 2002 *The Performance Prism. The Scorecard for Measuring and Managing Business Success* Prentice Hall.
[10] Juran J M and Godfrey A B 1998 *Juran's Quality Control Handbook* McGrawHill.
[11] Sousa S D, Aspinwall E M and Rodrigues A G 2006 Performance measures in English small and medium enterprises: survey results *Benchmarking – an Int. J.* **13** 120-134.
[12] Alarcón L F, Grillo A, Freire J and Diethelm S 2001 Learning from collaborative benchmarking in the construction industry. *Proceedings of 9th Annual Conference of the International Group of Lean Construction* March 1–10.
[13] Ahmad M and Dhafr N 2002 Establishing and improving manufacturing performance measures *Robotics and Computer-Integrated Manufacturing* **18** 171–176.
[14] Sousa S D and Aspinwall E 2010 Development of a performance measurement framework for SMEs *Total Quality Management & Business Excellence* **21** 475-501.
[15] Melnyk S A, Bititci U, Platts K, Tobias J and Andersen B 2014 Is performance measurement and management fit for the future? *Management Accounting Research* **25** 173-186.

[16] Neely A, Gregory M and Platts K 2005 Performance measurement system design - a literature review and research agenda *Int. J. of Operations & Production Management* **25** 1228-1263
[17] Nudurupati S S, Bititci U S, Kumar V and Chan F T S 2011 State of the art literature review on performance measurement *Computers and Industrial Engineering* **60** 279-290.
[18] Horta I, Camanho A and Costa J M 2012 Performance assessment of construction companies: A study of factors promoting financial soundness and innovation in the industry *Int. J. of Production Economics* **137** 84-93.
[19] Golam K, Razia S S, Rehan S and Solomon T 2018 Performance evaluation of employees using Bayesian belief network model *Int. J. of Mgmt. Science and Eng. Mgmt.* **13** 91-99.
[20] Rigby J, Dewick P, Courtney R and Gee S 2014 Limits to the Implementation of Benchmarking Through KPIs in UK Construction Policy: Insights from game theory *Public Management Review* **16**:6, 782-806.
[21] Toor S R, Ogunlana S O 2010 Beyond the "iron triangle": stakeholder perception of key performance indicators (KPIs) for largescale public sector development projects. *Int. J. Project Manage* **28** 228–236.
[22] Bititci U, T Turner and Begemann C 2000 Dynamics of performance measurement systems *Int. J. of Operations and Production Mgmt.* **20**(6), 692-704.
[23] CCI - Centre for Construction Innovation 2018 KPI Engine Available from: http://www.kpiengine.com/EngineHome/index.php
[24] BEC - Byggeriets Evaluerings Center 2018 Available from: http://www.byggeevaluering.dk/.
[25] [IcBench] IcBench 2018 Definição dos indicadores. Available from: http://www.icbench.net
[26] CII - Construction Industry Institute 2018 10-10 Metrics. Available from: http://www.construction-institute.org.
[27] Kagioglou M, Cooper R and Aouad G 2001 Performance management in construction: a conceptual framework *Construction Management and Economics* **19** 85-95.
[28] Bassioni H A, Price A D and Hassan T M 2004 Performance Measurement in Construction. *J. of Management in Engineering* **20** 42-50.
[29] Haas C T, Rankin J H, Fayek A R, Forgues D and Ruwanpura J 2012 Development and implementation of a benchmarking and metrics program for construction performance and productivity improvement *Canadian Journal of Civil Engineering* **39** 957–967.
[30] Ali H, Al-Sulaihi I and Al-Gahtani K 2013 Indicators for measuring performance of building construction companies in Kingdom of Saudi Arabia *J. of King Saud University – Engineering Sciences* **25** 125–134.
[31] Dillon W R, Madden T J and Firtle N 1994 *Marketing research in a marketing environment* (3rd Edition). McGraw-Hill Higher Education, IL. U.S.A.

IOP Conf. Series: Materials Science and Engineering **800** (2020) 012009 doi:10.1088/1757-899X/800/1/012009

Application of genetic algorithm method for soil nailing parameters optimization

F Benayoun[1*] D Boumezerane[2] S Rehab Bekkouche[3] and L Bendada[4]
[1] Department of Civil Engineering,University Larbi Ben Mhidi Oum El Bouaghi Algeria
[2] Engineering Division,University of the West of Scotland, High St,Paisley PA1 2BE,UK
[3] Department of Civil Engineering ,University 20 Aout 1955,Skikda,Algeria
[4] Department of Mechanic Engineering,University Larbi Ben Mhidi Oum El Bouaghi,Algeria
*E-mail: benayounfadila@yahoo.fr

Abstract. Soil nailing is a reinforcement technique used in constructing excavations and stabilizing slopes by reinforcing the ground in situ. The number of horizontal or inclined nails and their spacing is a key issue. The aim of this work is optimization of soil nailing parameters using genetic algorithm method. We look to obtain the optimal design and improve the parameters of soil nailing which affect stability. Analysis and parametric study of effect of algorithm configuration was carried out to search for optimum combination of the design variables to satisfy the required safety factor. The obtained results are discussed in comparison to existing literature.

1. Introduction

Although optimization is an elegant mathematical tool it was used quite early by researchers in many disciplines, geotechnical engineers have noticed the techniques quite late. Most advances in the development and application of classical optimization methods has taken place over the last 50 years. (Little and Price 1958) presented a computerized approach of slope stability analysis . An elementary beginning of application of optimization with (Baker and Garber 1978) to determine the critical slip surface and the corresponding minimum factor of safety.

Among the optimization methods, the genetic algorithm is a powerful stochastic optimization method based on the principles of genetics and natural selection. For near two decades, GA has been shown to be suitably robust for a wide variety of complex geotechnical problems : (Simpson and Priest 1993), (Pal et al 1996),(Deb et al 1999),(Goh A.T.C 1999) used genetic algorithm procedure to locate the critical circular slip surface, (Ponterosso et al 2000) produced a GA solution for the cost optimization problem of reinforcement layout for reinforced soil slopes, (McCombie and Wilkinson

2002),simple genetic algorithm is applied to the search for the minimum factor of safety in slope stability analysis.The first application of tree genetic programming (TGP) in the fields of geotechnical engineering is carried out by (Yang et al 2004) to analyze the stability of slopes. (Patra et al 2005) generalized method of computer based optimum design of soil nailed slopes is reported. (Zolfaghari et al 2005) and (Jianping et al 2008) have done several studies by GA for obtaining the critical slip surface. (Levasseur et al 2007) used a genetic algorithm to identify the soil parameters of a sheet pile wall retaining an excavation. (Prabir and Basudhar 2008) described methods that are available for optimal design and analysis of geotechnical engineering problems. (Torres and al 2009) discussed a characterization of different criteria for model classification while dealing with a system modeling. (Chao Li 2010) produced an efficient approach for locating the critical slip surface in slope stability analyses using a real-coded genetic algorithm.(Gandomi 2011) applied a LGP,GEP,MEP to the formulation of several complex geotechnical engineering problems. (C.hsein and al 2013) studied a reliability-based robust geotechnical design of spread foundations using multi-objective genetic algorithm. (Momeni and al 2014) developed an ANN-based predictive model enhanced with genetic algorithm (GA) optimization technique to predict the bearing capacity of piles [18] [25].(Hoseini.M et al 2016) used GA, PSO and multiple regression in the estimation of mechnical resistance value of soils. (Zhen- yu yin et al 2017) a comparative study of optimization techniques(genetic algorithms,particle swarm optimization and the artificial bee colony algorithm) are used for identifying soil parameters in geotechnical engineering .(Seyyed.M.H et al 2018) back analysis is carried out using genetic algorithm and particle swarm optimization in order to determine the soil strength parameters in an excavation project.

2. Genetique alghorithm optimization

Genetic algorithms (GA) were developed with the primary intention of imitating the processes that exist in nature. Basic principles of genetic algorithms were published in 1962 (Holland) and the mathematical framework for their development was published in 1975.

A genetic algorithm (GA) is classified as a global search heuristic, which is used in the calculation to find exact or approximate solutions to optimization and search problems.The genetic algorithm is implemented in a computer simulation in which a solution of the problem is represented using a genome or a chromosome. The genetic algorithm creates a population of solutions and applies genetic operators as mutation and crossover to evolve solutions in order to find the best solution.

The basic components common to almost all genetic algorithms are:

- a fitness function for optimization;
- a population of chromosomes;
- selection of which chromosomes will reproduce;
- crossover to produce next generation of chromosomes;
- random mutation of chromosomes in new generation.

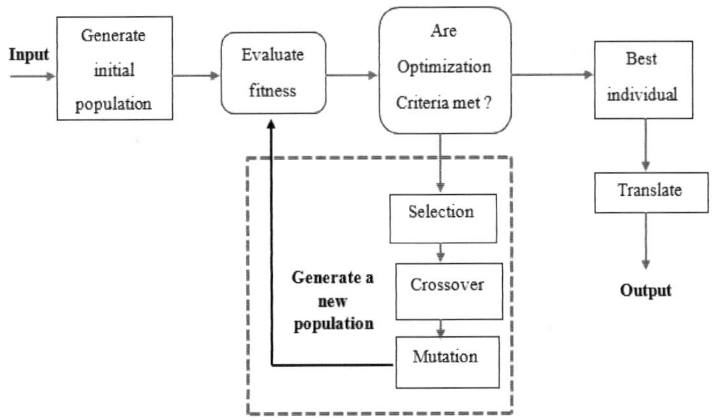

Fig 1. Basic structure of genetic algorithm [12]

3. Case study

Soil nailing is the technique used in slope and excavation to make it more stable with the use of passive inclusions, usually steel bars, termed as soil nails. It consists of passive reinforcement which is encased in grout to provide corrosion protection and improved load transfer to ground. As the excavation proceeds, the shotcrete, concrete or other grouting materials are applied on the excavation face to grout the reinforcing nails [6].

The studied nailed wall is inspired by the thesis [21] some results of modelization were used for the problem of optimization. The geometric model adopted is modeled with Plaxis 2D, a two-dimensional finite element program with a plane strain problem and long-term behavior is simulated using drained analysis conditions, Fifteen (15) -noded triangular elements are used for generating finite element mesh of appropriate density. Coarse mesh density is adopted globally, which is refined to fine density mesh around nails.

Figure 2 shows the simulated soil nail wall with dimensions and various parameters including in situ soil properties, mesh boundaries and conditions fixities.

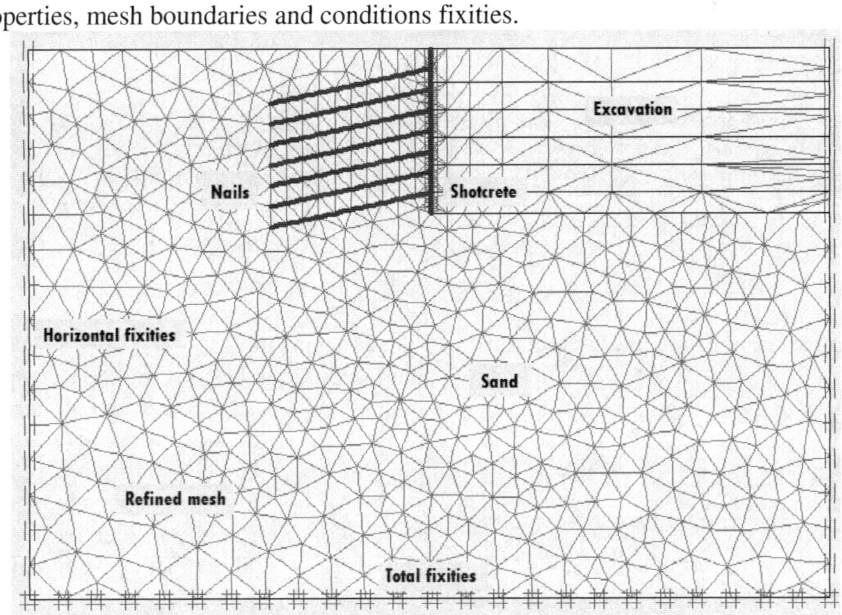

Fig 2. Geometric model.

The geometric model is 60m long and 40m deep, the excavation height is 12m.

According to exploration data, the field is composed of Sand characterized by these parameters:

Table 1. Soil properties.

Soil properties	Sand
Unit Weight, γ (KN/M^3)	18
Cohesion, (KN/M^2)	10
Friction angle, Φ	35
Dilatancy angle Ψ	5
Young's modulus, E (KN/M^2)	6.25 10^4
Poisson's ratio ν	0.3

The nail parameters used are axial stiffness EA and Bending stiffness EI (it is modeled as plate element).

Table 2. Characteristics of nails.

Axial stiffness EA (kN / m)	Bending stiffness EI (KN.m^2/m)
6.87*10^4	12.5

The shotcrete facing was considered to be 24 cm thick, and also modeled as a beam element with axial rigidity (EA) of 2.5 x 10^6 kN/m and flexural rigidity (EI) of 1.22 x 10^4 kNm^2/m. The height of the wall (H) was assumed to be constant and equal to 12 m. Wall inclination (β=90°), soil properties and diameter of nails (22mm) are considered constants also. In order to obtain optimal design for soil nailing wall, a genetic algorithm optimization was executed . Plaxis 2D is used to simulate the change of nail's length, inclination, and spacing nail to study the effect of these parameters on safety factor.

The following variables were used:

- 3 different nail lengths to wall height ratio : 0.75, 1, 1.2;
- 3 different vertical spacings 1.2 m, 1.5 m and 2 m;
- 3 different nail inclinations : 10°, 15°, 20° below the horizontal, because the inclination has an effect on the overall stability of the nailed wall.

In this way, 27 simulations were performed.

4. Mathematical modeling and optimization problem formulation

From mathematical point of view, optimization is about finding the best solution in a search space of a function, which describes the problem and is often called a fitness or objective function. The fitness function or objective function has some boundaries, which are the constraints of the problem. The optimization procedure has two phases. First phase is mathematical modeling where an objective multivariable function should be defined. In this phase, all constraints and bounds of the variables should be defined too. Second phase is searching for a global minimum of objective function. In the process of optimization, soil nailing safety factor is set as the objective function and used to provide a measure of how individuals have performed in the problem posed. The objective function is the function that the algorithm is trying to optimize, it is one of the most pivotal parts of the algorithm.

The objective function is defined as follows:

$$Y(x_1,x_2,x_3)=-1.56049\,x_1^2+0.0166667x_2^2-0.0008x_3^2+0.0525259x_1x_2+0.0079932x_2x_3+0.00153005x_1x_3$$
$$+3.95007x_1-0.349882x_2+0.0184172x_3-0.115709$$

The variables are :

- x_1 :length to height ratio L/H
- x_2 :Vertical spacing
- x_3 :nail inclination

Constraints:

- Bounds on length height ratio: $0.75 \leq X_1 \leq 1$
- Bounds on vertical spacing(m): $1 \leq X_2 \leq 2$
- Bounds on nail inclination(°): $10° \leq X_3 \leq 20°$

Genetic algortithm technique is applied to optimize soil nailing parameters. This function is a nonlinear problem.

As we mentioned in previous paragraph, the second phase is solving of mathematical model and minimize the objective function $Y(x_1,x_2,x_3)$ using GA toolbox in Matlab.

5. Parameters of Genetic Algorithm

Parameters have to be prechosen to control the search process, and they include the population size for each generation, the number of generations, the probabilities of genetic operators (reproduction, crossover and mutation), the selection method, the termination conditions, etc. The parameters used in the case study are as follows:

Table 3. Parameters of Genetic Algorithm.

Parameter	Settings
Population type	Double vector
Population size	100
Crossover fraction	0.8
Generations	100
Selection	Stochastic uniform
Mutation rate	0.05

The optimum parameters obtained with this model for different variables conditions are:

Variables	L/H	Sv (m)	I°
Optimal	0.75	1.99994	10°

The parameters obtained are in accordance with the Clouterre recommendations 1991: the Length to height ratio of Nails = (0.5 - 1.2).

The evaluation of generations and the best values obtained by GA are shown in figure 3.

IOP Conf. Series: Materials Science and Engineering **800** (2020) 012009 doi:10.1088/1757-899X/800/1/012009

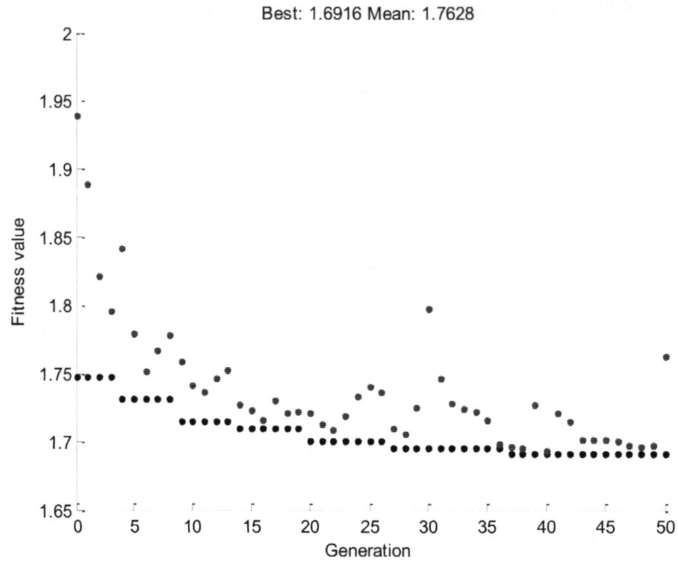

Fig 3. Variation of fitness values with generations.

Figure 3 the black point indicates the best value obtained from the objective function at each generation, shows little progress in lowering the fitness value, the blue point indicates the average value of the objective function at each generation, we have a number of points equal to the number of generations either for the mean or for the best, and for this cas we had 50 point for the best and 50 points for means.The fitness value is almost constant after generation 37, the final value obtained equal to 1.6916 expresses a very good result of minimization of safety factor.

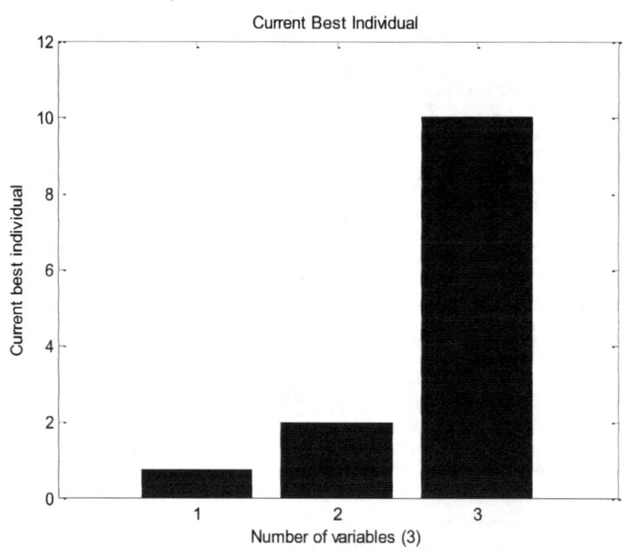

Fig 4. Current best individual with number of variables.

Figure 4 shows the number of best individuals that correspond to each variable in the optimized function. The best individual is inclination, then vertical spacing finally the height to length ratio.

6. Parmetric study

The parameters which condition the convergence of a genetic algorithm are:

- the size of the population of individuals
- the maximum number of generations
- the probability of crossing;
- the probability of mutation.

There are no parameters that are suitable for solving all the problems that can be posed to a genetic algorithm. However, some values are often used (defined in the literature) and can be good starting points for starting a search for solutions using a GA.

6.1. Selection type effect

Selection is the most important step it allows the individuals of a population to survive, to reproduce or to die. As a general rule, the probability of survival of an individual will be directly related to its relative effectiveness in the population (higher fitness).
Very strong selection will lead to highly fit individuals taking over the population, thus reducing the diversity needed for change and progress,on the other hand very weak selection may result in too slow evolution.We present a comparison using roulette wheel and uniform selection.

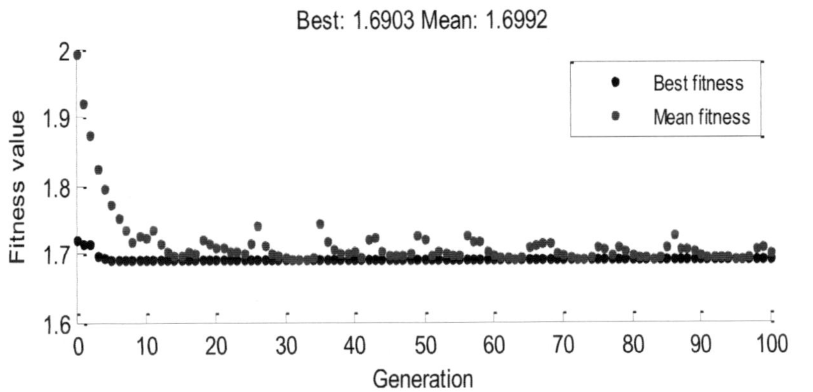

Fig 5. Variation of fitness values with generations (Roulette selection).

Roulette selection is the best known and the most used.With this method each individual has a chance of being selected proportional to his performance, so the more individuals are adapted to the problem, the more likely they are to be selected, In this case a rapid decreases and convergence is observed.

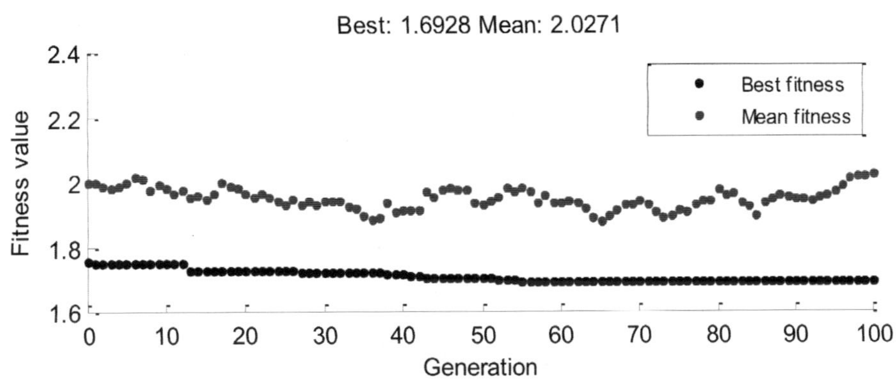

Fig 6. Variation of fitness values with generations (Uniform Selection)

In the uniform selection,parents are selected at random from a uniform distribution.This results is an undirected search.This selection is not a useful search because the selection is done randomly,each individual therefore has a 1 / P probability of being selected, where P is the total number of individuals in the population. In this case the best and mean fitness diverge at each generation.

7. Conclusion and perspectives

- This paper applies a methodology to arrive at optimal design parameters of soil nailing using GA optimisation;
- It has been demonstrated that the presented algorithm is able to find quickly the minimum safety factor and the optimum parameters. The optimum value for the process variables namely inclination vertical spacing and length to height ratio are found to be equal to 10°, 2.0m and 0.75 respectively;
- The results are satisfactory and in accordance with the recommendations Clouterre 1991.
- For a better optimization strategy the choice of the parameters of genetic algorithm is a really sensitive step (population size,mutation rate,crossover fraction ….etc);
- Multiobjective optimization techniques can be applied to solve this problem in future searches, as the shear resistance parameters also affect the safety factor value.
- The influence of the cost of realization of a soil nailed: cost of excavation, drilling, shotcreat ,steel …..etc poses a problem of optimization;
- This study will help engineers to choose reinforcement parameters of soil nailing that will produce economic and sable design.

References

[1] Ganesh R et al 2017 Modeling and optimization of furan molding sand system using design of experiments and particle swarm optimization, *mechanical engineering journals* doi: 10.1177/0954408917728636, pp 1–20

[2] Zhen yu yin et al Pierre-yves 2017 Optimization techniques for identifying soil parameters in geotechnicalengineering: comparative study and enhancement ,*International Journal for Numerical and Analytical Methods inGeomechanics* Doi: 10.1002/nag.2714

[3] Syed G A et al 2017 Evolutionary computations (GA and PSO) *in geotechnical engineering, Computational Water energy, and environmental engineering* 2017,6, http://www.scirp.org/journal/cweee issn online: 2168-1570 issn print:2168-1562, pp 154-179

[4] Marc F et al 2017 Réflexions et propositions pour le calcul optimisé des parois clouées, *Rev. Fr. Geotech. 2017,* 150, 1 ,EDP sciences,Doi:10.1051/getech/2017002.

[5] Mehdi H et al 2016 Estimation of soil mechanical resistance parameter by using particle swarm optimization, genetic algorithm and multiple regression methods,*sciencedirect soil & tillage research journal* homepage: www elsevier.com /locate /st i l l soil & tillage research 157, pp 32–42

[6] Arindam dey 2015 Issues and aspects of soil nailing *OIP -STC challenges and recent advances in geotechnical engineering research and practices,IIT guwahati ,conference paper ·*

[7] Kalehiwot N M et al 2015 Optimization of design parameters and cost of geosynthetic-reinforced earth walls using harmony search algorithm,*int. j. of geosynth. and ground eng.* 1:15 doi 10.1007/s40891-015-0017-springer international publishing

[8] Hyung-joon seo et al 2014 Optimization of soil nailing design considering three failure modes,*ksce journal of civil engineering* 18(2): copyright, korean society of civil engineers,doi 10.1007/s12205-014-0552-9 pissn 1226-7988, eissn1976-3808 www.springer.com/12205 geotechnical engineering, pp 488-496

[9] Lee I M 2014 Geotechnical optimization in underground construction geotechnical aspects of undergroundconstruction in soft groun*d –korean geotechnical society, yoo, park, kim & ban (eds),seoul, korea*, isbn 978-1-138- 02700

[10] Rabie M 2014 Performance of hybrid mse/soil nail walls using numerical analysis and limit equilibrium Approaches, *housing and building national research center hbrc journal* civil engineering dept., helwan university, cairo egypt http://ees.elsevier.com/hbrcj

[11] Rahmani R and Yusof R 2014 A new simple, fast and efficient algorithm for global optimization over continuous search-space problems. *appl math comput* 248(c), pp 287–300

[12] Dusan P and Miroslav R 2013 Using genetic algorithms for optimization ofturning machining process; *journal of engineering studies and research* – volume **19** no. 1

[13] Manish M R Saraswat M.R Ajay K.S 2013 Genetic algorithm for optimization using matlab *International Journal Of Advanced Research In Computer Science* volume **4**, no. 3, (special issue) issn n°.0976-5697

[14] Arokiasamy M et al 2013 Optimisation of shock absorber process parameters using failure mode and effect analysis and genetic algorithm. *journal of industrial engineering international , a springer open journal* 9:18 http://www.jiei-tsb.com/content/9/1/18

[15] Kushagra A and Sandeep CH 2012 Cost optimization of composite beams using genetic algorithm and artificial neural network, *International conference on computer technology and science iccts , ipcsit* vol. **47** iacsit press, singapore doi: 10.7763

[15] Darius M et al 2012 Optimization of grillages using genetic algorithms for integrating matlab and fortran environments *lietuvos ateitis science – future of lithuania,mechanics, material science, industrial engineering and management* issn 2029-2341 print / issn 2029-2252 doi:10.3846/mla.2012.93 mokslas , pp 564–568

[17] Alavi A H et al 2012 Linear and tree-based genetic programming for solving geotechnical engineering problem*s,*Book:Metaheuristics in Water Resources, Geotechnical and Transportation Engineering,*chapter 12 Elsevier, Editors: XS Yang et al*·Doi: 10.1016/b978-0-12-398296-4.00012-x ,pp 289-310

[18] Alavi A.H and Gandomi A.H 2011 A robust data mining approach for formulation of geotechnical engineering systems. *eng. computations* 28(3),pp 242–274

[20] Pucker T and Grabe J 2011 Structural optimization in geotechnical engineering: basics and application springer-verlag,acta geotechnica, doi 10.1007/s11440-011-0134-7,pp 41–49

[21] Ahmed A E 2010 Design charts for soil nailing. *a thesis submitted in partial fulfillment of the requirements for the degree of master of science in civil engineering faculty of engineering-shobra benha university* sc. in civil engineering

[22] Levasseur S et al 2009 Statistical inverse analysis based on genetic algorithm and principal component analysis:method and developments using synthetic data. *int. j. num. anal. meth. geomech.* 33, 1485–511

[23] Levasseur S et al 2008 Soil parameter identification using a genetic algorithm. *International journal for numerical and analytical methods in geomechanics* 32, no. 2. doi:10.1002/nag.614, pp 189–213

[24] Prabir k and Basudhar P.K 2008 Application of optimization and other evolutionary techniques in geotechnical engineering, *the 12th international conference of international association for computer methods and advances in geomechanics (iacmag)*, department of civil engineering, indian institute of technology kanpur, goa,india, pp 133-145

[25] Levasseur Set al 2007 Soil parameter identification using a genetic algorithm. *int. j. num. anal. meth. geomech.* 32,pp 189–213

[26] Tung-hsu hou et al 2007 Parameters optimization of a nano-particle wet milling process using the taguchi method,response surface method and genetic algorithm powder , *technology 173 www.elsevier.com/locate/powtec sciencedirect,pp 153–162*

[27] Patra C.R. and Basudhar P.K. 2005 Optimum design of nailed soil slopes,*Geotechnical and geological engineering* 23:springer,doi 10.1007/s10706-004-2146-7,pp 273–296

[28] Zolfaghari A.R et al 2005 Simple genetic algorithm search for critical non-circular failure surface in slope stability analysis.comput geotech 32(3), pp 139-152

[29] Yang C.X Tham l.g et al 2004 Two–stepped evolutionary algorithm and its application to stability analysis of slopes. *j. comput. civil eng. (asce)* 18, pp 145–53

[30] Mccombie P and wilkinson P 2002 The use of the simple genetic algorithm in finding the critical factor of safety in slope stability analysis. *comput. geotech.* 29, pp 699–714

[31] Brameier M and Banzhaf W 2001 A comparison of linear genetic programming and neural networks in medical data mining. *ieee trans. evol. comput.* 5(1), pp 17–26

[32] Goh A.T.C 1999 Search for critical slip circle using genetic algorithms *.Civil engineering and environmental systems* volume 17, 2000 - issue 3,pp 181-211

[33] Deb k 1999 Evolutionary algorithms for multi-criterion optimization in engieering designs *in engineering and computer science,* wiley, pp 135-161

[34] Banzhaf W et al *1998* Genetic programming–an introduction. on the automatic evolution of computer programs and its application. *dpunkt/morgan kaufmann, heidelberg/san francisco*

[35] Goldberg D.E 1994 Genetic and evolutionary algorithms come of age.Communications of the ACM, Volume 37 Issue 3, pp 113-119.

[36] Simpson A R and Priest S.D 1993 The application of genetic algorithms to optimisation problems in geotechnics. *comput. geotech.* 15,pp 1–19

[37] Goldberg D E 1989 Genetic algorithms in search,optimization,and machine learning,*addison-wesley boston*

[38] Baker R and Garber M 1978 Theoretical analysis of the stability of slopes,issn 0016-8505 | e-issn 1751-7656,volume **28** issue 4, pp 395-411

[39] Holland J 1975 Adaptation in natural and artificial systems, an introductory analysis with applications to biology, control and artificial intelligence MIT Press Cambridge, MA, USA ©1992 ISBN:0262082136

Coal ash use as a cement replacement in concrete production

S C Bostanci

European University of Lefke, Faculty of Engineering, Department of Civil Engineering, Lefke, Northern Cyprus, TR-10, Mersin, Turkey.

E-mail: sbostanci@eul.edu.tr

Abstract. Ordinary Portland cement (OPC) is the main contributor of concrete CO_2 emissions amongst concrete constituents. Use of alternative cementitious materials is highly encouraged to lower environmental footprint of concrete production. Local waste materials could be a feasible approach in terms of economic and environmental sustainability due to local availability. In this study, coal ash waste was used as a Portland cement replacement by 10% and 20% for the production concrete mix concrete mix with 28-day design strength of 45 N/mm^2. Developed mixes were tested against various engineering properties including workability, bulk density, compressive strength, sound permeability, thermal conductivity, water permeability and porosity. Results showed that coal ash incorporated mixes had slightly lower slump values with slightly lower fresh density. Hardened properties including bulk density and compressive cube strength were reduced slightly for CA mixes. Other hardened properties showed that CA mixes had more compact structures compared to conventional mix. Therefore, CA mixes showed enhanced sound permeability, thermal conductivity, water permeability and porosity.

1. Introduction

Global concrete production has reached to approximately 4.4 billion tonnes [1]. In parallel with this, utilization of waste materials has been encouraged by the science community to lower concrete environmental footprint [2-8]. Local waste materials are mainly preferred due to their availability for practical use. In addition, use of these materials has a potential to contribute environmental sustainability by reducing carbon footprint of concrete through reduced distance for collection and also economic sustainability having either lower or no cost. Coal ash (CA) is a waste material in North Cyprus that is obtained through coal burning. Coal is mostly consumed in restaurants. After consumption, CA is generally disposed to waste bins for municipal collection. There is no approach to collect CA waste and no specific treatment for CA disposal. CA is usually taken to landfill sites that are not pleasing from the environmental sustainability perspective. In addition, once CA is dumped in landfill sites, it may interfere to groundwater and airborne to surrounding site that could cause serious health problems. Concrete production could be an alternative solution for the CA disposal. This also encourages environmental, economic and social sustainability by lowering the amount of natural materials and enhancing built-environment for the society.

CA use in concrete as a replacement to either natural sand [9-10] and Portland cement [11-12]. Existing studies revealed that CA incorporation decreased concrete slump [10-11]. Singh [9] reported CA use decreased compressive strength considerably at 7 days while significant improvement was observed at 28 days. Previous studies [9-11, 13] claimed that use of CA reduced the compressive strength. Mangi [11] found that use of CA by 10% lowered compressive strength by 7% while Singh [10] found that 20% CA content lowered the compressive strength by 5% at 28 days. A study by Singh [9] reported increase in water absorption as CA content increased. There is no common agreement and insufficient data on the CA use as a cement replacement for optimum performance concrete from the hardened properties perspective.

This paper aims to investigate the effect of CA use in concrete production as a replacement to OPC. For this aim, CA was used in concrete with contents of 10% and 20%. A reference mix made with OPC and two CA incorporated mixes were produced to achieve 28-day target design strength of 45 N/mm^2. These concretes were tested against engineering (slump test, bulk density, compressive cube strength, sound permeability and thermal conductivity) and durability properties (water permeability and porosity).

2. Experimental programme

2.1. Materials

Three mixes, with 28-day target design strength of 45 N/mm^2, were examined in this study. First mix was reference mix made with only CEM I Ordinary Portland cement (OPC). In other two mixes, OPC was replaced with coal ash (CA) by 10% and 20% on mass basis. OPC used was 42.5N and complying with TS EN 197-1:2011. CA waste was collected from a local restaurant and sieved through 425 μm. Natural sand and gravel were used for all mixes. High range polycarboxylic ether-based versatile water-reducing admixture was used in conformity with TS EN 934-2.

2.2. Mix proportioning

Concrete mix designs were established based on BRE mix design document. Trial mixes were carried out then mixes were finalized. Control mix was designed for water/binder (w/b) ratio of 0.45 and to achieve target consistency of S3 workability class in conformity with TS EN 206:2013+A1 [14]. For CA incorporated mixes, both total aggregate and total cementitious contents were kept the same and replacements of OPC by CA were done on mass basis. The admixture was kept to 1% of the OPC content for all mixes. Target 28-day compressive cube strength sought was 45 N/mm^2. Control mix was denoted as OPC and then CA mixes were denoted as CA-10 and CA-20 for 10% CA and 20% CA incorporation respectively. Mix design proportions are provided in table 1. Casting and fresh testing of concrete were performed in conformity with TS EN 12350:2000 Part 1 and Part 2.

Table 1. Mix proportions of concretes.

Concrete Mix	Water (kg)	Portland cement (kg)	Coal ash (kg)	Aggregates (kg)		Admixture (kg)
				Fine	Coarse	
OPC	220	490	-	825	790	4.90
CA-10	220	440	50	825	790	4.40
CA-20	220	390	100	825	790	3.90

2.3. Test methods

Engineering properties examined include slump test, density, compressive strength, sound permeability, thermal conductivity, water absorption and porosity. Compressive strength values were determined through 150 mm by 150 mm cube samples at 1, 7, 14 and 28 days in conformity with TS EN 12390-3:2009. Sound permeability and thermal conductivity tests were conducted on square samples with dimensions of 500 mm by 500 mm with 40 mm thickness. Water absorption values were also established on 150 mm cube samples. Three samples were used and the average values were calculated for each test.

2.3.1. Sound Permeability Test. Sound permeability test was done on computer controlled test apparatus. Following casting, samples were covered with polythene sheet for 24 h and then cured in water tank for 28 days. Samples were then air-dried for a day and test was carried out. Sound permeability was established based on given sound on one side and received on the other side. Samples were covered with polystyrene foam during testing. Sound was applied for a duration of an hour then sound permeability in percentage was obtained according to given and received values.

2.3.2. Water Permeability Test. Water permeability test was carried out by apparatus (Fore A.Ş., Turkey) on 150 mm by 150 mm cube samples. Water permeability test was carried out at 28days. Similar to sound permeability test samples were covered with polythene sheet for 24 h and then water-cured until 28 days. Then, samples were oven-dried until a constant weight is achieved and left to cool down in the laboratory conditions prior to testing. Samples were placed on the test equipment with a water pressure through an area of 75 mm diameter at the centre of the bottom surface. The pressure was set to 5±0.5 bar throughout the test duration of 72 h.

3. Results and discussions

3.1. Fresh properties

3.1.1. *Slump Test.* Slump test results are provided in figure 1. It is seen from the results that slump values tend to decrease as the CA content increased. Similar findings were reported earlier [10-11, 14]. The reduction in slump value is more prominent for CA-20 mix (12%) compared to CA-10 mix (3%). Mangi [11] reported slump reduction (10%) when 10% CA content was used as a replacement to fine aggregates. It is also noteworthy to state that CA mixes had lower admixture content. Reduced water-reducing admixture might have had negative effect on dispersion of constituents.

IOP Conf. Series: Materials Science and Engineering **800** (2020) 012010 doi:10.1088/1757-899X/800/1/012010

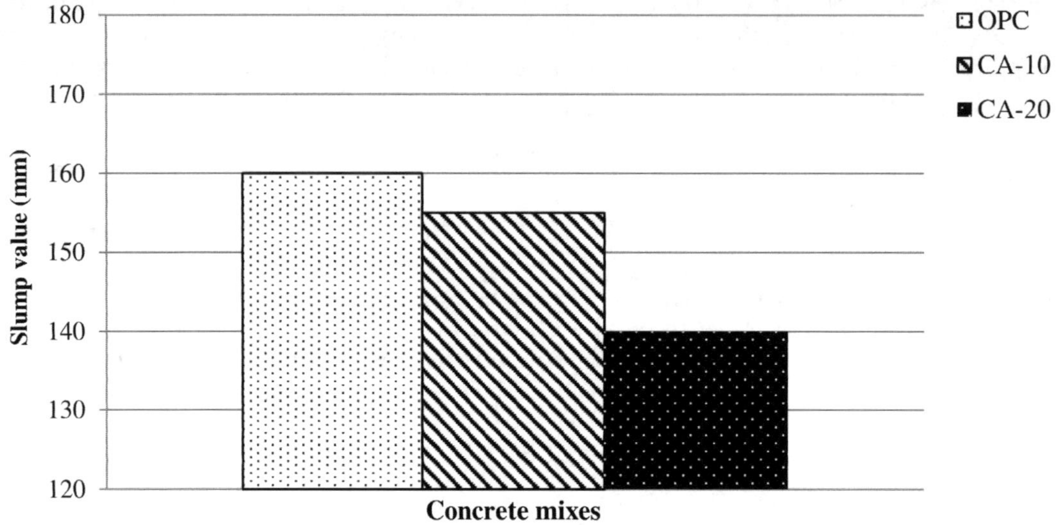

Figure 1. Slump test results.

3.1.2. Fresh density. Fresh density test results are given in table 2. Test results show that CA incorporation reduced the fresh density. Fresh density values were decreased by 3.7% and 4.9% for CA-10 and CA-20 respectively. Aggregate contents were kept the same for all mixes. Thus, this reduction is believed to be due to lower specific gravity of CA in comparison to OPC.

3.2. Hardened properties

3.2.1. *Bulk Density.* Hardened density test results are presented in table 2. The results showed similar trend as fresh density results. There is a significant reduction observed for CA incorporated mixes. Bulk density reductions were reported as 3.2 % and 4% for CA-10 and CA-20 mixes respectively. This result is in line with earlier study by Singh [10] who reported 2% reduction in density when 20% CA is used as fine aggregate replacement.

Table 2. Fresh and hardened density test results.

Mix	Fresh Density (kg/m^3)	Bulk Density (kg/m^3)
OPC	2355.0	2324.7
CA-10	2268.6	2249.1
CA-20	2240.6	2231.3

3.2.2. Compressive cube strength. The compressive cube strength results are given in figure 2. There is a gradual reduction observed for CA mixes at all ages, except for CA-10 mix at 1-day, in comparison to conventional concrete mix. These results are in agreement with previous researches [9-11, 13]. Strength losses were found as 14.6% and 19.2% at 7 days and 4.2% and 7.4% at 14 days for CA-10 and CA-20 mixes respectively. Strength losses at 28 days were observed as 4.7% for CA-10 mix and 11.8% for CA-20 mix. Earlier study by Singh [10] claimed 5% reduction in strength when 20% CA is used as fine aggregate replacement. Strength reduction reported for CA-10 mix is slightly lower than previous statement by Mangi [11] who reported 7% reduction in strength when 10% CA is used as OPC replacement. Lower strength results by CA mixes is believed to be due to lower CaO content of CA compared to OPC. The results suggest that CA did contribute to strength development after 7 days. This may be a potential indication for a pozzolanic activity by CA. However, post-28 day results need to be investigated for the possible evidence of pozzolanic activity of CA.

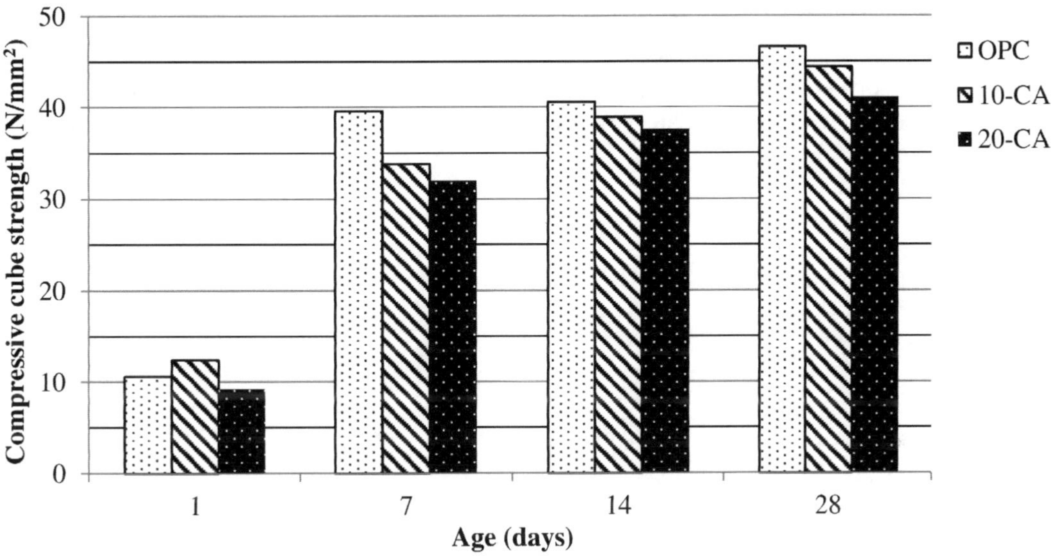

Figure 2. Compressive cube strength results.

3.2.3. *Sound Permeability*. Sound permeability test results are presented in table 3. It can be seen that CA addition reduced the sound permeability of concrete slightly. This indicates that CA has better inherent characteristics from the acoustic insulation point of view. The index of reduction of impact noise results were in agreement with the range (17-22 dB) identified earlier [15]. As green building tools favours acoustically-effective building materials for enhanced quality of the life of the occupants. The use of CA could be potentially used in applications where acoustic insulation is sought.

Table 3. Sound permeability test results.

Concrete Mix	Sound Permeability (%)	Standard Deviation	Index of reduction of impact noise (dB)
OPC	19.2	1.2	19.5
CA-10	18.6	1.3	18.1
CA-20	18.4	1.5	18.0

3.2.4. *Thermal Conductivity*. Thermal conductivity test results are revealed in table 4. It is seen from the results that CA use reduced thermal conductivity of concrete compared to conventional mix. Thermal conductivity of a material is influenced by material density, moisture content and ambient temperature [16]. As CA replaced OPC on mass basis and assumed to have lower specific gravity, the surface volume of CA particles is expected to increase and this leads to more compact concrete structure. This compact structure is believed to lower thermal conductivity. In addition, it is more pronounced with lower porosity values by CA mixes. The results may support the fact by earlier study [17] which claimed that higher admixture content may increase dissolved chemical ions in water and increments thermal conductivity. From the results, CA additive mixes showed a potential to be used for thermally-effective construction applications and provide thermal comfort for the buildings.

Table 4. Thermal conductivity test results.

Concrete Mix	Thermal conductivity (λ (W/mK))
OPC	2.3367
CA-10	2.1425
CA-20	2.1379

3.2.5. *Water Permeability*. Water permeability results are given in figure 3. It is seen from the results that water permeability is gradually reduced when CA is introduced compared to OPC mix. The result is in contrast with the earlier study [9]. The values obtained were slightly higher than previous study (4.68%-5.68%). Water permeability values were found as 5.8% and 6.1% for CA-10 and CA-20 mixes respectively while OPC mix had 13.2%. As CA mixes achieved lower compressive cube strengths, this reduction could be attributed to compact structure of CA mixes due to increased volume in paste content.

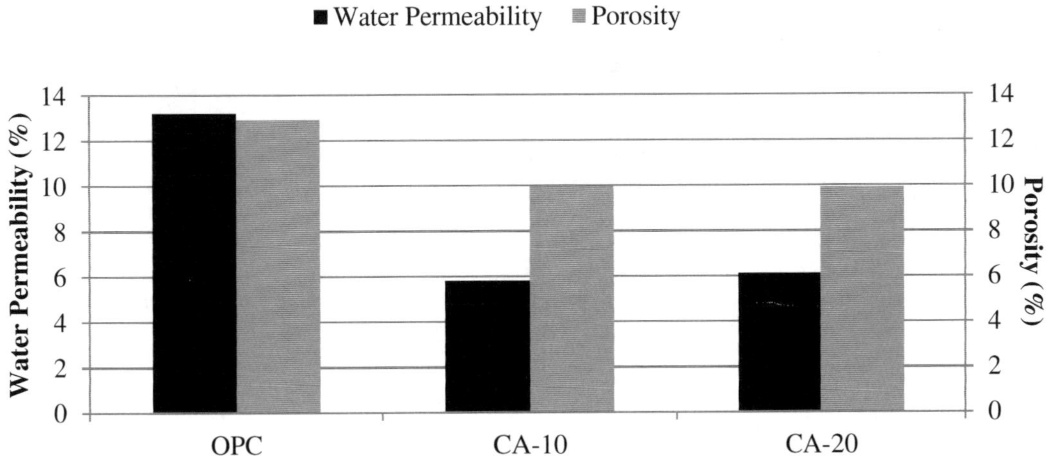

Figure 3. Water permeability and porosity test results.

3.2.6. *Porosity*. Porosity results of mixes developed are given in figure 3. Porosity values tend to decrease as CA content increased. Values were obtained as 12.93%, 9.94%, and 9.90% for OPC, CA-10 and CA-20 mixes respectively. The relation between the water permeability and porosity is given in figure 4. The results showed quite similar trend as water permeability. The relationship between water permeability and porosity showed an excellent correlation (R^2=0.998). The results were also in consistent with sound permeability and thermal conductivity test results indicating a compact structure that led to a denser structure for CA mixes.

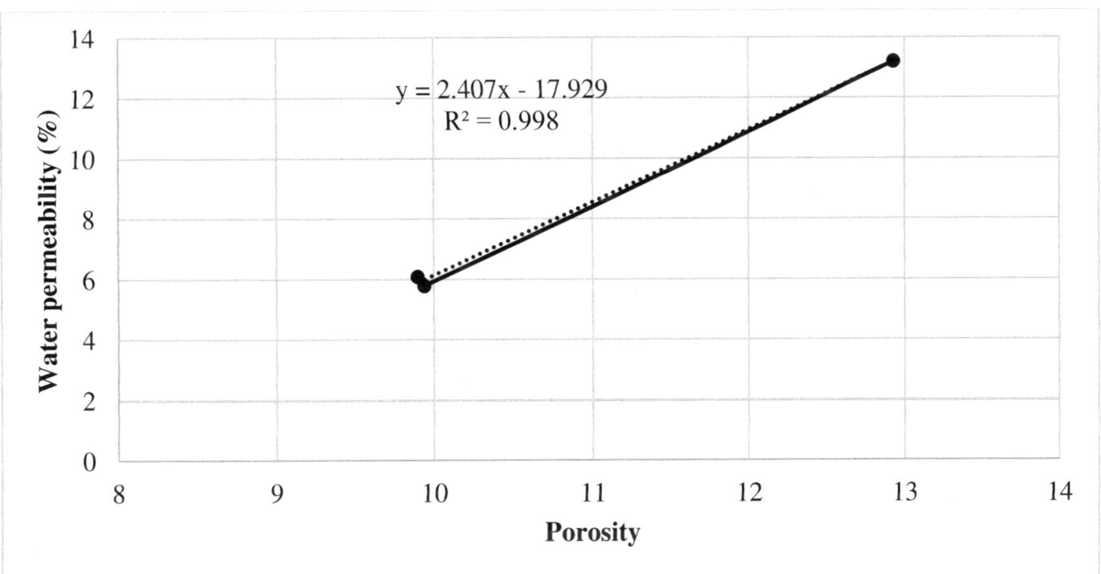

Figure 4. Relationship between water permeability and porosity.

4. Sustainability Considerations

The use of CA as a replacement to energy-intensive OPC, the main contributor to eCO_2 emissions amongst constituents, will potentially reduce environmental footprint of concrete. As North Cyprus has limited available natural resources, the use of recycled materials in order to promote sustainable development is inevitable. In common practice, North Cyprus construction sector imports most of the PC used which this releases eCO_2 emission through transportation. The implementation of recycled materials will then lower the extraction of raw materials needed and also will contribute to economic sustainability by lowering the OPC content used.

A study by Huising [17] pointed out that appropriate sustainable climate policies need to be put into action through social radical changes for carbon-effective economy. There is no specific treatment for CA disposal. If CA is not used in concrete production, it will be most probably sent to landfill. This, then, may lead to contamination of groundwater and could harm living habitats. Moreover, CA may expose a health risk due to airborne of CA. It was observed that CA could lead formation of upper respiratory related diseases and lung failure at long term exposure. The use of CA as a replacement to OPC in concrete will reduce the materials sent to landfill and lower the eCO_2 emissions considerably depending upon the replacement level while contribute to economic viability if obtained from a reasonable distance.

5. Conclusions and recommendations

This research focussed on using CA in concrete production as a replacement to OPC. It can be extracted from the results that CA could provide similar values as conventional OPC mix. The results revealed in this study could provide technical data on the fresh and hardened properties of CA concrete. The main conclusions can be drawn as;

- CA use reduces concrete workability which is believed to be due to reduced admixture content may have increased the friction between the particles.
- Fresh and bulk densities decreased as CA content increased due to the lower specific gravity of CA.
- Compressive strengths were gradually decreased as CA content increased (4.7% and 11.8%). Mechanical and durability related tests indicated CA mixes had compact structure. Therefore, this reduction could be due to CA addition might have weakened the bond between the cement paste and the aggregate.
- CA was observed to have better sound absorbing characteristics than OPC which could be used in acoustically-effective applications.
- Thermal conductivity values were lowered when CA is introduced. This could indicate CA mixes to be used in the energy-efficient construction applications for better thermal comfort.
- CA mixes showed lower water permeability and porosity compared to conventional mixes.
- Sound permeability, thermal conductivity, water permeability and porosity results suggest that CA mixes had more compact structure than conventional OPC mix.
- CA use in concrete could lead to environmentally friendly concrete production through reducing the need on the raw materials and lowering the potential eCO_2 emissions. In addition, use of no-cost CA waste could also enable economically-viable concrete production. CA use could support social sustainability by reducing the amount of CA sent to landfill and thereby contamination of natural resources and prevent health disorders for living habitat in the surrounding areas.
- Further research on the long term performance (>28days) of CA mixes is recommended to identify possible presence of pozzolanic reactions. CA with finer characteristics (<425 μm) is

also recommended for future studies which could enhance mechanical characteristics of concrete. Leaching tests should be also carried out to evaluate developed mixes from the health and safety point of view.

Acknowledgement

The author would like to thanks for individuals including Mrs. Tuğçe Mani, Mr. Metin Eray, Mr Kemal Sergen, Mr. Harbil Dağcın and Ede Isaac Ajogi and organizations including Boğaziçi Endüstri Madencilik Ltd., Nayat Yapı Ltd., Kascon Readymixed Concrete Company and Meril Restaurant for providing some of the materials used for the present work.

References

[1] Hilburg J 2019 *Archpaper.com* [online]. New York: The Architect`s Newspaper, LLC [cited 10th July 2019]. <https://archpaper.com/2019/01/concrete-production-eight-percent-co2-emissions/>.
[2] Aydin E 2018 *ACI Mater. J.* **115 (1)** 55-64.
[3] Onuaguluchi O and Eren Ö *Mater. Res.* **15(6)** 1029-1036.
[4] Özkan Ö and Yüksel İ 2008 *Constr. Build. Mater.* **22** 1288-1298.
[5] Çınar, M E and Kar F 2018 *Constr. Build. Mater.* **163** 734-741.
[6] Bostanci SC Limbachiya M C and Kew H 2018 *J. Clean. Prod.* **189** 176-196.
[7] Carsana M Frassoni M and Bertolini M 2014 *Cement Concrete Comp.* **45** 39-45.
[8] Erdoğan T Y 1997 *Admixtures for Concrete* (Middle East Technical University Press) 188 pp. Jalal M Nassir N and Jalal H 2019 *J. Clean. Prod.* **238** 117882.
[9] Singh M and Siddique R 2015 *J. Clean. Prod.* **91** 269-278.
[10] Singh M and Siddique R 2016 *J. Clean. Prod.* **112** 620-630.
[11] Mangi S A Ibrahim M H W Jamaluddin N Arshad M F and Jaya R P 2019 *Int. J. Eng. Sci. Technol.* **22** 515-522.
[12] Teixeira E R Camões A Branco F G Aguiar J B and Fangueiro R 2019 *Was. Man.* **94** 39-48.
[13] Rafieizonooz M Salim M R Mirza J Hussin M W Salmiati Khan R and Khankhaje E 2017 *Constr. Build. Mater.* **143** 234-246.
[14] Yogesh A and Siddique R 2014 *Constr. Build. Mater.* **54** 210-223.
[15] Dearnaulds V Costanzo E Carvalho A and Arlaud B 2005 *Proceedings of Twelfth International Congress on Sound and Vibration*, Lisbon, Portugal.
[16] Venkataraman M Mishra R and Militky J 2017 *J. Text. Eng. Fash.* **2(3)** 401-409.
[17] Huising D Zhang Z Moore J C Qiao Q and Li Q 2015 *J. Clean. Prod.* **103** 1-12.

IOP Conf. Series: Materials Science and Engineering **800** (2020) 012011 doi:10.1088/1757-899X/800/1/012011

Modification of Hydraulic Conductivity of Sandy Soil using Seawater and Alkaline Solutions

Osama Dawoud[1]

[1]Civil Engineering Department, Istinye University, Topkapi Campus,, Istanbul, Turkey

E-mail: osama.dawoud@istinye.edu.tr (Osama Dawoud)

Abstract. Altering soil characteristics by precipitation of the calcium carbonate between soil particles is attracting the interest of geotechnical engineers for a wide range of applications. However, current applied approaches still face challenges such as the high cost of materials, difficult handling of precursors, and hazardous by-products. The study described by the current paper investigated the modification of hydraulic conductivity of sandy soil by chemical precipitation of the calcium carbonate using seawater as the source of calcium. The designed experiment involved treatment of bench-scale soil columns by two readily mixed solutions. One of the solutions served as the source of calcium ions and the other contained carbonate ions. The current paper compares behavior of hydraulic reduction considering different resources. The results showed that the crystals formed in small sizes as a result of the instantaneous reaction of calcium precipitation; and thus, crystals transport with the solution as colloids. Accordingly, reductions in hydraulic conductivity were mainly governed by the filtration mechanisms; and the precipitated calcium carbonate was retained close to the injection point. Conducting treatment using seawater as the source for calcium ions induced 65% reduction in hydraulic conductivity at 5.5 g $CaCO_3$/g soil. This reduction is identical to the case when the artificial source was used. Further treatment by the artificial source of calcium reduced the hydraulic conductivity of sandy soil by 97.2%. These outcomes suggest that the chemically induced calcium carbonate precipitation from seawater can be employed by some geo-environmental applications which are conducted in the vicinity of the shore and require reduction of hydraulic conductivity.

Keywords: Calcium Carbonate Precipitation, Seawater, Alkaline Solution, Soil Conductivity

1. Introduction

Precipitation of calcium carbonate in between soil particles is one of the processes that recently acquired the interest of the scientific community and geotechnical engineers as a potential technique for soil improvement [1, 2] . However, the instantaneous occurrence of the chemical reaction hinders the in-situ precipitation, which drove researchers to focus on inducing precipitation by microbial activity (MICP). Thus, insufficient research can be found that consider chemical precipitation of calcium carbonate for geotechnical applications.

On the other hand, application of MICP on large scales still involves several pitfalls that could hinder its applicability. These pitfalls include the hazardous by-products, introduction of exogenous species to the environment, and the complications involved in materials delivery to the site.

The research described by the current paper employed an experimental approach in order to investigate the potential modification of hydraulic conductivity of the sandy soil using chemical precipitation of the calcium carbonate using seawater as the source of calcium ions. It was envisioned that such a process might be utilized for development of impermeable barriers in order to control seawater intrusion at the coastal aquifers.

1.1. Chemical Precipitation of Calcium Carbonate

The amendments that occur to the engineering characteristics of soil by precipitation of calcium carbonate are governed by a wide range of physico-chemical processes [3–7]. However, the following sections were restricted to the chemical factors that govern the precipitation process. Also, the previously reported studies on the soil behavior that contain calcium carbonate precipitation were briefly reviewed.

The chemical precipitation of calcium carbonate is an instantaneous irreversible reaction. It can be described by the equation:

$$Ca^{2+} + CO_3^{2-} \rightarrow CaCO_3 \downarrow \qquad (1)$$

However, the precipitation process in governed by several factors that influence the characteristics of the resulting mineral.

Calcium carbonate occurs in several mineral forms that include the stable forms of calcite and aragonite, metastable vaterite, and other amorphous forms [8]. The precipitation reaction of calcium carbonates results initially in the mineral forms of calcium carbonates that are thermodynamically unstable. This comports with Ostwald's Law of Stages, which states that 'the least stable phases with the highest solubility, precipitates first and subsequently transforms to the more stable ones' [9–11]. The calcite precipitation is a phased complex process, which is controlled by temperature, nucleation sites, ionic strength, pH, solution stoichiometry and degree of supersaturation [8, 12]. These factors control the essential processes necessary for the crystallization of calcite, which are the nucleation, the crystal growth, and the transformation process.

These processes govern the precipitation of calcium carbonate in the form of calcite, which is essential for the amendment of the engineering characteristics of soils. Other metastable and transformable forms do not contribute effectively to particles cementation. Moreover, they are vulnerable to dissolution in the pore water and being flushed out of the soil.

1.2. Hydraulic Conductivity

Previous studies reported differential behavior of the hydraulic conductivity of the soil that examines precipitation of calcium carbonate. Most of these studies describe the precipitation that is biologically induced.

A recent study revealed the actual sites of calcium carbonate precipitation within the soil pores by MICP [13] . The study showed that the precipitation occurs in the solution, and over the surface of soil particles. The solid particles that form in the solution transport by the fluid flow, and they could be intercepted by other soil particles.

As the concentration of calcite content increases by treatment, the flow paths get narrowed down to full clogging. The occurrence of this clogging depends on the treatment conditions. The rate of calcite precipitation and the growth rate of calcite crystals increases as the concentration of the treatment solution increases. The concentration also specifies the maximum sizes of calcite crystals that can form, which finally controls the behavior of the hydraulic conductivity of the bulk soil .

Such variable factors that control the bulk hydraulic conductivity may explain the differences in reduction of hydraulic conductivity that were reported by different researchers.

For example, a reduction of void ratio by 6% and 17% of the initial condition for 'heavily' and 'moderately' treated soils was reported [14] . The 'moderately' treated conditions were found to cause reductions not less than 31% in a different study [15] . It is expected that hydraulic conductivity would change according to the calcite content for each zone, and the bulk hydraulic conductivity of treated soil is dependent on the homogeneity of the property along the flow paths.

In fact, most of the measured values found in literature for hydraulic conductivity comes as part of the parameters measured for different experiments of different designs, purposes, and conditions. One of the studies that were focused on the hydraulic reductions by MICP treatment showed that the concentration of treatment solutions alter the ultimate reductions that can be achieved [16]. The low-concentration treatment was found to maintain the hydraulic conductivity of the soil with no significant changes, while abrupt and severe clogging was noticed when high-concentration treatment was employed. This was explained on the basis of the size of calcite crystals and their relation to the concentration of the solution. Large crystals seem to clog the pores much faster than the small ones that instead get uniformly precipitated over the surface of soil grains, which was demonstrated few years later [13].

The mechanism described for clogging of the pores by calcite precipitation indicates that a full clogging of soil is possible if the treatment conditions are set properly. This has been validated by a different study [17] where a model of sand was treated with MICP by permeation, and circulating the effluent back again to the model. A thin crust of calcite with low permeability 1.6×10^{-7} m/s at the surface of the model with thickness of 1.0 mm. However, the small depth of the crust gives an indication that clogging is more likely to happen close the injection point rather than to extend with a uniform distribution along the flow path.

Furthermore, the reduction in the hydraulic conductivity occurs according to two phases. The first phase was characterized by insignificant reductions in hydraulic conductivity that is accompanied by remarkable increase in calcium carbonate content in the soil. In the following phase, the hydraulic conductivity sharply decreases by any increase in the soil content [18].

1.3. Chemical precipitation of calcium from seawater

Calcium is of the most abundant ions in seawater, and its concentration is governed by the overall speciation of other ions. For example, oceans acidification causes dissolution of calcareous rocks and coral bleaching.

However, the precipitation of calcium carbonate might be affected by the presence of other ions in the seawater. Although the presence of unstable forms in natural systems is improbable, several studies have reported the presence of amorphous forms due to the presence of inhibitors of calcite crystallization [10] , which acts in different mechanisms. Some of these inhibitors are inorganic such as Mg^{2+}. These ions were found to incorporate the inhibition of calcite formation process by blocking active growth sites, or straining the local crystal lattice increasing its solubility of the formed crystals [19]. The inorganic orthophosphate ions were found to retard the dissolution of vaterite and crystallization of calcite by blocking the active sites by the adsorbed phosphate [20]. Also, it was reported that the rate of calcite precipitation slowed down by a factor of 2 due to the presence of dissolved sulfate in the solution [21]. Furthermore, other forms of precipitates such as magnesium hydroxide might appear if precipitation is induced by elevating pH magnitude. This precipitate does not contribute the binding process.

1.4. Concluding comments on previous studies

The soil amendment by chemical precipitation of the calcium carbonate was downgraded due to the practical issues facing its application. However, it is still worthy to investigation of its feasibility for applications where the seawater fills the soil pores as the seawater would represent the source of calcium. The soil salinity in these environments would be already high and the chemcial treatment would not pose extra contamination.

Many studies employed the soil columns for the sake of simplicity. These designs were found convenient for the current study. The concentrations of the calcium that were used for the current study comply with the magnitudes recommended by previous researchers for the

2. Design of the Experiment

The experiment design involved a controlled system for simultaneous injection of two chemical solutions in a column of a soil specimen, as shown in Figure 1.

Two separate tanks were used to store the chemical solutions. One tank contained the calcium-source solution, while the other was used to store the alkaline solution. Two tests were conducted using two different sources of calcium. For the first test, a solution of 0.1M $CaCl_2$ was used. For the second test, the seawater was used as the source of calcium. Seawater samples were collected directly from the sea in containers of 29L each. The Alkaline solution composed of 0.1M and 0.01M $NaHCO_3$ solution for Tests 1 and 2 respectively. The suction tubes from the two tanks are mixed at a tube junction that is fitted before the pump.

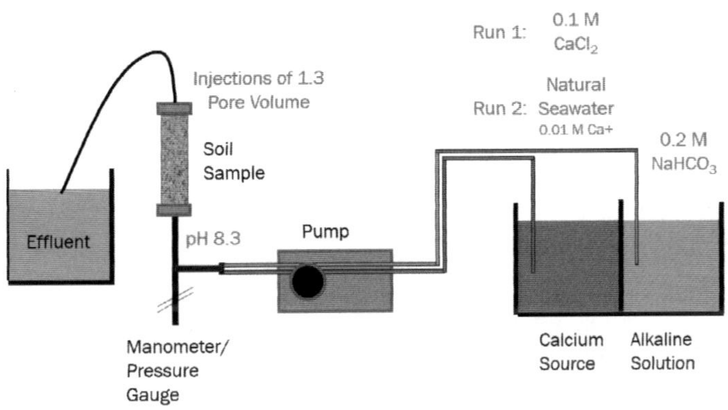

Figure 1. Setup of the experiment in the lab

A peristaltic pump was used for injection of the solutions. The pump was sufficient to provide the flow and pressure required for the experiment. However, as a centrifugal pump, the flow rate and the exerted pressure varied during the experiment. Thus, the pump was characterized in order to identify the pump curve. The flow rate was measured, and the hydraulic head was identified using the pump curve.

A timer was used to run the pump intermittently. The timer was designed and programmed explicitly for the current experiment. A retention time of 1 hour was given after each injection. This retention time was suggested to allow for any short-term potential transformation of the crystal forms of calcium carbonate.

A soil specimen of silica sand-medium grade was used for the test. The initial hydraulic conductivity of the specimens was measured as 2.13 m/s and 2.49 m/s for Tests 1 and 2, respectively. The column was 15 cm in height, and 5 cm in diameter. A nylon filter of 0.2 mm-diameter of openings was used at the top and the bottom surfaces of the soil specimen. The effluent of the soil specimen was collected in a dedicated disposal tank.

Concentration of calcium ion in seawater was determined in the lab using EDTA method. The ion concentration reported as 0.0103 M with 0.12% relative deviation.

3. Results and Discussion

The measured flow rate during each injection was transformed into a hydraulic head using the equation fitted to the pump curve. This helps calculation of the hydraulic conductivity of the specimen for every injection using Darcy's equation:

$$K = \frac{Q\,L}{A\,\triangle h} \tag{2}$$

Where:

Q is flow rate (cm^3/s);
L is the length of the soil column;
A is the cross-sectional area of the soil column (cm^2);
Δh is the head exerted by the pump (cm).

The calculated hydraulic conductivity from both tests is plotted together in Figure 2. For comparison reasons, the results of the two specimens were plotted against the cumulative number of moles of carbonate ion that were injected. This was applied since the available data is the number of injections. At the same time, the concentration of the chemicals used was different. The real indicator for the soil behavior is the amount of the precipitate. In the current case, it is presented by the amount of precipitate precursors that were injected since no approach is available for real-time measurement of the precipitate changes.

As shown in the figure, soil specimen in Test 1 (which employs $CaCl_2$ as calcium source) exhibited a gradual reduction in hydraulic conductivity by treatment. The general behavior matches the behavior described in literature for the reduction of the hydraulic conductivity by microbial calcite induced precipitation [18].

However, the visual inspection of the effluent shows that the effluent contained a precipitate of small particles that behave as colloids. It was suspected that these colloids exhibit sedimentation by gravity, and they are stable such that they do not transform into larger crystals. It reported that the calcite crystals that form as a result of chemical reaction are considerably smaller than the ones that are triggered biologically [22]. This probably due to the lack of preferential surfaces of precipitation. Furthermore, the calcium carbonate accumulated close to the injection surface causing clogging. After injection of 10.9 mole/g of soil of Ca^{2+} and $CO_3{}^{2-}$, the hydraulic conductivity of specimen from Test 1 dropped down to 2.7% of the original hydraulic conductivity. Figure 3 shows an optical microscopic scanning of the specimen from Test 1. The scanning was made to the specimen surface at the injection point. The precipitate appeared in white color filling the pores providing a visual evidence of the clogging that occurred.

35% reduction in the original hydraulic conductivity was attained after injection of 5.5 moles $CO_3{}^{2-}$/100 g of soil. This is comparable to the 28% reduction that was achieved using the artificial source of calcium ions.

However, the number of injections using the seawater was ten times the number of injections needed for Test 1 for the same amount of injected chemicals. This required huge amounts of water and a long time of the experiment; therefore, experiment was terminated at this stage. In fact, the effort and time required for treatment using seawater is an obstacle that has to be studied from a practical point of view.

4. Conclusion

The study described by the current paper investigated the reduction of soil hydraulic conductivity by chemical precipitation of the calcium carbonate. The experiment involved of injection of readily mixed solutions into a soil column. One of the solutions served as a source of calcium ions and the other contained carbonate ions. Two tests were conducted examining different sources of calcium. The first test employed two 0.1M solutions of $CaCO_3$ and $NaHCO_3$. The other test involved mixing a solution of 0.01M $NaHCO_3$ with seawater (as the source of calcium). During the test, it was noticed that crystals formed in small size, and they were prone to mobility by fluid flow. These observations suggested that the precipitate followed the mechanisms of colloids transport and sedimentation; and thus reductions in hydraulic conductivity were governed by

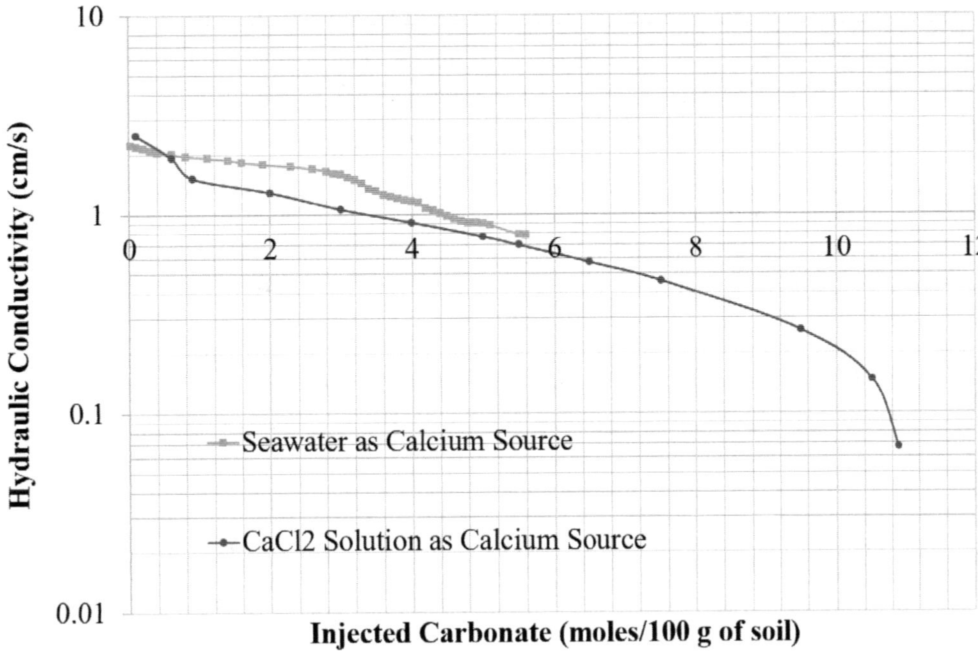

Figure 2. Changes in hydraulic conductivity by treatment

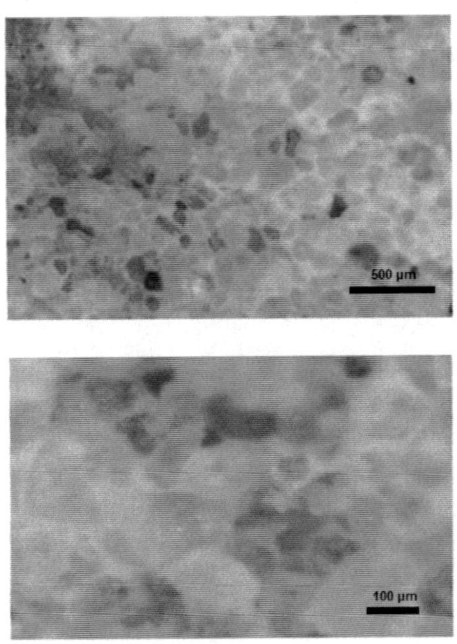

Figure 3. Opticalmicroscopic scanning of samples from Test No. 1 at injection surface after termination of the experiment

the filtration mechanisms. It was noticed that precipitate accumulated close to the soil surface at the injection point.

Employing seawater and artificial source of calcium ions induced reductions in hydraulic conductivity of 35%, and 28% of the original magnitude, respectively. These reductions were achieved after injection of 5.5 mole/100 g of soil. Both specimens exhibited the same general behavior of reduction of hydraulic conductivity by treatment.

Calcium content of seawater could be beneficial for geoenvironmental applications that focus on precipitation of the calcium carbonate. However, such application might be restricted to the vicinity of the shore. It could be utilized for applications such as impermeable barriers. The proposed technique offers several advantages against traditional methods. Using sweater would minimize the costs of treatment to a competitive level to other techniques and materials which are used for the same kind of soil treatment. The materials used by the current research are non-hazardous, and the applied treatment is easily reversible. However, the amount of injected water is considerably high. Therefore, this technique has to be assessed in terms of energy consumption and required time of treatment.

References

[1] Mitchell J K and Santamarina J C 2005 *Journal of geotechnical and geoenvironmental engineering* **131** 1222–1233

[2] Jiang N J, Tang C S, Hata T, Courcelles B, Dawoud O and Singh D N *Soil Use and Management*

[3] Achal V and Pan X *Applied Biochemistry and Biotechnology* 1–11 URL 10.1007/s12010-014-0842-1

[4] Cheng L, Cord-Ruwisch R and Shahin M A 2013 *Canadian Geotechnical Journal* **50** 81–90 URL 10.1139/cgj-2012-0023

[5] Gorospe C M, Han S H, Kim S G, Park J Y, Kang C H, Jeong J H and So J S 2013 *Biotechnology and Bioprocess Engineering* **18** 903–908 URL 10.1007/s12257-013-0030-0

[6] Kawano J, Shimobayashi N, Kitamura M, Shinoda K and Aikawa N 2002 *Journal of Crystal Growth* **237-239, Part 1** 419–423 URL 10.1016/S0022-0248(01)01866-8

[7] Nemati M and Voordouw G 2003 *Enzyme and Microbial Technology* **33** 635–642 URL 10.1016/S0141-0229(03)00191-1

[8] van der Weijden C H and van der Weijden R D 2014 *Journal of Crystal Growth* **394** 137–144 URL 10.1016/j.jcrysgro.2014.02.042

[9] Kralj D and Brečević L 1995 *Colloids and Surfaces A: Physicochemical and Engineering Aspects* **96** 287–293 URL 10.1016/0927-7757(94)03063-6

[10] Kralj D, Brecevic L and Nielsen A E 1990 *Journal of Crystal Growth* **104** 793–800 URL 10.1016/0022-0248(90)90104-S

[11] Kralj D, Brecevic L and Nielsen A E 1994 *Journal of Crystal Growth* **143** 269–276 URL 10.1016/0022-0248(94)90067-1

[12] Wolthers M, Nehrke G, Gustafsson J P and Cappellen P V 2012 *Geochimica et Cosmochimica Acta* **77** 121–134 URL 10.1016/j.gca.2011.11.003

[13] Wang Y, Soga K, Dejong J T and Kabla A J 2019 *G{é}otechnique* **69** 1086–1094

[14] DeJong J T, Mortensen B M, Martinez B C and Nelson D C 2010 *Ecological Engineering* **36** 197–210 URL 10.1016/j.ecoleng.2008.12.029

[15] Whiffin V S, van Paassen L A and Harkes M P 2007 *Geomicrobiology Journal* **24** 417–423 URL 10.1080/01490450701436505

[16] Qabany A A and Soga K 2013 *Geotechnique* **63** 331–339 URL 10.1680/geot.SIP13.P.022

[17] Stabnikov V, Naeimi M, Ivanov V and Chu J 2011 *Cement and Concrete Research* **41** 1143–1149 URL 10.1016/j.cemconres.2011.06.017

[18] Dawoud O, Chen C Y and Soga K 2014 Microbial Induced Calcite Precipitation for Geotechnical and Environmental Applications (American Society of Civil Engineers) pp 11–18 URL 10.1061/9780784413456.002

[19] Nielsen L C, Yoreo J J D and DePaolo D J 2013 *Geochimica et Cosmochimica Acta* **115** 100–114 URL 10.1016/j.gca.2013.04.001

[20] Katsifaras A and Spanos N 1999 *Journal of Crystal Growth* **204** 183–190 URL 10.1016/S0022-0248(99)00174-8

[21] Flaathen T K, Oelkers E H, ur R Gislason S and Aagaard P 2011 *Energy Procedia* **4** 5037–5043 URL 10.1016/j.egypro.2011.02.476

[22] Mitchell A C and Ferris F G 2006 *Geomicrobiology Journal* **23** 213–226 URL 10.1080/01490450600724233

IOP Conf. Series: Materials Science and Engineering **800** (2020) 012012 doi:10.1088/1757-899X/800/1/012012

Seismic Design of Mid-Rise Reinforced Concrete Structures According to TEC 2007 and TBEC 2018

S Derogar[1*], I Safkan[1], B Odabas[1]
[1] European University of Lefke, Faculty of Engineering, Department of Civil Engineering, Lefke, Northern Cyprus, TR-10 Mersin, Turkey

*Email: sderogar@eul.edu.tr

Abstract: Reinforced concrete (RC) structure are widely used structural system in buildings across the Turkey. The new Turkish standard on earthquake, TBEC 2018: Turkish Building Earthquake Code, was published on 2018 which is replaced for the old earthquake code, TEC 2007: Turkish Earthquake Code-2007. This paper aims to present a comparison between the seismic design provisions of these two codes for the design of mid-rise RC structures. The code provisions in both codes are evaluated and differences in terms of ductility and strength for the design of mid-rise RC structures are highlighted and presented. It was shown that the design base shear is higher when using TBEC 2018 code. A 4-story building is designed according to both codes and then the performance of the structure was evaluated. Due to differences on allowable drift limit, the R and Cd factors, the structure showed different behaviour when design according to both codes which are discussed in detail in this paper. A cost analysis was also carried out and compared for both buildings.

1. Introduction

Reinforced concrete moment-resisting frames are a widely used structural system in buildings across the world, especially in developing countries such as Turkey. Earthquake is one of the major problems in Turkey as the country is located on the Alp-Himalayas fault. Earthquakes, particularly on the north Anatolian faults are very dangerous as they are very close to the surface (about 5-30 km) [1]. Following the 1908 Messina earthquake (Italy), the Royal Government of Italy established a Committee in early 1909 to study the disaster and recommend earthquake disaster mitigation measures. The recommendation was adopted in Royal Decree No. 573 of April 29, 1915. Accordingly, the height of the buildings was limited to two stories, and the first story was assigned to be designed for a horizontal force equal to 1/8 the second-floor weight and the second story for 1/6 of the roof weight. This is believed to be the first known quantitative recommendation of design for seismic forces in the history of seismic codes [2].

Modifications to the design codes are normally accompanied by experiencing destructive earthquakes. Turkey experienced lots of catastrophic earthquakes which led thousands of casualties and are reported in the literature [1]. Main reason of the extent of the damage and losses are generally believed to be the construction malpractice and lack of adequate inspection of enforcement of the application of seismic design code [3].

Turkey adopted the first earthquake code in 1940 and since then the code was revised 9 times [3]. Turkey experienced various kinds of disasters, and earthquakes are the most hazardous kind of these disasters. Each disaster has been an experience for Turkey's disaster management. Hence, after these major disasters, new laws and new codes enacted or old ones were revised [1]. The Turkish earthquake code of 2007 [4] was the same as the Turkish earthquake code of 1998 except for a few parts and minor changes. Therefore, the scientific studies in earthquake engineering have not been reflected in the design rules since 20 years. The recent Earthquake Code of Turkey has been revised and published in 2018[5]. Almost 120 experts contributed to the whole document and all activities were coordinated by a permanent earthquake code committee consisting of 15 members. The new Turkish Building Earthquake Code is a comprehensive revision of the previous one dated 2007.

This paper aims to highlight the differences between the 2018 earthquake code and the 2007 earthquake code. For this, the earthquake performance analysis with different ground classes were performed considering a four storey reinforced concrete frame building. In performance evaluation, Equivalent Seismic Load Method which is suggested in Turkish Earthquake Code 2007 and Turkish Earthquake Code 2018 were adopted. Also a cost analysis were carried out for the designed structures according to both codes. All analysis were carried out using Prota Structure [6] and Sap2000 v.20, general-purpose finite element program [7] software programs.

2. Main changes/differences according to the earthquake codes of 2007 and 2018

In the earthquake code, 2007 [4]., "Analysis Requirements for Earthquake Resistant Buildings" were consisted of two parts where it is expanded into four sections in the new Turkish Earthquake Code of 2018. The most significant change that was introduced with the earthquake code of 2018 [5] regarding the calculation of earthquake ground motion are Turkey Earthquake Hazard Map (Turkey Earthquake Seismic hazards) [8]. Figure 1 (a & b) shows the change seismic hazard map of Turkey according to both Earthquake codes of 2007 and 2018 respectively.

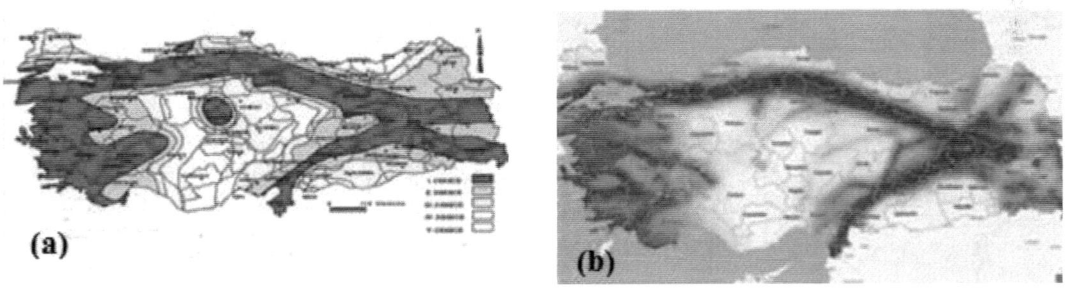

Figure 1: a) Earthquake Hazard Map of 2007, b) Earthquake Hazard Map of 2019

According to the earthquake code of 2007 [4], five different earthquake zones were defined in Turkey. However, with the new earthquake hazard maps, the concept of earthquake zones has completely disappeared. The S_s and S_1 value can be determine from the map for any location in Turkey.
In the earthquake code of 2007 [4]., the soil was divided into four different classes: Z1, Z2, Z3 and Z4. The determination of the soil class was carried out by A, B, C and D soil grouping, respectively. However, in the new earthquake code 2018 [5], soil classes and soil groups are combined from the best ground to worst ground ZA, ZB, ZC, ZD, ZE and ZF are defined. For the ZF soil class, which categorized as worst soil class, the new code requires test to be carried out on site-specific research and evaluation.

Classification of buildings are defined by introducing the importance factor. According to the Turkish earthquake code 2007 [4] structures such as schools, dormitories, prisons and museums were defined at the second level of importance. In the Turkish earthquake code of 2018 [5], the importance factor of these structures were increased to the first level and the coefficient was increased from 1.4 to 1.5. A parametric study on two identical building designed by using both codes of 2007 and 2018 were carried out to assess the effect of importance factor and it was shown that the change in importance factor can lead to an approximate increase of 7% on the calculated earthquake force of the building.

In the new earthquake code 2018 [5], the building performance levels are defined in 4 different categories as "continuous use (CU)", limited damage (LD), controlled damage (CD) and prevention of collapse (PC) which is shown in Figure 2. However, in the earthquake code 2007 [4], performance levels were defined only for the evaluation of existing buildings.

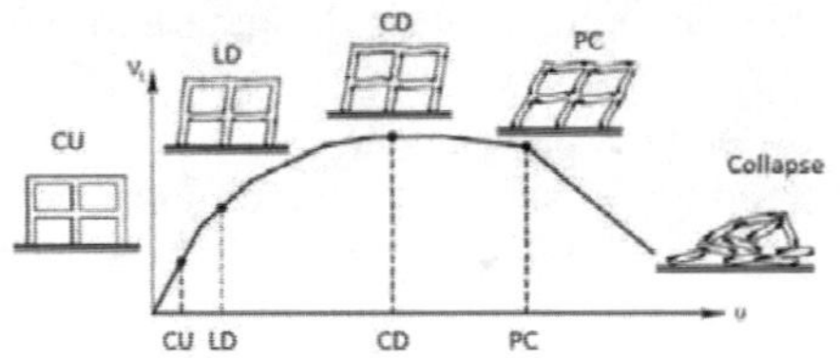

Figure 2. Building Performance Goals

The new earthquake code 2018 [5] introduced the over strength coefficient factor, "D", as a new parameter in the design of reinforced concrete structures. Those structures which are not expected to exhibit ductile behaviour, earthquake loads reduced by the behaviour coefficient are increased by the over strength coefficient. As a result of the application of the over strength coefficient, the shear force demands will increase for all column and beam members.

The most important feature that was introduced by the new earthquake code (2018) [5] was the necessity to use the effective stiffness coefficients. Accordingly, the effective stiffness coefficients were reduced by 70% for columns, 35% for beams and 50% for shear walls. Consequently, the horizontal displacement of the members could be increased and therefore, the dimension of the members need to be increased in order to ensure sufficient rigidity.

Modifications are provided in the new earthquake code 2018 on calculating the spectrum coefficients. Accordingly, Corner periods (TA, TB), which are important factors in determining the spectrum coefficients in the new earthquake code, dependent on SDS and SD1 values. Figure 3 compares the horizontal Elastic Design spectrum obtained for Istanbul-Gebze using earthquake code of 2018 [5] and earthquake code of 2007 [4]. As shown in Figure 3, higher spectral acceleration values are obtained according to Turkish earthquake code of 2018 [5] when compared to the Turkish earthquake code of 2007 [4]. As a result, the earthquake loads will be increased.

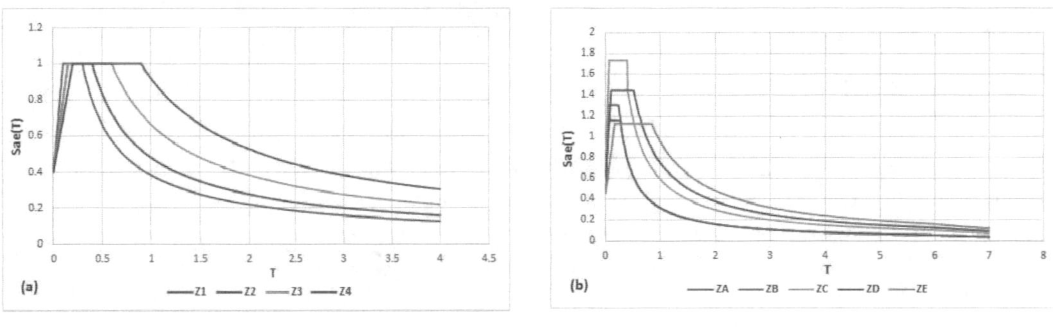

Figure 3. Horizontal spectral accelerations for different soil classes in Istanbul- Gebze, a) according to earthquake code 2007, b) according to earthquake 2018

Earthquake code of 2018 [3] proposed a major change on the calculation and limits of the effective inter-storey drift. In the new earthquake code, the displacement limit is determined separately depending on the contiguous or unified state of the walls and frames. According to this rule, it aimed the walls of the building do not receive any damage during the earthquake and the structure can exhibit more ductile behaviour. According to earthquake code 2007 [4], the drift limit was depending only on the floor height and seismic load reduction factor whereas in new earthquake code, the drift limit depends on the building important factor, the seismic load reduction factor and the building seismic period.

3. Design and cost analysis of R.C. structures according to earthquake codes of 2007 and 2018

As shown in Figure 4, a 4-storey reinforced concrete building was designed according to both earthquake code by using the "Prota Structure [6]" and "Sap2000 [7]" software program to examine the numerical differences that will arise from the design principles that differ in the 2007 [4] and 2018 [5] earthquake cedes. It was assumed that the project is located in Gebze district of Istanbul province. Earthquake performance analysis with different soil classifications were performed on this 4 storey reinforced concrete frame system building. In performance evaluation Equivalent Seismic Load Method which is suggested in Turkish Earthquake Code of 2007 [4] and Turkish Earthquake Code of 2018 [53] issues were used.

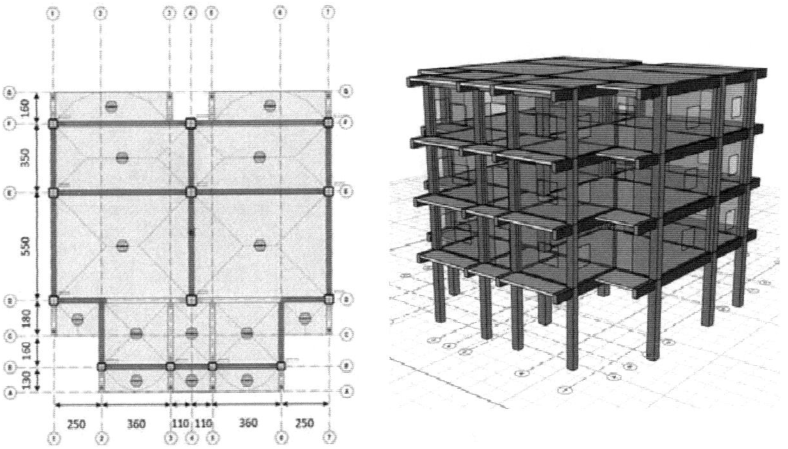

Figure 4. floor plan and view of the Building

According to Turkish Earthquake Code of 2007 [4], Gebze district belongs to Earthquake zone one.. The short and long periods corresponding to the spectral acceleration coefficients (S_s, S_1) were calculated by the new seismic hazard maps. The earthquake calculation parameters according to both earthquake codes are given in the Table 1. In this analysis, Concrete of class C25 and Steel S420 were assumed to be used for both design codes of 2007 [4] and 2018[5]. Structural members were designed according to TS 500 [9], whereas the loads were applied according to TS 498 [10].

Table 1, a) Required earthquake parameters for the design of building according to earthquake code of 2018 [5], b) Required earthquake parameters for the designed building according to earthquake code of 2007, [4]

(a) TS500-2000 , TS498 , TBDY 2018 **Earthquake Parameters**	
Latitude, Longitude	40.8005 / 29.6446
Short Period Map Spectral Acceleration (Ss)	1.445
Long Period Map Spectral Acceleration (S1)	0.394
Building Usage Class	BKS=3
Importance Factor	I=1
Behaviour Factor	R=8
Overstrength Coefficient	D=3
Building Height Classes	BYS=6
Earthquake Design Classes	DTS=1

(b) TS500-2000 , TS498 , TEC 2007 **Earthquake Parameters**	
Earthquake Zone	I.
Effective Ground Acceleration	0.4
Importance Factor	1
Behaviour Factor	R=8

No expansion joint was considered in the modelled structures, when Earthquake code 2018 was used. Table 2 shows the fundamental period of the structures when analysed according to earthquake codes of 2007 [4] and 2018 [5] respectively using different ground classes. It was observed that natural period of the structures are higher when the building was designed according to the earthquake code 2018 [5].

Table 2. Buildings time period values on different ground classes which are designed a) according to earthquake code of 2007 [4], b) according to earthquake code of 2018 [5]

a) Earthquake Code of 2007 [4]			**b) Earthquake Code of 2018** [5]		
Ground Class	X Direction (MOD 1)	Y Direction (MOD 2)	Ground Class	X Direction (MOD 1)	Y Direction (MOD 2)
Z1	0.499	0.498	ZA	0.614	0.579
Z2	0.499	0.498	ZB	0.614	0.579
Z3	0.499	0.498	ZC	0.614	0.579
Z4	0.499	0.498	ZD	0.614	0.579
			ZE	0.464	0.449

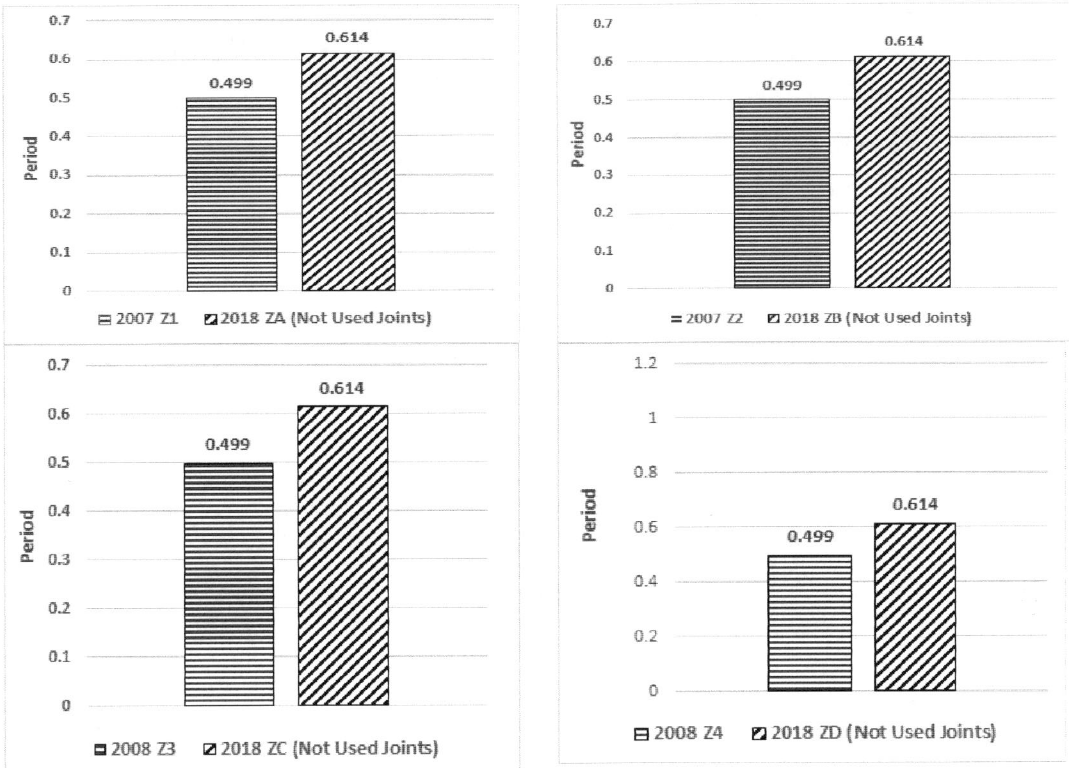

Figure 5. Comparison of natural time period values of the 4-story building on different soil classification according to earthquake codes of 2007 and 2018

As shown in Figure 5, incorporating the effective stiffness coefficients with the new earthquake code led the natural time period of buildings to be increased by approximately of 37%. For the buildings designed according to earthquake code of 2018, sections of structural members were increased depending on the ground classes. Due to the use of cracked stiffness coefficient increase in fundamental period was monitored.

Figure 6 represents the base shear force values obtained from the designed buildings in Gebze district according to earthquake codes of 2007 [4] and 2018 [5] considering different soil classifications. As shown in Figure 6, when the structure is designed according to earthquake code 2007 [4], the base shear force for Z2 ground class was increased by approximately 25%, and about 40% in ground class Z3 when compared to results of the structure designed for Z1 ground class. It was observed that the obtained base shear forces were the same when the structures were designed considering ground class Z3 and Z4.

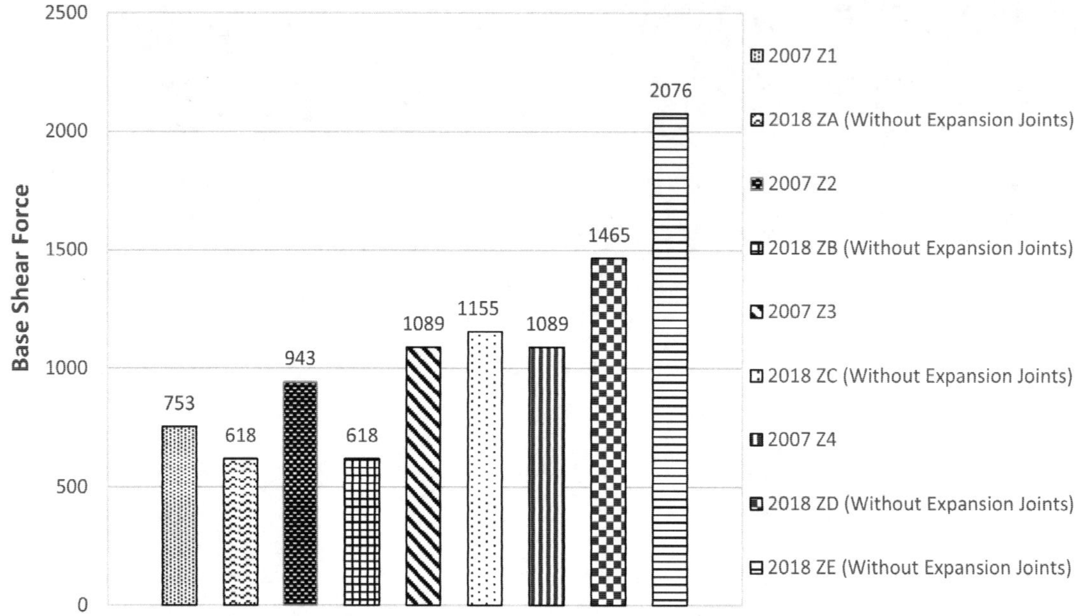

Figure 6. Comparison of calculated base shear force values on different ground classes according to earthquake codes of 2007 and 2018.

When the structures are designed according to earthquake code 2018 [5], I t was observed that the calculated base shear force of the building in the ZA and ZB classes were similar. This is due to the fact that the ZA and ZB ground class spectral acceleration values are very close to each other. It was noticed that the base shear forces of buildings designed with no expansion joints according to earthquake code of 2018 in ZA and ZB ground classes are less than those of buildings designed according to earthquake code 2007 for Z1 and Z2 ground classes. Low spectral acceleration values were obtained in the horizontal spectrum curve corresponding to the high period value and therefore, the values of the base shear forces are reduced.
The effective relative drift and storey drift results on different soil classes of the buildings designed according to the Turkish earthquake code of 2007 [4] and 2018 [5] are given in Table 3 (a & b) respectively.

Table 3-a: The effective inter-storey drift and storey drift values on different soil classes of the building designed according to earthquake code 2007 [4]

Storey Number	Storey Drift	Ground Zone	Storey Number	Inter-storey Drift
1	0.002358		1	0.006289
2	0.005349	Z1	2	0.007975
3	0.007712		3	0.0063
4	0.009089		4	0.003659
1	0.002956		1	0.007883
2	0.006703	Z2	2	0.009992
3	0.009665		3	0.007899
4	0.011461		4	0.004788
1	0.003412		1	0.009098
2	0.007738	Z3	2	0.011536
3	0.011156		3	0.009114
4	0.01314		4	0.005292
1	0.003412		1	0.009098
2	0.007738	Z4	2	0.011536
3	0.011156		3	0.009114
4	0.01314		4	0.005292

Table 3-b: The effective inter-storey drift and storey drift values on different soil classes of the building designed according to earthquake code 2018 [5]

Storey Number	Storey Drift	Ground Zone	Storey Number	Interstorey Drift
1	0.004097		1	0.003108
2	0.010745	ZA	2	0.005043
3	0.016656		3	0.004484
4	0.0215		4	0.003098
1	0.004097		1	0.003108
2	0.010745	ZB	2	0.005043
3	0.016656		3	0.004484
4	0.0215		4	0.003098
1	0.004097		1	0.003108
2	0.010745	ZC	2	0.005043
3	0.016656		3	0.004484
4	0.0215		4	0.003098
1	0.005189		1	0.004893

2	0.013608	ZD	2	0.007939
3	0.021095		3	0.00706
4	0.026268		4	0.004878
1	0.003306		1	0.00424
2	0.009197	ZE	2	0.007556
3	0.015027		3	0.00648
4	0.019744		4	0.00434

As given in Table 3, the storey drift and inter-story drift of the buildings designed according to earthquake code 2018, increased by about 55% in the Z2 ground class, and about 45% in the ground class Z3 when compared to the structures designed according to earthquake code 2007 [4]. The same spectral acceleration values were obtained for Z3 and Z4 period values corresponding to the resonance region in the horizontal spectral curve. Therefore, there were not any changes for Z4 ground class. Figure 7 represents the story drift and inter-story drift behaviour of the 4 story building designed according to earthquake codes of 2007 [4] and 2018 [5]. It was noticed that the structure designed on ZE ground class is more rigid than the structure designed on ZC and ZD ground classes and therefore, the building shows less storey drift values.

As shown in Figure 7, unlike storey drift values, the inter-storey drift value of the building designed in ZE ground class is higher than the inter-storey drift value of the building designed in ZC ground class. This is true as ZE ground class lamda (**λ**) value is greater than ZC ground class lambda (**λ**) value.

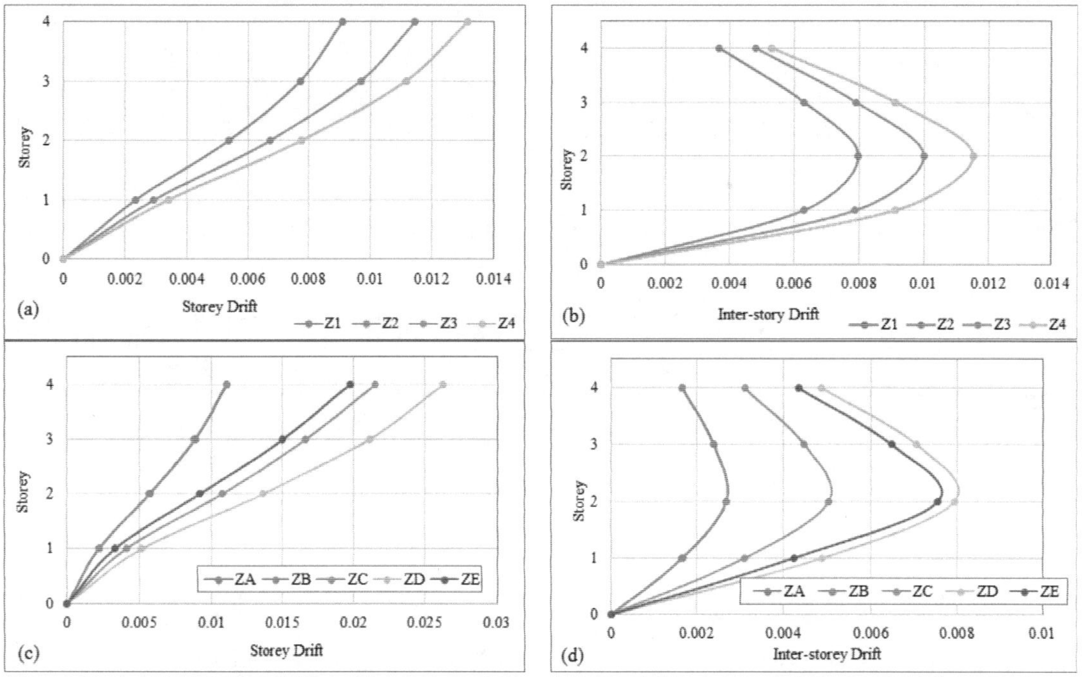

Figure 7. a) and b) Storey drift and inter-story drift values on different ground classes buildings designed according to earthquake code of 2007 respectively, c) and d) Storey drift and inter-story drift values on different ground classes buildings designed according to earthquake code of 2018 respectively.

A cost analysis was also carried out for the buildings designed according to earthquake code 2007 [4] and 2018 [5] considering different soil classifications. Only expenses on formwork, concrete and rebars were included for the cost analysis in this paper. The price list provided by approximate unit cost of Turkey

IOP Conf. Series: Materials Science and Engineering **800** (2020) 012012 doi:10.1088/1757-899X/800/1/012012

[11] was used for the calculation of the construction expenses. Figure 8 presents the results from cost analysis of the 4-story building considering different soil classifications designed according to earthquake codes 2007 [4] and 2018 [5] respectively. As discussed earlier, the new code considers a reduction in the stiffness of the members and this leads to a higher expense on the construction price of buildings designed with earthquake code 2018 [5].

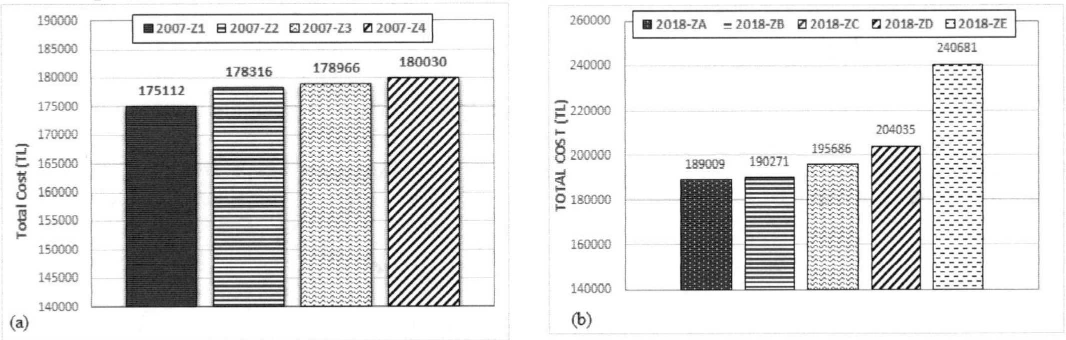

Figure 8. Total cost analysis on different ground classes buildings designed a) according to earthquake code 2007, b) according to earthquake code 2018.

4. Conclusion

This paper aimed to highlight the differences between Turkish earthquake codes of 2007 [4] and 2018 [5]. A mid-rise structure, 4-story building, considering different soil classifications, was selected and designed according to earthquake code 2007 [4] and 2018 [5]. Equivalent Seismic Load Method which is suggested in both Turkish Earthquake Code 2007 [4] and Turkish Earthquake Code 2018 [5] was adopted in this paper. A detailed comparison was carried out on the design aspects and cost of the construction of the selected structure when designed according to both earthquake codes of 2007 [4] and 2018 [5]. All analysis were carried out using Prota Structure [6] and Sap2000 [7] software programs. The following conclusions are provided:

It was shown that the increase in importance factor from 1.4 to 1.5 led the calculated earthquake force to be increased by approximately 7%.

The adoption of effective stiffness coefficient in earthquake code 2018 [5], caused the increase in horizontal displacement of the structure and automatically the member dimensions were increased to ensure sufficient rigidity.

Higher spectral acceleration values were obtained according to Turkish earthquake code 2018 [5] when compared with the Turkish earthquake code 2007 [4].

According to the earthquake code 2018 [5], depending to the period of the structure, the seismic load reduction factor values are decreased. Consequently, the earthquake loads are increased.

When the structures were designed according to earthquake code of 2018 [5] with no expansion joints, the base shear forces of the structures in ZA and ZB ground classes are less than those of structures in Z1 and Z2 ground classes of earthquake code 2007 [4]. Low spectral acceleration values were obtained in the horizontal spectrum curve corresponding to the high period value and therefore the values of the base shear forces are reduced.

The adoption of effective stiffness coefficients in the design of structures according to earthquake code 2018 [5], caused a dramatic increase in fundamental period of structure when compared to earthquake code 2007 [4]. It was also noticed the depending on this increase, storey drift and inter-storey drifts were increased significantly depending on the period of the structure.

Unlike storey drift values, the inter-storey drift value of the building designed in ZE ground class is higher than the inter-storey drift value of the building designed in ZC ground class. This is because ZE ground class lambda (λ) value is greater than ZC ground class lambda (λ) value.

The adoption of effective stiffness coefficients in structures designed according to the Turkish earthquake code of 2018 [5] led to dramatic increase in the building period. Considering the increase in period of the structure and to ensure adequate rigidity of the buildings, the dimension of the structural members are increased and therefore, the building costs increased significantly.

References

[1] Mokarrami A 2009 Iran Deprem Yönetmeliğin Türk Deprem Yönetmeliği İle Karşılaştırılması, Istanbul Technical University, MSc Thesis, June 2009.
[2] Hampshire S Zanaica L Bucur C 2013 Comparative Study of Codes for Seismic Design Of Structures, Mathematical Modelling In Civil Engineering, **9 (1).**
[3] Soyluk A Harmankaya S Y 2012 The History of Development in Turkish Seismic Design Codes International Journal of Civil & Environmental Engineering IJCEE-IJENS **12 (01)**, February 2012.
[4] TEC 2007, Deprem Bölgelerinde Yapılacak Yapılar Hakkında Yönetmelik, Bayındırlık ve İskan Bakanlığı, Ankara, 2007.
[5] TBEC 2018 Turkish Earthquake Code: Specifications for building design under earthquake effects. Ankara, Turkey.
[6] Prota Structure, Ver 2019, Prota Software Informatics and Engineering, Ankara, Turkey.
[7] CSI, SAP2000, Ver 20.1, Integrated Finite Element Analysis and Design of Structures, Computers and Structures, Inc., Berkeley, CA.
[8] Turkey Earthquake Hazard Maps Interactive Web Applications (https://www.afad.gov.tr)
[9] TS 500, 2000, Turkish Standard on Requirements for design and construction of reinforced concrete structures, Ankara, Turkey, ISC 91.080.40
[10] TS 498, 1987, Turkish standard on Design loads for Buildings, Ankara, Turkey, ICS 91.040
[11] Approximate unit cost of construction, 2019, public work s and Housing ministry. Ankara, Turkey.

IOP Conf. Series: Materials Science and Engineering **800** (2020) 012013 doi:10.1088/1757-899X/800/1/012013

Influence of fillers on structure and behaviour of gypsum mortars

M Doleželová[1], J Krejsová[1], L Scheinherrová[1], A Vimmrová[1*]

[1]Czech Technical University in Prague, Faculty of Civil Engineering, 166 29 Prague 6, Czech Republic

*E-mail: vimmrova@fsv.cvut.cz

Abstract. Gypsum mortars with different types of lightweight fillers were studied. The importance of gypsum as one of the most environmentally friendly building binder increases nowadays and new ways of its utilization are searched. Using of fillers in the gypsum is generally not necessary, because gypsum does not shrink, but they can be used for economic reasons or to improve the thermal properties or fire resistance of the gypsum material. We studied the structure and properties of the gypsum mortars with three types of lightweight fillers (expanded perlite, expanded clay aggregate, recycled PUR) and compared them with the gypsum mortar with siliceous sand. The SEM analyses were performed together with the mechanical tests. The roughness of the fillers surface and its influence on the structure and properties was studied by confocal laser scanning microscopy. It was found, that the type of the filler has principal impact on the microstructure of gypsum mortars and their properties.

1. Introduction

The significance of gypsum, as one of the most environmentally friendly building binders is currently increasing and new ways of its utilization are searched. Generally, the fillers are not necessary in the gypsum-based materials, because gypsum (contrary to cement-based materials) does not shrink during setting [1]. Nevertheless, the fillers can be used to improve e.g. the fire-resistant properties of gypsum, because the gypsum paste without fillers shrinks significantly at high temperatures [2]. Lightweight fillers can be used in order to improve the thermal resistance of the gypsum materials.

As a lightweight filler perlite [3] or vermiculite [4] are used most often, also expanded polystyrene, either new [5] or recycled [6] can be used. Other waste polymer materials were tested, e.g. polyurethane foam from automobile and construction industry [7] or ground rubber from used tires [8] or from pipe foam insulation [9]. Mostly the impact of the filler on the properties of gypsum material was studied and all authors state, that by addition of any lightweight aggregate the thermal properties of gypsum material were improved, but the mechanical properties usually decreased significantly. The impact of the filler on the structure of the materials was not studied often and also direct comparison of different types of fillers used in the same gypsum material was not performed yet.

We studied three types of fine lightweight aggregates – perlite, ceramic aggregate and waste ground polyurethane and their influence on the structure and properties of gypsum mortar. The results were compared with the gypsum mortar containing standard sand.

IOP Conf. Series: Materials Science and Engineering **800** (2020) 012013 doi:10.1088/1757-899X/800/1/012013

2. Materials

All mortars were prepared from the commercial gypsum plaster (calcium sulfate hemihydrate, CaSO4·1/2H_2O), calcined from flue gas desulfurization product (CaSO4·2H_2O), type A2 according ČSN EN 13 279-1:2009 [10] (producer Saint-Gobain Construction Products CZ, branch RIGIPS, Czech Republic).

As a lightweight aggregate the expanded perlite EP 150 (producer PERLIT, Czech Rep.), expanded clay aggregate Liapor (producer LIAS Vintířov, Czech Rep.) and crushed waste polyurethane foam from used white goods (producer DAXNER Technology, Czech Rep.) were used. The standardized sand CEN, ČSN EN 196-1 [11] was employed for the preparation of the reference material. The mineral composition of raw materials (when possible), obtained by XRD analysis is given in the Table 1. The granulometry of the aggregates can be seen in Figure 1 and their bulk densities (provided by the producers) are in Table 2. Set retarder Retardan-200 P (producer SIKA, Germany) was used in all mortars. There can be seen, that even if the maximal size of particles in all aggregates was declared as 2 mm, the crushed PUR has the finest granulometry with particles smaller than 0,5 mm, while expanded perlite has most of the particles between 0.5 and 2 mm.

Table 1. Mineral composition of raw materials.

		Gypsum	Standard sand	Expanded clay	Expanded perlite*
Mineral		[wt. %]			
Bassanite	$CaSO_4 \cdot 1/2H_2O$	89			
Anhydrite	$CaSO_4$	6			
Calcite	$CaCO_3$	4			
Muscovite	$Kal_2(AlSi_3O_{10})(OH)_2$	1			traces
Quartz	SiO_2		100	6	
Mullite	$Al_6Si_2O_{13}$			38	
Anorthite	$CaAl_2Si_2O_8$			23	
Hercynite	$Fe^{2+}Al_2O_4$			21	
Diopside	$CaMgSi_2O_6$			6	
Hematite	Fe_2O_3			4	
Magnesite	$MgCO_3$			2	

* Glassy material – mostly amorphous

Table 2. Bulk density of used aggregates.

Aggregate		Standard sand	Expanded clay	Expanded perlite	Crushed PUR foam
Bulk density	[kg/m^3]	2575	1050	80	180

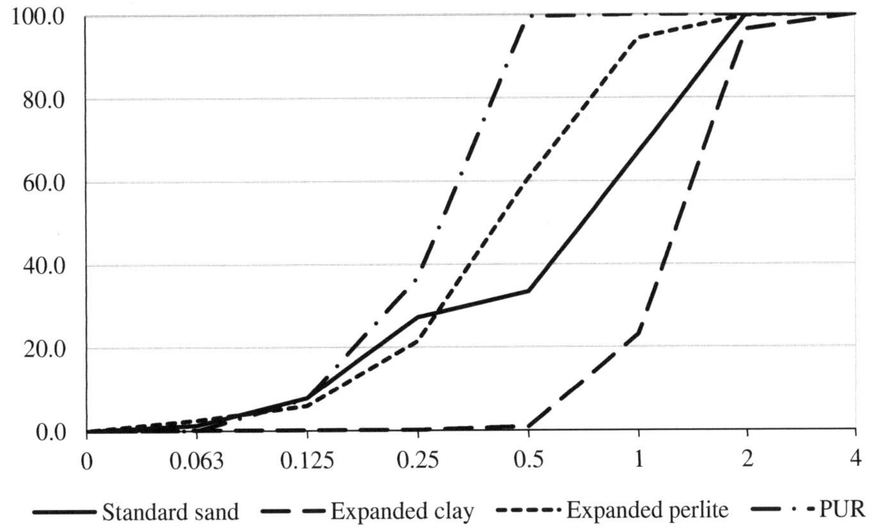

Figure 1. Particle size distribution of used aggregates.

2.1. *Composition and preparation of mortars.*

The composition of the mortars was designed so that the volume of gypsum and aggregate was the same in all mortars. As an initial composition the reference mortar GR with standard sand was chosen and its composition was designed according ČSN EN 196-1 [11]. The mass ratio between the gypsum powder and standard sand was 1: 3. Amounts of 450 g of gypsum and 1350 g of sand were used for one batch of reference mortar. The mass of each aggregate in other mortars was calculated from its bulk density ρ_V and the volume of standard sand V_R in mortar GR (Eq. 1) in order to maintain the same volume of aggregate in all mortars.

$$m = \rho_V . V_R \tag{1}$$

The amount of water was determined for a flow tests diameter value of 165 ± 5 mm and there can be seen, that the mortar with PUR needed substantially higher amount of water than mortars with inorganic aggregates. The amount of retarding agent was 0.02% from dry gypsum weight in all mortars. Composition of tested materials is in Table 3.

Table 3. Composition of tested mortars.

Material	Aggregate type	Gypsum	Aggregate	Retardan	Water
			[g/batch]		[ml/batch]
GR	standard sand	450	1350.0	0.09	270
GEC	expanded clay	450	550.5	0.09	320
GEP	expanded perlite	450	41.9	0.09	260
GPUR	crushed PUR foam	450	94.4	0.09	400

The materials were prepared according to ČSN EN 13454-2 [12]. First, the required amount of water was poured into a bowl. Then dry gypsum was mixed with the retarding agent and dry mixture was poured into the water. The bowl was immediately placed in an automatic mixer. The total mixing time was 4 minutes, which included mixing for 30 s at low speed, 30 s of adding aggregates at low speed, 30 s mixing at high speed, pause 90 s during which the mixture was manually wiped from the bowl wall and finally mixing for 60 s at high speed. Then the gypsum mixture was poured into a mold and was firstly manually and then mechanically compacted. From each material the three prismatic samples 40 x 40 160 mm were prepared. The surfaces of the samples were levelled using a metal knife. The hardened samples were tipped out of the molds 60 - 90 min later and stored at 20 ± 5 °C and relative humidity of 50 ± 5%.

3. Methods

The *particle size distribution* was determined according to EN 933-2 [13]. Standard sieves with apertures of 0.063 mm, 0.09 mm, 0.125 mm, 0.25 mm, 0.5 mm, 1.0 mm, 2.0 mm and 4.0 mm were used.

The *microstructural morphology* of the gypsum mortars and surfaces of the aggregate grains was studied by scanning electron microscopy (SEM) using a Phenom XL electron microscope. The samples were dried and they were not coated or polished.

The *grain and fracture surface roughness* was determined by confocal laser scanning microscopy (CLSM) as the three-dimensional arithmetical mean roughness value SRa [μm] by the Phenom XL. The three-dimensional roughness is an analogy to the arithmetical mean deviation of the profile Ra (two-dimensional roughness) according to ISO 468 [14]. The SRa value of the typical grain surface of each aggregate was determined according Eq. 2.

$$SRa = \frac{1}{l_1 l_2} \int_0^{l_2} \int_0^{l_1} |f(x,y)| dx dy \quad (2)$$

where f(x,y) is height of the assessed profile and l1, l2 are measured lengths.

The *bulk density* of mortars was calculated from the mass and dimensions as a ratio of mass and volume. The dimensions of samples were measured by the digital calliper and their mass was determined by weighing. Volume of samples was calculated from dimensions.

The *flexural and compressive strength* were determined according EN 13454-2 [12] on standard test samples. The experiment was performed as a common three-point bending test. The measurements were carried out 7 days after mixing. The compressive strength was measured on the halves of the specimens left over from the bending test.

4. Results and Discussions

The shape and quality of the surface of the aggregate grains were studied by the SEM. From each aggregate several grains was chosen and the surface roughness SRa was measured 10 times for each aggregate at 20 μm wavelength and the mean value was calculated. The surface of typical grain of each aggregate type and the mean surface roughness are given in the Table 4. There can be seen that the roughest surface has the standard sand, because it was crushed at production process in order to obtain the required particle size. The crushed PUR differs significantly from other materials and it has also the smoothest surface of all aggregates. The values of the surface roughness correspond to the visual observations. There can be also seen, that particles of expanded clay and perlite are very porous, while sand and PUR particles have not any pores inside.

Table 4. Shape and surface roughness of aggregates.

Aggregate		standard sand	expanded clay	expanded perlite	crushed PUR
SEM	500 x magnif.		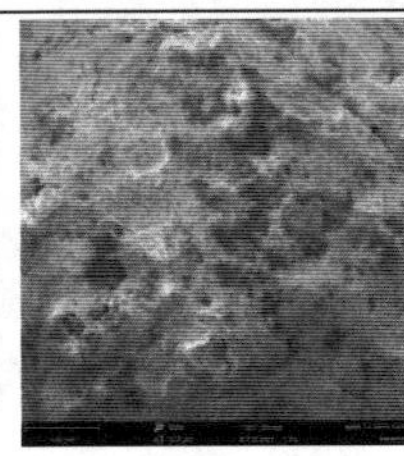	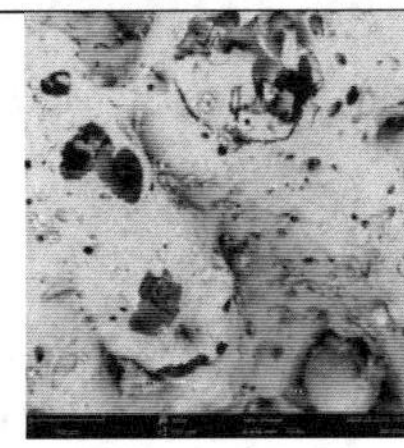	
SRa	[μm]	0.591	0.557	0.474	0.351

SEM images of the mortars with the different fillers are in the Figure 2. The images taken at 250x magnification show the overall structure of the mortars. There can be seen that the microstructures of mortars differs significantly. While in the mortars with well-defined aggregates (i.e. sand and expanded clay) the boundary between the aggregate particles and gypsum can be seen distinctly, in the mortars with less regular particles (perlite and PUR) the crystals of gypsum grows into the surface of the particles. At magnification 3000 x can be also seen, that the size of gypsum crystals in particular mortars differs. The largest gypsum crystals are in the mortar with PUR (length approx. 13 μm, thickness 2 μm) and they are separated from the particle surface and loosely packed and distinct interfacial transition zone (ITZ) is formed. The smallest particles can be seen in the mortar with expanded clay (length approx. 8 μm, thickness 1 μm) and the structure is less porous. The densest structure is formed in the mortar with perlite, where the gypsum crystals are shorter (length approx. 10 μm, thickness 1.5 μm) and they are not as smooth as crystals in other mortars. There can be said, that the biggest crystals and the most distinct ITZ were formed in the mortars with nonporous particles, regardless of the surface roughness. Also, the amount of water in the mixture did not play any important role in the crystal formation.

The values of bulk density, strength and roughness of fractured surface are given in Table 5. The fracture surface roughness was measured on the broken surface after bending test at the wavelength 50 μm.

Table 5. Physical properties of mortars.

		standard sand	expanded clay	expanded perlite	crushed PUR
compressive strength	[MPa]	12.7	12.5	9.4	2.9
flexural strength	[MPa]	5.9	4.3	3.9	1.8
bulk density	[kg/m^3]	1886	1107	647	838
fracture surface roughness	[μm]	1.46	1.72	1.74	1.47

IOP Conf. Series: Materials Science and Engineering **800** (2020) 012013 doi:10.1088/1757-899X/800/1/012013

Aggregate	250x	3000x	Structure
standard sand	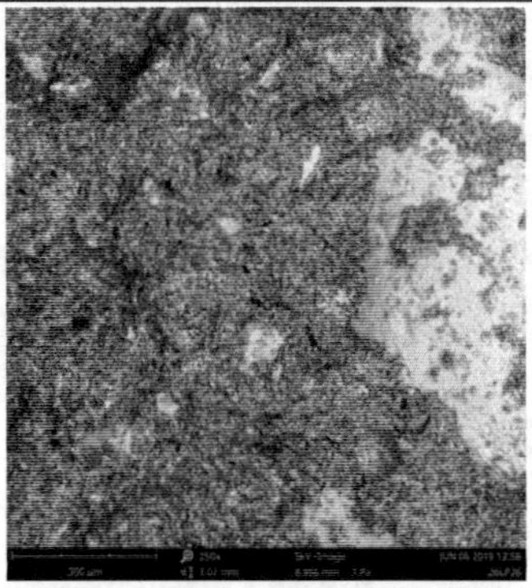		· large, well formed crystals · smooth crystals · porous structure · distinct ITZ
expanded clay	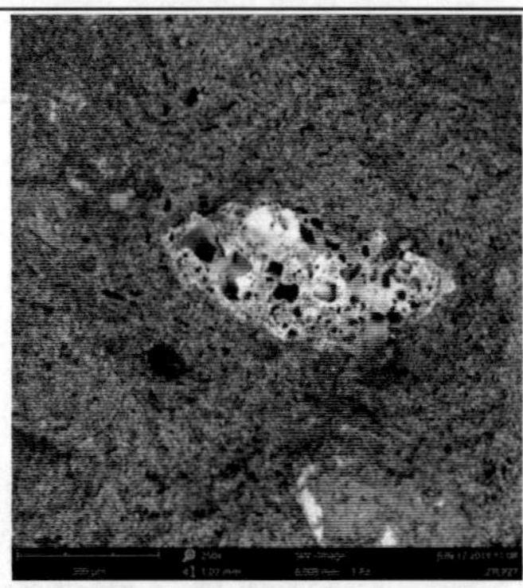	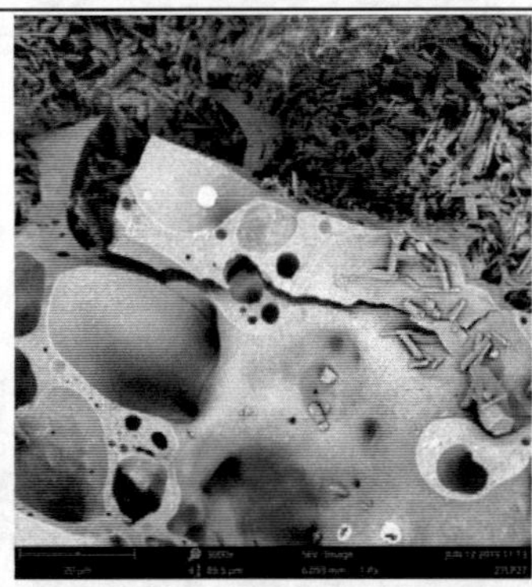	· smallest, well formed crystals · less porous structure · less distinct ITZ
expanded perlite	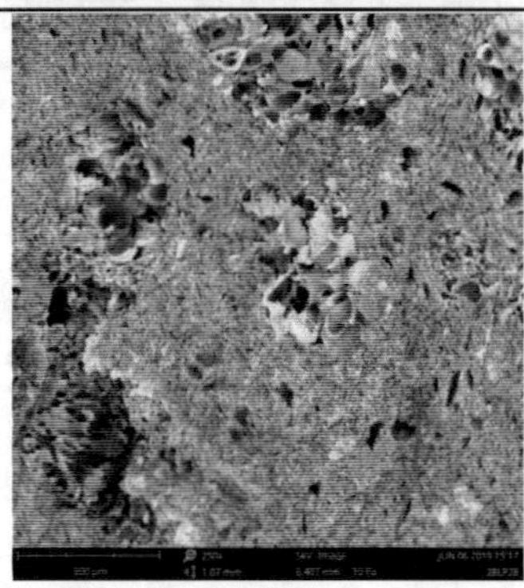	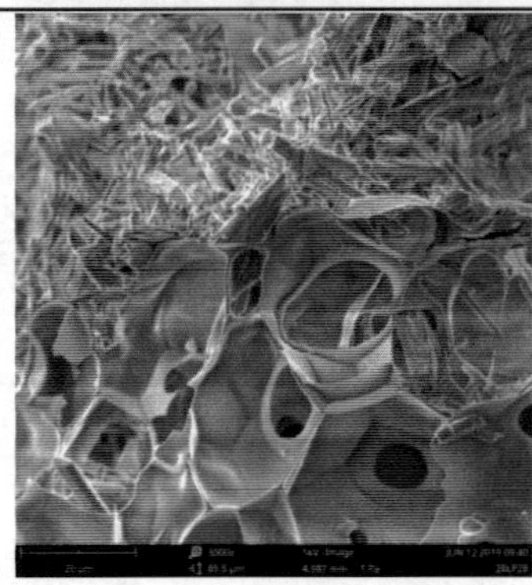	· medium size, frayed crystals · densely packed · no visible ITZ
crushed PUR		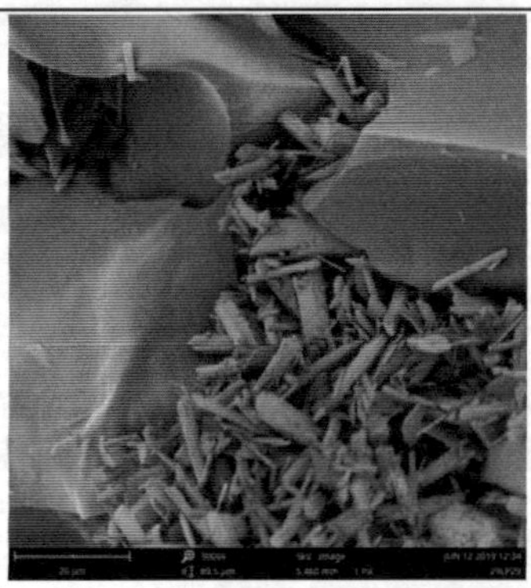	· largest crystals · distinct ITZ

Figure 2. The SEM images of microstructure of mortars.

There can be seen, that the values of strength are related to the surface roughness of aggregate particles (Fig.3). The rougher the particles are the higher the strength is. This phenomenon is known in the materials with cement binder [15] and was observed by the authors also in the gypsum mortars [16]. It can be seen that even if the bulk density of mortar with expanded clay decreased to 60% of the bulk density of mortar with standard sand, the compressive strength stayed nearly the same and fracture strength decreased 14% only. On the opposite side the strength of mortar with PUR decreased significantly, because of very smooth PUR particles.

Unlike the strength the fracture surface roughness does not correspond with the roughness of particles. The fracture surface roughness depends more on the porosity of particles. It can be explain by the fact that the nonporous particles are stronger and therefore not easily broken and the fracture line goes through the gypsum paste only, which is relatively smooth. In the mortars with weaker porous particles the fracture line goes through the broken particles also and these broken particles has rougher surface than gypsum paste and therefore the fracture surface roughness is bigger

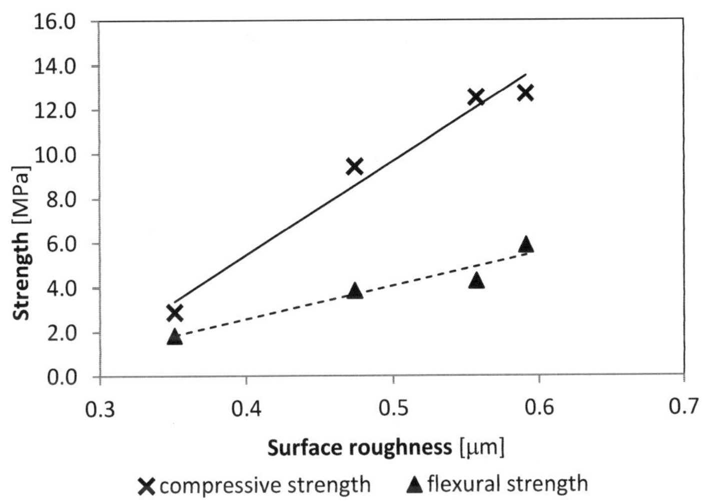

Figure 3. Relationship between the particle surface roughness and strength of mortars.

5. Conclusions

The influence of the different types of the lightweight aggregates was studied. It was found, that the structure of gypsum was changed by the presence of different types of particles and the type of particles had also the influence on the physical properties.

There conclusions can be drawn:

- The rougher the surface of aggregate particles is, the higher the strength of gypsum mortars.
- The size and shape of gypsum crystals in the vicinity of particles depends on the surface roughness of aggregates. In the mortars with smoother particles the gypsum crystals are larger and better defined and there is more pores between them.
- The fracture surface roughness depends on the strength of the aggregate particles. Stronger particles do not break during the bending and the fracture goes through the smoother gypsum only, while the weak porous particles breaks and the fracture goes through the rougher particles also.

Acknowledgments

This research was supported by the Czech Science Foundation, Project No. 19-08605S.

References

[1] Karni J, Karni E. 1995 Gypsum in construction: origin and properties. Mater Struct. **28** 92 doi:10.1007/bf02473176

[2] Doleželová M, Scheinherrová L, Krejsová J, Vimmrová A. 2018 Effect of high temperatures on gypsum-based composites. *Constr Build Mater.* **168** 82 doi:10.1016/j.conbuildmat.2018.02.101

[3] Demir I, Serhat Baspinar M. 2008 Effect of silica fume and expanded perlite addition on the technical properties of the fly ash–lime–gypsum mixture. *Constr Build Mater.* **22** 1299 doi:10.1016/j.conbuildmat.2007.01.011

[4] Gencel O, del Coz Diaz J, Sutcu M et al. 2014 Properties of gypsum composites containing vermiculite and polypropylene fibers: Numerical and experimental results. *Energy Build.* **70** 135 doi:10.1016/j.enbuild.2013.11.047

[5] Sayil, B. and Gürdal, E. 1999 The physical properties of polystyrene aggregated gypsum blocks. In: 8th International Conference on Durability of Building Materials and Components. Vancouver: NATIONAL RESEARCH COUNCIL CANADA, 496

[6] San-Antonio-González A, Del Río Merino M, Viñas Arrebola C, Villoria-Sáez P. 2015 Lightweight material made with gypsum and extruded polystyrene waste with enhanced thermal behaviour. *Constr Build Mater.* **93** 57 doi:10.1016/j.conbuildmat.2015.05.040

[7] Gutiérrez-González S, Gadea J, Rodríguez A, Junco C, Calderón V. 2012 Lightweight plaster materials with enhanced thermal properties made with polyurethane foam wastes. *Constr Build Mater.* **28** 653 doi:10.1016/j.conbuildmat.2011.10.055

[8] Serna Á, Río M, Palomo J, González M. 2012 Improvement of gypsum plaster strain capacity by the addition of rubber particles from recycled tyres. *Constr Build Mater.* **35** 633 doi:10.1016/j.conbuildmat.2012.04.093

[9] Jiménez Rivero A, de Guzmán Báez A, García Navarro J. 2014 New composite gypsum plaster – ground waste rubber coming from pipe foam insulation. *Constr Build Mater.* **55** 146 doi:10.1016/j.conbuildmat.2014.01.027

[10] ČSN EN 13 279-1 2014 Gypsum binders and gypsum plasters. Definitions and requirements. Czech Standardization Institute. Prague

[11] ČSN EN 196-1 Methods of testing cement –Part 1: Determination of strength 2016 Czech Standardization Institute. Prague

[12] ČSN EN 13454-2 Binders, composite binders and factory made mixtures for floor screeds based on calcium sulfate - Part 2: Test methods 2019 Czech Standardization Institute. Prague

[13] ČSN EN 933-2 Tests for geometrical properties of aggregates. Determination of particle size distribution. Test sieves, nominal size of apertures. 1997 Czech Standardization Institute. Prague

[14] ISO 468. Surface roughness. Parameters, their values and general rules for specifying requirements. 1982 International Organization for Standardization

[15] Mindess S, Young J, Darwin D 2003 *Concrete* (Upper Saddle River, N.J.: Prentice Hall)

[16] Krejsová J, Doleželová M, Pernicová R, Svora P, Vimmrová A 2018 The influence of different aggregates on the behavior and properties of gypsum mortars. *Cem. Concr. Compos.* **92** 188 doi:10.1016/j.cemconcomp.2018.06.007

IOP Conf. Series: Materials Science and Engineering **800** (2020) 012014 doi:10.1088/1757-899X/800/1/012014

Use of Slab composite with galvanized steel deck in reinforced concrete frame

M Eissa * M Celikag

[1]Department of Civil Engineering, Faculty of Engineering, Eastern Mediterranean University, Famagusta, North Cyprus, via Mersin 10 Turkey
*Email: mahmoud6cyprus@gmail.com

Abstract. Slabs composite with galvanized steel deck and in-situ concrete are commonly used in steel framed structures. This approach proven to reduce structural weight and increase the efficiency of structural design. Furthermore, the steel deck acts as permanent formwork and also as positive reinforcement after concrete hardening. The study reported in this paper concentrates on the use of composite slab with galvanized steel deck in reinforced concrete (RC) framed buildings. This requires the assessment of the system under flexural loads and slip forces that develops between the RC beam and the composite slab in shear transferring zone. Four-point static and push-off tests were used to find the flexural capacity of system and the amount of slip between the beam and the slabs. Results of the numerous experimental tests and analytical studies demonstrate that the composite slab can be implemented in RC frames since both composite beam flanges and RC beam flanges conducted by the same beam stem manifested almost the same flexural behavior and had roughly similar load-carrying capacity in static test. Moreover, no slip was developed between the RC beam and the composite slab in all four test cases with different reinforcement set up. This can be considered as an important advantage of this system.

1. Introduction

Over the last few decades' composite slab application has been frequent and mostly accounted for the supremacy of steel frames in multi-story buildings. The interpretation of composite slab, which is formed by using in-situ concrete with galvanized steel deck, is acceptable when the deck develops a satisfactory shear bond with the concrete. In such situations the deck is not only used as a permanent formwork but it also sustains the construction loads during the execution, including the fresh concrete during casting process and acts as tension reinforcement for the composite slab. The slab weight dominates the overall weight of the building. The composite slab with a thin galvanized deck and approximately 25 percent less concrete than other types of reinforced concrete slabs, is lighter and it can develop the same stiffness. As a result, the structural members become smaller in cross section and lighter too and hence the loads that are transferred to the foundations are reduced leading to smaller size foundations. The light weight galvanized sheets reduce the transportation time and the time in which crane spends to transfer the material. Moreover, the application of the composite slab is quite simple when compared to other slabs since it often only needs shrinkage and temperature reinforcement after placing the sheets on beams. Reduced workmanship and material consumptions make composite slab a favorite alternative in steel building industry. The purpose of this paper is to investigate the possibility of providing these advantages to the reinforced concrete buildings through investigating the flexural capacity of the RC beam flange with composite slab. For this purpose, four points load static test was used, the results were compare with the traditional reinforced concrete solid slab and the slip that usually occurs between the composite slab and the steel beam (in this paper between the composite slab and RC beam) in the shear transferring zone was measured using push-off test as per the requirements of Euro-code.

Literature review did not reveal any evidence of investigation into this kind of composite slab with galvanized deck being used with RC beams in RC buildings. Hence, there is a need to understand such slabs individual behavior and also behavior as part of an RC frame. To do that one needs to review the understanding of such slab behavior within a steel framed building and hence try to understand how it would behave as part of a RC framed building. Nie and Wang [1] and Nie and Cai [2] both claimed almost the same results when they investigated composite slab with steel beam under four points static test and they concluded that including the slip impact has remarkably enhanced the deflection prediction. The slip effect developed an additional flexural moment which caused a depletion in the elastic flexural capacity. In addition to that composite beam failure mode relies on the shear connectivity degree. In fact, the main concern in composite structure is the slip between the concrete and the steel which requires shear connector to transfer the shear loads between the two materials. The literature is rich of such manuscripts that discusses this topic. Zona and Ranzi [3] has done a parametric study on slip demand in composite slab and they concluded that, for a given shear connection degree, the design parameters that influence shear connection demand are construction succession (propped construction shows slip demand always larger than unpropped construction), the length of the span (longer spans are the most censorious in terms of slip demand), and the shape of the steel section (sections with equal flanges are less censorious). They also found out that the shear connection distribution has a crucial effect as the non-uniform shear connection distributions with more connectors near the supports are more effective in limiting the slip demand, mostly for the full shear connection designs. Furthermore, the thickness of the slab, dead load to live load ratio, and concrete compressive strength, can have indispensable influence on the slip demand even though their overall importance is limited. The slip capacity test in this paper has been done in accordance to EN 1994-1-1 2004 [4] Annex B.2. and all specimens in both tests has been prepared in accordance to Euro code design requirements.

2. Material properties, sample preparation and load set-up

C25 concrete was used in static test and low strength concrete C5 were used in push-of test. S420 rebar were used for the RC slab and beams and Almetsan ALDECK 70/915, with S300 yield strength, a deck depth of 60 mm, thickness of 1 mm with an embossment distributed on the web of the sheet, were used. A total of sıxteen specimens were tested in two different test set-up as follows:

- Four-point load static test were carried out to assess the behavior of the new system under flexural load. Then the behavior of new system was compared to the traditional reinforced concrete solid slab system to assess the efficiency of the new system.
- Push-of tests are standard for finding out slip between steel beams and composite slab with steel deck. Hence, twelve push-of test specimens were used to investigate if slip occurs between the composite slabs and the reinforced concrete beam of the new system.

2.1. Static test

The tests were done by using a simply supported reinforced concrete (RC) beam carrying a continuous RC slab to primarily investigate the flexural behavior of both of them under four-point load (two loads applied on the slab and two reactions from the supports). The overall length of the specimens was 3700 mm with a pure bending (pure bending is the distance between the two line loads) of 250 mm. 150 mm thickness was used for both solid slab and composite slab. Both slabs were carried by a 25/45 cm reinforced concrete beam. The reinforcement in solid slab was Ø8/17 in the short direction and Ø8/25 in the long direction with Ø8/25 at the negative moment zone to make it similar to the composite slab, since steel mesh were used. On the other hand, the composite slab had only a mesh reinforcement with Ø8/25 in both directions to prevent concrete shrinkage cracking. Finally, the RC beam has 3Ø12 rebar at top and bottom for flexure and Ø8/20, as shear reinforcement. Figure 1 displays one of the specimens before loading. The load has been applied gradually so that the failure load can be achieved within an hour. Low Voltage Displacement Transducers (LVDTs) were placed in the mid-span of the beam and the two slabs to monitor the deflection throughout the test period.

Figure 1. Static test load set-up and specimen dimensions

2.2. Push-off tests

The push test is required for the composite beam to investigate the effectiveness of the new type of shear connecter to guarantee the composite action between the concrete and the steel component. In this new system the test was carried out to analyze the developed stresses and internal forces between the reinforced in-situ concrete beam and the composite slab with galvanized deck. Push test was carried out by using four different test cases in conformity with Euro code 4, Annex B.2. Figure 2 shows specimen dimensions and load set-up. In Case1 (SC-FR) the beam, and slab had reinforcement and there was a proposed connector. In Case2 (SC-SR) beam and slab had reinforcement but without connectors. In Case3 (SC) beam had reinforcement but slab had no reinforcement and connector. In Case4 (SC-Con) beam had reinforcement with connector and slab had no reinforcement. Thus the tests were carried out in order to investigate which component will influence the slip capacity and the failure mode.

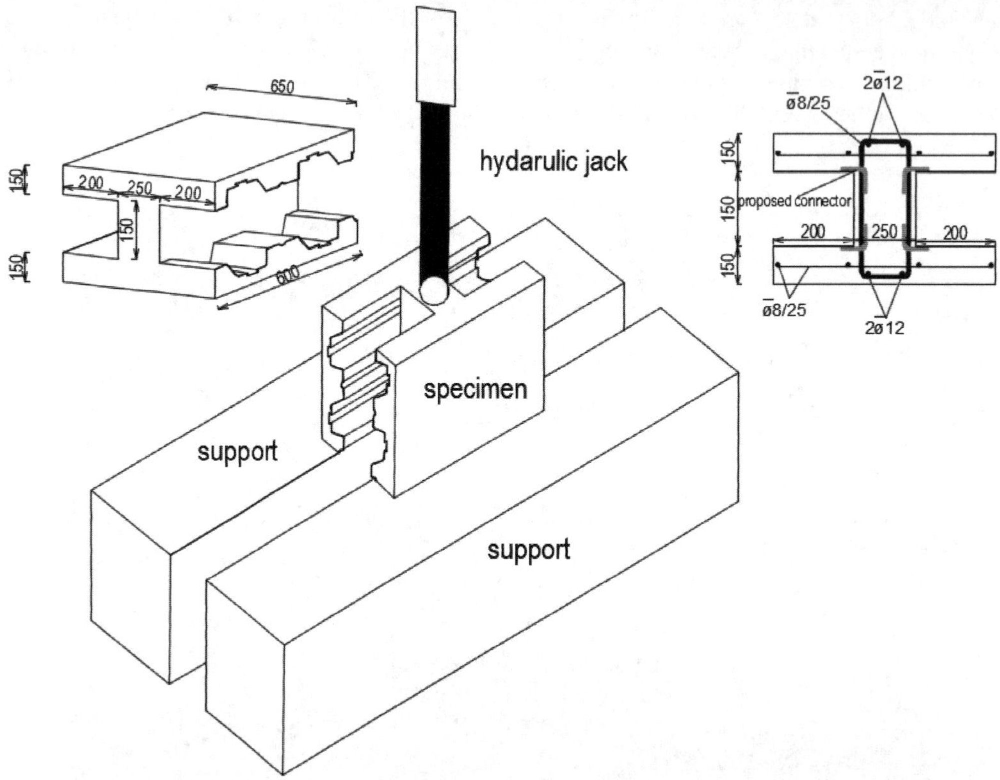

Figure 2. Specimen dimensions and load set-up for push-off test.

The load has been applied in a way that the failure would not occur in less than 15 minutes with LVDTs were replaced at the bottom of the beam and in the mid-span of the beam where the connectors were embedded.

3. Results and discussions

3.1. Static test

Composite and solid slab supported by RC beam behaved in a similar manner, with small differences, such as. the initial elastic stiffness of the solid slab was higher than the composite one; the elastic load was about 10% higher with elastic mid-span deflection being around 30% lower. On the other hand, although the failure load was almost the same, the total mid-span deflection corresponding to failure load was 25% more for the composite slab whereas the average crack width was considerably lower for composite slab. The solid slab specimen attained less total mid-span deflection than the composite one. This could be due to composite slab having galvanized flexible steel sheet that made the overall slab to have more flexible behavior. Figure 3 compares the mid-span deflections of the composite slab specimen and solid slab one.

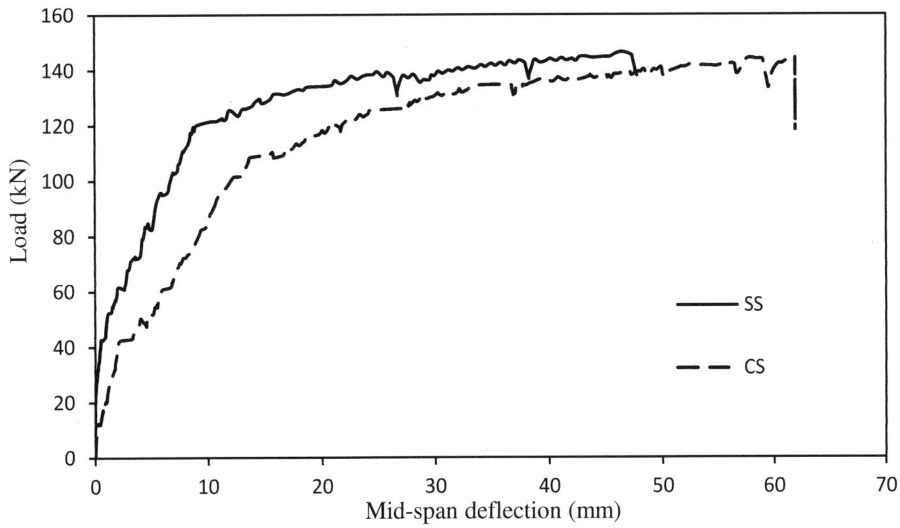

Figure 3 Load versus mid-span deflection of composite and solid slab specimens.

3.2. Push-off test

Euro code 4 requires 6 mm slip for ductile behavior, all cases had brittle behavior. Ductile failure does not exist in this system as the connector does not play any role in either load carrying capacity or total slip. This behavior has been predicted with an understanding of the nature of reinforced concrete structures. This can be an advantage of the new system since slip capacity is one of the major concerns in designing composite beam with slab composite with galvanized steel deck.

Despite the fact that proposed connectors had slightly increased the load carrying capacity of the specimen. However, more investigations should be done in future with different connectors and different types of concrete to try to understand the behavior and level of ductility of such systems. Figure 4 shows load versus longitudinal slip for a specimen from each series.

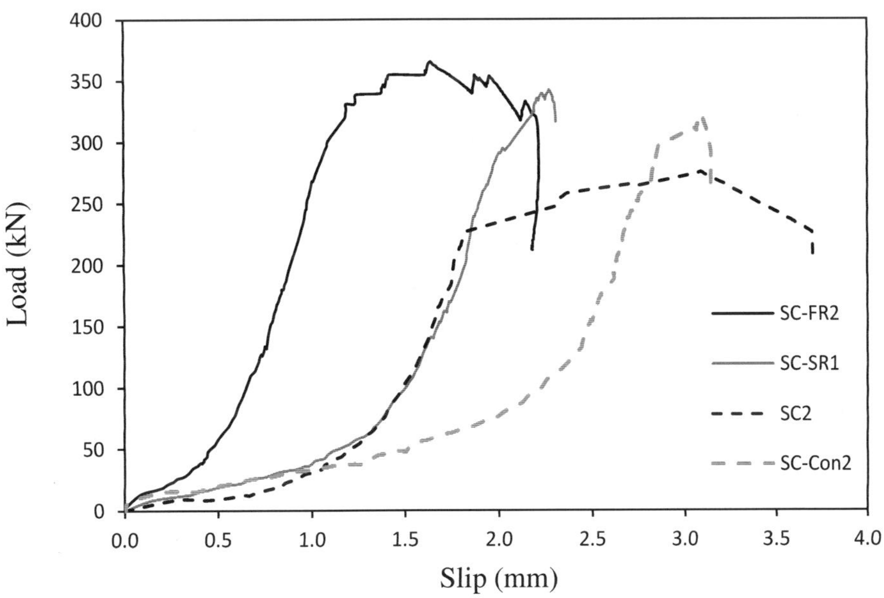

Figure 4 Load versus longitudinal slip for a specimen from each series.

4. Conclusion

The conclusion drawn from the work done can be summarized as follows.

- Composite slab and reinforced concrete solid slab carried by the same RC beam showed the same flexural behavior and had almost identical load carrying capacity in static test. Yet composite slab showed more total deflection than solid slab but with less crack width.
- The four cases used for push-off tests indicated that no slip occurred between the slabs and the beam since all of them has been casted together so the shear forces would be transferred by the concrete itself. Thus usage of this system may provide some advantages since no slip was developed. However, all test cases showed brittle behavior in accordance to Euro code 4 evaluation. It should be mentioned that although slab reinforcement is only for shrinkage and temperature it played an important role in increasing the system load carrying capacity and causing the specimen to have a ductile failure mode.
- In summary the composite slab which consist of galvanized steel deck and in-situ concrete can be applied easily on reinforced concrete beam under gravity loads.

References

[1] Nie, J., Cai, C., & Wang, T. (2005). Stiffness and capacity of steel–concrete composite beams with profiled sheeting. *Engineering Structures*, *27*(7), 1074-1085.

[2] Nie, J., Fan, J., & Cai, C. (2008). Experimental study of partially shear-connected composite beams with profiled sheeting. Engineering Structures, 30(1), 1-12.

[3] Zona, A., & Ranzi, G. (2014). Shear connection slip demand in composite steel–concrete beams with solid slabs. Journal of Constructional Steel Research, 102, 266-281. doi: 10.1016/j.jcsr.2014.07.018

[4] EN 1994-1-1:2004. Eurocode 4 design of composite steel concrete structures, part 1.1: general rules and rules for buildings. Brussels: European Committee for Standardization (CEN); 2004.

IOP Conf. Series: Materials Science and Engineering **800** (2020) 012015 doi:10.1088/1757-899X/800/1/012015

Performance of encased single stone column in layered soil

R E Farah*, Z Nalbantoglu

Department of Civil Engineering, Eastern Mediterranean University, Gazimağusa, North Cyprus via Mersin 10 Turkey.

*Email: rowad.farah@cc.emu.edu.tr

Abstract. Stone columns were found to be effective, economical and widely-used ground improvement technique in soft grounds. It is important to understand the behavior of stone columns when they are installed in layered soils. This paper presents results from a series of small scale laboratory model tests carried out in unit cell tank to investigate the behavior of floating and end bearing stone columns in layered soils with and without encasement, consisting of loose sand overlying a soft soil. Tests were conducted with loading only the stone column area to estimate the limiting axial capacity of stone column. Laboratory tests were carried out on a column of 10 cm diameter surrounded by layered soil. Test results indicated that the bearing capacity of layered soil has been improved in all cases of stone column applications. The non-encased floating stone columns gave better results than non-encased end bearing stone column. The inclusion of geotextile encasement resulted in further improvement of layered soil and the encased floating stone column had superior bearing capacity improvement.

1. Introduction

Stone columns have been used in many difficult foundation sites throughout the world to increase the bearing capacity and the rate of consolidation, to reduce the total and differential settlements, and to improve slope stability of embankments and also to improve the resistance to liquefaction [1].

It is well established that stone columns derive their bearing capacity from the lateral confining pressure from the surrounding soils [2-4]. When stone columns are installed in very soft soils or in layered soils where the top layer is very soft, they may not derive significant bearing capacity owing to low lateral confinement, which leads to excessive bulging. Improvement in the bearing capacity of the stone column can be achieved by the encasement of the stone column in geotextiles [5]. Increased confinement of the stone column reduces lateral bulging, thereby increasing the bearing capacity of the stone column [6-7].

Previous studies investigated the performance of stone columns in the settlement and bearing capacity of soft soils [1, 8-9]. There have been very limited studies performed on the behavior of stone columns in layered soils [10-11]. However, in reality, foundations are usually set in layered soils and understanding the behavior of stone columns in layered soils is very important for civil engineers [11]. In the literature, the condition of layered soils was mostly considered to be a soft soil overlaying a very soft soil, each of varying thicknesses [10-11]. However, the performance of stone columns in loose sand overlaying soft soil was not considered. In the present study, the performance of stone columns in layered soil consisting of loose sand overlaying soft soil was studied in a small-scale laboratory model test tank with the single stone column at the center and the layered soil surrounding it.

For the enhancement of weak soils where the depth of the hard layers lies at deeper layers, the choice of floating stone columns may be one of the best options [12]. Literature review points out that most of the conducted researches on the behavior of stone columns were at most on the condition of end bearing stone columns [13-14]. There are very limited studies conducted on the vertical stress-settlement behavior of floating stone columns [13-14]. Hence, in the present study, small scale laboratory model tests have been conducted to examine the performance of both floating and end bearing stone columns with and without geotextile encasement in layered soil.

2. Experimental investigation

2.1Materials used

A clayey silty soil was used in the present study, which was taken from Famagusta, North Cyprus. The specific gravity of this soil was found to be 2.65. The liquid limit, plastic limit and plasticity index of the soil were found to be 58%, 30% and 28%, respectively. As per the Unified Soil Classification System (USCS), the soil was classified as silt with high plasticity (MH). To analyze the worst soil condition in the laboratory, soil specimens at different water contents were prepared and unconfined compressive strength, UCS tests were performed on those specimens. The variation of UCS with increasing water content values was shown in Figure 1. The values in Figure 1 indicated that the water content corresponding to 33 kPa was found to be 33%. According to Das [15], the consistency of the soil at 33 kPa was described to be soft. Throughout the study, the water content of the soil in the model test tank was preserved at 33%. The bulk density of the soil at the same water content was 1.81 g/cm3.

A sandy soil which was used in the study was taken from the Bedis Beach in Famagusta, North Cyprus. The average particle size of the sand material (D_{50}) was 0.22 mm. The sand used has a coefficient of uniformity (C_u) of 1.29 and a coefficient of curvature (C_c) of 1.06 and was classified as poorly graded sand (SP) as per Unified Soil Classification System (USCS). The specific gravity, maximum dry density (ρ_{dmax}) and minimum dry density (ρ_{dmin}) of the sand were found to be 2.65, 1.55 g/cm3 and 1.46 g/cm3, respectively. The friction angle of the sand was found to be 31°.

The stone columns formed were of 10 cm diameter and constructed by crushed stone aggregate with particle size varying in the range of 1mm to 5mm. The average particle size of the aggregate material (D_{50}) was 3.0 mm. The crushed stone aggregate used has a coefficient of uniformity (C_u) of 1.67 and a coefficient of curvature (Cc) of 0.99 and was classified as poorly graded sand (SP) as per Unified Soil Classification System (USCS). The specific gravity, maximum dry density (ρ_{dmax}) and minimum dry density (ρ_{dmin}) of the aggregate were found to be 2.48, 1.61 g/cm3 and 1.49 g/cm3, respectively. The friction angle of the aggregate at 70% relative density was found to be 45° and the corresponding density was 1.57 g/cm3.The grain size distribution of soft soil, sand and the crushed stone aggregate used in this experimental investigation were depicted in Figure 2.

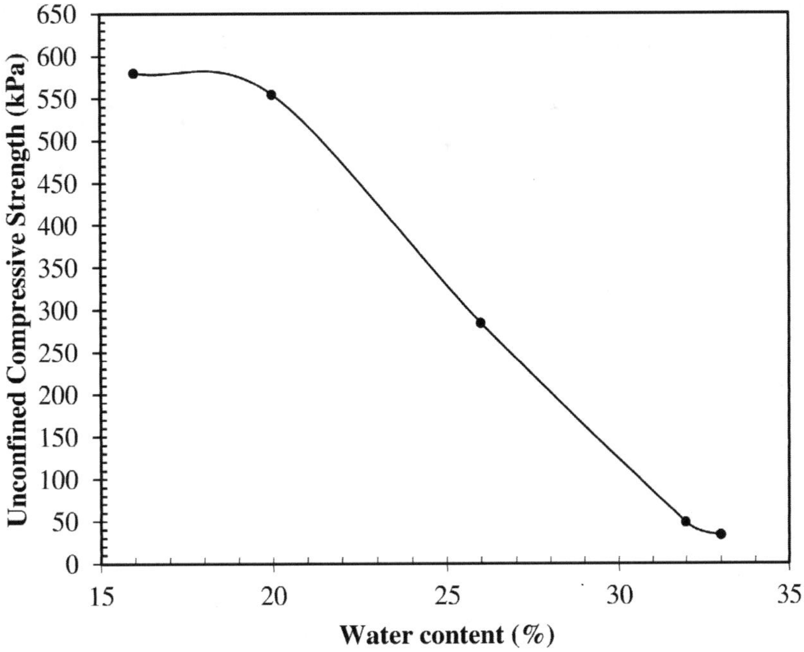

Figure 1. Unconfined compressive strength (UCS) of soil with changing water content.

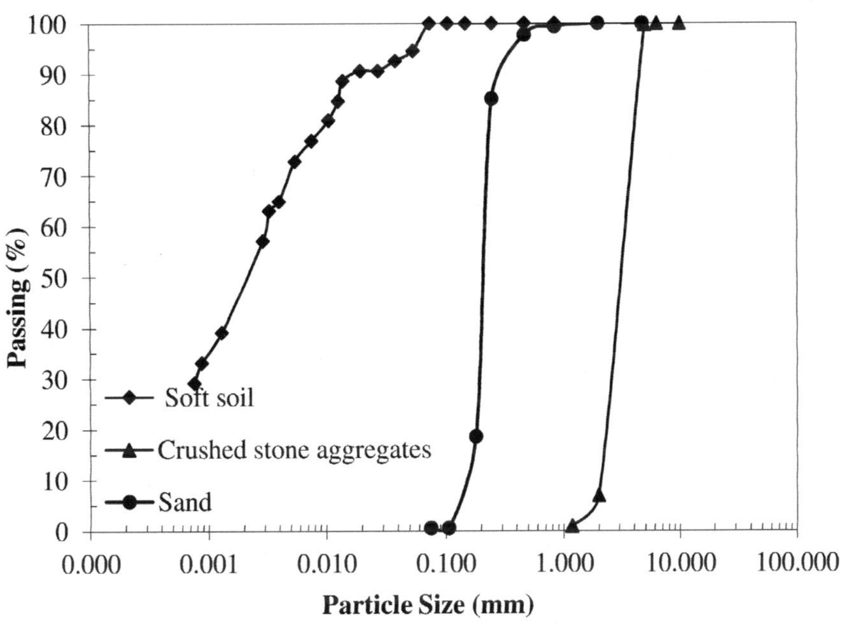

Figure 2. Particle size distribution of soft soil, sand and crushed stone aggregates used in the study.

In this study, non-woven geotextile was used as an encasing material in the stone columns. The geotextile material had a tensile strength of 7 kN/m. The properties of the non-woven geotextile utilized in this study were illustrated in Table 1.

Table 1. Properties of the non -woven geotextile [16-22].

Test	Standard	Value	Unit
Weight	EN ISO 9864	500	gr/m^2
Thickness (2kPa)	EN ISO 9863-1	2.7	mm
Tensile strength (longitudinal transverse)	EN ISO 10319	7	kN/m
Break extension (longitudinal-cross)	EN ISO 10319	min. 60	%
Static puncture	EN ISO 12236	2040	N
Dynamic puncture	EN ISO 13433	10.04	mm
Water permeability, VH50	EN ISO 11058	0.034	l/s* m^2
Visible pore size, O90	EN ISO 12956	0.070	mm

2.2 Experimental setup

Two types of stone columns were considered in the study: floating stone column with L= 50 cm and end bearing stone column with L= 75 cm (at the base of model test tank). All stone column diameters were 10 cm. A steel rod of 2 cm diameter was used for the compaction of stone column material to reach the required density. The load was applied through a 10 cm diameter steel circular plate and 38 mm thickness. Two linear variable differential transformers, LVDT were used to measure the settlement of the loading plate. A circular steel test tank of 40 cm diameter and 80 cm high was used in the present study. The layered soil in the model test tank consisted of 35 cm soft soil bed at the tank bottom and 35 cm of loose sand was placed on top of it.

The soft soil was placed in the test tank and the soil was consolidated with an in-situ overburden pressure. Then, the sand layer was laid over the consolidated soil layer. For the placement density of the sand layer, minimum dry density of 1.46 g/cm3 was used.

2.3 Details of test series

Three series of model tests were carried out on unreinforced layered soil, non-encased stone columns and encased stone columns. After the preparation of the soil specimens in the test tank, the vertical load was applied over the area of the stone column by a constant rate of 1.2 mm/min up to 30 mm vertical settlement of the footing. The loading period was maintained short to simulate the undrained condition during construction.

3. Results and discussion

3.1 Vertical stress-settlement behavior of stone column

In the present study, in order to reach the clear ultimate load in the load-settlement curves of stone columns, the loading of the stone column was continued until 30 mm settlement and the vertical stress corresponding to this settlement value was considered to be the ultimate vertical stress of the stone column. The behavior of non-encased and encased floating and end bearing stone columns in layered soil were studied with 10 cm diameter stone columns.

3.1.1 Behavior of non-encased and encased floating stone columns

Figure 3 presents the vertical stress-settlement curves of single stone column with and without encasement. The figure indicated that the ultimate vertical stress of the non-encased floating stone column resulted in a significant amendment of layered soil bearing capacity. The ultimate vertical stress value of non-encased floating stone column reached 124 kPa at 30 mm settlement, while the ultimate vertical stress value of encased floating stone column increased further and reached 240 kPa at 30 mm settlement. In case of encased floating stone column, the increase in the bearing capacity of the soil was due to the radial elongation of the geotextile under applied loads; this was achieved by the tensile strength and stiffness of the geotextile material [8].

3.1.2 Behavior of non-encased and encased end bearing stone columns

From Figure 3, it can be seen that the non-encased end bearing stone column resulted in a considerable improvement in the ultimate vertical stress of soil. The ultimate vertical stress reached 114 kPa at 30 mm settlement. The encased end bearing stone column exhibited little improvement to the bearing capacity of soil and the ultimate vertical stress value of the encased end bearing stone column reached 150 kPa at 30 mm settlement.

3.1.3 Behavior of floating and end bearing stone columns

Figure 3 indicated that the non-encased floating stone column with shorter length (L/d=5) gave better improvement to the bearing capacity of soil than the non-encased end bearing stone column with longer length (L/d=7.5). In soft soils, an increase in the L/d ratio of stone column within the range of 6-10 increases the ultimate vertical stress of the stone columns [23-24]. In the case of non-encased floating and end bearing stone columns in layered soil, an increase in L/d ratio resulted in a reduction in the ultimate vertical stress. Thus, the findings in layered soil were contradicting the findings obtained for soft soil.

From Figure 3, it can be seen that the encased floating stone column (L=50 cm) resulted in a significant improvement in the ultimate vertical stress of soil compared to encased end bearing stone column (L=75 cm). The reason for this behavior was that for shorter encased stone column, the stone column was densified in a shorter time then the time required for the elongation of geotextile started and the soil-column interface shear resistance contributed to the bearing capacity of the stone column. While, in longer encased stone column, the densification of stone column material took longer time to transfer the load along the stone column length. Due to the limited elongation of the geotextile, soil-column interface shear resistance was not attained and thus, the geotextile did not contribute to the increase the bearing capacity of the stone column.

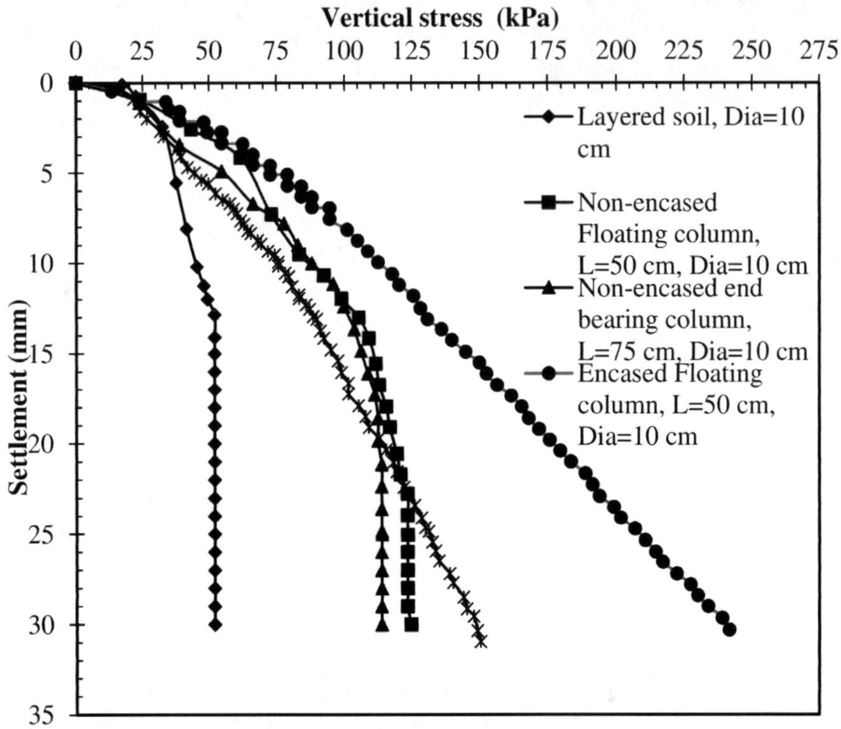

Figure 3. Vertical stress-settlement curves of single stone columns with and without encasement.

4. Conclusions

Based on the small-scale laboratory model test results, the following conclusions can be drawn from the study:

1. The ultimate vertical stress and settlement were improved in all cases of stone column applications in layered soil.
2. The encased floating stone column gave the maximum performance in the improvement of bearing capacity among all stone column applications.
3. The non-encased floating stone column gave better results of improving bearing capacity of soil than the non-encased end bearing stone column. The geotextile encasement of stone columns gave further improvement to the bearing capacity of the soil.

References

[1] Alamgir M, Miura N, Poorooshasb HB and Madhav MR 1996 *Computers and Geotechnics.* **18(4)** 267-90.
[2] Greenwood DA 1970 *In Inst Civil Engineers Proc.*
[3] Hughes JM, Withers NJ and Greenwood DA 1975 *Geotechnique.* **25**(1) 31-44.
[4] Barksdale RD and Bachus RC 1983 *Design and construction of stone columns* vol 2, appendixes Federal Highway Administration (Washington, DC, USA).
[5] Raee E, Hataf N, Barkhordari K and Ghahramani A 2019 *International Journal of Civil Engineering.* **17(6)** 673-85.
[6] Debnath P and Dey AK 2017 *Geotextiles and Geomembranes.* **45(6)** 653-64.
[7] Nazariafshar J, Mehrannia N, Kalantary F and Ganjian N 2019 *International Journal of Civil Engineering.* **17(2)** 253-63.

[8] Murugesan S and Rajagopal K 2009 *Journal of Geotechnical and Geoenvironmental Engineering*. **136(1)** 129-39.
[9] Shivashankar R, Babu MD, Nayak S and Manjunath R 2010 *Geotechnical and Geological Engineering*. **28(5)**695-706.
[10] Shivashankar R, Babu MD, Nayak S and Rajathkumar 2011 *Geotechnical and Geological Engineering*. **29(5)** 749.
[11] Prasad S, Vasavi C and Sai K 2017 *International journal of civil engineering and technology* **8(8)** 453-462.
[12] Dash SK and Bora MC *2013 Geotextiles and Geomembranes*. **41** 26-35.
[13] McCabe B 2009 *Proceedings of ICE geotechnical engineering*.
[14] Ng KS. 2017 *Journal of GeoEngineering*. **12(2)** 89-96.
[15] Das, BM 2008 *Advanced Soil Mechanics* (Taylor and Francis, London and New York).
[16] ISO E 9864: 2005 Geosynthetics--Test Method for the Determination of Mass per Unit Area of Geotextiles and Geotextile-Related Products, European Committee for Standardization, Brussels, Belgium.
[17] ISO E 9863-1: 2005 Geosynthetics–Determination of thickness at specified pressures–Part 1: Single layers. CEN, Brusel.
[18] ISO E 10319: 2008 Geosynthetics Wide-width Tensile Test. International Organization for Standardization, Geneva, Switzerland.
[19] ISO E 12236: 1996 Geosynthetics—Static puncture test (CBR test).
[20] ISO E 13433: 2006. Geokunststoffe—Dynamischer Durchschlagversuch (Kegelfallversuch) (Geosynthetics—Dynamic perforation test (cone drop test)), Deutsches Institut für Normung eV, Berlin.
[21] ISO E 11058: 1999–Geotextiles and geotextile-related products–determination of water permeability characteristics normal to the plane, without load. European Committee for Standardization, Brussels. International Organization for Standardization, Geneva, Switzerland.
[22] ISO E 12956: 2010 Geotextiles and Geotextile Related Products—Determination of the Characteristic Opening Size.
[23] McKelvey D, Sivakumar V, Bell A and Graham J 2004 *Proceedings of the Institution of Civil Engineers-Geotechnical Engineering*. **157(3)** 137-49.
[24] Ali K, Shahu JT and Sharma KG *In Indian geotechnical conference* p 620-628.

IOP Conf. Series: Materials Science and Engineering **800** (2020) 012016 doi:10.1088/1757-899X/800/1/012016

Oxidation of reactive black 5 solutions with sulfate radicals activated by iron

N Gamadze, S Pirgalıoğlu*

Engineering Faculty, European University of Lefke , Lefke / Mersin 10 Turkey

*E-mail:spirgalioglu@eul.edu.tr

Abstract
Textile Effluents are problematic effluents which cannot be treated with conventional biochemical treatment methods. One of the most promising methods for treatment of non-biodegradable wastewaters is advanced oxidation. In this study, effect of sulfate radical based advanced oxidation methods were investigated for treatment of an azo dye; Reactive Black 5 (RB-5). Sulfate radicals were activated with the addition of Iron(II) and Iron (III) Sulfate to the reaction medium. Effect of pH and persulfate dosage on RB-5 removal were also investigated. It was observed that runs which were conducted with Iron (II) ions yielded better results both in terms of RB-5 removal and reaction rate. It was recorded that RB-5 removal and removal rate both improved with increased persulfate dosage while optimum initial pH was found to be unadjusted pH which was measured as 3. RB-5 removal over 99% was observed in the run conducted at pH 3 with a persulfate dosage of 4 mM and $FeSO_4$ (Fe(II)) dosage of 1 mM.

1.Introduction

Textile effluent containing dyes are known to be toxic and non - biodegradable. With the stability of the modern dyes, the use of conventional biological treatment methods for industrial wastewater is proved to be ineffective [1] Due to the fact that conventional biochemical treatment methods do not effectively treat the textile effluents, physical and chemical methods can be considered as alternative treatment methods. Advanced oxidation is a chemical treatment method relying on generation of hydroxyl or sulfate radicals and oxidation of targeted pollutants with these radicals. [2-4] sulfate radicals are obtained by activation of persulfate anion (PS) thermally or chemically such as transition metal ions [5]. Iron (II) and Iron (III) ions are amongst these transition metals. Metal activation takes place as shown in equation 1 [6]

$$S_2O_8^{2-} + Me^{n+} \rightarrow SO_4^{\cdot-} + Me^{(n+1)+} + SO_4^{-2} \quad (1)$$

When Fe^{2+} is used as activator reactions below take place equation 2 shows sulfate radical formation. While equation 3 shows undesired radical consumption. [6]

$$S_2O_8^{2-} + Fe^{2+} \rightarrow SO_4^{\cdot-} + Fe^{3+} + SO_4^{-2} \quad (2)$$

$$SO_4^{\cdot-} + Fe^{2+} \rightarrow Fe^{3+} + SO_4^{-2} \quad (3)$$

On the other hand, reaction mechanism for Fe^{3+} can be given as below [7]:

$$S_2O_8^{2-} + Fe^{3+} \rightarrow SO_4^{\cdot-} + Fe^{2+} + SO_4^{-2} \quad (4)$$

It is also reported that Fe^{3+} ions can react with reaction intermediates to produce Fe^{2+} and activation of PS is carried out with Fe^{2+} ions in some cases resulting a decrease in decolorization rate with the runs conducted with Fe^{3+} [6].

Reaction equations show that increase in iron concentration favor sulfate radical formation while over a certain iron concentration sulfate radical scavenging occurs. [6] On the other hand increase in PS concentration is expected to increase sulfate radical formation leading to increased oxidation rate of the targeted pollutant.

Advanced oxidation of an azo dye; Reactive black 5 (RB-5) with sulfate radical activated by Fe^{2+} and Fe^{3+} was studied. Oxidation of RB-5 with various advanced oxidation methods were studied [8-10] although there is limited information about oxidation of RB-5 with sulfate radicals activated by iron ions. Effect of activator, PS dosage and pH of the medium was observed on decolorization of RB 5. Reaction kinetics were also investigated.

2. Experimental

2.1 Materials

Potassium persulfate (ACS reagent ≥99%), Reactive Black 5 (Dye content 50%), sodium hydroxide, sulfuric acid, ferric and ferrous sulfate were purchased from Sigma-Aldrich.

2.2 Analytical

RB5 Concentration was measured by using Rayleigh UV 1601, UV-visible spectrophotometer at 590 nm. pH of the reaction medium was monitored by WTW multi 3430 multimeter. Iron concentration was measured by Raleigh WFX-130B atomic absorption spectrophotometer (AAS).

2.3 Decolorization experiments

Oxidation reactions were conducted in a 250 mL batch reactor for 90 minutes samples were taken with certain intervals and analyzed immediately with UV-Vis spectrophotometer. Iron concentrations throughout reaction period and after precipitation was monitored by using AAS. Initial dye concentration was selected as 50 $mg.L^{-1}$ (0.05 mM), Iron (II) and Iron (III) concentration was selected as 1mM. Parameters that were investigated were; Activator type (Fe^{2+} and Fe^{3+}), PS concentration and solution pH. PS concentrations were selected as: 0.5, 1, 2, 4 mM and medium pH was adjusted to be 3, 4, 7 and 10. This procedure was followed in presence of both Fe^{2+} and Fe^{3+}. Selected concentrations correspond to molar ratios of; 1\ 10-20-40-80 \ 20 in terms of Dye \ PS \ Iron.

3. Results and discussions

Experiments without pH adjustment (pH = 3) were conducted in order to investigate effect of PS dosage and activator type on decolorization rate. The rest of the experiments which were conducted to observe effect of pH on decolorization rate were conducted with the selected PS dosage in the first set of experiments.

3.1 Effect of activator on decolorization rate

Decolorization rate of RB-5 was observed to be higher when Fe^{2+} was used as the activator for all PS dosages. Figure 1 represents the decolorization of RB-5 in sulfate radical oxidation experiments activated by Fe^{2+} and Fe^{3+}. Having higher decolorization rates with presence of Fe^{2+} was expected since oxidation mechanism with Fe^{3+} involves production of Fe^{2+} followed by activation of PS with Fe^{2+} ions [6]. It is also worth to note that after a reaction period total decolorization values were observed to be similar. Comparison of RB-5 decolorization that were conducted at different PS dosages are presented in Figure 2. It was observed in this study that increase in PS dosage increased both decolorization rate and RB-5 removal. Increase in reaction rate with increasing PS dosage was expected since sulfate radical generation rate is dependent on PS concentration. It can also be reported that increasing PS concentration did not have any inhibiting effect on oxidation reaction for the studied PS concentration range. It is also worth to note that almost 100% dye removal was observed with the runs conducted with 4mM PS and Fe^{2+} as activator (30 minutes of reaction time). On the other hand, 96% dye removal was observed at the end of 90 minutes of reaction time for the runs conducted with presence of Fe^{3+} and 4mM PS.

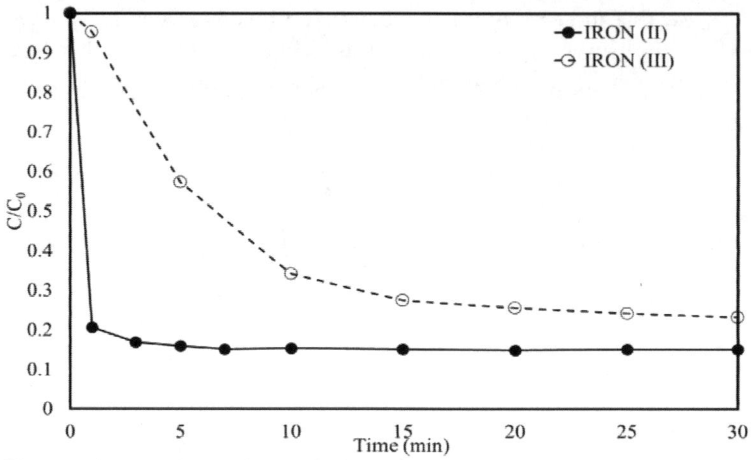

Figure 1. Decolorization of RB-5. PS Concentration: 0.5 mM, Fe Concentration: 0.1 mM, Dye Concentration: 0.05 mM. Activator: Fe^{2+} and Fe^{3+} pH: unadjusted (pH = 3).

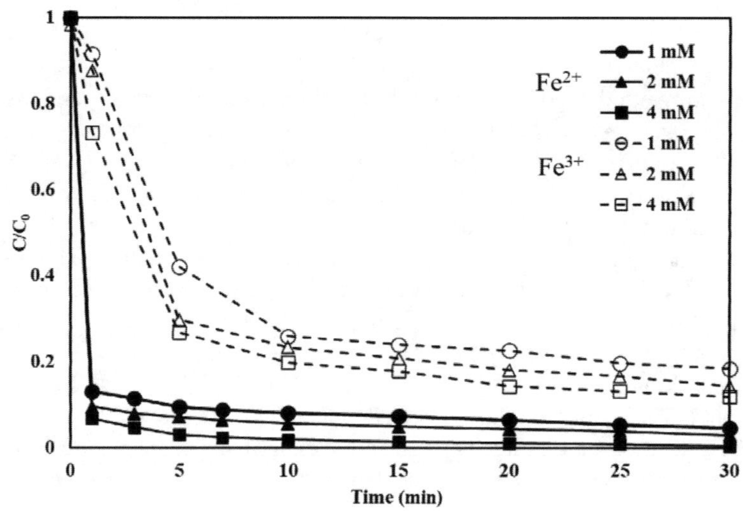

Figure 2. Comparison of activator type with different PS concentrations. Fe Concentration: 0.1 mM, Dye Concentration: 0.05 mM. Activator: Fe^{2+} and Fe^{3+} pH: unadjusted (pH = 3)

3.2 Effect of pH on decolorization rate

Effect of medium pH was also investigated by adjusting pH of dye solution through addition of 0.1 M NaOH solution. Reactions were started by addition of the required amount of PS. It is well known that addition of OH^- ions will result iron hydroxide formation which is insoluble in water. Increasing pH didn't just affect the reaction mechanism it also affected the dissolved iron amount in the medium which decreased the amount of the activator. Results obtained from the pH adjustment experiments are given in Figure 3.

It can be observed from the figure 3 that increasing pH of the medium decreased both RB-5 decolorization rate and dye removal. no significant color removal was observed in the run conducted with presence of Fe^{3+} ions at pH = 10. This can be explained by precipitation of iron ions as $Fe(OH)_3$.

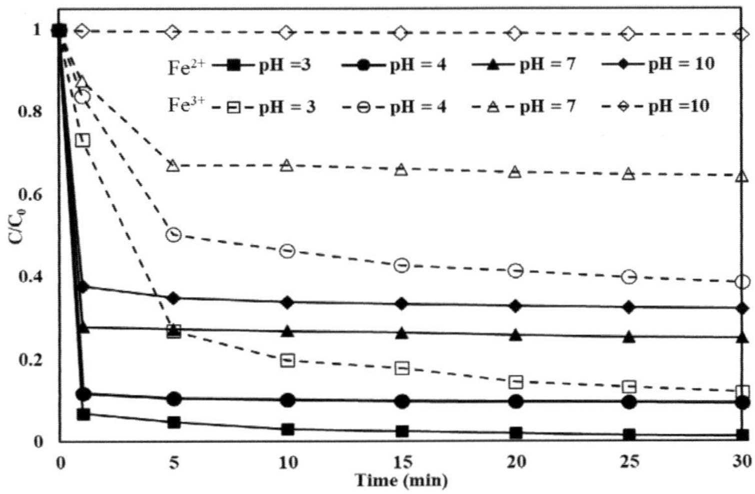

Figure 3. Effect of pH on decolorization of RB-5. PS Concentration: 4 mM, Fe Concentration: 0.1 mM, Dye Concentration: 0.05 mM. Activator: Fe^{2+} and Fe $^{3+}$

4.Conclusions

Iron (II) sulfate addition was found to be more effective than iron (III) sulfate for activation of sulfate radicals. Iron activated sulfate radical oxidation was found to be most effective at acidic conditions. Increase in decolorization rate and RB-5 removal was observed with the increase of PS concentration. TOC removal and byproduct formation should also be investigated in order to understand reaction mechanism better.

References

[1] Ghaly AE, Anathashankar R. Alhattdom, Ramakrishinan, "Production, Characterization and Treatment of textile effluents: A Critical Review", Journal of Chemisrty and Engineering Process 182. (2014)

[2] Yang Shiying, Wang Ping, Yang Xin, Wei Gang, Zhang Wenyi, Shan Liang, "A novel advanced oxidation process to degrade organic pollutants in wastewater: Microwave activated persulfate oxidation"' Journal on Environmental Science 21,1175-1180 (2009)

[3] P. Hu, M. Long, "Cobalt-catalyzed sulfate radical-based advanced oxidation: A review on heterogeneous catalysts and applications", Applied Catal. Built Environment", 181, 103–117(2016).

[4] S. Bekkouche, S. Merouani, O. Hamdaoui, M. Bouhelassa," Efficient photocatalytic degradation of Safranin O by integrating solar-UV/TiO_2/persulfate treatment: Implication of sulfate radical in the oxidation process and effect of various water matrix components" Journal of Photochemisrty. 345, 80-90 (2017)

[5] Xiang-Rong Xu, Xiang-Zhong Li, "Degradation of azo dye Orange G in aqueous solutions by persulfate with ferrous ion", Separation and Purification Technology 72 105–111 (2010)

[6] S. Rodriguez, L. Vasquez, D. Costa, A. Romero, A. Santos "Oxidation of Orange G by persulfate activated by Fe(II), Fe(III) and zero valent iron (ZVI)", Chemosphere 101 86–92 (2014)

[7] Tunlawit Satapanajaru, Maneekarn Yoo-iam, Pinnaree Bongprom, Patthra Pengthamkeerati, "Decolorization of Reactive Black 5 by persulfate oxidation activated by ferrous ion and its optimization", Desalination and Water Treatment, 56, (2015)

[8] Ming Chang,Yang-hsin Shih, "Synthesis and application of magnetic iron oxide nanoparticles on the removal of Reactive Black 5: Reaction mechanism, temperature and pH effects", Journal of Environmental Management 224, 235-242 (2018)

[9] Saltuk Pirgalıoğlu Tülay A. Özbelge "Comparison of non-catalytic and catalytic ozonation processes of three different aqueous single dye solutions with respect to powder copper sulfide catalyst", Applied Catalysis A: General, 363,157-163, (2009)
[10] Mustapha Mohammed Bello, Abdul Aziz Abdul Raman, Anàm Asghar, " Activated carbon as carrier in fluidized bed reactor for Fenton oxidation of recalcitrant dye: Oxidation-adsorption synergy and surface interaction", Journal of Water Process Engineering 33, (2020)

IOP Conf. Series: Materials Science and Engineering **800** (2020) 012017 doi:10.1088/1757-899X/800/1/012017

An experimental study on the behavior of fiber-reinforced concrete flexural members under cyclic loading

H R Gatabi[1], M Celikag[1*], H A Bengar[2]

[1]Department of Civil Engineering, Faculty of Engineering, Eastern Mediterranean University, Famagusta, North Cyprus, via Mersin 10 Turkey

[2]Department of Civil Engineering, University of Mazandaran, Babolsar, Iran

*Email: murude.celikag@emu.edu.tr

Abstract. An experimental work was performed to address the cyclic performance of reinforced concrete flexural members containing steel, PVA, and Forta-Ferro fibers. Four full-scale reinforced concrete flexural members having the same geometry were subjected to a one-point bending test with slow cyclic loading. One of the four specimens was a control specimen with 0% fiber volume fraction (V_f), and the other three contained 1.5, 0.6, and 1.5% V_f of steel, Forta-Ferro, and PVA fibers, respectively. The ultimate resistance, ductility, stiffness, dissipation of energy was studied, and the influence that different fibers had on the cyclic performance was evaluated. In general, the cyclic performance of the reinforced concrete flexural members improved by adding fibers, leading to higher ultimate displacement and maximum strength value.

1. Introduction

One can trace back the procedure of adding fibers to materials with brittle behavior to improve their ductility to ancient times [1]; for example, Egyptians added to straws to mud bricks [2]. The fiber types utilized nowadays across different industries have increased significantly; for instance, the fiber types used most commonly in concrete structures are steel, polymeric, and carbon fibers.

Extensive research has been focused on steel fiber-reinforced concrete (SFRC) [3-9]. The performance of beams reinforced with steel fibers was investigated by [10]. They tested the beams under uniaxial tension to obtain the post-cracking behavior, whereby they defined characteristic stress vs. crack opening. Fibers can change the post-cracking softening behavior seen in concrete to a post-cracking hardening behavior, which allows the formation of extensive cracking prior to failure; however, in post-cracking softening, a strength reduction occurs after initial cracking, which prevents the formation of further cracks.

Furthermore, reinforcing cementitious materials with PVA fibers is also advantageous in terms of durability criterion when subjected to freeze-thaw or wet-dry cycles compared to plain concrete, with the crack width being limited to 100 μm [11, 12]. Short cut PVA (polyvinyl alcohol) fibers show good performance in improving the flexural and tensile behavior since they possess high strain and tensile strength capacities together with unique strain hardening property, which limits the crack width and allows for multiple fine cracks [13].

Forta-Ferro fibers are a type of synthetic fibers composed of 100 percent virgin copolymer/polypropylene. They are in the form of a twisted bundle of non-fibrillated monofilaments, as well as some fibrillated network fibers. Through their use, high quality and efficiency are achieved for reinforced concrete. Using this fiber type leads to less shrinkage in fresh and hardened concrete and improves the impact resistance, fatigue resistance, and stiffness of hardened concrete. By incorporating an integrated network of these fibers in concrete, the long-term durability and structural properties are improved, and secondary or thermal cracking in concrete is controlled. Other properties of Forta-Ferro fibers include their noncorrosive nature and 100 percent resistance in alkaline environments [14].

The present was work aimed at investigating the performance of reinforced concrete beams containing fibers throughout the concrete mix under cyclic loading through an experimental program. This study focused on the performance of flexural concrete members reinforced with longitudinal and transverse steel bars built-in full scale under cyclic loading. Here, four beam specimens comprising of one fibreless, one with 1.5% steel fibers, one with 0.6 Forta-Ferro fibers, and one with 1.5% PVA fibers were manufactured in full scale and exposed to the slow cyclic one-point bending loading. To be able to study the effect of fibers on the cyclic performance of the specimens, their capacity curve, cyclic response, ductility, ultimate deflection, and loading and unloading stiffness, as well as damage propagation and energy absorption and dissipation were evaluated as some of the important parameters.

2. Experimental procedure

The following sections give more information regarding the specimens and their constituents, test set-up, and applied loading history.

2.1. Materials and specimens

In this work, four reinforced concrete flexural members were manufactured in the laboratory, among which one was fibreless and the other three each contained a specific volume fraction of a different fiber type including one with 1.5% steel fibers, one with 0.6% Forta-Ferro fibers, and one with 1.5% PVA fibers. Apart from the inclusion of fibers and their content, the three fiber-reinforced specimens had the same properties to eliminate the probability of the scattering of the results. Table 1 shows the mechanical properties of the fibers used in the specimens. The specimens had a length of 2440 mm and a cross-section of 270×270 mm. For the top and bottom longitudinal reinforcement, 4Φ14 deformed bars were used, and for transverse reinforcement, Φ8@56mm were used. The corresponding yield strength of the longitudinal and transverse rebar were 4560 and 3850 kg/cm^2, respectively. Fig. 1 gives some details about the specimens. The concrete cover thickness was 50 mm. Also, the concrete binder materials were Portland cement type II. For obtaining the compressive strength of the concrete, compression tests were conducted on three cubic 15×15×15 cm specimens of each mixture design after curing for 28 days. The specimens containing 1.5% steel fibers, 0.6% Forta-Ferro fibers, 1.5% PVA fibers and the one without fiber had compressive strength values of 29.4, 28.85, 29.7, and 30.8 MPa, respectively.

Table 1. Mechanical properties of fibers

Fiber	Density (gr/cm³)	Tensile Strength (N/mm²)	Length (mm)	Diameter (mm)	V_f (%)
Steel	7.85	1100	30	0.500	1.5
Forta-Ferro	0.91	660	54	-	0.6
PVA	1.30	1300	8	0.085	1.5

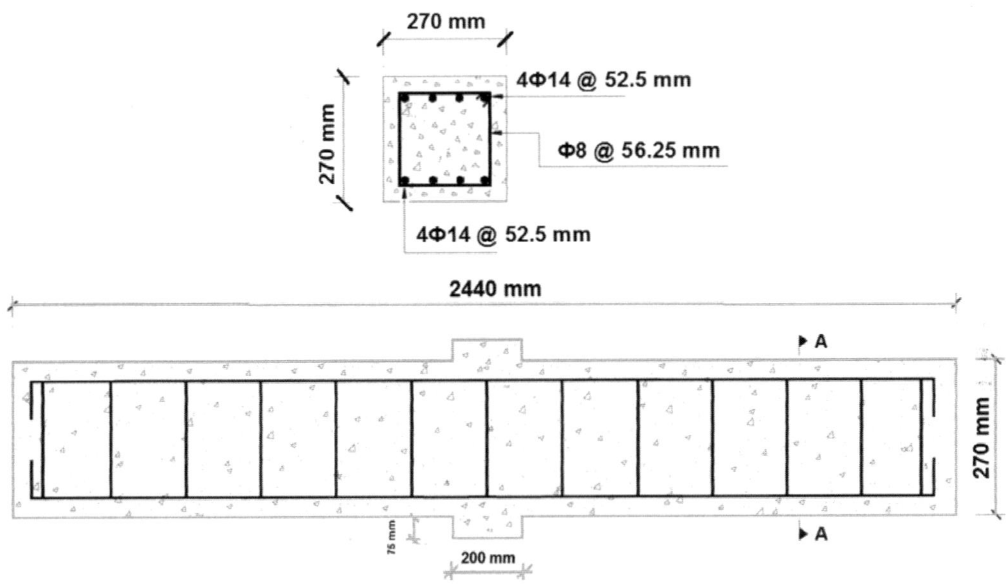

Fig. 1. Dimensions and reinforcement detail of specimens

2.2. Test setup and experiments

The simply-supported beam specimens were subjected to the one-point flexural testing under slow cyclic loading applied by a universal machine. The equipment and devices used in the test setup included a reaction frame, a load cell for measuring the load, and two Low Voltage Displacement Transducers (LVDTs) for measuring the mid-span deflection of the specimens. The experimental setup is schematically shown in Fig. 2.

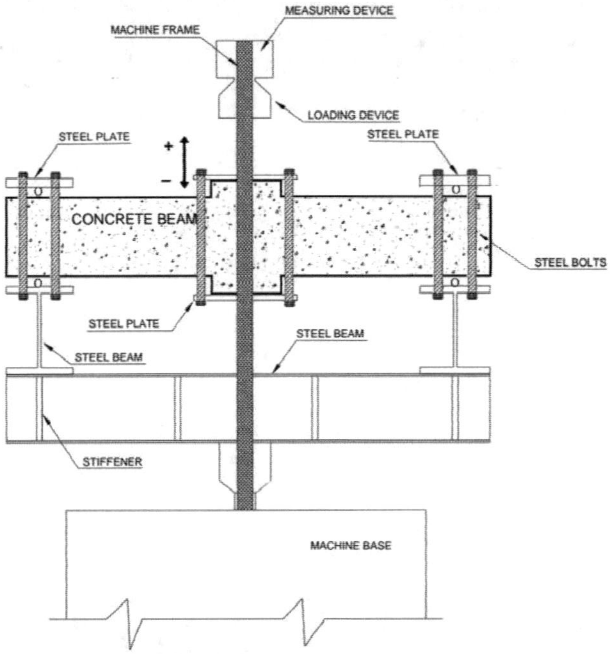

Fig. 2. Schematic view of the test setup

2.3. Loading history

During the test, the specimens were subjected to cyclic loading in a displacement-controlled manner. Each cycle was repeated three times, with the first one initiated at ± 2 mm. In total, 14 drift ratios were applied to the specimens, which included 0.20, 0.25, 0.35, 0.50, 0.75, 1.00, 1.40, 1.75, 2.20, 2.75, 3.5, 4.5, 6.0, and 8.0%. Fig. 3 demonstrates the cyclic loading regime conducted compliant to the ACI Committee 374.1-05 [15].

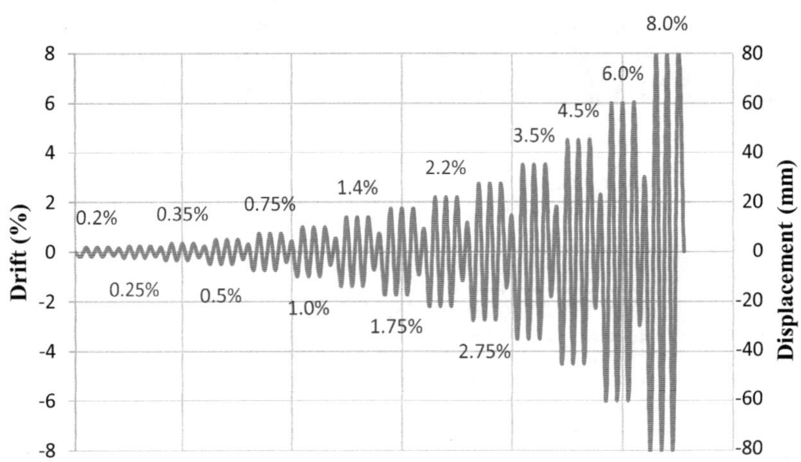

Fig. 3. Cyclic loading protocol

3. Results and discussion

This section gives the results obtained from the experiments. In this work, the hysteresis behavior of specimens was investigated, and based on the obtained results, their peak and ultimate strength, capacity curve, ultimate displacement, and ductility, as well as energy dissipation, were investigated. A comparison was made among the results of different specimens, including fibreless and fiber-reinforced ones, to specify the influence of fibers on the cyclic performance of reinforced concrete flexural members.

3.1. Hysteresis behavior

The hysteresis behavior of the specimens is represented in the form of load vs. the mid-span displacement at the top and bottom surfaces of the steel beam. The cyclic performance and failure mechanism of the specimens are demonstrated in Fig. 4.

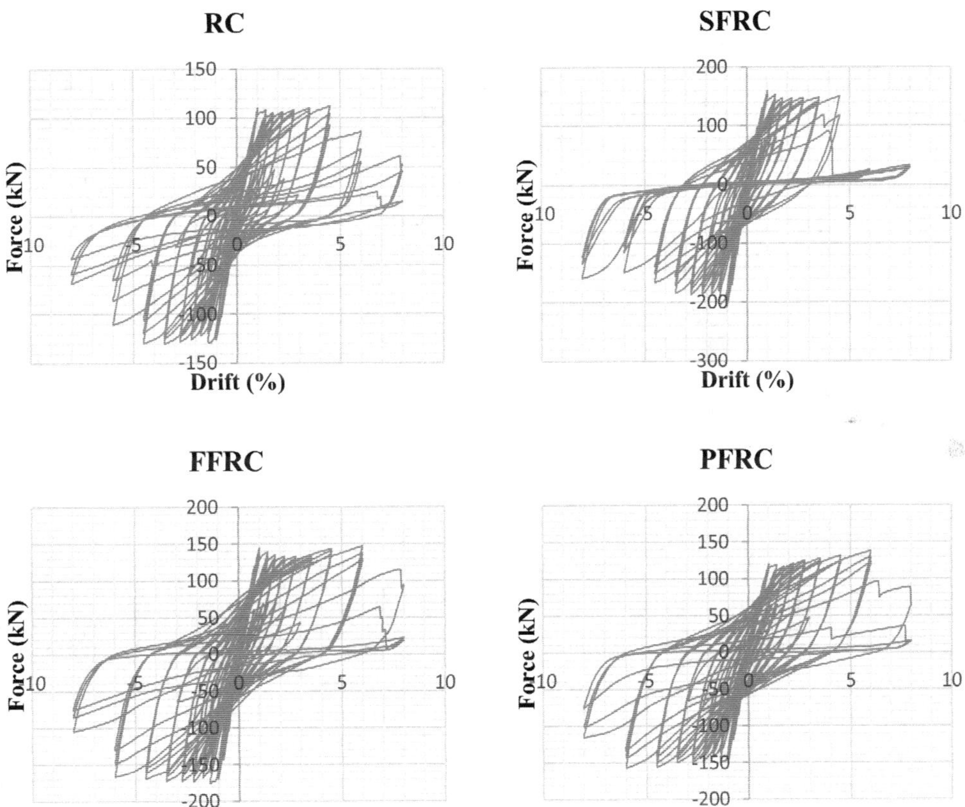

Fig. 4. Hysteretic behavior

The failure of the flexural specimen without fibers was characterized by concrete crushing in the compressive zone as well as extensive cracking with branching. However, flexural specimens containing steel, Forta-Ferro, and PVA fibers failed through a major crack, commonly located in the middle region of the flexural members (Fig. 5). Fig. 4 shows that in the case of using fibers in the concrete mix, an increase was seen in the ultimate displacement and maximum strength relative to the fibreless specimen. In addition, in the specimens without fibers, the damage was more extended, and cracks propagated more in the entire length compared to the fiber-containing specimens. In addition, in the specimens containing fibers, a major crack located in the middle region occurred.

The capacity curves derived from the hysteresis curves are compared in Fig. 6. According to the mean values, as well as the assessment of the specimen results, these curves were derived. Furthermore, Table 2 lists the ultimate displacement and maximum strength values in the positive and negative loading range for the tested specimens.

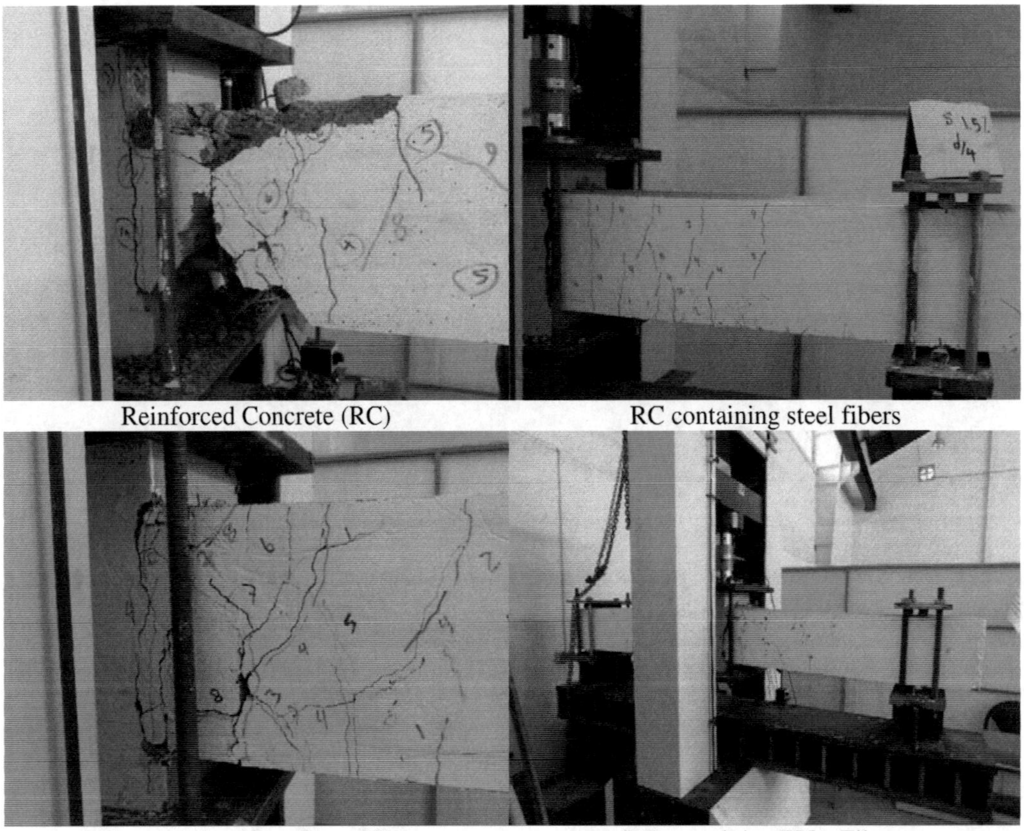

Reinforced Concrete (RC) RC containing steel fibers

RC containing Forta-Ferro Fibers RC containing PVA Fibers

Fig. 5. Damage and crack observation for specimens

Fig. 6 shows that using fibers leads to increased maximum strength and ultimate displacement in the negative and positive loading ranges. In order to better compare the capacity curves of the specimens, it is easier to use idealized form of the tri-linear force-displacement diagrams of the specimens considering a considerable stiffness change for defining the elastic stiffness and yielding displacement (y) as well as a strength reduction of 20% from the highest obtained ultimate strength (Table 2). In general, when fibers were used, it was observed that the maximum strength, ultimate strength, and maximum displacement were all increased.

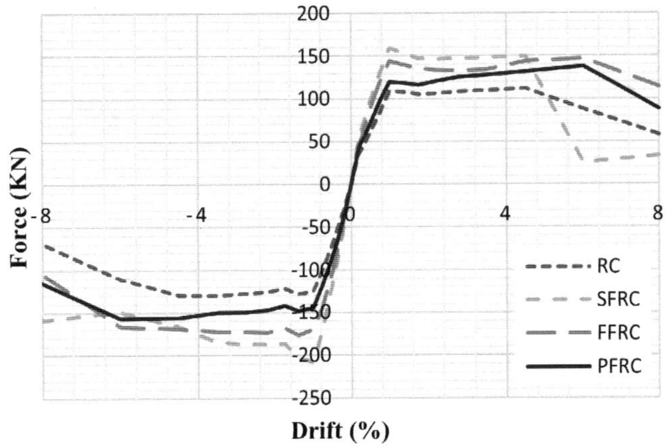

Fig. 6. Capacity curves of the four tested specimens

Table 2. Characteristic values of capacity curves of specimens

Specimen	Peak load (kN)		Average peak load (kN)	Displacement at yielding point (mm)		Displacement at 20% drop in peak load (mm)		Ductility factor	
	Pull (+)	Push (-)		Pull (+)	Push (-)	Pull (+)	Push (-)	Pull (+)	Push (-)
RC	112.00	129.35	120.67	11.0	12.0	58.83	63.77	5.34	5.31
SFRC	159.2	207.61	183.4	9.8	11.6	47.7	44.8	4.86	3.86
FFRC	146.6	176.49	161.54	9.5	11.2	78.1	68.1	8.22	6.08
PFRC	137.61	156.71	147.16	9.2	9.9	71.2	75.1	7.73	7.58

3.2. Stiffness degradation

Fig. 7 shows the slope of a line connecting peak points in each cycle approximates the cyclic stiffness of the flexural member. Eq. (1) was used to calculate the cyclic stiffness at different cycles (Fig. 7), and for this purpose, only the first reversal cycle was considered.

$$K_i = \frac{F_i^+ - F_i^-}{d_i^+ - d_i^-} \quad (1)$$

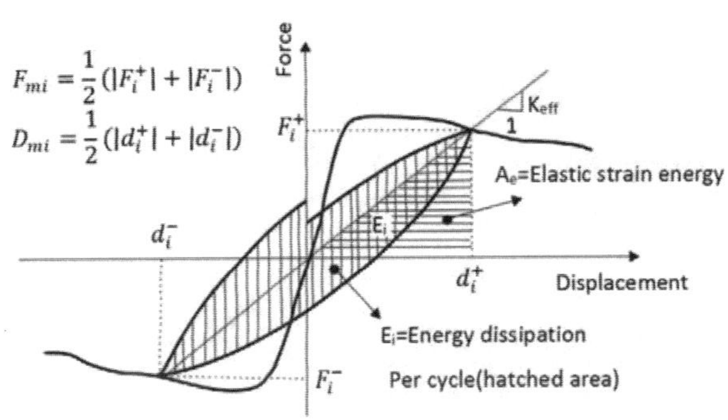

Fig. 7. Description of loading secant stiffness

Fig. 8 shows a degradation for the peak-to-peak secant stiffness (slope) of all the specimens that sees a degradation. This is believed to have been a result of the nonlinear concrete deformations, flexural and shear cracking, cover fracture, and the reinforcement slippage.

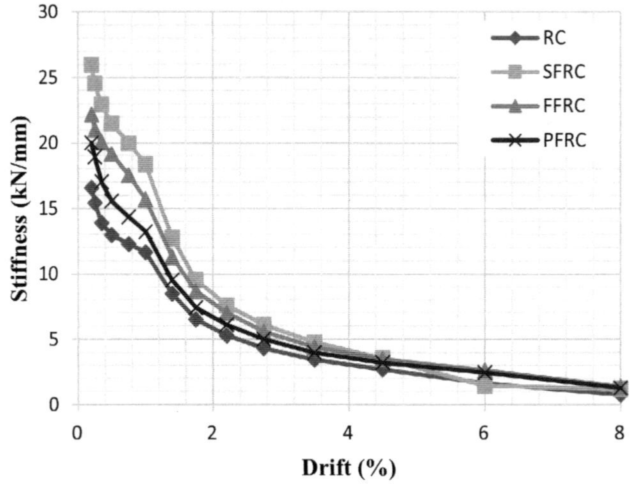

Fig. 8. Changes in loading secant stiffness of the four specimens

According to Fig. 8, the steel fiber reinforcement (SFRC), Forta-Ferro fiber reinforcement (FFRC), and PVA fiber reinforcement (PFRC) specimens showed initial stiffness values of 25.98, 22.18, 20.02, respectively and these are 57.16, 34.18, and 21.11% greater than that of the specimen without fiber, respectively. This difference became even higher at a drift of 0.75%, such that the secant stiffness of the SFRC, FFRC, and PFRC specimens were around 62.75, 42.78, and 50.5% higher than that of the fibreless specimen, respectively. On the other hand, the initial stiffness of the FRC specimens is more stable in a way that the initial cyclic stiffness of these specimens has a lower decline compared to that of the specimen without fibers. In addition, four specimens had closely spaced stiffness values at large displacements, namely drifts of above 4.5%.

3.3. Energy absorption and dissipation

The resistance of a structure to seismic loads is dependent on the ability of the structure to absorb and dissipate the energy of an earthquake. The area inside the hysteresis loop specifies the energy dissipation, and as this area increases, more seismic energy is absorbed and transformed into the hysteresis energy. In Fig. 9, the hysteresis energy and displacement curves of different beams are compared. In general, at small displacements, there is no considerable difference among hysteresis energy values; however, at large displacements, this difference among hysteresis energy values was increased. This is attributed to the addition of fibers. The specimens containing steel, Forta-Ferro, and PVA fibers showed around 7.21, 50.51, and 37.12% increase in energy dissipation relative to the specimen with 0% fiber volume fraction (V_f). Up to 8% drift, the cumulative dissipated energy of the specimens RC, SFRC, FFRC, and PFRC were 96866, 103853, 145793, and 132821 kN.mm, respectively.

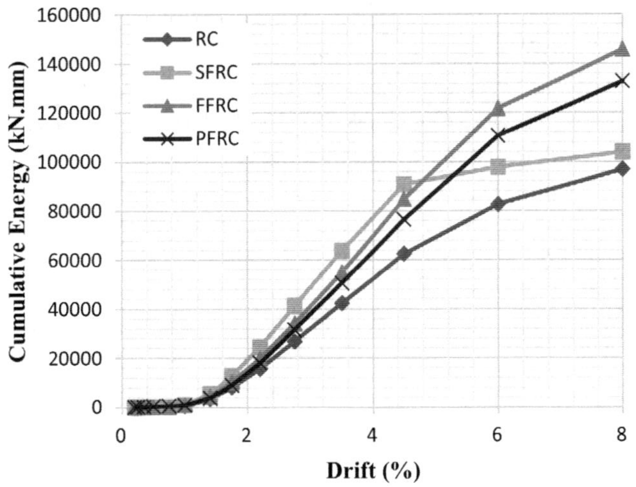

Fig. 9. Cumulative energy of the specimens compared with each other

3.4. Discussions

Based on test findings, all the beams experienced the flexural failure mode. Nevertheless, when the fibers were present in the concrete, damage in the specimens was concentrated to the maximum displacement area, while cracks propagated along the entire length of the flexural member in the fiberless beam, with the final failure occurring at the location of load application. The steel, Forta-Ferro, and PVA fibers bridge across the two faces of micro-cracks, preventing their further widening. As the displacement increased, all of the fibers crossing the mouth of the crack at the mid-span started to elongate, then yielded, and eventually ruptured at the ultimate state of the flexural members. This phenomenon led to increased, ultimate loading capacity. Including fibers in the concrete matrix that were considerably affected the stiffness and stability of the specimens at loading and unloading increments. When the fibers were added to the mixes, more energy was absorbed and dissipated, as expected.

4. Conclusion

This experimental study attempted to investigate the influence that steel, Forta-Ferro, and PVA fibers have on the cyclic performance of reinforced concrete flexural members. The main conclusions obtained are given as follows:

1. In general, the maximum strength and displacement increased by the addition of fibers.
2. When the various fibers were present to the concrete mix, the cracking was limited to the main crack localized at the mid-span of the flexural members, and also, cracks were fewer compared to the case of the fiberless specimen.
3. Adding fibers, in general, led to a higher secant stiffness during cycles of loading and unloading. This means that via the addition of steel, Forta-Ferro, and PVA fibers in the concrete mixes, damage to the specimens decreased, and the stability of the cyclic behavior increased relative to the specimens without fibers.
4. At smaller displacements, no considerable difference was seen among the hysteresis energy values of the beams; however, at greater displacements, this discrepancy became significant.

References

[1] Felekoğlu, B., Türkel, S., & Altuntaş, Y. (2007). Effects of steel fiber reinforcement on surface wear resistance of self-compacting repair mortars. Cement and Concrete Composites, 29(5), 391-396.

[2] Ranjbaran, F., Rezayfar, O., & Mirzababai, R. (2018). Experimental investigation of steel fiber-reinforced concrete beams under cyclic loading. International Journal of Advanced Structural Engineering, 10(1), 49-60.

[3] Sahoo, D. R., Bhagat, S., & Reddy, T. C. V. (2016). Experimental study on shear-span to effective-depth ratio of steel fiber reinforced concrete T-beams. Materials and Structures, 49(9), 3815-3830.

[4] Sahoo, D. R., Maran, K., & Kumar, A. (2015). Effect of steel and synthetic fibers on shear strength of RC beams without shear stirrups. Construction and Building Materials, 83, 150-158.

[5] Sahoo, D. R., Solanki, A., & Kumar, A. (2014). Influence of steel and polypropylene fibers on flexural behavior of RC beams. Journal of Materials in Civil Engineering, 27(8), 04014232.

[6] Özcan, D. M., Bayraktar, A., Şahin, A., Haktanir, T., & Türker, T. (2009). Experimental and finite element analysis on the steel fiber-reinforced concrete (SFRC) beams ultimate behavior. Construction and Building Materials, 23(2), 1064-1077.

[7] Nataraja, M. C., Dhang, N., & Gupta, A. P. (1999). Stress–strain curves for steel-fiber reinforced concrete under compression. Cement and concrete composites, 21(5-6), 383-390.

[8] Vandewalle L et al (2003) Final recommendation of RILEM TC 162-TDF: Test and design methods for steel fiber reinforced concrete sigma-epsilon-design method. Materials and Structures 36:560–567

[9] Ganesan, N., Indira, P. V., & Abraham, R. (2007). Steel fiber reinforced high performance concrete beam-column joints subjected to cyclic loading. ISET Journal of Earthquake Technology, 44(3-4), 445-456.

[10] Casanova, P., & Rossi, P. (1997). Analysis and design of steel fiber reinforced concrete beams. Structural Journal, 94(5), 595-602.

[11] Yu, J. H., Chen, W., Yu, M. X., & Hua, Y. E. (2010). The microstructure of self-healed PVA ECC under wet and dry cycles. Materials Research, 13(2), 225-231.

[12] Şahmaran, M., Lachemi, M., & Li, V. (2010). Assessing the durability of engineered cementitious composites under freezing and thawing cycles. In Recent Advancement in Concrete Freezing-Thawing (FT) Durability. ASTM International.

[13] Li, V. C., & Kanda, T. (1998). Innovations forum: engineered cementitious composites for structural applications. Journal of Materials in Civil Engineering, 10(2), 66-69.

[14] Arafa, M., Arafa, M., Alqedra, M., & Almassri, H. G. (2013). Effect of forta-ferro fibers on fresh and mechanical properties of ultra-high performance self-compacting concrete. Effect of forta-ferro fibers on fresh and mechanical properties of ultra-high performance self-compacting concrete, 1.

[15] ACI Committee 374.1-05. Acceptance criteria for moment frames based on structural testing and commentary. Farmington Hills, Michigan, USA: American Concrete Institute; 2005.

IOP Conf. Series: Materials Science and Engineering **800** (2020) 012018 doi:10.1088/1757-899X/800/1/012018

Properties of high-strength concrete containing well graded rubber particles

A Habib*, U Yildirim, O Eren

Civil Engineering Department, Eastern Mediterranean University, Famagusta. North Cyprus

*E-mail: ahed.habib@gmail.com

Abstract. Discarding of old rubber tires is indeed a very serious environmental problem all over the world represented by a high risk for uncontrolled fires and other environmental and health hazards. It was estimated that every year about one billion tires get to the end of their life span. Recently, the construction industry has taken up the challenge to incorporate recycled materials in concrete mixtures by means aggregate replacement. Nowadays, many researches are focused on investigating rubberized concrete as a structural material due to its enhanced properties such as ductility, energy dissipation and damping ratio. Previous studies have suggested the use of fine rubber particles rather than coarse ones when high strength concrete is targeted despite the fact that using coarse ones provide enhanced energy dissipation, damping ratio and vibration behavior. This study is intended to address the effects of utilizing significant amount of well graded fine and coarse rubber aggregates on the properties of high strength concrete. On the basis of the investigations some mechanical, durability and dynamic tests will be conducted on concrete with different rubber replacement percentages. The results of the experimental works have shown that it is possible to develop high strength concrete when well graded fine and coarse rubber particles is used to replace 25% percent of the natural aggregates. Furthermore, the vibration behavior of the concrete mixture was improved considerably when high content of rubber aggregates was added into concrete.

1. Introduction

Disposal of old rubber tires causes very serious environmental problems every year [1, 2, 3]. Nowadays, studies on the application of non-biodegradable wastes as aggregates replacement in concrete are becoming very common all over the world aiming to incorporate the principles of sustainability in the construction activities [4]. Recently, many investigations on the use of recycled rubber particle in concrete by means of aggregate replacement were conducted trying to develop what so-called 'rubberized concrete (RBC)' [5, 6, 7].

Previous investigations on RBC have confirmed a significant reduction in the strength capacity of concrete as the rubber content is increasing [8, 9, 10, 11]. Nevertheless, developing of high strength rubberized concrete using fine rubber particles was reported to be possible when up to 12.5% of the fine aggregate is replaced [4]. Furthermore, the resistance of RBC against water permeability is decreased due to its increased porosity [12]. On the other hand, adding rubber to concrete improves its ability to highly dissipates energy which when used in structural members can positively influence the vibration behavior of the building [13].

This study is intended to take the research in the field of the RBC one step further by studying the effects of replacing natural aggregates by well graded fine and coarse rubber particles on the behavior of concrete. As a part of the study some of the mechanical, durability and dynamic characteristics of RBC mixtures will be reported at two rubber replacement percentages.

2. Materials and Methods

2.1. Materials Properties

In order to produce a concrete that has the characteristics of high-strength concrete the materials with the properties shown in Table 1 were used.

Table 1. Properties of the materials used in concrete mixtures.

Material	Properties	Standard
Cement	Blast-furnace Slag Cement CEM II/B-S 42.5 N with a specific gravity of 3.15	ASTM C595
Silica Fume	It has a 95 % SiO_2 content	ASTM C1240
Water	Tap water free from contamination	ASTM C1602
Admixtures	High range water reducer Polycarboxylic ether-based superplasticizer GLENIUM 27 with specific gravity of 1.026 and brown color	ASTM C494
Steel Fiber	Hooked end steel fiber with a length of 50 mm, slenderness of 80 and tensile strength of 1225 MPa	ASTM A820
Fine Aggregate	Crushed stones with a maximum particle size of 5 mm distribution as shown in Figure 1.	ASTM C33
Coarse Aggregate	Crushed limestone rock with a nominal maximum size of 10 mm and distribution as shown in Figure 1.	ASTM C33
Rubber	A mix of fine and coarse rubber particles from old car tiers with a specific gravity of 1.06 and particle size distribution as shown in Figure 1.	ASTM C330

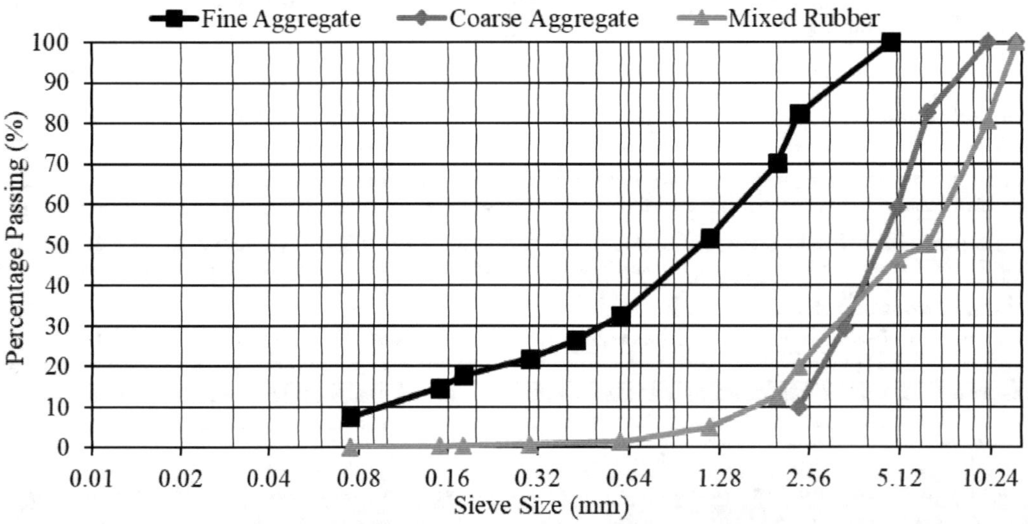

Figure 1. Aggregate particle size distribution.

2.2. Rubber Pre-Treatment Using NaOH Solution
The aim of this treatment is to enhance the surface roughness of each rubber particle which will result in improved concrete properties. In fact, the process was started by washing the rubber aggregates in tap water to remove any impurities and dust, soaking them in the 10% NaOH solution for half an hour, then the rubber were washed by water until their pH became 7 (to prevent any possible negative impact on the concrete durability), finally, they were left away to get air-dried.

2.3. Mix Proportioning
The concrete mix proportion shown in Table 2 was designed to achieve high-strength concrete after adding the rubber aggregates. Thus, based on the findings of the previous investigations over 50% reduction was expected when 25% of the aggregates will be replaced by recycled rubber particles. Therefore, to overcome this reduction in the strengths of concrete both silica fume and steel fiber were incorporated in the mixture.

The rubber replacement percentages in this study were selected as 15% and 25% which were based on the results of old researches that recommended high content of rubber aggregate.

In fact, the process of concrete production was handled in controlled laboratory conditions. Due to the high amount of binding materials required to produce high-performance concrete that includes significant amount of rubber aggregate the production stage followed the requirements of ultra-high-performance concrete standard ACI 239-18 in terms of mixing duration. As a result, concrete mixing procedure took in total 20 minutes. Thereafter, ASTM C1758 standard practice was followed for making and curing of the concrete samples.

Table 2. Mix proportions used in this study.

Material	Control Mix (ZRBC)	15% Rubberized Mix (15RBC)	25% Rubberized Mix (25RBC)
Cement (kg)	1000	1000	1000
Water (kg)	180	180	180
Fine Aggregate (kg)	448	381	336
Coarse Aggregate (kg)	672	571	504
Rubber Aggregate (kg)	0	70	117
Admixture (kg)	50	50	50
Silica Fume (kg)	300	300	300
Steel Fiber (kg)	78	78	78

2.4. Experimental Program
As part of this study, experimental tests will be conducted to evaluate the effects of adding fine and coarse rubber aggregates on the mechanical, durability and dynamic behavior concrete.

In general, the guidelines shown in Table 3 were used to carry the mechanical and durability tests on concrete specimens

On the other hand, to obtain the dynamic characteristics of each mixture a free vibration test was performed with the acceleration values being recorded using an accelerometer sensor with a range of ±2g and a sampling frequency of 200 Hz. Generally, the procedure of testing each cantilever beam as shown in Figure 2 was started by applying the supporting conditions, setting up the sensor location and imposing vibration on the specimen by impact loading. Then, the recorded data were undergoing a pre-processing stage to reduce the noise emerged from the surrounding environment and improve their reliability. The followed method is called moving average filter [14]. Thereafter, Excel 2016 and SeismoSignal software package [15] were used to produce the fast Fourier transform (FFT) that converted the recorded data from the time domain to the frequency domain and to calculate the natural frequency of each specimen. Finally, the expressions below were used to calculate the dynamic modulus of elasticity of each sample.

$$f_n = \frac{1}{2\pi}\sqrt{\frac{k}{M}} \quad (1)$$

$$\Rightarrow \; E = \frac{4\pi^2 {f_n}^2 q L^4}{12I} \quad (2)$$

Table 3. Standards to be followed in this research.

Test	Specimen Size in (mm)	Remarks
Density of Hardened Concrete	150*150*150	ASTM C642
Compressive Strength	150*150*150	ASTM C39
Splitting Tensile Strength	150*150*150	ASTM C496
Static Modulus of Elasticity	200*100	ASTM C469
Water Penetration	150*150*150	DIN 1048

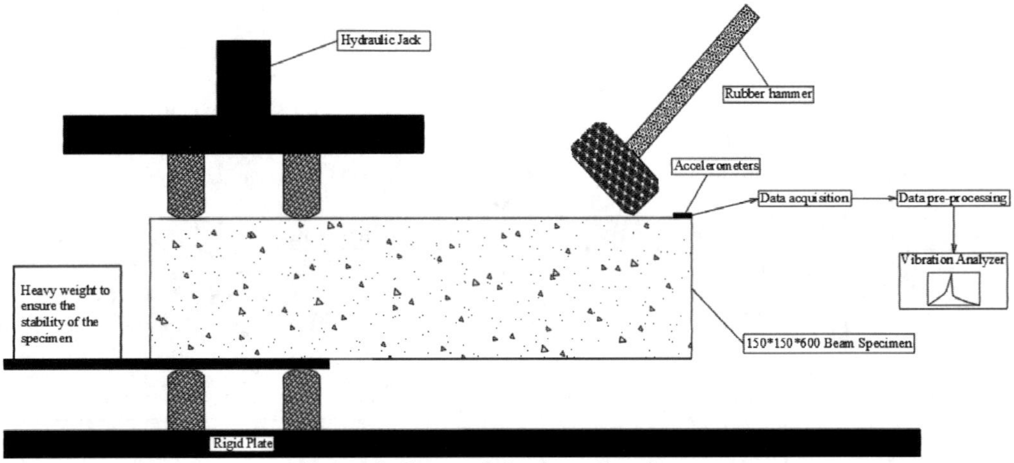

Figure 2. Setup of the free vibration test

3. Results and Discussions

3.1. Bulk Density

The bulk density for each concrete mix was obtained using the method provided in ASTM C642 by taking the mass of the specimen (in kg) divided by its volume (in m^3). Thereafter, the average of six different specimens were found as shown in Table 4. In fact, it was observed that concrete faces a decrease in its density related directly to the increase in rubber content, this change is attributed to the significant difference between the specific gravity of rubber and natural stone particles.

3.2. Compressive Strength

The average 28 days compressive strength of each mixture is shown in Table 4. In general, the results indicated a significant reduction in the strength corresponds to 29% and 43% for 15RBC and 25RBC respectively. This drop is attributed to many reasons including replacing higher strength and load-carrying capacity aggregate by lower one [13] and the weaker bond between the rubber aggregates and cement past in comparison to the natural aggregate one which lead to rapid rupture of concrete [16].

3.3. Splitting Tensile Strength

Similar to compressive strength the 28 days splitting tensile strength of concrete, Table 4, was affected by the rubber replacement, in which the observed reduction were 4% and 17% for 15RBC and 25RBC respectively.

3.4. Static Modulus of Elasticity

The variation in the static elastic modulus of RBC with respect to the control mixture is shown in Table 4. This difference is due to the reduction in the compressive strength of the RBC specimens.

3.5. Depth of Water Penetration

This test was conducted to examine the durability behavior of RBC in comparison to the control mixture. In general, the depth of water penetration, Table 4, was very small in all cases (below 5 mm) due to using high amount of silica fume while the pattern of variation showed an increase in the depth as more rubber particles are being added to the concrete.

Table 4. Results of the experimental tests.

Test	ZRBC	15RBC	25RBC
Bulk Density (kg/m^3)	2312	2191	2139
Compressive Strength (MPa)	96.77	68.5	55.07
Splitting Tensile Strength (MPa)	5.32	5.1	4.39
Static modulus of elasticity (GPa)	53.6	38.2	36.6
Depth of water penetration (mm)	1.25	3.5	4.5
Dynamic Modulus of Elasticity (GPa)	60.06	44.66	42.93

3.6. Dynamic Modulus of Elasticity

Figure 3 shows the results of the Fast Fourier transform (FFT) of each concrete mix. As can be seen there, the frequency of concrete decreased as the rubber replacement percentage is increasing. This reduction is mainly attributed to the drop in the specimen's modulus of elasticity.

The dynamic modulus of elasticity for each specimen was calculated using the formula mentioned previously chapter and the results are shown in Table 4. Moreover, the pattern of reduction in the dynamic modulus of elasticity is very close the one for static modulus of elasticity as shown in Figure 4 due to their direct relation.

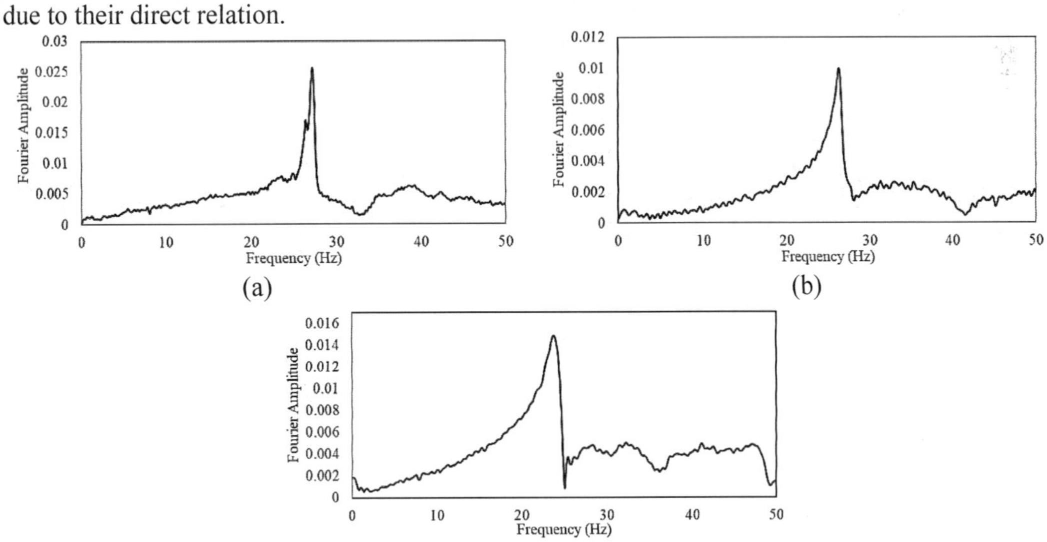

Figure 3. Fast Fourier transform graph of (a) ZRBC, (b) 15RBC and (c) 25RBC.

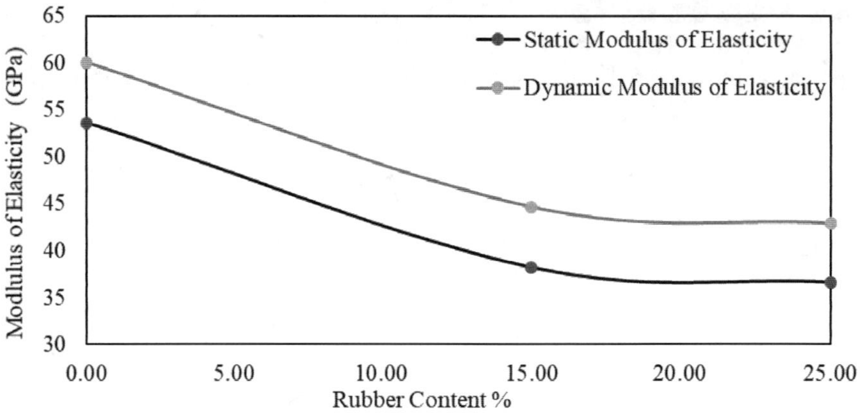

Figure 4. Concrete modulus of elasticity (GPa) for different rubber replacement percentages.

4. Conclusion

This study has focused on the properties of high-strength rubberized concrete. As a part of this study, experimental tests were conducted to comprehensively examine the effect of using well graded fine and coarse rubber particles as aggregates replacement on the some of the mechanical, durability and dynamic properties of high-strength concrete. On the basis of the above statements the following conclusions are drawn:

- Adding rubber to concrete significantly reduces its mechanical properties in comparison to conventional concrete.
- Using rubber aggregates in concrete reduces its bulk density considerably.
- Replacing 25% of the natural aggregates by well graded fine and coarse rubber particles reduced its compressive strength by almost 40%.
- Incorporating rubber in a concrete mix reduces its resistance to water absorption in comparison to the normal one.
- Both static and dynamic modulus of elasticity are reducing as the rubber content in concrete is being increased.

References

[1] Pelisser, F., Zavarise, N., Longo, T., & Bernardin, A. (2011). Concrete made with recycled tire rubber: Effect of alkaline activation and silica fume addition. *Journal of Cleaner Production.*

[2] Yung, W., Yung, L., & Hua, L. (2013). A study of the durability properties of waste tire rubber applied to self-compacting concrete. *Construction and Building Materials.*

[3] Aslani, F. (2015, 9 18). Mechanical Properties of Waste Tire Rubber Concrete. *Journal of Materials in Civil Engineering, 28*(3), 04015152.

[4] Thomas, B., & Chandra Gupta, R. (2016). Properties of high strength concrete containing scrap tire rubber. *Journal of Cleaner Production.*

[5] Eldin, N., & Senouci, A. (1993). Rubber-Tire Particles as Concrete Aggregate. *Journal of Materials in Civil Engineering.*

[6] Eldin, N., & Senouci, A. (1994). Measurement and prediction of the strength of rubberized concrete. *Cement and Concrete Composites.*

[7] Topçu, B. (1995). The Properties Of Rubberized Concretes. *Cement and concrete research, 25*(2), 304-310.

[8] Eldin, N., & Senouci, A. (1992). Engineering properties of rubberized concrete. *Canadian Journal of Civil Engineering*.

[9] Fattuhi, N., & Clark, L. (1996). Cement-based materials containing shredded scrap truck tyre rubber. *Construction and Building Materials*.

[10] Khatib, Z., & Bayomy, F. (1999). Rubberized Portland Cement Concrete. *Journal of Materials in Civil Engineering*.

[11] Zheng, L., Huo, X., & Yuan, Y. (2008). Strength, Modulus of Elasticity, and Brittleness Index of Rubberized Concrete. *Journal of Materials in Civil Engineering*.

[12] Su, H., Yang, J., Ling, T., Ghataora, G., & Dirar, S. (2015). Properties of concrete prepared with waste tyre rubber particles of uniform and varying sizes. *Journal of Cleaner Production*.

[13] Xue, J., & Shinozuka, M. (2013). Rubberized concrete: A green structural material with enhanced energy-dissipation capability. *Construction and Building Materials*.

[14] Smith, S. W. (1997). *The scientist and engineer's guide to digital signal processing*.

[15] Antoniou, S., Pinho, R., & Bianchi, F. (2018). SeismoSignal. Seismosoft.

[16] Bisht, K., & Ramana, P. (2017). Evaluation of mechanical and durability properties of crumb rubber concrete. *Construction and Building Materials*.

[17] ASTM. (2013). ASTM C642 Standard Test Method for Density, Absorption, and Voids in Hardened Concrete.

[18] ASTM. (2019). ASTM C595 Standard Specification for Blended Hydraulic Cements.

[19] ASTM. (2017). ASTM C496 Standard Test Method for Splitting Tensile Strength of Cylindrical Concrete Specimens.

[20] ASTM. (2017). ASTM C494 Standard Specification for Chemical Admixtures for Concrete.

[21] ASTM. (2014). ASTM C469 Standard Test Method for Static Modulus of Elasticity and Poisson's Ratio of Concrete in Compression.

[22] ASTM. (2018). ASTM C39 Standard Test Method for Compressive Strength of Cylindrical Concrete Specimens.

[23] ASTM. (2017). ASTM C330 Standard Specification for Lightweight Aggregates for Structural Concrete.

[24] ASTM. (2018). ASTM C33 Standard Specification for Concrete Aggregates.

[25] ASTM. (2015). ASTM C1758 Standard Practice for Fabricating Test Specimens with Self-Consolidating Concrete.

[26] ASTM. (2018). ASTM C1602 Standard Specification for Mixing Water Used in the Production of Hydraulic Cement Concrete.

[27] ASTM. (2015). ASTM C1240 Standard Specification for Silica Fume Used in Cementitious Mixtures.

[28] ASTM. (2016). ASTM A820 Standard Specification for Steel Fibers for Fiber-Reinforced Concrete.

[29] DIN. (1975). DIN 1048-5: Testing concrete, testing of hardened concrete (specimens prepared in mould).

[30] ACI. (2018). *ACI 239R-18: Ultra-High Performance Concrete: An Emerging Technology Report.* USA: Farmington Hills.

IOP Conf. Series: Materials Science and Engineering **800** (2020) 012019 doi:10.1088/1757-899X/800/1/012019

Integrating AHP and GIS as a decision-making tool for the optimal allocation of wind farm: A case study of Syria

M Habib [1, *], A Matouk [2]

[1]Civil Engineering Department, Al-Balqa Applied University, Amman, Jordan

[2]Architectural Engineering Department, Al-Jazira Private University, Damascus, Syria

[*] E-mail: maan.habib@gmail.com

Abstract: Nowadays, renewable energy is commonly applied to ensure a sustainable development and minimize the reliance on fossil fuels. Wind is considered as one of the most significant inexhaustible sources of energy for power generation through turbines. However, choosing an appropriate site of wind farm and designing a turbine array to fit it are a key issue to success as expected. Geographic Information Systems (GIS) technology has been widely accepted in a number of areas as a decision support system based on analyzing geospatial data to facilitate reaching desired results in shorter time and less cost. The process of establishing the best-suited location of wind parks require analysis of spatial multi-criteria within a GIS environment using the Analytic Hierarchy Process (AHP). This research focuses on defining the main factors that influence wind energy development and presents GIS-based land suitability analysis model to determine the optimal allocation of wind farm in Syria. As a part of this study, the procedure of designing geodatabase for wind power plants is illustrated that includes conceptual, logical, and physical design phases. The results show efficiency of the integrated technique in site suitability evaluation of the presented case.

1. Introduction

Global warming as an essential form of climate change is a major factor leading to a higher incidence of natural disasters. It has also an impact on population health status of affected communities that gives rise to humanitarian crises [1]. This phenomenon is resulted from the carbon dioxide releases due to many human activities such as burning of fossil fuels [2]. Nowadays, renewable energy is considered as a suitable option to produce clean energy that seek to minimize greenhouse gas emissions and to solve the problem of its increasing demand around the world [3]. The green energy is environmentally friendly and its resources are available free of charge all over the world [4]. The use of wind power is quickly growing as the one of those commonly applied renewables that relies on mean value of wind speed and variation of this speed [5]. Determination of the ideal site of wind farms plays a vital role in wind power development to achieve a satisfying turbines performance [6].

Geographic Information Systems (GIS) is a type of decision-making support system, which is used for collecting, processing, analyzing and visualizing geospatial data and information [7, 8]. It is a powerful tool for spatial siting of wind parks based on multi-criteria decision analysis (MCDA) by comparing the alternatives within the framework of the determined criteria [9]. The most popular and practical approach to deploy MCDA is the AHP that is defined as an organised process to build weighted factors to split the decision-making method into a simple steps [10]. The AHP provides a ranking of options with regard to decision-makers to fulfil the objective that can be hierarchically performed for the case study [11]. The extensive use of AHP for providing solutions to a multi-criteria issue is attributed to the fact that it takes into account both tangible and intangible criteria [12].

Accordingly, identifying an appropriate site of wind farm is complex and difficult process, where the evaluation is based on selecting the most suitable location among the other alternatives [13]. It contains different factors such as economic, environmental and physical restrictions, which should be described correctly during the decision-making stages with reference to the characteristics of the project. Furthermore, the estimation of weights to these various parameters is very critical and should be performed using AHP [14]. The main objective of this paper is to design and implement a geodatabase for wind farms of a pilot study, which addresses the application of a GIS in analyzing data related to land suitability selection for determining the optimal allocation of wind power stations. Several GIS functions are utilized including spatial multi-criteria assessment model using ArcGIS 10.2. Figure 1 indicates a schematic diagram for the general methodology of the framework in this research.

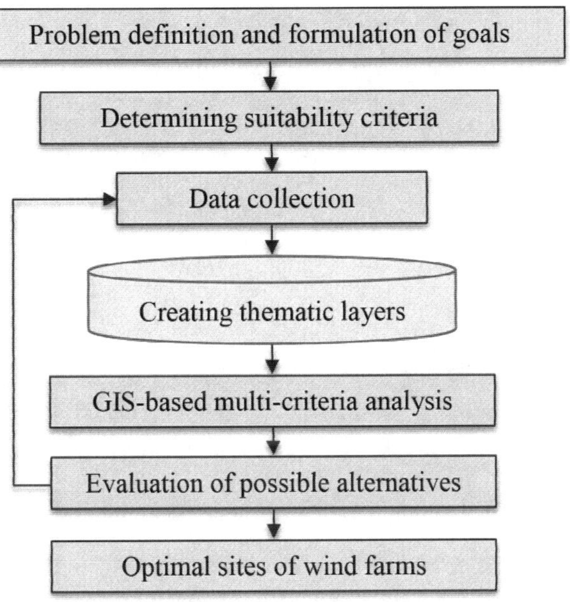

Figure 1. Schematic flowchart methodology of wind farm site selection

2. Materials and Methods

As previously mentioned, wind power is the most common environmentally friendly and alternative resource of renewable energy to generate electricity through turbines, which should be erected especially in windy sites to reach the maximum performance and efficiency. The increasing wind energy investment is related to land-use planning, which has two different goals, economic benefit and nature preservation, both linked to the concept of sustainable development [15]. Therefore, choosing a suitable location of wind farm requires taking into account a set of factors and balancing several purposes. This section focuses on the description of the study area, data collection and identification of decision criteria..

2.1 Study area

The case study is situated in the Middle East in the region of Syria that stretches northward from the western end of Asia and consists of the capital (Damascus) and thirteen governorates, as shown in Figure 2. The land has four distinct seasons and a climate that is well suited for variety of activities. Syria is bordered on the north by Turkey, the south by Jordan, the east by Iraq and the west by the Mediterranean Sea and Lebanon. Its area is 184,479 Km^2, located at 32° 30′ N - 37° 30′ N and 35° 30′ E - 42° 30′ E. The terrain surface throughout Syria consists of various landforms such as mountain ranges, deserts,

IOP Conf. Series: Materials Science and Engineering **800** (2020) 012019 doi:10.1088/1757-899X/800/1/012019

plains, water bodies and coastal features. Land-use within the country contains rural and agricultural land, power generation and distribution, built-up areas and so on.

The uprising and conflict that had begun in March 2011 affected migration patterns in the country, where much of the population was expelled to other parts to be internally displaced or outside to become refugees. This war has transformed Syrian cities into scenes of apocalyptic devastation where buildings and infrastructure had bombed or entirely destroyed. In the positive view, it is the best time to start using clean and sustainable energy in a right way to serve the country well into the future by decreasing greenhouse gas emissions and reducing fuel consumption.

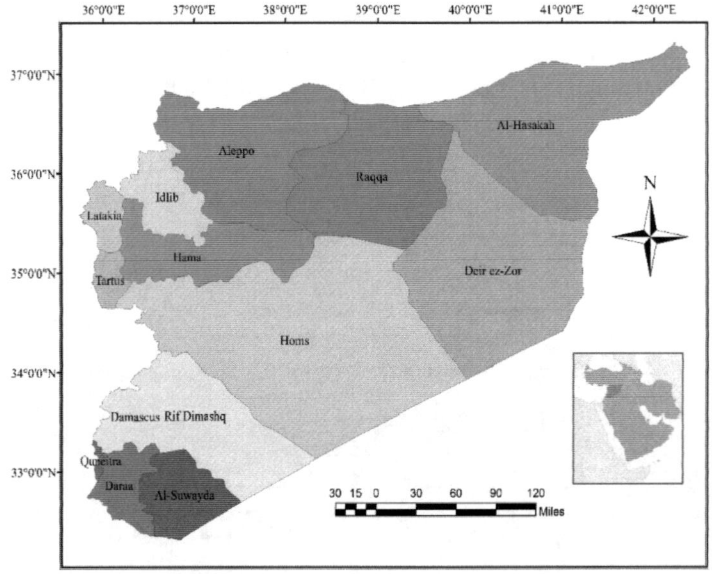

Figure 2. Administrative divisions of the test area.

2.2 Data collection and preparation

Data collection remains the most expensive and time-consuming aspect of setting up a major GIS facility. The creation of accurate geodatabases is a very important part of GIS and vital to decision making, since good decisions require good information derived from data. The key aspect of GIS functionality helps the users integrating and analyzing data from many sources with difference in scale, accuracy, resolution but does not provide information about the effects of combining these various levels of data uncertainty on the input maps and the output maps resulting from spatial querying and analysis [16]. As the high level of the positional accuracy will be the popular means of all spatial data, they should be defined in a proper coordinate reference system [17]. A rectangular coordinate system is one of the most popular in representing points on a map. Georeferenced data is captured from various sources (GPS, remote sensing, surveying, scanned topographic map, etc) that usually reveals the data having different coordinate systems. Therefore changing the coordinates must be applied to unify data references. The conformal mapping system used for Syria in the study is UTM zone 37 N that is based on WGS 84 datum. The main sites of Syria that have the potential for wind farm are represented using ESRI's ArcGIS 10.2, where the GIS layers illustrated in Figure 3.

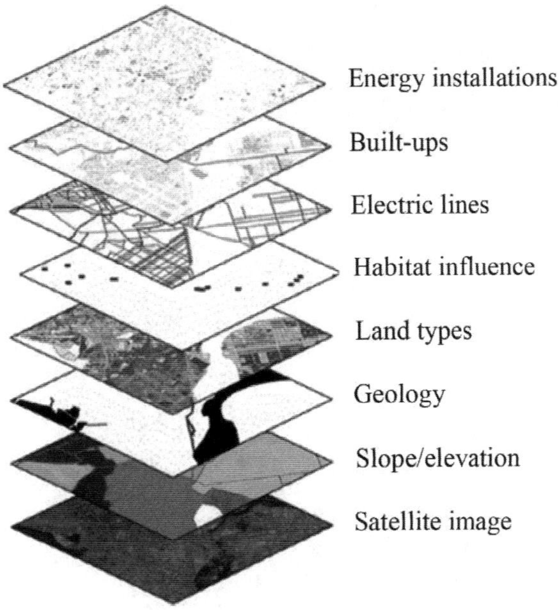

Figure 3. GIS layers for wind park sites [18].

2.3 Identification of decision criteria

Choosing an appropriate site of wind power plants does not rely only on a wind speed, but indeed it is a multi-criteria decision issue and the relative significance of each parameter must be performed correctly. By taking into consideration each of the defined criteria, it must also be verified that final determined wind farm locations are not on restricted areas. the criteria should meet the national legislation according to wind farm development and integrate local conditions such as infrastructure and place characteristics [19]. Figure 4 presents the hierarchy structure of the site suitability criteria for wind parks.

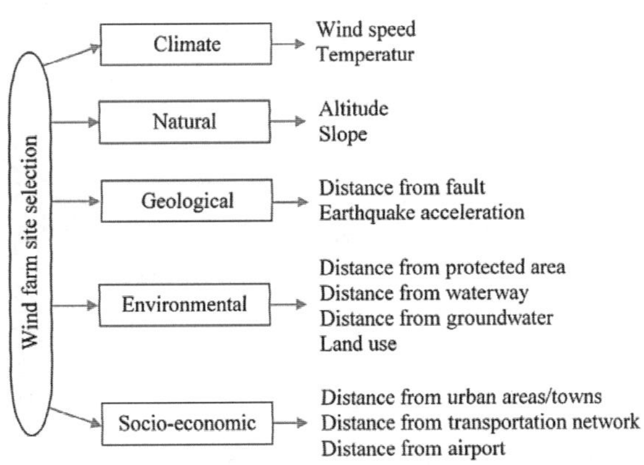

Figure 4. Wind farm selection criteria.

Annual average wind speeds in the region is considered as the most significant criteria to convert the wind energy into electrical power. Figure 5 represents the map of annual average wind speed in Syria that was produced using kriging interpolation method within GIS environment. Digital Elevation Model (DEM) is useful to avoid highlands when assigning locations of wind power plants as a high altitude will increase the investment cost [20]. The DEM surface and streams for the test area are demonstrated in Figure 6. The restriction criteria, data structures and feature types are listed in Table 1, where this data is reclassified to produce the raster dataset as shown in Figure 7. The exclusionary zones are deemed as unsuitable to place wind parks due to realistic considerations and regulations.

Table 1. Constraints on wind farm siting

Criterion	Data structure	Buffer (m)	Feature type
Urban area/town	Vector	1000/750	Polygon
Waterways	Vector	400	Polyline
Protected regions	Vector	1000	Polygon
Airports	Vector	12000	point
Railways	Vector	250	Polyline
Highways	Vector	500	Polyline

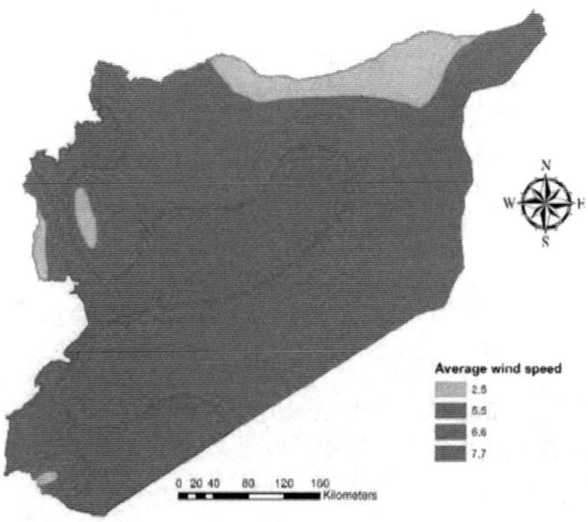

Figure 5. Distribution of annual average wind speed (m/s) in Syria.

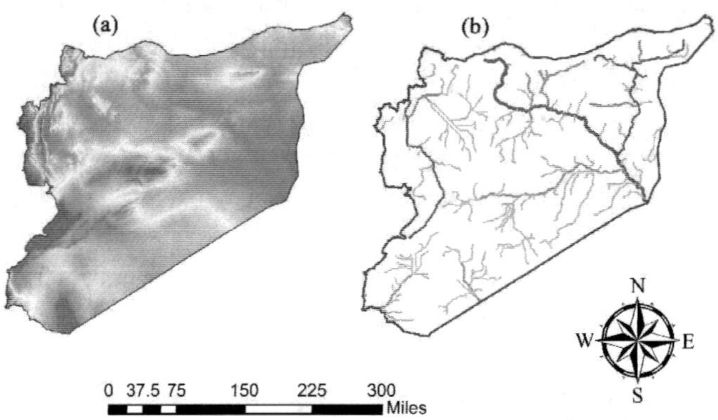

Figure 6. (a) DEM and (b) waterways for the study area.

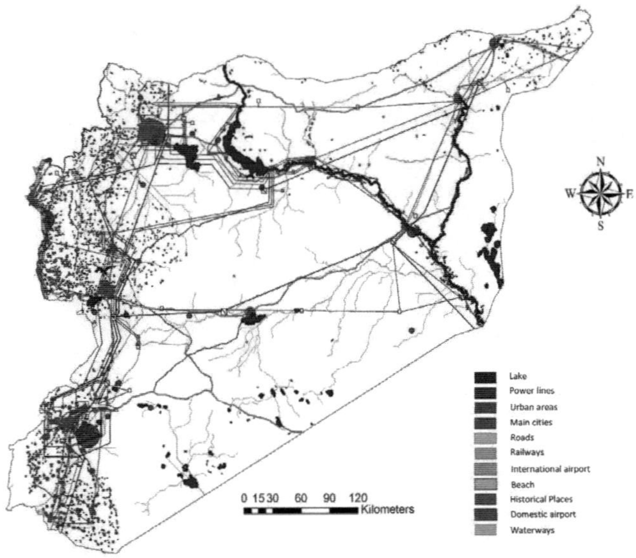

Figure 7. Map of buffer zones around the restricted features in Syria.

3. Results and discussions

As stated previously, the suitability analysis is applied to rank and score sites relied on defined criteria. After the restricted areas were excluded from the country (Figure 7), it is possible to determine wind farm sites by evaluating variety of decision factors according to the specific requirements of planners. Table 2 describes criteria that are used in the current study with their justifications of choice, data structure and feature type. Each criterion is represented in GIS as a feature class (layer) for the whole area. It is important that different formats of data are standardized and changed into forms, which can be compared with each other. All of these criteria are converted into raster format by reclassifying to alternative values for creating suitability model as shown in Figure 8.

Table 2. Decision factors used in this research

Criterion	Reason of selection	Data structure	Feature type
Wind energy potential	Decisive for wind energy success	Vector	Polygon
Slope	Affect construction costs and maintenance	Vector	Polygon
Land use	Decreasing the environmental influence of development on socio-ecological systems	Vector	Polygon
Proximity to power lines	Reducing the cost of energy transmission to the grid	Vector	Polyline
Proximity to main roads	Facilitating access for logistics and construction	Vector	Polyline

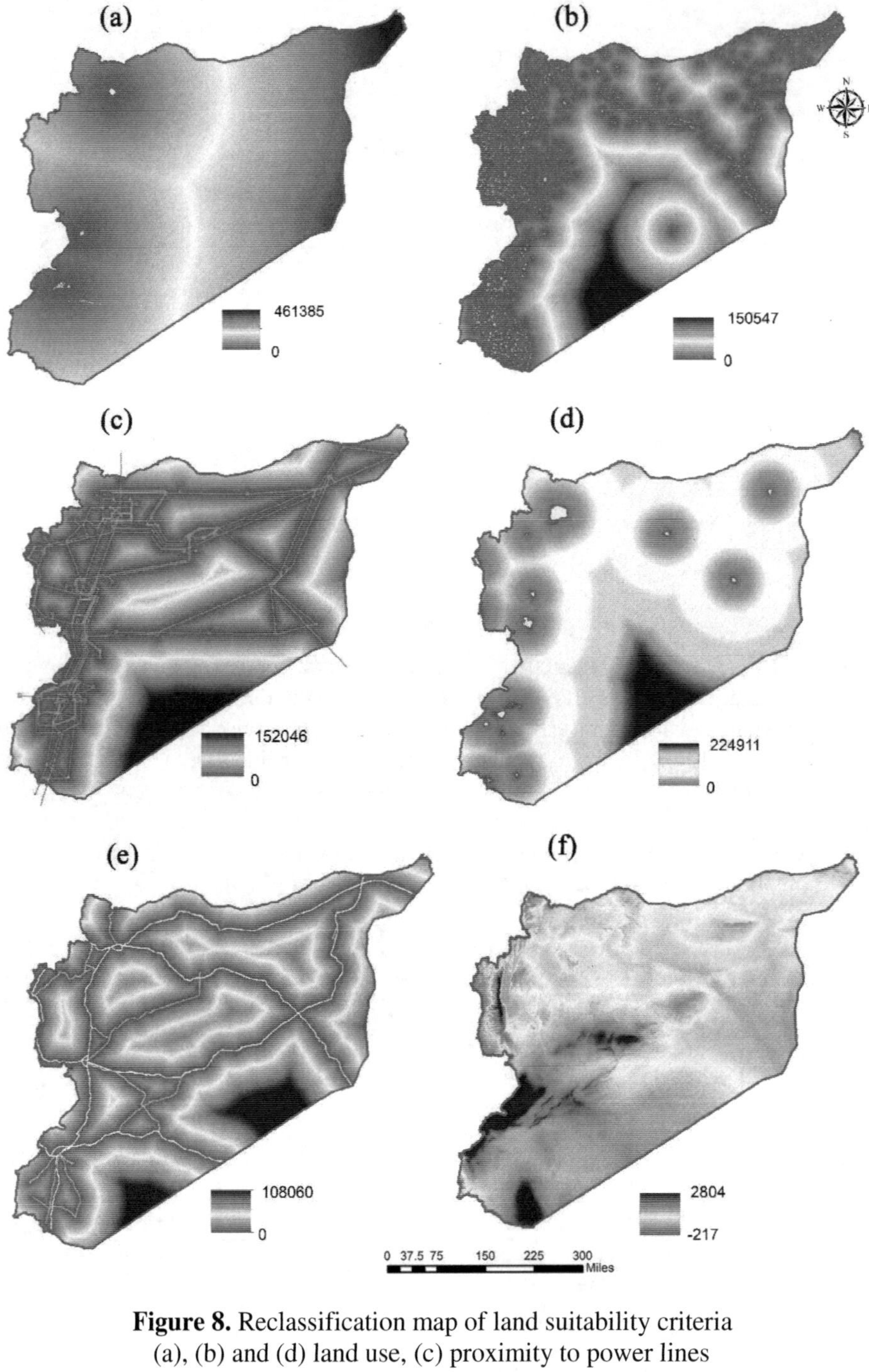

Figure 8. Reclassification map of land suitability criteria (a), (b) and (d) land use, (c) proximity to power lines (e) roads distances and (f) slope.

Land suitability assessment of wind power is extremely related to the opinions of different stakeholders. Therefore, this research focuses on some criteria are deemed out of the scope of this study. Furthermore, in order to build accurate suitability map for wind energy development, it should be associated with a weight to each criterion or GIS layer in a systematic and comprehensive procedure. AHP uses a pairwise comparison method to evaluate criteria relative significant and compute their

IOP Conf. Series: Materials Science and Engineering **800** (2020) 012019 doi:10.1088/1757-899X/800/1/012019

weights. The wind energy potential is given high weight since it is essential in location selection of wind turbines as listed in Table 3. Weighted overlay analysis is a useful tool for handling multi criteria issues such as wind power plants suitability model. However, the resulted map of appropriate sites of wind parks in Syria. Figure 9 is distributed into four classes where the most convenient areas are placed in the southwestern and in the middle part of the country.

Table 3. Assigned weights of criteria

Criterion	Weight %
Wind energy potential	37.5
Slope	22.5
Land use	20
Proximity to power lines	10
Proximity to main roads	10

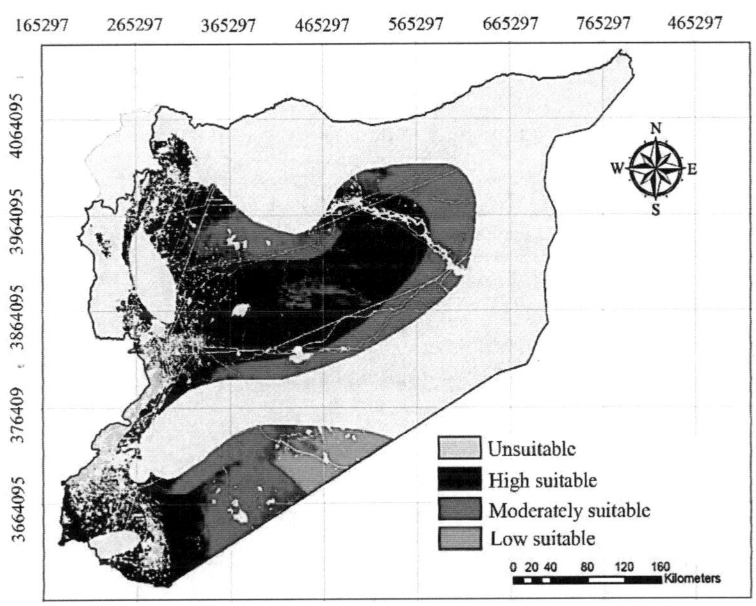

Figure 9. Suitability map for development of wind power stations in Syria.

4. Conclusions

Wind energy is commonly utilized in recent years for increasing the usage of renewable energy resources as an alternative of fossil fuels. Choosing a proper site of wind parks is an essential problem to get a high performance of wind power turbines from technical and economic point of view without harming the environment and the society. A proposed methodology is introduced to define appropriate places of wind power stations, which are related to some assumptions, requirements, and constraints. Climate, natural, geological, socio-economic, and environmental are the main factors contribute to develop site selection issue. Although GIS has effective tools for analysing geospatial data to identify the best locations, but this application not capable to overcoming the problem of inconsistency in expert opinions when attempt to rate each of criteria that are included in a suitability analysis. The AHP approach is widely used in multi-criteria decision making to estimate the weights to the various factors. Applying this technique with GIS in the study have proved to be a powerful technique for wind farm site selection. The resulted map of wind farm in Syria shows that the majority of the areas with high appropriate are situated the southwestern and in the middle part of the country. Finally, efforts should be made to promote using clean and sustainable energy as a post-conflict plan in Syria that contributes to reduce greenhouse gas emissions and decrease fuel consumption. In addition, commitment with environmental standards and principles of spatial planning to develop wind energy will help in avoiding their future negative impacts on land use.

5. References

[1] Berry, H. L., Waite, T. D., Dear, K. B., Capon, A. G., & Murray, V. (2018). The case for systems thinking about climate change and mental health. Nature Climate Change, 8(4), 282.

[2] Ghazali, D. A., Guericolas, M., Thys, F., Sarasin, F., Arcos Gonzalez, P., & Casalino, E. (2018). Climate change impacts on disaster and emergency medicine focusing on mitigation disruptive effects: an international perspective. International journal of environmental research and public health, 15(7), 1379.

[3] Ariza-Montobbio, P., & Farrell, K. N. (2012). Wind farm siting and protected areas in catalonia: planning alternatives or reproducing one-dimensional thinking?. Sustainability, 4(12), 3180-3205.

[4] Azad, A. K., Rasul, M. G., Islam, R., & Shishir, I. R. (2015). Analysis of wind energy prospect for power generation by three Weibull distribution methods. Energy Procedia, 75.

[5] Gul, M., Tai, N., Huang, W., Nadeem, M. H., & Yu, M. (2019). Assessment of wind power potential and economic analysis at hyderabad in pakistan: powering to local communities using wind power. Sustainability, 11(5), 1391.

[6] Tegou, L. I., Polatidis, H., & Haralambopoulos, D. A. (2009). Wind turbines site selection on an isolated island. WIT Transactions on Ecology and the Environment, 127, 313-324.

[7] Awange, J. L., & Kiema, J. B. K. (2013). Fundamentals of GIS. In Environmental Geoinformatics (pp. 191-200). Springer, Berlin, Heidelberg.

[8] Burrough, P. A., McDonnell, R., McDonnell, R. A., & Lloyd, C. D. (2015). Principles of geographical information systems. Oxford University Press.

[9] Malczewski, J., & Rinner, C. (2015). Multicriteria decision analysis in geographic information science (pp. 90-93). New York: Springer.

[10] Mahdy, M., & Bahaj, A. S. (2018). Multi criteria decision analysis for offshore wind energy potential in Egypt. Renewable energy, 118, 278-289.

[11] Saaty, T. L. (1990). Decision making for leaders: the analytic hierarchy process for decisions in a complex world. RWS publications.

[12] Aras, H., Erdoğmuş, Ş., & Koç, E. (2004). Multi-criteria selection for a wind observation station location using analytic hierarchy process. Renewable Energy, 29(8), 1383-1392.

[13] Ali, S., Lee, S. M., & Jang, C. M. (2017). Determination of the most optimal on-shore wind farm site location using a GIS-MCDM methodology: Evaluating the case of south korea. Energies, 10(12), 2072.

[14] Solangi, Y., Tan, Q., Khan, M., Mirjat, N., & Ahmed, I. (2018). The selection of wind power project location in the southeastern corridor of pakistan: a factor analysis, AHP, and Fuzzy-TOPSIS application. energies, 11(8), 1940.

[15] Van Lier, H. N. (1998). The role of land use planning in sustainable rural systems. Landscape and Urban Planning, 41(2), 83-91.

[16] Devendran, A. A., & Lakshmanan, G. (2014). A Review on accuracy and uncertainty of spatial data and analyses with special reference to urban and hydrological modelling. ISPRS Annals of Photogrammetry, Remote Sensing & Spatial Information Sciences, 2(8).

[17] Habib, M. (2008). Proposal for developing the Syrian stereographic projection. Survey Review, 40(307), 92-101.

[18] Shaheen, M., & Khan, M. Z. (2016). A method of data mining for selection of site for wind turbines. Renewable and Sustainable Energy Reviews, 55, 1225-1233.

[19] Hansen, H. S. (2005). GIS-based multi-criteria analysis of wind farm development. In Proceedings of the 10th Scandinavian Research Conference on Geographical Information Sciences, 13th-15th June.

[20] Azizi, A., Malekmohammadi, B., Jafari, H. R., Nasiri, H., & Parsa, V. A. (2014). Land suitability assessment for wind power plant site selection using ANP-DEMATEL in a GIS environment: case study of Ardabil province, Iran. Environmental monitoring and assessment, 186(10), 6695-6709.

IOP Conf. Series: Materials Science and Engineering **800** (2020) 012020 doi:10.1088/1757-899X/800/1/012020

A framework for pavement crack detection and classification

A Habib*, M M Kunt

Civil Engineering Department, Eastern Mediterranean University, Famagusta. North Cyprus

*E-mail: ahed.habib@gmail.com

Abstract. Pavement damage detection is indeed a very important process for the management of roads. Nowadays, scholars are focusing on finding a simple and accurate way to detect road cracks aiming to increase its life span and improve its safety and quality. However, due to some factors including cost of implementation and level of experiences required small communities and developing countries are yet unable to adapt this idea. Therefore, developing a feasible method is becoming extremely important for improving the quality and safety of pavements. Thus, this study is intended to propose a framework for pavement crack detection and classification through the use of inexpensive set of sensors and the applications of artificial neural networks. As part of the project, difficulties that are going to be faced by local agencies in small communities will be highlighted and possible solutions will be suggested. In general, this framework is expected to be helpful to government entities in developing plans and taking actions toward providing a suitable pavement distress mitigation strategy.

1. Introduction

Deterioration of the pavement is the system through which distresses develop under the combined impacts of road traffic and environmental factors. Serviceability, security and road quality are significantly affected by deterioration of pavements. Because of its extensive use, roads deteriorate with time and must, therefore, be maintained to ensure that safety, efficiency and durability requirements are met. Common deterioration of pavement involves cracking, surface defects, deformation and structural failure [1]. The most common types of cracking are fatigue cracking, longitudinal cracking, transverse cracking, block cracking, slippage cracking, reflective cracking and edge cracking [2]. The combination of fatigue and cracking block may be regarded as alligator cracking. These cracks are mainly caused by high load applied or excess moisture. Therefore, it is important to continuously check the pavement condition in order to provide a solid data for a pavement management system [3]. However, local governments in small cities often lack experienced staff for effectively using a fully implemented pavement management system [3]. Furthermore, available techniques for pavement crack detection are considered to be very expensive especially for small countries such as the case of Turkish Republic of North Cyprus (TRNC). For this reason, finding a way to reduce this cost and complexity is essentially towards an effective management of pavements.

Thus, the aim of this paper is to propose a framework empowered by a low-cost tool that can be used by small communities and to highlight the main difficulties and their solution that might be faced during the implementing of this methods.

IOP Conf. Series: Materials Science and Engineering **800** (2020) 012020 doi:10.1088/1757-899X/800/1/012020

2. Materials and methods

In this section, components of the framework for pavement damage detection empowered by a low-cost tool is going to be discussed.

2.1. Data collection

In this study a data collection circuit will be developed based on the use of Raspberry Pi Zero, Figure 1, as the main computer that works for data acquisition and data storing. An acceleration sensor MPU6050, Figure 2, with a data sampling rate of 60 Hz and range of ±8g will be used for measuring the vibration in the vertical direction. Furthermore, a Global Positioning System (GPS) GY-NEO6MV2 sensor, Figure 3, will be used for defining the location of the crack. In general, the total cost of the circuit and the sensors is considered low-cost system which allows small communities to get benefit of this method even at small budgets.

In order to collect the data, the device must be mounted on any vehicle or bicycle. Thereafter, the Raspberry Pi Zero will work as a data acquisition system through a python coded program that will read the output of the sensors and will store it in the memory card.

Figure 1. Raspberry Pi Zero

Figure 2. Chip of the MPU 6050 sensor

Figure 3. Chip of the GY-NEO6MV2 GPS sensor

2.2. Data Pre-Processing

In fact, the recorded data must undergo a pre-processing stage to reduce their noise emerged from the uncontrollable environment in the pavement. The suggested method in this study is called the moving average filter which is composed of taking the average of a certain number of outputs and subtract that value from the readings of the sensor [4]. In addition, to removing the surrounding noise the suggested method will also remove the readings caused by the Earth's gravity as the observed accelerations are in vertical direction.

2.3. Data processing

This stage includes both the crack detection and classification. At the beginning, the collected data mainly the vibration one will be divided into number of cases in which each case will represent a sudden change in the vibration input. This is going to be done by selecting all the values that cross a specified acceleration limit such as 10 m/s^2. Thereafter, 15 readings before and 15 readings after each of these values will be defined as a crack that will be classified separately.

Classifying each crack is going to be undertaken through the use of Artificial Neural Network (ANN) **Figure 4** because of its capability to learn and capture the functional relationships among the hard description of data [5]. In order to develop the ANN system part of the recorded data will be used to train the system and rest will be to verify and test it.

In fact, the use of an accelerometer provides an easy way for identifying the vibration behavior of each change in the pavement texture such as cracks or speed humps. Furthermore, using a reliable ANN system can help in classifying each of the texture changes into the correct crack type. Table 1 shows examples of different types of changes in the pavement surface. It is clear that each one has unique characteristics. For example, the alligator crack can be defined by consecutive fluctuation of a similar peak values. On the other hand, the pothole can be observed by a sudden change in the vibration behavior that have a single peak value, while the speed hump is identified by a negative peak directly followed by a high value of a positive peak and its long duration. Using the same procedure, it is possible to specify the characteristics of each crack. These characteristics can be the duration of the fluctuation in vibration, the standard deviation in of the vibrations, and the number of positive and negative peak values.

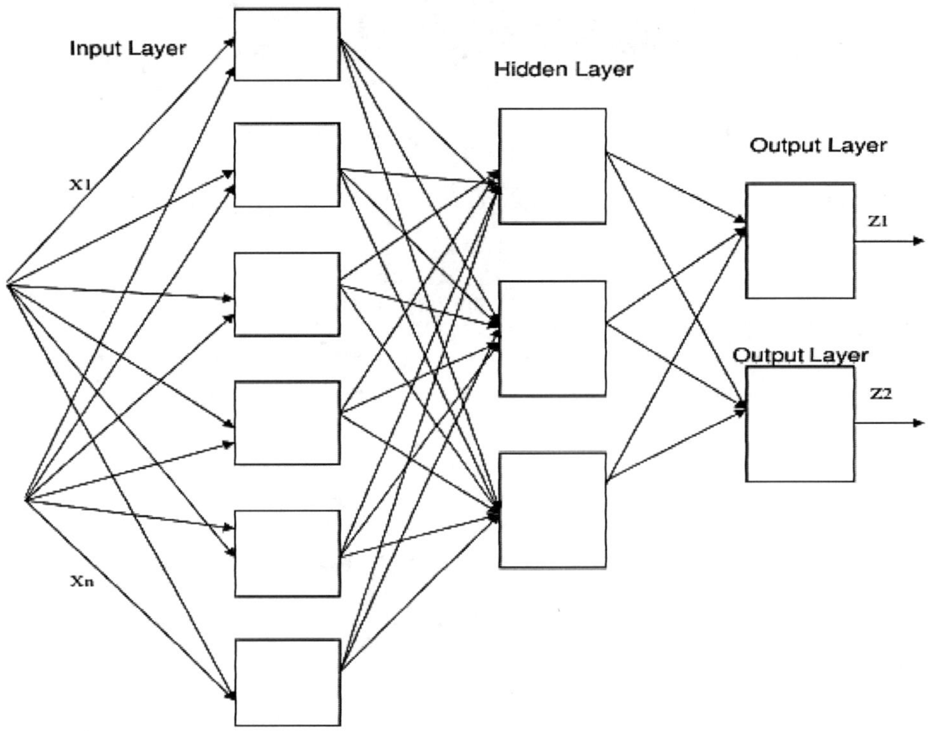

Figure 4. An artificial neural network [6].

IOP Conf. Series: Materials Science and Engineering **800** (2020) 012020 doi:10.1088/1757-899X/800/1/012020

Table 1. Vibration behavior of various pavement texture changes.

Type	Photo	Vibration Behavior
Alligator Crack	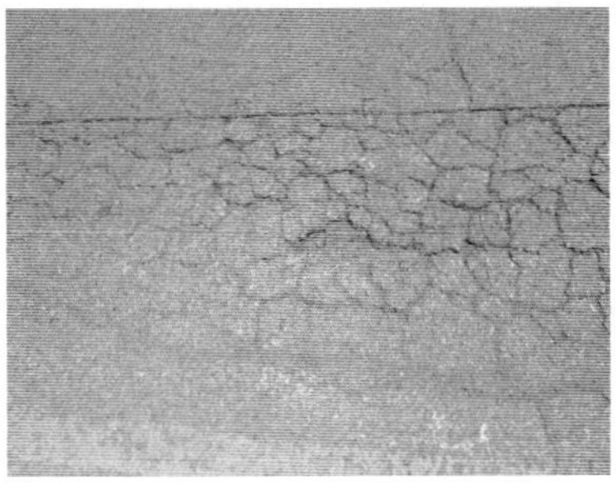	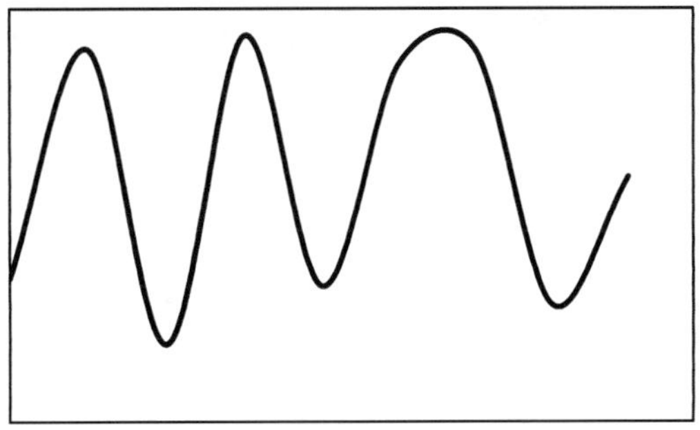
Pothole Crack		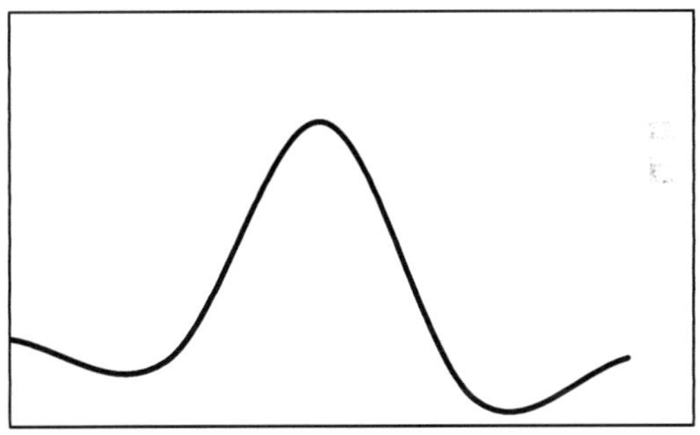
Speed Humps		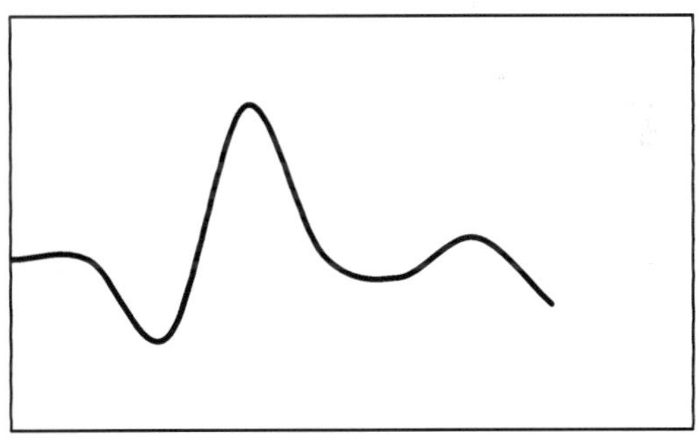

2.4. GIS Database

The importance of any GIS is its capability in data modelling, which is a set of constructs for describing and representing selected aspects of the real world in computer [7]. It is a procedure of designing and developing a data system by taking all the information that would be needed to support various business processes of the organisation [8]. Data model is created to depict the structure of the data handled in information systems that is demonstrated in entity-relationship diagrams (ERD). Three basic styles of data modelling, Figure 5, are generally used in practice today [9]:

- Conceptual data models: the process of constructing a model of the data, creation of ERD, used in organization and independent of all physical considerations.
- Logical data models: the method of representing data architecture and organization in a graphical way, mapping ERD to tables, without any regard to the physical data storage device.
- Physical data models: the internal schema database design.

In this stage, each crack will be defined by its type and location on a designed geodatabase that composed of a digitized roads network such as the one shown in Figure 7.

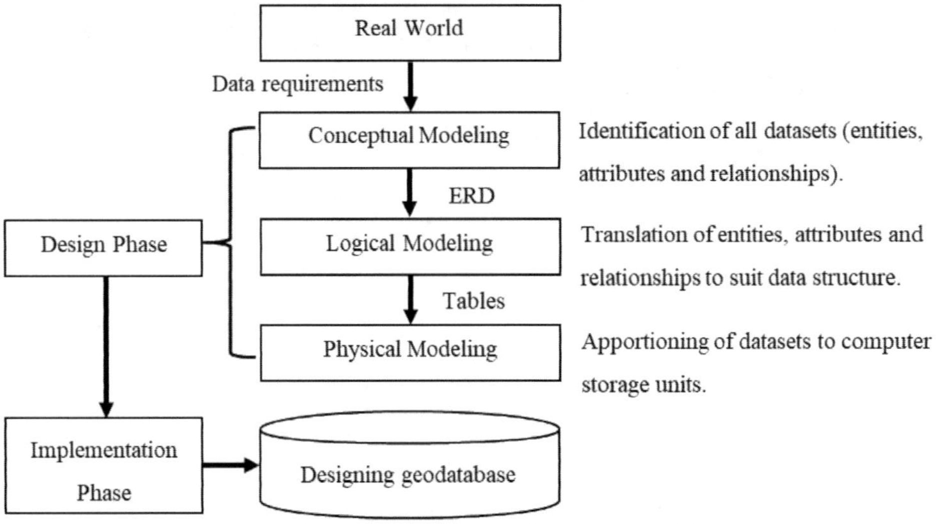

Figure 5. Levels of abstraction in data modelling [10]

Figure 6. An example of a digital road network

3. Proposed Framework

3.1. Pavement Crack Detection and Classification Framework

As mentioned previously, this paper is proposing a framework that can be used for pavement crack detection and classification.

The framework proposed in this study, as depicted in Figure 7, is composed of four main stages which are data collection, their pre-processing and processing and storing them in a GIS database for possible mitigation plans.

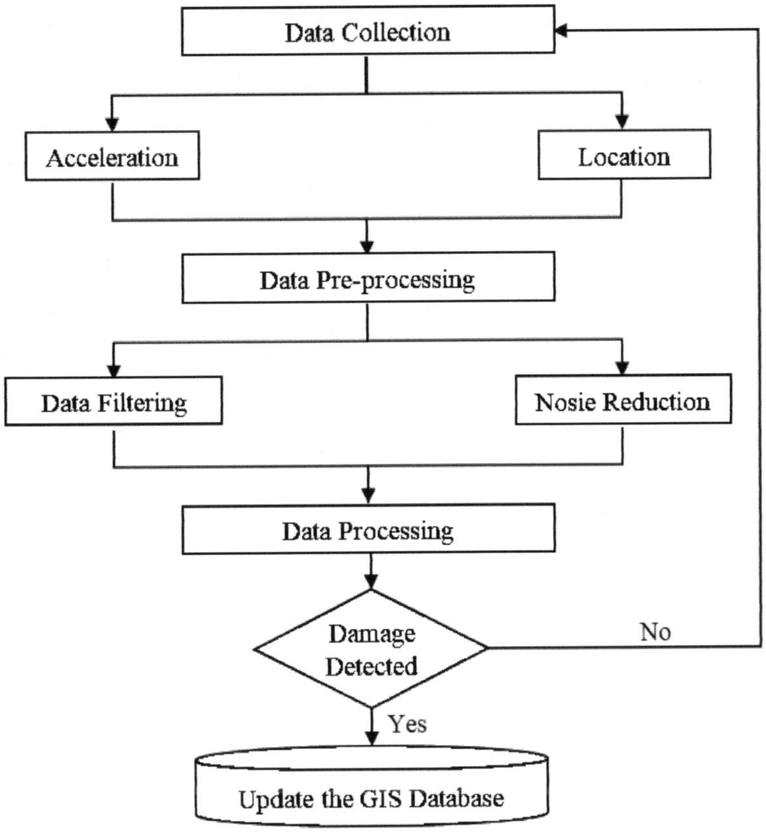

Figure 7. General framework for pavement crack detection and classification

3.2. Challenges for Small Communities

In fact, the major difficulty that is expected to be faced by small communities is the lack of an accurate GIS database. Previous study [11] have illustrated how a GIS map can be collected, edited, and verified from other map sources such as locally maintained road ones. Furthermore, another challenge would be the cost of continuous data collection which can be overcome by licensing local agencies who run vehicle or bicycle renting services to work as a data provider for the government by introducing this low-cost device to their business.

4. Conclusion

This study contributes to the pavement damage detection and classification in small communities with the developed framework. As a part of this study, a proposed framework that can be implemented with a low-cost tool mounted to a vehicle or a bike was discussed. Furthermore, some of the challenges that are expected to be faced by these communities were mentioned and possible solutions were suggested.

References

[1] Aziz, M. M., Mustafa, K. H., Hainin, M. R., Zaini, A. A., & Warid, M. N. (2016). A Review on Crack Relief Layer (crl) in Roads and Highways. Jurnal Teknologi, 45-52.

[2] Adlinge, S. S., & Gupta, A. K. (2013). Pavement deterioration and its causes. International Journal of Innovative Research and Development, 2(4), 437-450.

[3] Kunt, M. M. (2008). The Benefits of Introducing a Pavement Management System in Small Cities: The Case of the City of Famagusta. Washington D.C.: Transportation Research Board 87th Annual Meeting.

[4] Smith, S. W. (1997). The scientist and engineer's guide to digital signal processing.

[5] Gunawan, F. (2018, 11 1). Detecting road damages by using gyroscope sensor. ICIC Express Letters, 12(11), 1089-1098.
[6] Daniel, G. (2013). Principles of artificial neural networks . World Scientific.
[7] Longley, P. A., Goodchild, M. F., Maguire, D. J., & Rhind, D. W. (2005). Geographic information systems and science. John Wiley & Sons.
[8] Ponniah, P. (2007). Data modeling fundamentals: a practical guide for IT professionals. John Wiley & Sons.
[9] Singh, S. P., & Singh, P. (2014). Modeling a geo-spatial database for managing travelers' demand. International Journal of Database Management Systems.
[10] Akinyemi, F. O. (1999). Database design and development for census mapping in Nigeria. Cartography in the Service of Government, Nigeria, Lagos: Nigerian Cartographic Association, 191-202
[11] He, Z., Qin, X., Wang, H., & Comes, C. (2017). Implementing practical pavement management systems for small communities: a south Dakota case study. Public Works Management & Policy, 22(4), 378-391.

IOP Conf. Series: Materials Science and Engineering **800** (2020) 012021 doi:10.1088/1757-899X/800/1/012021

Viscoelastic properties of EVA interlayer used in laminated glass structures

T Hána[2]*, M Eliášová[1], M Vokáč[2] and K V. Machalická[2]

[1]Department of Steel and Timber Structures, Faculty of Civil Engineering, Czech Technical University in Prague, Thákurova 7, 166 29 Praha, Czech Republic
[2]Klokner Institute, Czech Technical University in Prague, Šolínova 7, 166 29 Praha, Czech Republic

*E-mail: tomas.hana@cvut.cz

Abstract. Laminated glass is getting used more extensively in a current architecture due to its transparency, aesthetic impression, and post-breakage behaviour. Examples of such structures are balustrades, stairs, or facade panels. These are usually loaded in bending. Polymeric interlayer embedded between glass plies has no flexural stiffness, but it can itself transfer shear stress. The rate of this transfer depends on the shear stiffness of interlayer which is time and temperature dependent parameter. Producers of interlayers often do not specify this quantity thus engineers rather neglect the shear interaction of glass plies in perpendicularly loaded laminated glass panels. This paper provides the values of shear stiffness modulus of common interlayer EVASAFE® by Bridgestone™ in time and temperature domain. This shear stiffness is expressed through Maxwell model whose parameters are based on the series of Dynamic mechanical thermal analysis (DMTA) performed on single lap small-scale specimens in Klokner institute CTU in Prague. Results show that shear stiffness of this interlayer is able to provide significant shear coupling of glass plies in broad temperature range of short-term perpendicularly loaded laminated glass structures.

1. Introduction

Laminated glass is used in a current architecture even as a load bearing element. It is a composition of glass plies bonded with polymeric interlayer. Typical applications of these structures are balustrades, roof panels, stairs, or canopies. The direction of applied load is usually perpendicular to central plane and stress-state and deflections are significantly affected by the shear stiffness of a particular interlayer [1]. This parameter is time and temperature dependent quantity [2] and it is not stated in technical sheets of most of interlayers. Increase of temperature as well as long-term loading induce molecular movement in a polymeric material resulting in reduction of its shear stiffness [2]. Therefore, laminated glass structure loses its bending stiffness and main tensile stress in glass increases. Drop of temperature and short-term loading have opposite effects [3]. To design load bearing laminated glass in bending taking shear interaction of glass plies into account, an engineer is referred to national standards, e.g. CNR-DT 210 [4], or to the current European draft prEN 16612 [5]. The mechanical tests of interlayers can be performed on small-size samples of a single interlayer [6], on small laminated glass specimens [7], or on large specimens [8]. They can be basically sorted into two groups: (i) static long-term creep or relaxation experiments, and (ii) dynamic tests called Dynamic mechanical thermal analysis (DMTA) where the specimen is cyclically loaded in various frequencies and temperatures. Static tests do not require complicated evaluation methods, but they cannot be usually performed in broad time and temperature range. Contrary DMTA results need to be thoroughly evaluated, but they enable to express material stiffness in long relaxation times and broad temperature range. It is more appropriate to perform the experiments of an interlayer in shear or torsion. From the shear or torsion test, the shear modulus G is determined with more precision in comparison with tensile tests where the conversion using Poisson ratio is necessary [9]. This paper is focused on viscoelastic description of ethylene-vinyl acetate interlayer EVASAFE® made by Bridgestone™ based on series of displacement controlled DMTA experiments performed on small-scale single lap laminated glass specimens. Particularly, Prony series of constructed Maxwell model are provided and relaxation shear modulus G at reference temperature 20 °C in time is illustrated. Response of the constructed model was further compared to the response of EVASAFE® in static single lap displacement-controlled shear tests performed in various loading rates and temperatures using the identical specimens and testing equipment as in DMTA. All experiments and analyses were performed in Klokner Institute CTU.

2. Theory of Viscoelasticity

Response of viscoelastic material subjected to shear cyclic loading in linear viscoelastic region is described by Boltzmann superposition principle [10]

$$\tau(t) = \int_0^t G(t-s)\,\dot{\gamma}(s)\,ds, \tag{1}$$

where $\tau(t)$ is the shear stress in time in [MPa], s is the time of the loading step from the interval <0; t> in [s], $\dot{\gamma}(s)$ refers to the engineering shear strain rate in [s^{-1}] at the certain time s, and $G(t)$ is the relaxation function in [MPa] of a viscoelastic material in the time domain. This function is usually considered in the form of the generalized Maxwell model [11] according to the equation

$$G(t) = G_\infty + \textstyle\sum_{i=1}^{n} G_i\, exp(-t/\tau_i), \tag{2}$$

where G_∞ is the equilibrium shear stiffness modulus in [MPa] in long relaxation times, G_i denotes the shear stiffness of spring in [MPa] together with the relaxation time τ_i in [s] of the i-th Maxwell element, and n is the number of Maxwell elements in the Maxwell model. Relaxation time τ_i expresses the ratio η_i/G_i, where η_i [Pa.s] is the viscosity of the dashpot of i-th Maxwell element. The structure of Maxwell model is shown in figure 1.

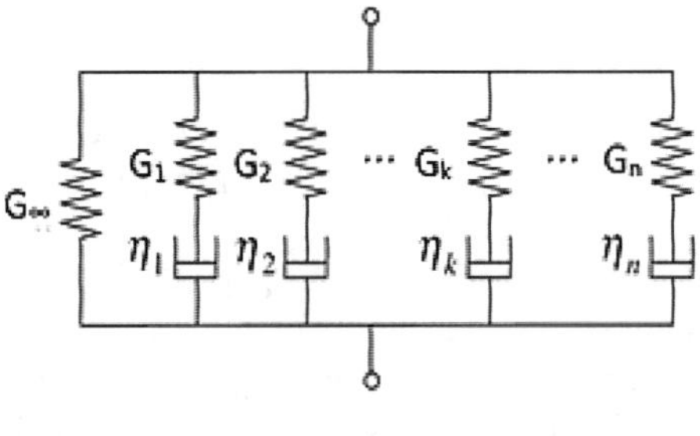

Figure 1. Maxwell model with its Prony parameters.

To include the temperature effect into the Maxwell model, the relaxation time $\tau_i(T_{ref})$ of i-th element determined for the reference temperature is modified as $\tau_i(T) = \alpha_T(T)\cdot\tau_i(T_{ref})$ for thermorheologically simple material, where α_T denotes the shift factor for temperature T [12]. This quantity may be obtained from the following Williams-Landel-Ferry (WLF) equation

$$log\,\alpha_T = -C_1(T - T_{ref})/(C_2 + T - T_{ref}), \tag{3}$$

where constants C_1 and C_2 are free non-dimensional WLF parameters [13]. WLF equation provides reliable results of extrapolated relaxation times for temperatures above Glass transition temperature T_g. For lower temperatures, the horizontal shift factor α_T should be expressed using different formulas [12]. In DMTA the interlayer is loaded by cyclic deformation according to the equation (4), where $\gamma(t)$ denotes shear strain in time t [s], γ_0 is the shear strain amplitude [-], and ω denotes the angular frequency [rad/s]. If shear strain $\gamma(t)$ is implemented into Boltzmann superposition principle in equation (1), its integration leads to the shear stress in the interlayer $\tau(t)$ through the complex modulus G^* [MPa] with its real part G' (storage modulus) and imaginary part G'' (loss modulus) [10] as shown in equation (5).

$$\gamma(t) = \gamma_0 \cdot \sin \omega t \tag{4}$$

$$\tau(t) = G^* \cdot \gamma(t) = \gamma_0 \cdot (G' \cdot \sin \omega t + G'' \cdot \cos \omega t) \tag{5}$$

General viscoelastic loop of stress-strain dependence with its important points as well as stress and strain courses in time observed during DMTA are displayed in figure 2. Viscoelastic materials during cycling exhibit time delayed response to the applied strain characterized by their phase shift δ [rad] as shown in figure 2.

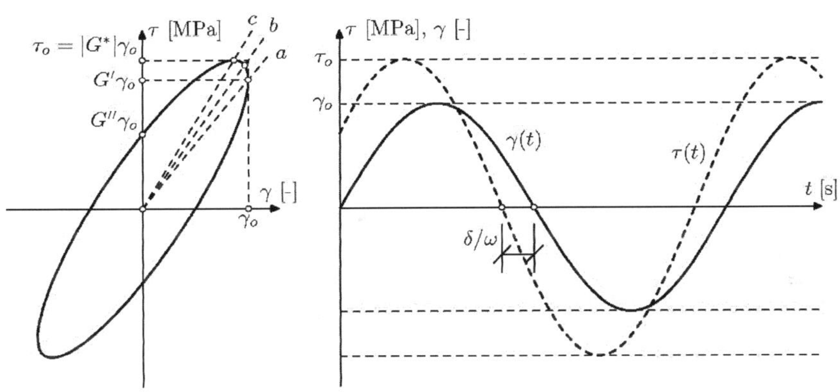

Figure 2. Shear stress-strain diagram and their courses in time during DMTA.

DMTA is usually performed at a certain range of angular frequencies and temperatures. Several curves of complex modulus G^* are obtained in that range depending on a testing equipment, see the example in figure 3. For thermorheologically simple materials, horizontal shifts of measured relations for the value log $\alpha_T(T)$ are performed to get a smooth Master curve at reference temperature T_{ref}, which enables to express the complex modulus G^* (resp. shear stiffness of material G) in wide frequency (resp. time) domain. This procedure is called Time-Temperature-Superposition-Principle (TTSP) as a sufficient approximation for engineering purposes [14]. As soon as Master curve $G^*(\log \omega)$ for a chosen reference temperature is completed, it is possible to approximate this curve by the response of the Maxwell model loaded in harmonic oscillatory loading by angular frequency ω according to the equation (4). Shear response $\tau(t)$ of that model is then expressed from the equation (5) by the equation (6) [11], where n denotes the number of Maxwell elements. Relaxation curve of fitted model is then expressed by the equation (2).

$$\tau(t) = \gamma_0 \cdot (G_\infty \cdot \sin \omega t + \sum_{i=0}^{n} \frac{G_i . \omega^2 . \tau_i^2}{1 + \omega^2 . \tau_i^2} \cdot \sin \omega t + \sum_{i=0}^{n} \frac{G_i . \omega . \tau_i}{1 + \omega^2 . \tau_i^2} \cdot \cos \omega t) \tag{6}$$

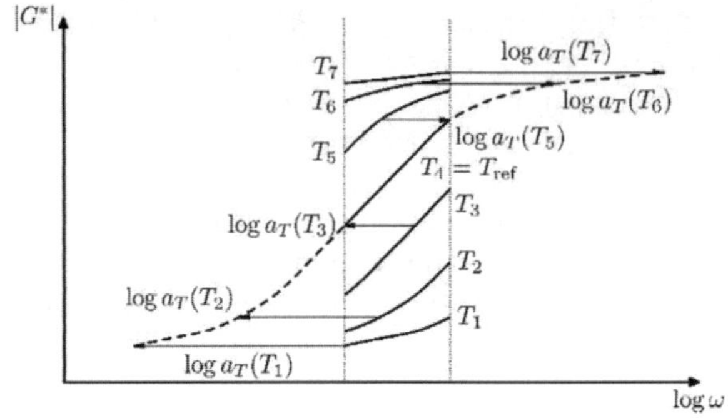

Figure 3. TTSP – horizontal shift of measured relations in DMTA and construction of Master curve.

3. Test Set-up

To find the viscoelastic parameters, DMTA tests of six single lap small glass specimens EVASAFE® in the range of temperatures -5 °C to +50 °C and testing frequencies between 0.05 Hz and 4.95 Hz were completed in Klokner Institute CTU in Prague. Experiments were performed in hydraulic testing system MTS with climatic chamber TIRA TEST T250/1. Shear strain of interlayer γ was controlled according to the equation (4). The shear area A of the interlayer was 50 x 50 mm with its thickness 0.81 mm. Engineering shear strain γ of the interlayer was determined from its thickness h and mutual displacement of glass panes u measured by displacement sensors MMR 1011 as $\gamma = u/h$. Shear stress τ acting on the interlayer was calculated from the applied force F and shear area A as $\tau = F/A$. Temperature in a climatic chamber was controlled by sensors Pt100 glued directly on the sample surface. As it has been stated, static displacement-controlled tests of the identical interlayer had been performed using identical testing equipment, specimens, and MTS system. Stress-strain relations obtained from those tests will serve for the evaluation of EVASAFE® Maxwell model. Detailed description of performed DMTA tests and static tests including methods of results evaluation has been specified by Hána et al. [15].

Figure 4. Test arrangement during DMTA and static displacement-controlled tests.

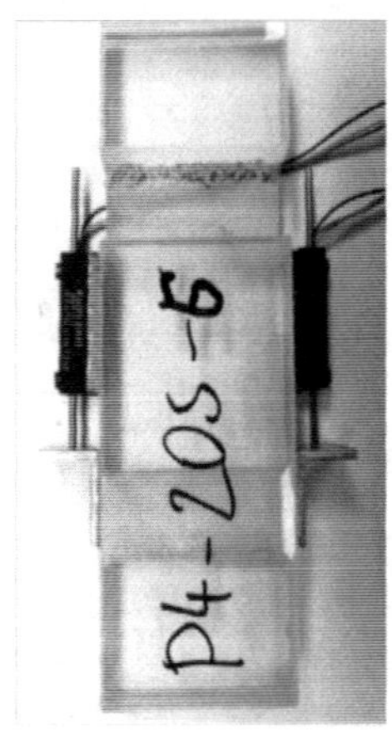

Figure 5. Testing single lap specimen with displacement transducers MMR 1011.

4. Results and Discussions

DMTA results, particularly storage modulus-frequency relations of representative specimen are shown in figure 6. Loss modulus was much lower than storage modulus ($G'' \ll G'$), thus it was not considered in master curving process. In the sense of regression analysis, WLF parameters C_1 and C_2 and the appropriate master curve at reference temperature 20 °C were constructed, see figure 7.

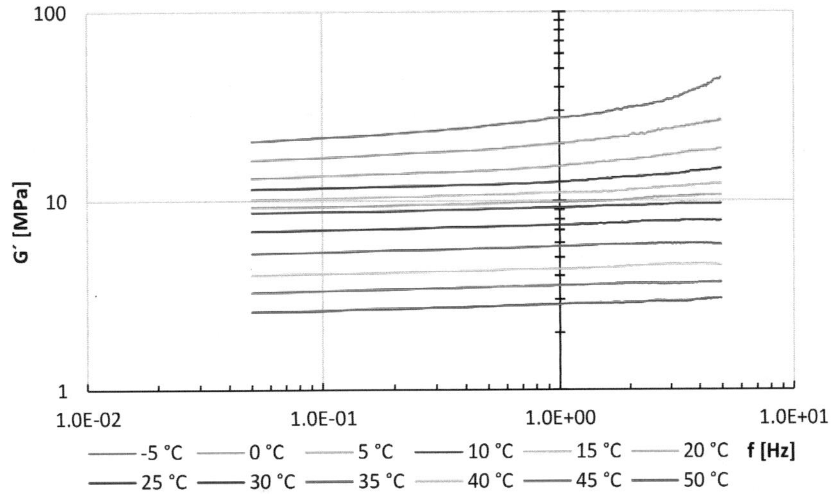

Figure 6. Measured storage modulus G' of representative specimen EVASAFE at testing temperatures.

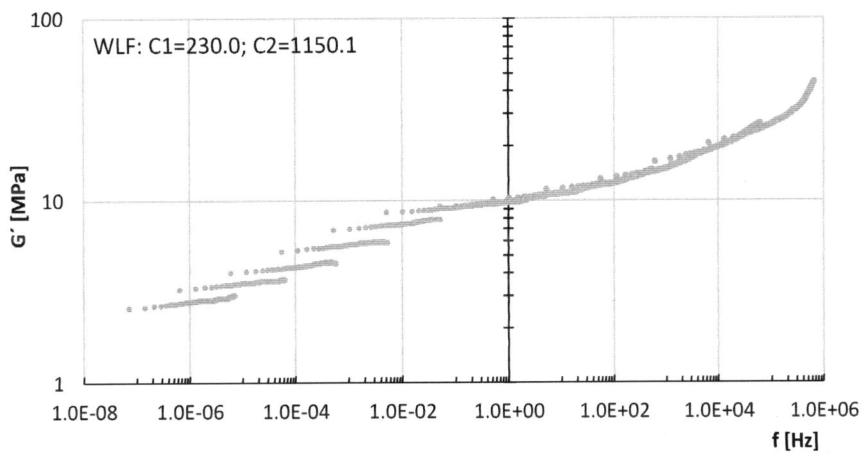

Figure 7. Constructed Master curve at reference temperature 20 °C for representative specimen.

Master curves of another testing specimens are of similar course. Glass transition temperature of ethylene-vinyl acetate is below 0 °C [16] thus master curving process using WLF equation is justified [12]. It is visible that a clear Master curve emerges in the range of high stiffness, while at the range of low stiffness it shows a discontinuity. It reflects the fact that EVASAFE® shows signs of thermorheological complexity and vertical displacements and rotations of the experimental curves are also necessary to get a clear Master curve. Similar discontinuity of Master curve in low frequencies of another EVA-based cross-linked interlayer evguard® has been revealed by Schuster et al. [16]. Based on the WLF constants and Master curve, generalized Maxwell model was constructed. To reduce the number of unknown Prony parameters, relaxation times τ_i were chosen. Homogenous distribution of relaxation times in frequency (time) domain was adopted [17]. It was chosen to use 22 relaxation times10^{-10+k}, where $k = 1{:}\,22$, to cover the range of testing frequencies. Fitting the shear moduli $\{G_i\}$, and G_∞ then leads to linearized problem. The objective function to be minimized is given by the equation (7)

$$F(\{G_i\}, G_\infty) = \sum_{j=1}^{m} (G'(\omega_j) - G_j')^2 + \sum_{j=1}^{m} (G''(\omega_j) - G_j'')^2, \tag{7}$$

where m is the number of all experimental measurements, G_j' and G_j'' are the measured storage and loss moduli for frequency ω_j, and $G'(\omega_j)$ with $G''(\omega_j)$ are the storage and loss moduli given by Maxwell model for frequency ω_j. Minimizing condition of this function based on calculus takes the form

$$\frac{\partial F}{\partial G_\infty} = 0; \frac{\partial F}{\partial G_i} = 0; i = 1, 2, \dots, n. \tag{8}$$

After solving the set of equations (8), the set of $\{G_i\}$, and G_∞ with the corresponding τ_i was obtained, see table 1. When these are used as input parameters in the equation (2), the shear modulus G(t, T = 20 °C) in time domain is obtained, see figure 8. To express the shear modulus in other tested temperatures, relaxation times according to TTSP can be modified as $\tau_i(\mathrm{T}) = \alpha_T(\mathrm{T}) \cdot \tau_i(\mathrm{T_{ref}})$.

Table 1. Prony parameters of investigated interlayer EVASAFE®.

WLF model at reference temperature T_{ref} = 20 °C					
C_1	230.0			C_2	1150.1
G_∞ = 0.362 MPa					
τ_i (s)	G_i (MPa)	τ_i (s)	G_i (MPa)	τ_i (s)	G_i (MPa)
1.00E-09	5.899	1.00E-02	0.365	1.00E+06	0.159
1.00E-08	2.882	1.00E-01	1.670	1.00E+07	0.194
1.00E-07	0.789	1.00E+00	0.063	1.00E+08	0.220
1.00E-06	0.136	1.00E+01	0.284	1.00E+09	0.060
1.00E-05	1.412	1.00E+02	1.322	1.00E+10	0.985
1.00E-04	17.767	1.00E+03	0.653	1.00E+11	0.141
1.00E-03	0.071	1.00E+04	1.260	1.00E+12	0.396
		1.00E+05	1.003		

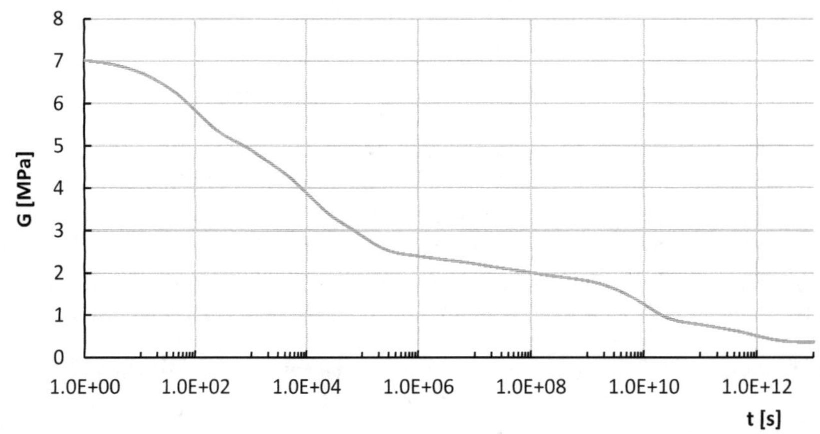

Figure 8. Relaxation function for the reference temperature 20 °C.

Short-term shear modulus at 0 °C – G(1s, 0 °C) = 18.2 MPa ensures monolithic response of laminated panel in bending. According to Galuppi and Royer-Carfagni [18] for the shear modulus values in range G = 1 – 10 MPa, the assumption of significant shear interaction of the individual glass plies in laminated panel is justified. Short-term shear moduli of tested interlayer, G(1s, 20 °C) = 7.0 MPa; G(1s, 40 °C) = 3.9 MPa; G(1s, 60 °C) = 2.1 MPa fulfill this assumption. Contrary, shear interaction of glass plies in long-term loaded laminated glass panel (G_∞ = 0.36 MPa) would be limited [18].

To verify the presented model, its shear stress response to the constant shear strain rate in tested temperatures was calculated and compared to stress-strain relations of already mentioned static displacement-controlled shear tests. In static tests, thickness of the interlayer was also 0.81 mm, temperatures were kept as 0; 20; 40; 60 °C and loading strain rates of the interlayer $\dot{\gamma}$ based on set MTS crosshead displacement were calculated as 0.0407 s^{-1} (for displacement 2.0 mm/min), 0.0102 s^{-1} (for displacement 0.5 mm/min), and 0.0025 s^{-1} (for displacement 0.125 mm/min). Representative results from static tests are in figure 9. If relaxation function of the constructed Maxwell model from the equation (2) is given into the integral equation (1), its evaluation results in the equation (9) which reflects the evolution of shear stress in constant strain rate. This way enabled to obtain shear stress-strain relations for tested temperatures in figure 10. When comparing both figures, one can see that static tests provide similar relations as Maxwell model in the range of engineering shear strains to 20%. This value is far over that observed in intact laminated glass panel (normally up to 1% [11]). For higher strains, Maxwell model is stiffer which might be a cause of exceeding the limits of linear viscoelasticity (usually between 20% and 50% [12] for softer polymeric material) or lower real shear strain rate of the interlayer.

$$\tau(t,T) = \dot{\gamma}\cdot\{G_\infty\cdot t + \alpha_T(T)\cdot\sum_{i=1}^{n} G_i\cdot\tau_i[1-\exp(-t/(\alpha_T(T)\cdot\tau_i))]\} \tag{9}$$

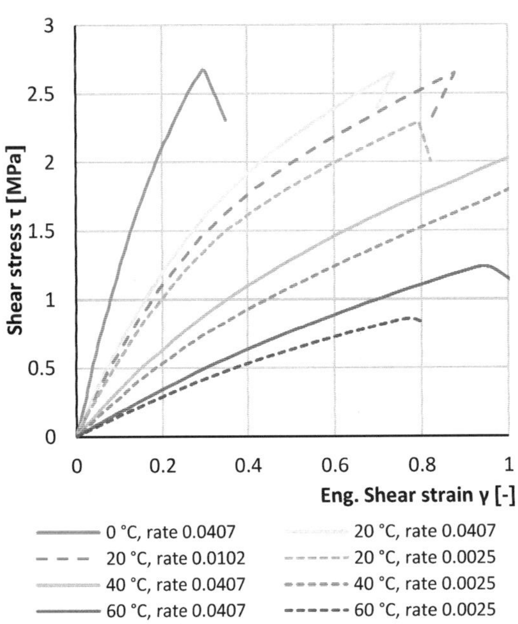

Figure 9. Stress-strain relations from static shear tests in various strain rates.

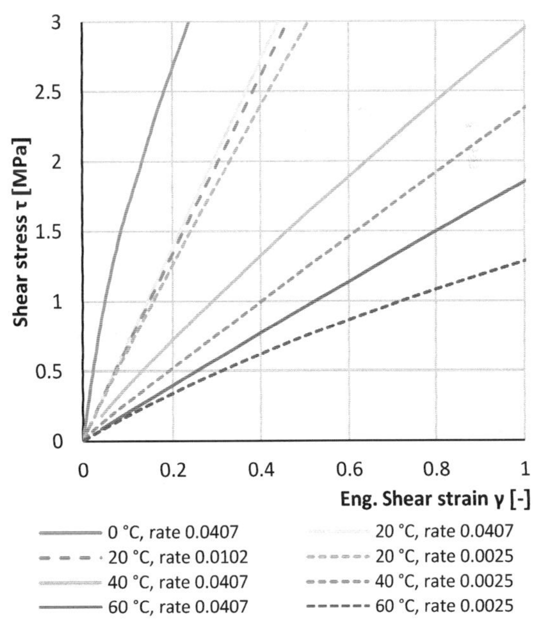

Figure 10. Response of constructed Maxwell model to the applied various strain rates.

5. Conclusion

This paper provided Prony parameters of the constructed WLF and Maxwell model of polymeric interlayer EVASAFE® based on the series of DMTA shear tests performed on small single lap laminated glass specimens at CTU in Prague. Experiments showed that tested interlayer exhibits signs of thermo-rheologically complex behaviour. Short-term relaxation modulus G based on constructed model ensures significant shear interaction of glass plies in a laminated panel in the range of 0 – 60 °C. Prony parameters were further used to calculate the theoretical stress-strain relations of the constructed Maxwell model loaded by various constant shear strain rates at various temperatures. These strain rates were identical with those from static displacement-controlled shear tests calculated on the set value of MTS crosshead displacement (2.0 mm/min, 0.5 mm/min, and 0.125 mm/min). Static shear tests were performed on the same testing specimens and interlayer as DMTA. Presented model provided similar stress-strain relations with static shear tests for engineering strains to 20% at all tested temperatures. For higher strains, response of Maxwell model to the applied load was stiffer. To validate the presented Maxwell model, sets of relaxation experiments of laminated glass with EVASAFE® at various temperatures are necessary. Experimentally verified models of polymeric interlayers is the way how to design safe and economic glass structures.

Acknowledgement
This research was supported by International Mobility of Researchers in CTU number CZ.02.2.69/0.0/0.0/16_027/0008465. The support is gratefully acknowledged.

References
[1] Serafinavicius T, Kvedaras A K and Sauciuvenas G 2013 *Mech. of Comp. Mater.* **49** 437-446
[2] Ferry J D 1990 *Viscoelastic Properties of Polymers* (New York: Wiley)
[3] Hána T, Eliášová M, Machalická K and Vokáč M 2017 Determination of PVB interlayer's shear modulus and its effect on normal stress distribution in laminated glass panels *IOP Conf. Ser.: Materials Science and Engineering. 3rd Int. Conf. on Innovative Materials, Structures and Technologies (Riga)* vol 251 (Bristol: IOP Publishing Ltd)
[4] CNR-DT 210 2013 *Guide for the Design, Construction and Control of Buildings with Structural Glass Elements* (Roma: National Research Council of Italy)
[5] prEN 16612 2017 *Glass in Building–Determination of the Lateral Load Resistance of Glass Panes by Calculation* (Brussels: European Committee for Standardization)
[6] Hooper P A, Blackmann B R K and Dear J P 2012 *J. Mater. Sci.* **47** 3564-3576
[7] Biolzi L, Cagnacci E, Orlando M, Piscitelli LR and Rosati G 2014 Long term response of glass-PVB double-lap joints *Composites: Part B* **63** 41-49
[8] Shitanoki Y, Bennison S J and Koike Y 2014 *Polymer Testing* **37** 59-67
[9] prEN 16613 2013 *Laminated Glass and Laminated Safety Glass–Determination of Interlayer Mechanical Properties* (Brussels: European Committee for Standardization)
[10] Lakes R 2009 *Viscoelastic Materials* (Cambridge: Cambridge University Press)
[11] Schneider J, Langer S and Schuster M 2015 Viscoelastic Properties of Laminated Glass Interlayers–Theory and Experiments *Glass Performance Days 2015* (Tampere: Tammerprint Oy) pp 143-147
[12] Brinson H F and Brinson L C 2015 *Polymer Engineering Science and Viscoelasticity* (New York: Springer Science+Business Media)
[13] Williams M L, Landel R F and Ferry J D 1995 The Temperature Dependence of Relaxation Mechanisms in Amorphous Polymers and Other Glass-forming Liquids *J. Am. Chem. Soc.* **77** 3701-3707
[14] Schwarzl F R 1990 *Polymermechanik* (Berlin: Springer Verlag)
[15] Hána T, Vokáč M, Machalická K and Eliášová M 2018 Material properties of polymeric interlayers under static and dynamic loading with respect to the temperature *Challenging glass 6–Conf. on Architectural and Structural Applications of Glass* (Delft: TU Delft) 417-426
[16] Schuster M, Kraus M, Schneider J and Siebert G 2018 Investigations on the thermorheologically complex material behaviour of laminated safety glass interlayer ethylene-vinyl-acetate. *Glass Struct. Eng.* **3** 373-388

[17] Barrientos E, Pelayo F, Noriega Á, Lamela M J, Fernández-Canteli A and Tanaka E 2018 *Mech. Time-Depend. Mater.* **23** 193-206
[18] Galuppi L, Royer-Carfagni G 2013 *Composites: Part B* **55** 109-118

Design and verification of properties of heavy-weight concrete for the production of weights

R Hela and L Bodnarova*

Brno University of Technology, Faculty of Civil Engineering, Veveri 95, Brno, Czech Republic

*E-mail:bodnarova.l@fce.vutbr.cz

Abstract. The paper deals with the design of composition and verification of the properties of heavy-weight concrete for the production of weights. Heavy weights from fine-grained heavy-weight concrete can be used as a substitute for harmful lead weights (sinkers) for fishing purposes. Lead fishing weights contaminate water, posing a risk to aquatic animals and birds. The production of fishing sinkers from heavy-weight concrete seems to be a possible substitute. A significant reduction in the use of lead fishing sinkers in the European Union is currently under preparation. The formulas of heavy-weight concrete from fine aggregate (fractions max. 2mm) were designed and tested. The consistency of concrete was designed for casting. The optimum vibration frequency of concrete compaction was determined at 75 Hz, resulting in a smooth, colourable surface. The density of the dried concrete after 28 days was 4000 kg.m^{-3} using a magnetite aggregate with an apparent density of 5.07 Mg/m^3.

1. Introduction

Heavy-weight concretes are concretes with a density above 2600 kg.m^{-3}[1], achieved by using heavy aggregate with a volumetric weight over 3000 kg.m^{-3} or via barite cement. Heavy-weight concretes are used as shielding concretes for radiation protection in nuclear power plants, hospitals or laboratories. Moreover, heavy-weight concrete elements are also used to load structures, such as the foundations of dynamically loaded machines, and as different weights, such as crane weights. Fishing weights (fishing sinkers) made of heavy-weight concrete are a prospective application of heavy-weight concrete as these concrete fishing sinkers can act as substitutes for widely used non-ecological lead fishing sinkers. Lead sinkers weights contaminate water, posing a risk to aquatic animals and birds. Recently, significant restrictions on the use of lead weights for fishing are being prepared by the European Union. Fishing sinkers made of heavy-weight concrete may be a possible substitute.

2. Ecotoxicity of Lead Weights

Because of its low melting temperature (327 °C) and high density (11 340 kg.m^{-3}), lead is an ideal material for casting fishing sinkers—if we leave out its toxicity. The problem with lead lies in its negative impacts on health; it seems that even low-level exposure can affect health. Lead is very toxic, and, when ingested, it threatens many body organs and functions. Lead poisoning has a negative impact on the functions of CNS, kidneys, immune system, heart and blood vessels. When ingested directly, e.g. from lead paint, contaminated water, food, etc., lead quickly gets into the blood and spreads throughout the body. Long-term exposure to lead can cause slackness of the musculoskeletal system (e.g. fingers, ankles, etc.), a slight increase in blood pressure, anaemia, reduced male and female fertility and even miscarriage. For young children, low-level or long-term exposure to lead affects brain functions and causes blood disorders. However, thanks to research on lead harmfulness, this metal is restricted in several countries, even for fishing. The EU has examined the impact of fishing lead sinkers on the environment and human health, finding that lost lead weights have an impact on the environment, for, if a weight is lost in water, the lead is abraded and dissolves with time. If a sinker gets lost outside the aquatic environment, it slowly corrodes, resulting in toxic substances, especially lead sulphides and oxides. It can also lay down sediments, thus being preserved for a long time. The substitution of lead weights is difficult because lead can be shaped very well; moreover, it is heavy and, given the alternatives, cheap. Alternatives include other metallic elements (tin, bismuth, antimony, steel, brass, tungsten, etc.); however, they have their disadvantages, too. For example, zinc is also toxic, copper has a negative impact on invertebrate animals, elements like tungsten and gold are very expensive, and steel is subject to corrosion. Non-metallic alternatives include resin and polypropylene, but these are disadvantaged by their low bulk density (compared to metallic elements) and their price. Other substitutes can be plastic stone or heavy-weight concrete [2].

3. Components of Heavy-weight Concrete

The components of heavy-weight concrete are very individual, depending on the concrete's usage. For protection against radiation, when objects are implemented in relatively great volumes, it is necessary to prevent the concrete from overheating via the generating heat of hydration. Thus, there is a need to choose the correct type of cement (cements of higher strength class generate more heat of hydration, and cements with the R designation, which stands for high initial strength, also generate more heat of hydration than cements with the N designation—normal initial strength). It is also necessary to treat hardened concrete well, especially after high-temperature concreting. Fillers of such concretes include steel slugs, synthetically produced corundum or natural aggregates: baryte, steel ore–magnetite, hematite, etc. Heavy-weight aggregates complicate the mixing of fresh concrete, for a mixer absorbs a lower volume of filling because of the heavy weight and worse homogenization). For these very same reasons, it is possible to transport lower volumes of concrete compared to standard concrete. In addition, it is necessary to check that there is no dusting of heavy aggregate grains during concreting [3].

3.1. Aggregates

A heavy-weight aggregate can be natural (rocks) or industrial. There are several general aggregate requirements:

- Guaranteed grain size distribution, density, content of crystalline water
- Defined chemical composition
- An aggregate must be of such character as to not threaten the strength and compactness of concrete
- During storage and mixing an aggregate, there should not be excessive abrasion
- The surface of an aggregate shall not reduce the bond strength of mortar and concrete
- An aggregate shall not contain substances that can degrade concrete or steel

3.1.1. Widely used aggregates. Baryte contains more than 85% $BaSO_4$. Its density is within 4000 and 4300 kg.m^{-3}. The compressive strength is about 48 MPa in average. It is not very resistant to abrasion, and it is mainly used for massive constructions that need to eliminate the impact of radiation.
Magnetite, an iron ore, contains up to 65% Fe. Its density is about 4600 and 4800 kg.m^{-3}. The compressive strength can be up to 220 MPa.
Limonite contains α-FeOOH a Fe_2O_3 and 6 to 11% crystalline water. Its density is within 3400 to 3650 kg.m^{-3}. The compressive strength can be up to 300 MPa.
For steel and cast iron, it is possible to use steel cuttings, waste from machining or cast iron slugs. Its advantage is its especially high density (about 7850 kg.m-3).

Corundum, with a chemical formula of Al_2O_3, can be natural or synthetic. With a density of 3600 up to 4300 kg.m^{-3}, it is extremely hard (9 on the Mohs scale). Corundum is used for concretes subjected to abrasion, e.g. floors.
According specific requirements, combinations of aggregates can also be used [1], [3].

3.2. Colour pigments
When using heavy-weight concrete for the elimination of radiation or for protection against dynamic effects, high quality is not expected, compared to architectural concrete. When using heavy-weight concrete for, e.g., weight, certain appearance requirements can arise. One such requirements can be a colour different from standard concrete; in this case, it is possible to use colour pigments. The following requirements exist for inorganic pigments:

- Colour stability in contact with concrete and weathering
- Minimal impact on concrete strength, concrete setting and hardening
- Thermal stability
- Good covering ability ensured by proper granulometry and limited aggregation of particles
- Minimal content of soluble salts (SO_4^{2-}, Cl^-, SiO_3^{2-})
- Limited content of oxides devaluing the colour (SiO_2, Al_2O_3)
- Good dispersion of particles with the size of 0.1 μm to 0.2 μm.

Pigment particles tend to gather—flocculate, thereby decreasing the colouration. The intensity of colouration logically depends on a pigment dose. A common maximal effective dose is about 6 to 9% of the cement weight [4].

4. Fine-grained Heavy-weight Concrete
For using heavy-weight concrete for other than shielding reasons, the composition requirements are quite different from standard heavy-weight concretes. The aim of this research is to suggest and test fine-grained, heavy-weight concrete for weights production. Fine-grained concrete was chosen because of its comparatively small dimensions of weights as well as the availability of narrow fractions of a heavy-weight aggregate. An example of standard density, fine-grained concrete is the production of floor screed, where fine-grained concrete recipes are used. Requirements for final product and workability differ from those of a heavy-weight aggregate.

4.1. General requirements for fine-grained, heavy-weight concrete:

- As low a water factor as possible, but sufficient for good casting (the higher the factor, the more significant the segregation will be)
- Using a superplasticizer with the best effect
- Higher amounts of cement in relation to the greater specific surface of fine aggregate
- Minimal content of air both in hardened and fresh concrete (the content of air in concrete is the main factor reducing its density)
- The best possible compacting
- If a high-quality surface is required, there are considerations of quality moulds, release agents, sufficient amounts of fine proportions of rigid particles, and the use of suitable water (e.g. unrecycled)

5. Experimental Set-up
The design of a concrete recipe and the test of concrete properties consisted of these steps:

1. Selection of a heavy-weight aggregate
2. Optimisation of fine fractions and water amount to reach the maximum density
3. Setting parameters of the concrete-compacting process
4. Verification of selected superplasticizers with simultaneous effects of set accelerators
5. Selection and testing of suitable releasing agents to achieve a smooth, poreless surface

6. Experimental verification of concrete parameters—defining the density of fresh and hardened concretes and defining compressive strength after 12, 24 hours and 28 days.

5.1. Selection of a heavy-weight aggregate and test of its properties
For the production of heavy-weight concrete, magnetite was chosen, with Sweden as the country of origin. For the determination of particle size distribution, the sieving method was carried out according to EN 961-1 [5]. Results are stated in Fig. 1. Apparent particle density was defined according to EN 1097-6 [6]. Apparent particle density ρ_a was 5.07 Mg/m^3.

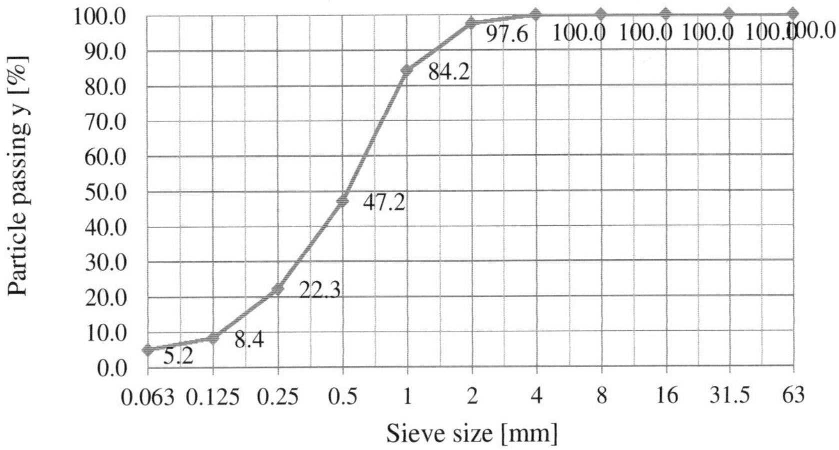

Figure 1. Particle size distribution curve.

5.2. Optimisation of fine fractions and water amount to reach the maximum density
A probationary recipe of concrete was designed. The recipe was further adjusted, and it was necessary to find an optimal amount of fine proportions and water amount to secure the maximal density of concrete.

Table 1. The basic recipe of heavy-weight concrete.

Concrete components	Amount per 1 m^3 (kg)
Cement CEM I 52.5 R	530
Crush magnetite sand 0-2 mm, Sweden	3340
Water	144
Superplasticizer based on modified polycarboxylic ether (PCE)	5.3 (1 % z m_c)
Liquid accelerating admixture (does not contain any added chloride ions)	21.2 (4 % z m_c)
Pigment – solution (shade of red)	10.2

Properties of superplasticizer: Homogenous liquid of orange or light-brown colour, specific gravity at 20 °C 1.06 g.cm^{-3}, pH 4-7, CL-ion content not more than 0.1 %wt.

Properties of accelerating admixture: Colourless free-flowing liquid, relative density 1.26 g.cm^{-3} at 25°C, pH ≥ 6.0 at 25°C, chloride ion content < 0.2%.

The optimal dose of fine proportions and water was specified experimentally. The components of concretes were homogenised in a mixer with forced circulation. For compacting, the researcher used a vibrating table with a frequency converter. Because of magnetism of the used aggregate, a vibrating table with electromagnetic clamping of moulds was not used for compacting. Concrete was filled into polymer moulds.

At first, the desired amount of water was determined under constant frequency, as it was crucial to find the minimal amount of water needed at the highest possible density of concrete. The goal was to reach the maximal compactness. Several probationary batches with different doses of water per 1 m^3 of concrete (140, 145, 150, 155, 160, 165 and 170 kg per 1 m^3) were made, and 150 kg of water per 1 m3 of concrete was defined as an optimal dose. With either a lower or higher dose of water, the weight of a mould was lower; therefore, the weight of fresh concrete was also lower. In this phase, 40 $kg.m^{-3}$ of stone dust was added to the mixture to achieve a high-quality surface for the concrete weight. Experimentally, it was specified that an amount higher than 40 $kg.m^{-3}$ increases the amount of water needed for proper workability. However, the higher amount of water was inadmissible because it can lead to a worsening of the concrete surface thanks to increased porosity, thus decreasing density. To ensure as fast a demoulding as possible, a hardening accelerator was used.

Table 2. The optimised recipe of heavy-weight concrete

Concrete components	Amount per 1 m^3 (kg)
Cement CEM I 52.5 R	530
Crush magnetite sand 0-2 mm, Sweden	3340
Filer amphibolite	40
Water	150
Superplasticizer based on modified polycarboxylic ether (PCE)	5.3 (1 % z m_c)
Liquid accelerating admixture (does not contain any added chloride ions)	21.2 (4 % z m_c)
Pigment – solution (shade of red)	10.2

5.3. Setting parameters of the concrete-compacting process

The compacting of concrete was done on a vibrating table. After defining the amount of water and fine proportions, it was necessary to define the optimal vibrating frequency, during which the best compacting can be reached. After previous experiences with the compaction of fine-grained, heavy-weight concrete, it was obvious that the ideal frequency would be between 67 Hz to 78 Hz (determined according to engine speed, monitored via an indicator on the vibrating table). During each experiment, a vibrating table was running at a constant frequency. The engine speed of a vibrating table was changed within each experiment (4000, 4300, 4500 and 4700 rpm). The definition of an ideal frequency was made according the maximal density of fresh concrete, calculated from the ratio of density of fresh concrete and the volume of a mould. A vibrating frequency of 75 Hz (4500 rpm) was selected as an optimal frequency [7].

Figure 2. Heavyweight concrete fishing weights.

5.4. Experimental verification of concrete parameters

The properties of fresh and hardened heavy-weight concrete were specified. Results for compressive strength and density of concrete are shown in Fig. 3 and Fig. 4.

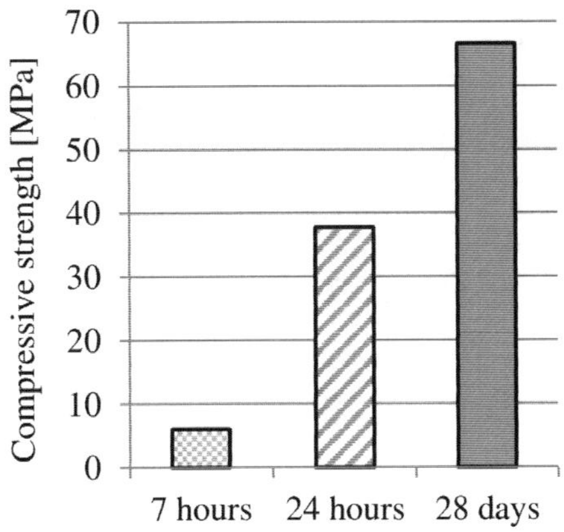

Figure 3. Compressive strength of heavy-weight concrete after 7 hours, 24 hours and 28 days.

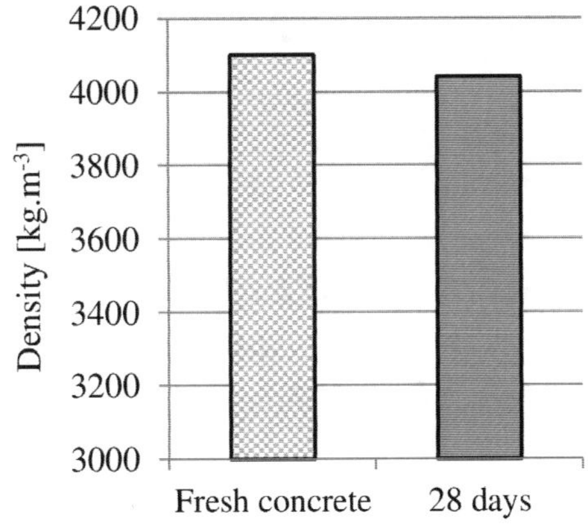

Figure 4. Density of fresh and hardened heavy-weight concrete.

6. Conclusion

The goal of this experiment was the design and verification of heavy-weight concrete properties for the production of weights. Weights made of fine-grained, heavy-weight concrete can be used as a substitute for harmful lead fishing sinkers. The recipe design of heavy-weight concrete was made to reach the maximal compactness and density of concrete. Recipes of heavy-weight concretes from fine aggregates up to 2 mm were designed. Concretes were designed with consistency for casting with requirements of a perfect smooth surface and the possibility of coloration. The water dose for reaching the maximal compactness was experimentally specified. When using an accelerator, the compressive strength after seven hours was 6 MPa, allowing for the quick demoulding of products. The optimal compacting frequency was specified as 75 Hz. Thanks to the reduction of air content in concrete, its density increased. After 28 days, the density of the dried concrete was 4000 kg.m^{-3} with use of the magnetite aggregate with an apparent particle density of 5.07 Mg/m^3.

Acknowledgement

FV10680 Development of prefabricated components from HWC and HPC for special applications in precision engineering and radiation protection.

References

[1] EN 206:2013+A1:2016 Concrete - Part 1: Specification, performance, production and conformity
[2] Scheuhammer AM, Money SL, Kirk DA, Donaldson G 2003 Lead fishing sinkers and jigs in Canada: Review of their use patterns and toxic impacts on wildlife *Occasional Paper of Canadian Wildlife Service* **108** 1-45
[3] Topçu I B 2003 Properties of heavyweight concrete produced with barite *Cement and Concrete Research* **33** 815-822
[4] Kotatkova J and Reiterman P 2014 Coloured Concrete with Focus on the Properties of Pigments *Advanced Materials Research* vol 1054 ed P Reiterman pp 248-+
[5] EN 961-1:2012 Tests for geometrical properties of aggregates - Part 1: Determination of particle size distribution - Sieving method
[6] EN 1097-6:2014 Tests for mechanical and physical properties of aggregates - Part 6: Determination of particle density and water absorption
[7] Jobanek J 2017 Development of mix design for heavy concrete weights *Bachelor thesis* supervisior R Hela Brno (Brno, University of Technology Faculty of Civil Engineering) p 60

IOP Conf. Series: Materials Science and Engineering **800** (2020) 012023 doi:10.1088/1757-899X/800/1/012023

Evaluation of carbonation depth evolution tendencies of reinforced concrete buildings located in coastal and inland areas of north cyprus

Q Al Haj Houseen and P Akpınar*

Near East University, Civil Engineering Department, Lefkosa, Mersin 10, Turkey

*pinar.akpinar@neu.edu.tr

Abstract. Carbonation is a critical concrete durability problem that is known to have potential adverse effects both on the concrete cover and steel reinforcing bars. This study involves detailed experimental investigations on the carbonation progress status of 149 samples from 39 existing buildings in North Cyprus that are located in coastal and inland areas. It is known that the varying exposure conditions in coastal and inland areas are likely to affect the durability performance of concrete buildings in different manners. Concrete core samples that were originally extracted from structures for compressive strength determination, were later tested by using Phenolphthalein indicator in order to determine the extent of carbonation occurred throughout their service duration. Besides studying their current carbonation states, the expected carbonation depth at the end of a 50-years of service life, as well as time needed to carbonate 30 mm-depth concrete cover were also investigated by using the Fick's law. Results showed that 41% of the inland samples have been exhibiting satisfactory carbonation resistance performance according to Eurocodes, where this rate was as high as 71% for the samples obtained from coastal areas of North Cyprus.

1. Introduction

Concrete is one of the most commonly used and dependable construction material all around the world [1], due to its high compressive strength in hardened state and other advantages such its ability to be casted into any desired form while it is in fresh state, its fire resistance and it's affordability in the aspects of the costs of constituents materials as well as the technology required for its manufacture [2,3]. However, concrete's durability is a critical issue to be considered in order to ensure its performance throughout its designed service life. The service lives of structures are defined by codes and regulations to vary between 50 to 100 years according to their designed service functions [4].

Corrosion of steel bars is known as one of the most critical durability threats faced by reinforced concrete structures. Reinforcement corrosion is known to yield damages both on the steel bars and the concrete surrounding them. Hence, in the case of reinforcement corrosion occurrence, costly and elaborated repairing activities or demolition of the structure might become inevitable [5-6]. Carbonation of concrete is a durability problem that is known to be likely to facilitate the initiation of reinforcement corrosion. During carbonation, CO_2 gas from the atmosphere permeates into concrete through its pores and is converted into a weak carbonic acid. Later on, this carbonic acid is known to further permeate in concrete and yield a reduction in the overall pH in the concrete microstructure. In an undamaged concrete, pH is known to be as high as 12 or more, and this state is known to provide a natural protection to embedded steel bars Once the carbonation problem occurs, the naturally high pH value of concrete may drop drastically even below the value of 10, and in this case, steel corrosion may also start besides the ongoing changes in concrete due to continuing carbonation [2,3]. The initial quality of concrete is expected to play a critical role on the rate of carbonation progress in concrete during its lifetime. Concrete having high compressive strength values are known to have higher density and hence lower permeability [2], yielding a lower rate of carbonation progress in concrete. Characteristic compressive strength of concrete is known to be mainly dependent on the mix design; the types and quantities of constituents materials used in the manufacture of concrete [7, 8]. Beside the compressive strength and the initial quality of concrete, environmental conditions such as external CO_2 concentration, relative humidity, and temperature are all known to affect the extent of carbonation problem in concrete [9, 10]

This study aims to evaluate carbonation progress tendency in reinforced concrete structures located in coastal and inland areas of North Cyprus. Coastal and inland areas of Cyprus are expected to yield differences in the carbonation progress tendencies of concrete buildings, mainly due to the varying relative humidity and temperature exposure conditions. The methodology for handling and testing the samples and the obtained results and discussions are presented in the following sections.

2. Methodology

2.1. Acquisition and preparation of concrete samples

The concrete samples used in this study were originally extracted from the buildings by the Cyprus Turkish Chamber of Civil Engineers in North Cyprus, in order to be tested to verify their current compressive strength performance according to EN ?????. These extracted core samples were delivered to Civil Engineering Department Laboratory of Near East University, for additional investigations on their current level and the further tendency of carbonation. Carbonation depth of a total of 149 concrete samples obtained from 39 existing structures have been determined experimentally and the results have been analysed. These 149 samples were taken from buildings located in Nicosia (inland area) and in coastal cities such as Kyrenia and Famagusta. Their ages were reported by the Cyprus Turkish Chamber of Civil Engineers to be varying between 3-50 years.

Samples were first grouped according to their origin (i.e. coming from coastal or inland areas) and then they were split vertically into two halves carefully, regarding to minimize any excess damage causing inacceptable irregularity on the inner surface where the phenolphthalein solution was to be applied for carbonation depth measurements.

2.2. Carbonation depth measurements by phenolphthalein indicator solution

Phenolphthalein solution is an pH indicator solution widely used in concrete carbonation depth determination studies. Phenolphthalein solution used in this study was prepared by following the RILEM recommendations that are described in the related literature [11] where 1 g of phenolphthalein powder was added into and dissolved in 70 ml of Ethyl alcohol, and then the solution was completed to 100 ml by addition of distilled water.

As a next step, phenolphthalein solution was sprayed evenly on the inner surface of split concrete core samples. When phenolphthalein is applied, the areas with high pH, as in the natural state of concrete, turn in to pink color; where the areas that have pH lower than 10 show no color change, indicating the occurrence of carbonation in concrete [12] , as can be seen in Figure 1.

IOP Conf. Series: Materials Science and Engineering **800** (2020) 012023 doi:10.1088/1757-899X/800/1/012023

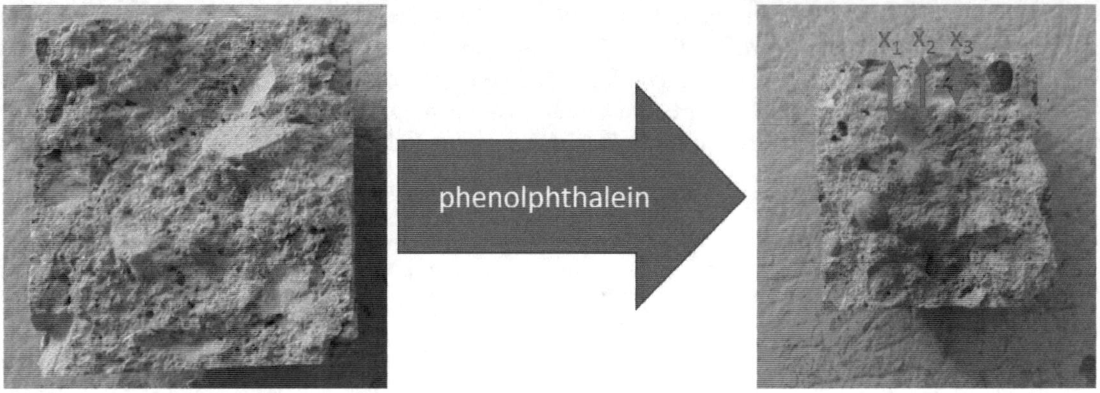

Figure 1. Color changes observed after applying phenolphthalein solution on the sample and taking an average depth measurement.

After the color changes could be observed within few seconds, the carbonated area's (i.e. remaining in grey color) depth was measured to the possible highest precision with the aid of a calliper, by taking average of three measurement for the sample of concern, as indicated in equation 1 given below:

$$d = \frac{x_1+x_2+x_3}{3} \quad (1)$$

2.3. Carbonation tendency calculations for studied structures

Carbonation is a time-dependent durability of concrete, whose progress in concrete could be expressed by Fick's Law as indicated in the below Equation 2 [13]:

$$k = \frac{D_{carbonation}}{t^{0.5}} \quad (2)$$

Where;

D carbonation = depth of carbonation from the surface of concrete
k = coefficient of carbonation
t = exposure duration

Initially, the coefficient of carbonation (k) is individually calculated for each concrete element provided that its actual age and corresponding carbonation depth is known. Once the coefficient of carbonation (k) is known, then calculation the depth of carbonation at a certain concrete age(i.e. carbonation exposure duration) as shown in equation 3, or calculating the time required for reaching a critical concrete carbonation depth, as shown in equation 4, becomes possible.

$$D_{(carbonation)} = k \times t^{0.5} \quad (3)$$

$$t = \left(\frac{D_{carbonation}}{k}\right)^2 \quad (4)$$

Using equations 3 and 4, the depth of concrete carbonation at the end of 50 years of service life, as well as the time required to reach 30mm of carbonation (i.e. conventional concrete cover thickness in which the steel bars are embedded) have been calculated for each studied structure.

2.4. Carbonation Depth Performance Satisfaction According to Eurocode EN1992-1-1
General Rules and Rules for Buildings specified in EN1992-1-1 [14], defines the safe limit of carbonation as maximum 30mm at the end of a 50-years of service life period for a concrete building. If the carbonation depth progresses up to only this level at the end of the service life of the building, it is expected that the embedded steel bars would not be affected of the resultant reduced alkalinity and therefore, corrosion would not be initiated. Performance of the structures studied in this work has been evaluated considering this criterion of EN1992-1-1 [14].

3. Results and Discussions
149 extracted core samples coming from 39 different existing structures from inland and coastal areas of North Cyprus have been examined by using phenolphthalein indicator solution and based on the observed results, Fick's law equation was employed to study the carbonation depth progress tendencies of the studied structures.

The concrete samples which were observed to have less than 1 mm carbonation, as well as the samples which were observed to be carbonated throughout the entire studied surface were eliminated, since these two findings indicate potential local anomaly regarding the individual specific sample of concern. Figure 2. a, b, and c represent the location distribution of structures, and the ratio of employed and eliminated structures at coastal and inland areas.

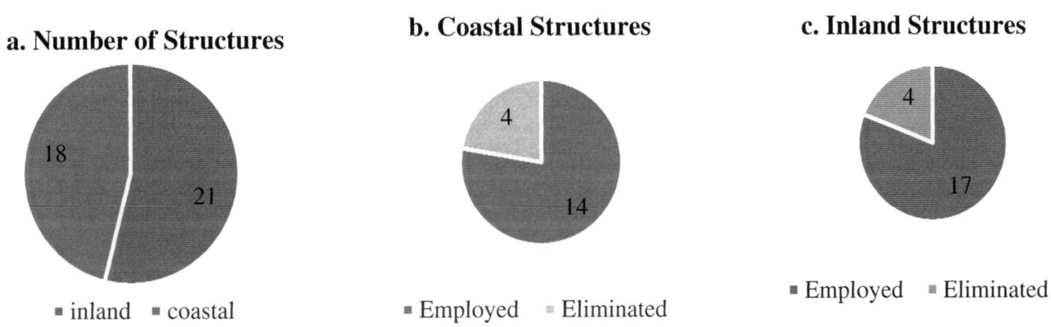

Figure 2. (a) general distribution of inland to coastal structures (b) proportional of eliminated structures within inland (c) proportional of eliminated structures within coastal.

Table 1 and 2 provide detailed results on the carbonation status and progress tendencies of studied structures located in inland and coastal areas of North Cyprus, respectively. The eliminated results for inland and four coastal structures have been reported in Tables 1 and 2 as grey-highlighted.

When the carbonation coefficient (k) values estimated by Fick's law for inland structures that are shown in Table 1 are studied, it is observed that the k values are varying between 1.16 and 14.67 mm/year$^{0.5}$. When the values of k reported in Table 1 was further observed, it can be seen that these significantly varying k values do not necessarily follow any structure age relation neither. For instance, the youngest inland structure, which is represented with structure no-2 that is only four years old was observed to have the highest reported carbonation coefficient, where other inland structures 19 and 20 that are both forty-two years-old are observed to have very varying k values of 1.77 ad 7.10 mm/year$^{0.5}$, respectively. This finding may potentially be due to the variance of initial concrete quality and characteristic strength of each structure that could not be measured or reported in this study, due to the limitations in acquiring satisfactory information on the exact initial state of buildings. It is expected that if the characteristic strength of concrete was low and initial quality of concrete was relatively poor in the sense of having higher permeability, such a building would experience faster progress of carbonation even though its age could be young.

Table 1. Carbonation depth and compressive strength information for inland structures.

Structure No.	Age of Sample (years)	Carbonation Depth (mm)	Carbonation Coefficient; K (mm/year$^{0.5}$)	Estimated Carbonation Depth at 50 Years Age (mm)	Estimated Carbonation Time for 30 mm Cover (years)
1	44	42.00	6.33	*44.77**	22.45
2	4	29.33	14.67	*103.71**	4.18
3	46	45.50	6.71	*47.44**	20.00
4	28	31.50	5.95	*42.09**	25.40
5$^+$	16	-	-	-	-
6	44	26.20	3.95	27.93	57.69
7	44	26.13	3.94	27.86	57.98
8	8	5.23	1.85	13.08	262.89
9	33	41.23	7.18	*50.75**	17.47
10	50	22.27	3.15	22.27	90.71
11	10	38.60	12.21	*86.31**	6.04
12$^+$	44	-	-	-	-
13	18	41.47	9.77	*69.11 b*	9.42
14$^+$	28	-	-	-	-
15	49	65.43	9.35	*66.09**	10.30
16	8	3.28	1.16	8.21	667.89
17	37	19.40	3.19	22.55	88.48
18	11	16.73	5.05	*35.68**	35.36
19	42	11.46	1.77	12.50	287.96
20	42	46.03	7.10	*50.23**	17.84
21$^+$	43	-	-	-	-

$^+$ *Grey-highlighting refers to the case of eliminated structures due to local anomalous measurements.*
* *Asterisks refers to the case of samples that do not satisfy the carbonation criterion in EN1992-1-1.*

Besides the carbonation coefficient, the depth of carbonation expected to be experienced at the end of 50 years, as well as the time to carbonate 30 mm concrete cover thickness were also determined and presented in Tables 1 and 2. It was observed that 41% of samples obtained from inland structures would not exceed 30 mm in 50 years of service life; therefore, these could be classified as conforming the acceptable carbonation level criterion defined in the related European Norms.

When the carbonation coefficient values determined for coastal structures that are presented in Table 2 are studied, it was observed that k values are varying between 0.69 and 6.97 mm/year$^{0.5}$. Similar, to what is observed in Table 1 for inland structures, the k value showed no detectable relation with respect to the age of the coastal structures, as it could be clearly seen in the cases of structures 22-26 that are all four years old but having varying k values. Similarly, this finding is expected to be due to the differences in the initial qualities of concrete buildings. Since exact data regarding the initial states of the structures could not be obtained in this study, further experimental investigations that could also control such missing parameters, are necessary in order to be able draw more solid conclusions on this matter.

Table 2. Carbonation depth and compressive strength information for coastal structures.

Structure No.	Age of Sample (years)	Carbonation Depth (mm)	Carbonation Coefficient; K ($mm/year^{0.5}$)	Estimated Carbonation Depth at 50 Years Age (mm)	Estimated Carbonation Time for 30 mm Cover (years)
22	4	3.50	1.75	12.37	293.88
23	4	7.37	3.68	26.05	66.34
24	4	1.38	0.69	4.88	1890.36
25	4	2.25	1.13	7.95	711.11
26	4	6.17	3.08	21.80	94.67
27	44	33.83	5.10	*36.07* *	34.59
28	44	46.27	6.97	*49.32* *	18.50
29	3	10.58	6.11	*43.19* *	24.12
30	14	8.47	2.26	16.00	175.77
31	22	25.53	5.44	*38.49* *	30.37
32	4	1.73	0.87	6.13	1198.22
33+	46	72.70	-	-	-
34+	23	73.00	-	-	-
35	44	18.57	2.80	19.79	114.88
36+	3	0.00	-	-	-
37+	26	0.00	-	-	-
38	44	25.32	3.82	26.99	61.78
39	44	6.92	1.04	7.38	826.96

+ *Grey-highlighting refers to the case of eliminated structures due to local anomalous measurements.*
* *Asterisks refers to the case of samples that do not satisfy the carbonation criterion in EN1992-1-1*

When the values for the depth of carbonation expected to be experienced at the end of 50 years for coastal structures that are presented in Table 2 were examined, it was observed that 71% of studied coastal samples would not exceed 30 mm in 50 years of service life; hence, these samples are conforming the acceptable carbonation level criterion defined in the related European Norms[14].

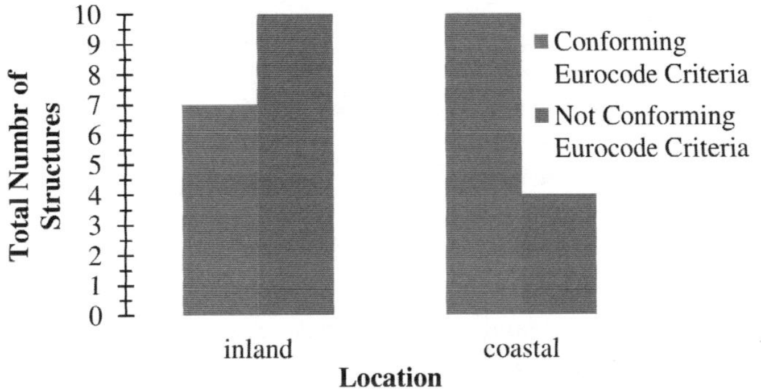

Figure 3. Number of inland and coastal structures conforming and non-conforming EN1992-1-1 [14]. criterion for acceptable concrete carbonation level.

The exact number of structures located in inland and coastal areas of North Cyprus that conform EN1992-1-1criterion for carbonation are illustrated in Figure 3.

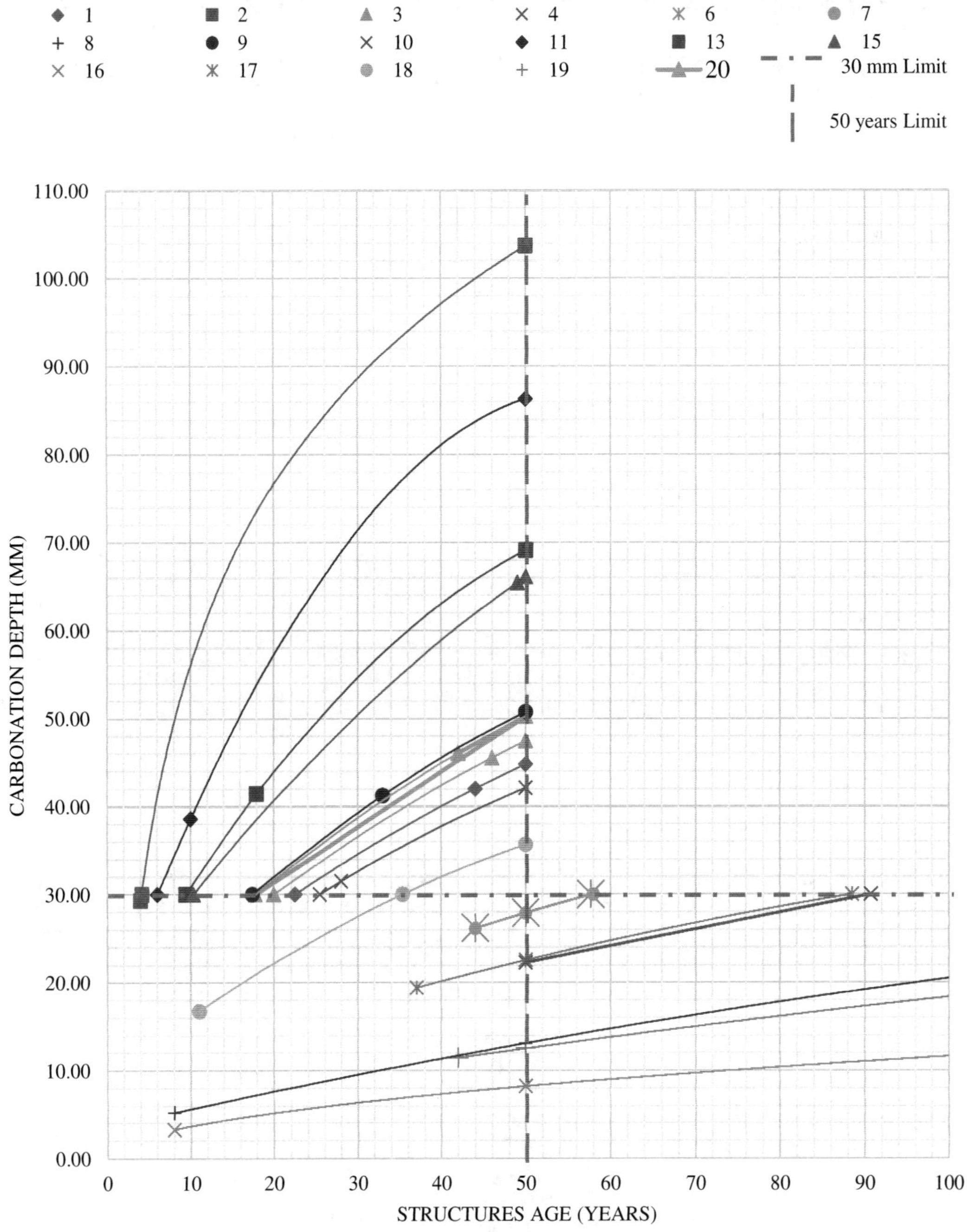

Figure 4. Carbonation depth for inland structures at current age, at 50 age years and the time required to reach 30 mm of carbonation depth.

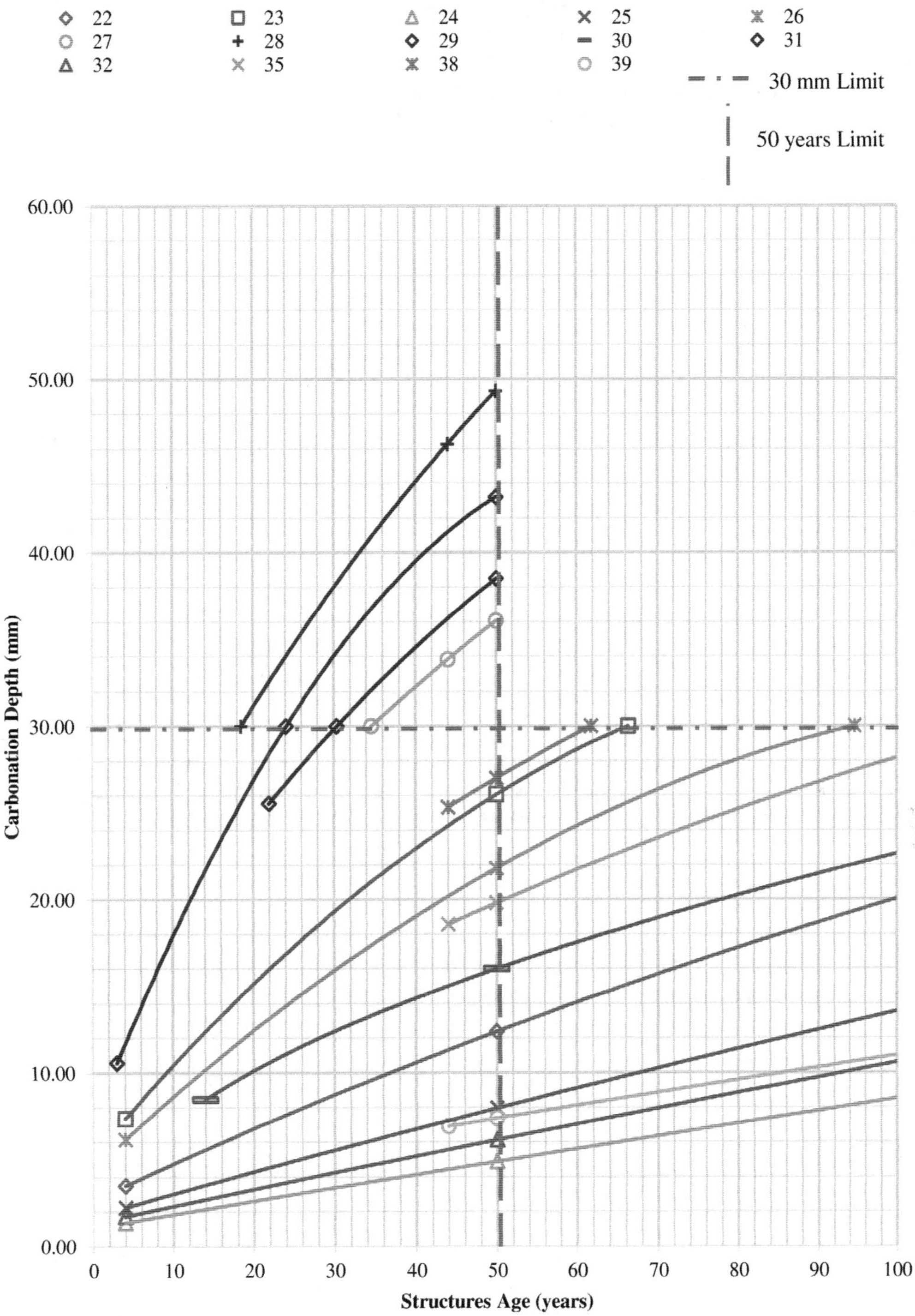

Figure 5. Carbonation depth for coastal structures at current age, at 50 years age and the time required to reach 30 mm of carbonation depth

IOP Conf. Series: Materials Science and Engineering **800** (2020) 012023 doi:10.1088/1757-899X/800/1/012023

Figures 4 and 5 illustrates collectively the estimations presented in Tables 1 and 2 for inland and coastal structures. The numbers given in the legends of figures 4 and 5 represent the structure numbers that were previously defined for inland and coastal structures. For an individual structure; the first point indicates its current age and corresponding carbonation depth; where the following points are the time to reach to a 30 mm carbonation depth and the total carbonation depth to be experienced at the end of 50 years of service life. The carbonation progress tendency lines connecting these tree points for each building was observed to be of a different characteristics (e.g. linear or parabolic) according to the determined coefficient of carbonation in both graphs.

Figures 4 and 5 serve to illustrate more clearly the differences in the actual carbonation depths observed for the current ages of the structures (i.e. the first data point for each structure) in the cases where the structures of same age located in similar conditions (e.g. all coastal or all inland). Besides the possible variations in the initial concrete quality being a potential reason for this observed carbonation progress performance difference, it should be kept in mind that in some cases the exact location of the concrete element from which the test cores were extracted could be influential. It is known that even within a specific building, the carbonation progress observed could vary if the element to be tested is an internal or an external element. Also, for external concrete elements, their state of being sheltered or unsheltered is known to yield different levels of carbonation progress in concrete [2]. Lack of these specific information on the samples acquired is observed to cause difficulties in clear interpretation of the observed carbonation progress tendencies of the structures.

Nevertheless, the fact that the carbonation coefficient values for coastal structures are significantly lower in general than the values of inland structures, as well as the fact that 71% of the coastal structures meet the acceptable concrete carbonation level in European Norms, where only 41% of inland structures could meet the same criteria, indicate that coastal structures are less prone to carbonation attack than inland structures in North Cyprus. Similar findings indicating that the coastal structures experience less severe carbonation occurrence are also reported by Alexander et al. [15], in which the carbonation depth progress of different localities of South Africa was investigated. This finding might be due to the effect of relatively higher relative humidity in the coastal areas that could fill the concrete porosity yielding a slower progress of carbonation [2]. However, the exact factors affecting the ultimately observed carbonation depths in studied structures should be further verified with additional experimental studies.

4. Conclusive remarks

The carbonation progress tendency of concrete structures located in inland and coastal areas of coastal structures were evaluated by carrying out studies on 149 concrete samples acquired from 39 existing reinforced concrete structures of varying ages. The conclusive remarks attained are as the following:

I. Concrete carbonation coefficient values determined by Fick's law for structures in coastal areas of North Cyprus are observed to be generally lower than the values determined for inland structures; indicating a slower progress of carbonation for coastal structures.

II. 71% of the coastal structures are estimated to yield less than 30 mm of concrete carbonation at the end of a 50 years of service life indicating conformity to EN1992-1-1 criterion; where this value is only 41% for inland structures of North Cyprus. When the results obtained for the studied 39 structures are taken into consideration, it is observed that coastal structures are less prone to concrete carbonation.

III. The observed cases of structures having the same age and being located in the same type of areas yielding varying carbonation depths and carbonation coefficient values are expected to be due to varying initial qualities of concretes, as well as potentially varying localized status of concrete elements tested. Further specific information and further experimental studies are required to verify the exact influencing factors yielding different carbonation tendencies for the buildings located in same areas with similar ages.

Acknowledgments
Authors would like to acknowledge Cyprus Turkish Chamber of Civil Engineers for their support and collaboration throughout this study.

References

[1] Moreno E I 2013 Carbonation Coefficients from Concrete Made with High-Absorption Limestone Aggregate *Advances in Materials Science and Engineering* **2013** 1–4 https://doi.org/10.1155/2013/734031
[2] Neville AM 2005 *Properties of Concrete* (New York: John Wiley & Sons)
[3] Monteiro P J M and Mehta P K 2005 *Concrete: Microstructure, Properties, and Materials-3rd ed.* (USA: McGraw-Hill Professional)
[4] Domone P L J and Illston J M 2010 *Construction materials: their nature and behaviour -4th ed.* (Milton Park, Abingdon, Oxon ; New York: Spon Press)
[5] Zhao H, Sun W, Wu X and Gao B 2017 The effect of the material factors on the concrete resistance against carbonation *KSCE Journal of Civil Engineering* **22(4)** 1265–74 https://doi.org/10.1007/s12205-017-0988-9
[6] Elmoaty A E M A 2018 Four-years carbonation and chloride induced steel corrosion of sulfate-contaminated aggregates concrete *Construction and Building Materials* 163 539–56 https://doi.org/10.1016/j.conbuildmat.2017.12.128
[7] Akpinar P and Khashman A 2017 Intelligent classification system for concrete compressive strength *Procedia Computer Science* **120** 712-8 https://doi.org/10.1016/j.procs.2017.11.300
[8] Khashman A and Akpinar P 2017 Non-destructive prediction of concrete compressive strength using neural networks *Procedia Computer Science* **108** 2358-62 https://doi.org/10.1016/j.procs.2017.05.039
[9] Akpinar P and Uwanuakwa I D 2016 Intelligent prediction of concrete carboration depth using neural networks *Bull Transilv Univ Brasov, Ser III Math Informatics, Phys* **9(2)** 99–108
[10] Malami S I, Akpinar P and Lawan M M 2018 Preliminary investigation of carbonation problem progress in concrete buildings of north Cyprus *MATEC Web Conf.* **20306007** https://doi.org/10.1051/matecconf/201820306007
[11] Neville A M 2006 *Concrete: Neville's Insights and Issues* (Thomas Telford)
[12] Houst Y F and Wittmann F H 2002 Depth profiles of carbonates formed during natural carbonation *Cement and Concrete Research* **32 (12)** 1923-30
[13] Monteiro I, Branco F A, Brito J d and Neves R 2012 Statistical analysis of the carbonation coefficient in open air concrete structures *Construction and Building Materials* 29 263–9 https://doi.org/10.1016/j.conbuildmat.2011.10.028
[14] EN 1992-1-1 2014 *Eurocode 2: Design of concrete structures - Part 1-1 : General rules and rules for buildings* (London: BSI Standards)
[15] Alexander M G, Mackechnie J R and Yam W 2007 Carbonation of concrete bridge structures in three South African localities *Cement and Concrete Composites* **29(10)** 750–9. https://doi.org/10.1016/j.cemconcomp.2007.06.005

IOP Conf. Series: Materials Science and Engineering **800** (2020) 012024 doi:10.1088/1757-899X/800/1/012024

Estimation of compaction characteristics of soils using Atterberg limits

A Hussain[1] and C Atalar[2*]

[1] M Sc., Omar Al-Mukhtar University, Algubba, Libya
[2] Prof.Dr. NEU Earthquake and Soil Research and Evaluation Center, Near East University, Nicosia, North Cyprus

*Email: cavit.atalar@neu.edu.tr

Abstract. Soils are mostly used construction materials. Naturally occurring soils when are used as a construction material their engineering properties need to be determined and they mostly need to be compacted for the improvement of their engineering properties. The determination of these engineering properties becomes a vital process for the successful design of any geotechnical structure. Testing the properties of compaction, optimum water content (OWC), and maximum dry density (MDD), requires high amount of work and consumes a large amount of time in the laboratory. In this paper, an attempt to develop predictive models between liquid limit, and compaction test parameters is made. To achieve this purpose, 8 samples of North Nicosia Kythrea group soils in North Cyprus were subjected to Atterberg limits and compaction laboratory tests. The soils were tested using standard Proctor compaction tests. Stepwise multiple linear regression analyses were carried out on the experimental data and predictive models were developed in terms of liquid limit (LL). Their compaction results were used to validate the proposed models. According to the findings, the R2 values of the models suggested are over 75% and the error variation between the predicted and experimental compaction characteristics values was found to be below ±1. It has been shown that these models will be useful for a preliminary design of earthwork projects which involves Nicosia soils in Cyprus.

1. Introduction

Compaction of soil is the artificial improvement of the mechanical properties of the soil .The soil is densified by the removal of pore spaces and the particles are rearranged. Since the soil particles are closely packed together during this process, the void ratio is reduced thus making it difficult for water or other fluid to flow through the soil. The type of soil and the grain sizes of the soil play a significant role in the compaction process as a reduction in the pore spaces within the soil increases the bulk density. Soils with higher percentages of clay and silt have a lower density than coarse-grained soils since they naturally have more pore spaces. The compaction curve obtained in the laboratory tests or field compaction represents the typical moisture-density curve which explains the compaction characteristics theory.

Soil compaction investigations started during the 20th century due to the automobile invention along with the paved roads. Since then, many efficient and cost effective methods came up; different compaction methods were used for different type of soils. Proctor, a pioneer in soil compaction established this fact in 1933. It was also established that the moisture content affected the degree of compaction for any compaction method used. Soil compaction only affects the volume of air and has no effect on the water content or the size of the solids. Air ratio in the void ratio is removed completely effective during the compaction process, but in practice, it is not. The diminution of pore spaces lead to a rearrangement of the soil particles making them more intense.

The importance of this property is the subject of a good estimate in the construction of earth dams and other projects to fill the ground. And is a vital process and is employed through construction projects such as: highways, and pavements, railways etc. The main goals of soil compaction are: reduction in permeability of the compacted soil, increase in the shear strength of the soil, and to reduce the subsequent settlement of the soil mass under working loads.

In the laboratory, soil compaction is conducted using Procter compaction device test. And in the field are compaction of soil by different equipment with different compression energy. The characteristics of the compaction test of an optimum water content (OWC), and maximum density (ρdmax), which is used to determine the shear strength and bearing capacity of the natural soil. Proctor test which is a laborious and time consuming method. Therefore, there is a need of correlating the compaction characteristics with consistency limits as they are easy to determine. Plastic limit bears a good correlation with OMC than any other consistency limits.

2. Previous work

Alluvial soils and overconsolidated (OC) clays constitute the most soils of Cyprus. Alluvial soils Holocene to recent in age containing gravel, sand, silt and clay are widespread in the Mesaoria plain, especially at Nicosia, Famagusta, and Limassol and at the east and west coast. They comprise loose - medium dense gravel and sand, and soft - firm silt and clays. The alluviums mostly contain low amounts of clay size material. The alluviums also contain high amount of montmorillonite. These alluviums show relatively high apparent strength in their dry state. However, with saturation their strength decreases. These clayey soils have low to medium swelling potential. The silting of the ancient harbours created very soft soils especially at the east and west coastline of the Mesaoria plain. Old stream beds filled with alluviums are found at the coastline and at the interior [1].

The Kythrea group mostly contains turbiditic rocks. It consists, from bottom to top, gravel, conglomerates, greywacke, marl, and mostly abyssal turbidites with a shallow environmental chalk, marl, limestone, and gypsum finishing. Alternation of sandstone-siltstone-marl-claystone is widespread within it. The group is only observed in North Cyprus and has a complete coverage of the northern and southern slopes of Kyrenia range from east to west. It consists of Mia Milia, Skylloura, Lapatza (Pre-evaporitic) and Lapatza (evaporitic) formations. The clayey units of several m thickness to tens of meters thickness in the different formations of the group exhibit different swelling potential. Pre-evaporitic and Skylloura formations exhibit high to very high, Mia Milia intermediate to high swelling potential [2].

Many researchers have made attempts to predict compaction test parameters from several factors such as soil classification data, index properties, and grain size distribution. Gurtug and Sridharan studied the compaction behaviour and prediction of its characteristics of three cohesive soils taken from the Turkish Republic of Northern Cyprus and other two clayey minerals based on four compaction energy namely, standard Proctor, modified Proctor, Reduced standard Proctor and Reduced modified Proctor to develop relationship between maximum dry unit weight and optimum water content and plastic limit with particular reference to the compaction energy [3].

They proposed the Equations 1 and 2 below.

$$OMC = 0.92PL \quad (1)$$

$$MDD = 21.61 - 0.26\,OMC \quad (2)$$

Recently, Sridharan and Nagaraj conducted a study of five pairs of soils with nearly the same liquid limit but different plasticity index among the pair and made an attempt to predict optimum moisture content and maximum dry density from plastic limit of the soils. They developed with the following Equations 3 and 4 [4].

$$OMC = 0.92PL \tag{3}$$
$$MDD = 21.46 - 0.23PL \tag{4}$$

Sivrikaya et al correlated maximum dry unit weight and optimum water content of 130 fine-grained soils using standard Proctor compaction test with a plastic limit. They developed the following Equations 5 and 6 [5].

$$OMC = 0.94PL \tag{5}$$
$$MDD = 21.97 - 0.27\,OMC \tag{6}$$

Sivrikaya's work as summarized by Sivrikaya and Soycan showed the following correlations using standard Proctor data with index properties of fine-grained soils [6].

$$OMC = 0.92PL \tag{7}$$
$$MDD = 20.90 - 0.21\,PL \tag{8}$$
$$MDD = 21.84 - 0.27\,OMC \tag{9}$$

Most recently, Jyothirmayi et al used nine types of fine-grained soils like black cotton soil, red clay, china clay, marine clay, silty clay etc. which were taken from different parts of Telengana and Andhra Pradeshin, India to propose a correlation Equation 10 using plastic limit (PL) in order to determine the compaction characteristics namely, optimum water content (OMC) of these soils[7].

$$OMC=12.001e^{0.0181PL} \tag{10}$$

3. Methodology

The soil samples were obtained from depths of between 1 to 10 metres from the North Nicosia soils. In total, 8 samples were collected and they were subjected to particle size analysis test, Atterberg limit tests, and compaction tests. The tests were performed in accordance to American Society for Testing and Materials (ASTM) standard specifications to determine the physical and compaction properties of the soils.

Mechanical sieve analyses were performed on each soil sample according to ASTM D6913-04 to determine the grain size distribution. Sieve analysis was conducted using U.S. Sieve sizes; #40, #60, #100, and #200. A sample of the soil was dried in the oven at a temperature of 105oC - 110oC for overnight. The whole specimen sample was allowed to cool and the test was done [8].

The Atterberg limits (plastic and liquid limit) were determined on all the 8 samples using distilled water as the wetting agent. The liquid limit test was done on the soil fraction passing through the U.S. No. 40 (0.425mm) sieve in accordance with ASTM D4318-10. This method involves finding the moisture content at which the groove cut in the wet sample with a standard grooving tool closes In accordance with ASTM D4318-10 procedure, the plastic limits were determined on the soil fraction passing the U.S. No. 40 sieve. This method involves finding the moisture content at which the wet soil just begins to crumble or break apart when rolled by hand, into threads of 3mm diameter [9].

Standard Proctor tests were conducted manually on the soil samples. Standard Proctor test was performed on 8 soil samples. This was used to determine the maximum dry unit weight and optimum moisture content of the soil. Compaction of the soil was done using the mechanical energy obtained from an impacting hammer. The mechanical energy is a function of hammer weight, height of the hammer drop, the number of soil layers, and number of blows per layer. The parameters of the standard Proctor tests in accordance to ASTM D 698-12 [10].

Stepwise multiple linear regression analyses were carried out on the experimental data and predictive models were developed in terms of liquid limit (LL). Their compaction results were used to validate the proposed models.

4. Results and discussions

The test results of sand, silt, clay contents, Atterberg limits and compaction characteristics are given in the Table1.

Table1. Laboratory test results for regression analysis of standard Proctor compaction test.

TEST NO	SAND	SILT	CLAY	LL	PL	PI	OMC%	MDD (KN/m3)	Soil Type
92	48.02	14.98	37	48.8	18.5	30.3	17.5	17.26	CL
93	30	16	54	60	22.6	37.4	19	16.42	CH
94	17.73	20.27	62	76.8	25.7	51.1	22.5	15.76	CH
95	54.32	12.68	33	48.5	19.5	29	18	17.76	CL
97	67.75	7.25	25	43.2	20.3	22.9	15.5	19.25	OL
98	21.35	19.65	59	63	21.7	41.3	19	16.36	CH
99	21.35	28.65	50	49.2	18	31.2	18	17.82	CL
100	22.55	19.45	58	65.7	23.9	41.8	21	15.63	CH

Graphs are plotted to correlate optimum moisture content with liquid limit and maximum dry density with optimum moisture content as shown in Figure 1 and Figure 2 respectively.

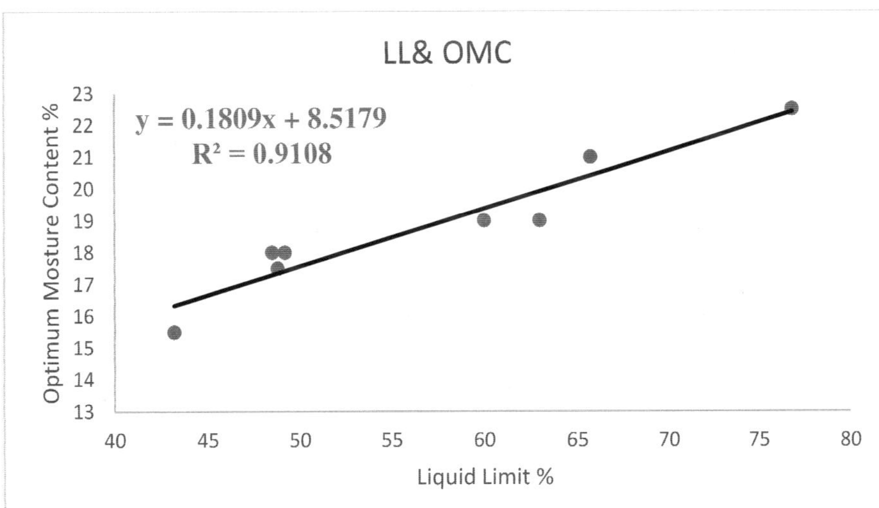

Figure 1. Scatter plot and best-fit curve of liquid limit and OMC.

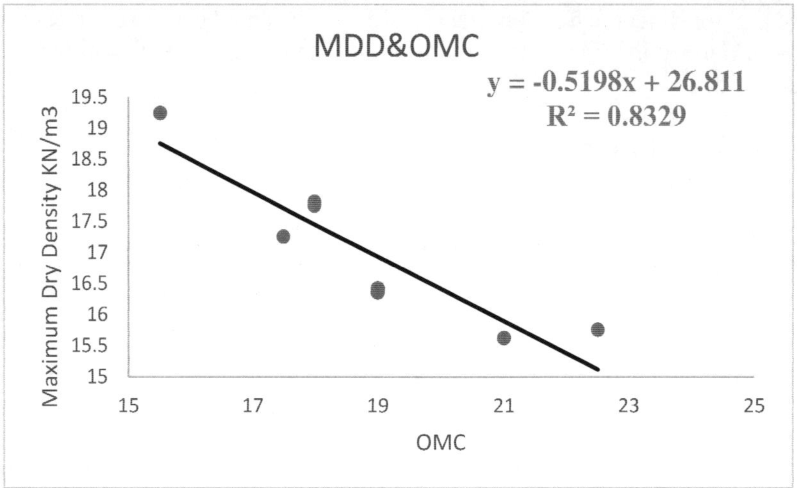

Figure 2. Scatter plot and best-fit curve for OMC and MDD

The equation obtained by the relation is,

$$OMC = 0.1809\ LL + 8.5179 \quad (11)$$

The equation obtained between the OMC with MDD is,

$$MDD = 26.811 - 0.5198\ OMC \quad (12)$$

The measured and predicted OMC and MDD are shown in Table 2.

Table 2. Validation of standard Proctor compaction parameters models.

Optimum Water Content OMC%			Maximum DryUnitWeight MDD(KN/M3)		
Measured	Predicted	Abs. Error	Measured	Predicted	Abs. Error
17.5	17.3458	0.15418	17.26	17.7145	0.4545
19	19.3719	0.3719	16.42	16.9348	0.5148
22.5	22.411	-0.089	15.76	15.1155	0.6445
18	17.2916	0.70845	17.76	17.4546	0.3054
15.5	16.3328	0.83278	19.25	18.7541	0.4959
19	19.9146	0.9146	16.36	16.9348	0.5748
18	17.4182	0.58182	17.82	17.4546	0.3654
21	20.403	0.59697	15.63	15.8952	0.2652

5. Conclusion

Based on the analysis of laboratory data, it is obtained that the correlation between optimum moisture content with liquid limit and maximum dry density with optimum moisture content for fine grained soils. The correlations based on the investigation in this study are given below:

$$OMC = 0.1809\ LL + 8.5179 \qquad R^2 = 0.911$$

$$MDD = 26.811 - 0.5198\ OMC \qquad R^2 = 0.833$$

The results showed that these proposed models had R^2 values greater than 75% and the variation of error between the experimental and the predicted values of compaction characteristics was less than ±1.

It has been shown that these models will be useful for a preliminary design of earthwork projects which involves Nicosia soils in Cyprus.

References

[1] Atalar C and Das B M 2018 *Proc. Int. Conf. on XVI Danube - European Conference on Geotechnical Engineering* (Skopje: Macedonia) 6 pages

[2]Atalar C and Das B M 2009 *Proc. 2nd Int. Con. on New Developments in Soil Mechanics and Geotechnical Engineering*, Nicosia, North Cyprus, pp 360-367

[3] Gurtug Y and Sridharan A 2004 Compaction Behavior and Prediction of its Characteristics of Fine-grained Soils with Particular Reference to Compaction Energy. *Journal of Soil and Foundation*, Japanese Geotechnical Society **5**, pp 27-36

[4] Sridharan A and Nagaraj H B 2005 Plastic limit and compaction characteristics of fine-grained soils. Journal of Ground Improvement, *Indian Institute of Science* pp 17-22

[5] Sivrikaya O, Togrol E and Kayadelen C 2008 Estimating compaction behavior of fine-grained soils based on compaction energy, *Canadian Geotechnical Journal* pp 877-887

[6] Sivrikaya O and Soycan YT 2009 Estimation of compaction parameters of fine-grained soils using Artificial neural networks. *Proc. 2nd international conference on new developments in soil mechanics and geotechnical engineering*. Near East University, Nicosia, North Cyprus pp 406-412

[7] Jyothirmayi K, Gnanananda T, and Suresh K 2015 Prediction of Compaction Characteristics of Soil using Plastic Limit. *International Journal of Research in Engineering and Technology*. Volume: **04** Issue: 06, pp 253-256

[8] ASTM International. 2009 ASTM D6913-04(2009) e1, Standard Test Methods for Particle-Size Distribution (Gradation) of Soils Using Sieve Analysis, West Conshohocken, PA. USA

[9] ASTM D4318-10. 2010 Standard Test Methods for Liquid Limit, Plastic Limit, and Plasticity Index of Soils, Annual Book of ASTM Standards, ASTM International, West Conshohocken, PA. USA

[10] ASTM D698-12 2012 Standard Test Methods for Laboratory Compaction Characteristics of Soil Using Standard Effort (12 400 ft-lbf/ft3 (600 kN-m/m3)), ASTM International, West Conshohocken, PA, USA

IOP Conf. Series: Materials Science and Engineering **800** (2020) 012025 doi:10.1088/1757-899X/800/1/012025

Two dimensional determination of peak acceleration distributions during earthquakes

Y İpek[1], M Akgün[2*], Ö C Özdağ[3], T Gönenç[2], E Pamuk[5], H Dindar[1] and C Atalar[4]

[1] M Sc., The Graduate School of Natural and Applied Sciences, Dokuz Eylül University, Buca - İzmir, Turkey
[2] Prof.Dr., Doç.Dr., The Department of Geophysical Engineering, Dokuz Eylül University Engineering Faculty, Buca - İzmir, Turkey
[3]Lecturer, Aegean Implementation and Research Centre, Dokuz Eylül University, Balçova - İzmir, Turkey
[4]Prof.Dr., NEU Earthquake and Soil Research and Evaluation Center, Near East University, Nicosia, North Cyprus
[5] Dr., General Directorate of Mineral Research and Exploration, Çankaya/Ankara, Turkey

*Email: mustafa.akgun@deu.edu.tr

Abstract. In this study, the two dimensional peak acceleration values between the bedrock and the ground surface is examined by using the soil bedrock model formed for a profile in the Bornova Plain in the east of İzmir Bay. For this purpose, two dimensional soil bedrock model was created for profile using microgravity and Extended Spatial Autocorrelation (ESPAC) method results. The S-wave velocities obtained from the ESPAC method were used to determine the density for microgravity modelling. After the density determination process, microgravity data on the profile were obtained by forward Talwani modelling. The obtained model is divided with gridding the data at regular intervals. In the study area, a recorded earthquake data used for deconvolution by soil-bedrock model known strong ground motion station BYN , and then transfer to the interface of bedrock in addition to obtain required bedrock data. Then, the calculated site- specific bedrock earthquake data was transferred from the bedrock surface of the grid model to the soil surface and layer boundaries. By combining these peak acceleration values for the profile, the two dimensional peak acceleration distribution for the profile was calculated.

1. Introduction

Turkey has suffered a great deal with the 1999 Kocaeli earthquake. After this event, the earthquake – soil- structure common interaction has gained more importance. In order to minimize earthquake damages and consciously build structure design, we need to determine the parameters that will be used for earthquake –soil- structure in a healthy way. Some parameters in earthquake engineering change the parameters of earthquake waves as they travel between the layers between the engineering bedrock and the soil surface. The most important of these parameters are the changes in earthquake duration, frequency content, and amplitude. These changes can be defined by using the density and depth of the engineering bedrock and the P and S velocity values, densities and layer thicknesses of the soil forming layers [2].

In order to provide a practical way to design earthquake resistant structures, these changes are defined in the Eurocode 8 regulation for an average depth of 30 m. In some cases, the same regulation uses the on-site design description, especially where the soil is thicker than 30 meters and / or S-velocity values are less than 250- 300 m /s [1]. In some circumstances it will be necessary to calculate the soil transfer function, which will define the joint effect of the layers between the engineering bedrock and the soil surface. The soil transfer function gives us detailed information about a depth the earthquake wave will be affected. The engineering bedrock and soil of the region should be studied. The geological units in the study area of the engineering bedrock should be investigated and defined by in-situ studies to determine the density of this layer and its depth from the soil surface. In order to define the effects of earthquake waves on focusing and repetitive reflections, the topography of the interface between the engineering bedrock and the soil should be investigated in small scale lenticular structures and thin layers that may have high S velocities and density differences in the soil.

Soil dynamic parameters can be determined with higher accuracy by using multidisciplinary methods together. Determining the dynamic parameters of the soil is a necessity before construction. Earthquake-resistant building design is one of the most important tasks in reducing the damages of earthquakes. Determining the soil characteristics of the structure, revealing how the soil and the structure behave during an earthquake, and constructing the structure in accordance with the soil characteristics constitute an important step in the fight against earthquake. In this study, microgravity, SPAC, and 1D equivalent static earthquake load analyzes were used to determine 2D peak acceleration distribution on a profile for İzmir Bornova Plain.

2. Study Area and Geological Features

The basin of the İzmir and its region contains the Upper Cretaceous Bornova Melange [4]. Limestone mega-olistoliths older than the matrix of the mélange are found in a random order within the matrix of the Bornova Melange. These limestones are known as Işıklar limestone in Altındağ and at its region [5]. Bornova Melange (complex) is composed of platform-type limestone and diabase blocks and conglomerate lens / channel fillings within the matrix consisting of sandstone / shale-calcareous shale intercalations [3]. The Neogene aged lacustrine sediments overlie the Bornova Melange with an angular unconformity. Yamanlar volcanics cover the existing units unconformably. The Quaternary aged soil area unconformably overlies all existing units.

3. Applications

Applications in the study area can be divided into 5 stages. These stages are;

- Microgravity Modeling (Determination of Bedrock)
- Deconvolotion of Earthquake Data
- Model Gridding (1000 m)
- Convolotion of Earthquake Data
- 2D Acceleration Distribution

Figure 1. Illustration of the Study area.

Scintrex CG-5 gravity device was used for gravity measurements at approximately 500 m sampling intervals depending on the terrain and urbanization conditions along 10 km profiles in the Northwest-Southeast direction. Within the scope of measurement planning, fixed station with absolute gravity value determined in Dokuz Eylül University Campus was used as the main base station. All measurements are finalized by connecting to this station. A total of 7 profile measurements were taken. In order to obtain healthy tilt angle values, low standard deviation values and low error values, 5-15 repetitive readings with a minimum duration of 60 seconds were performed. Within the scope of this study, calculations will be made on the results of 1 profile in the westernmost part of the area.

Stage I was obtained in 2017 and published by Pamuk et al. 2017[6]. As a result of the microgravity and SPAC measurements, the models for the A-A model section were made and the calculations made in this study were performed on this model [6]. The microgravity Bouguer Anomaly Map (Raw) and the microgravity Bouguer Anomaly Map (residual) are shown in Figure 2.

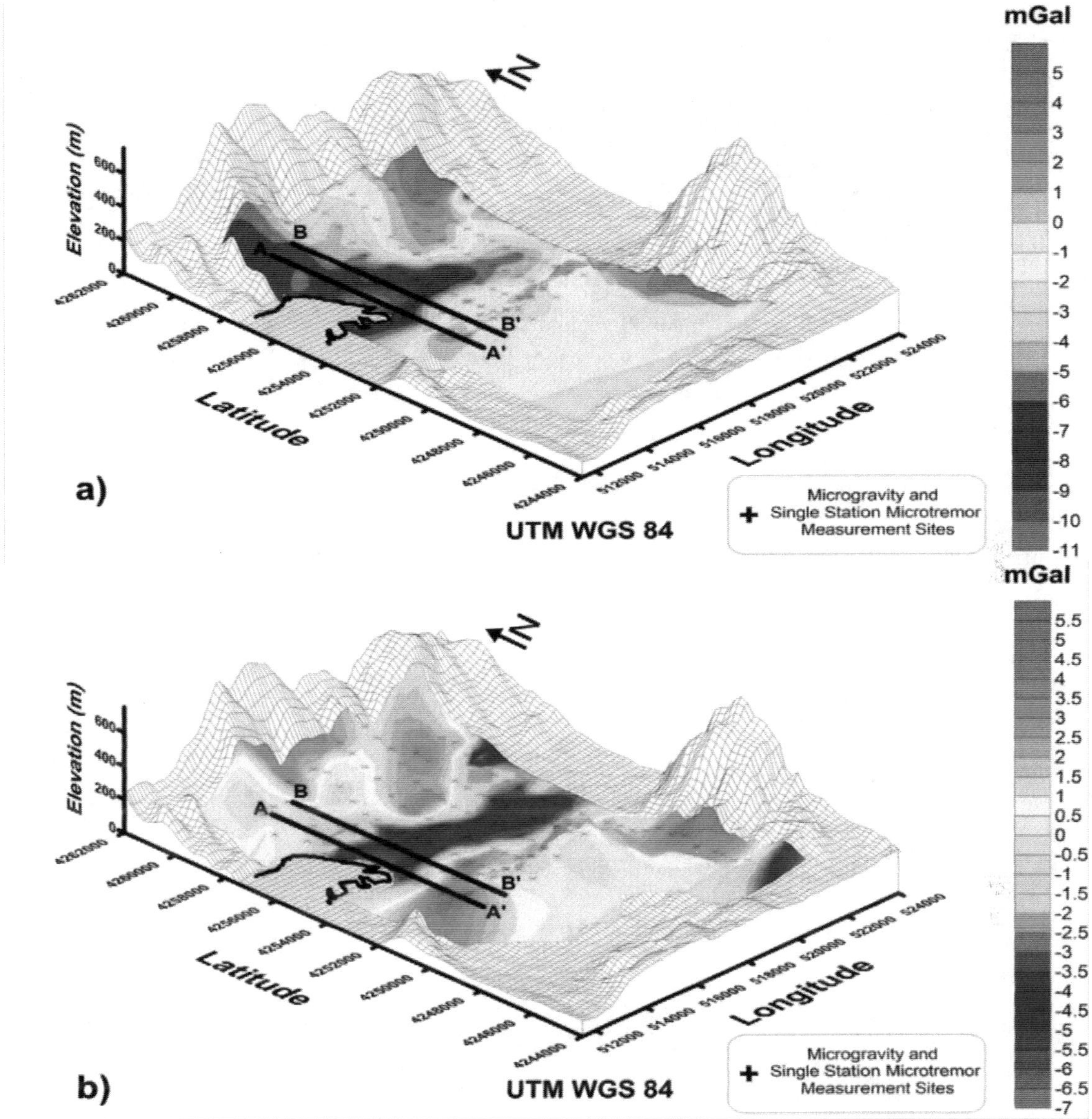

Figure 2. a) Microgravity Bouguer Anomaly Map (Raw), **b)** Microgravity Bouguer Anomaly Map (residual)

After the completion of the modeling steps, the Mw = 6.2 Midilli earthquake at 12:28:37 GMT on 12.06.2017, which was recorded at the BYN strong motion station located in the middle of the A-A' profile, was selected for the calculations. Primarily, the soil model for the substation is determined (Figure 3).

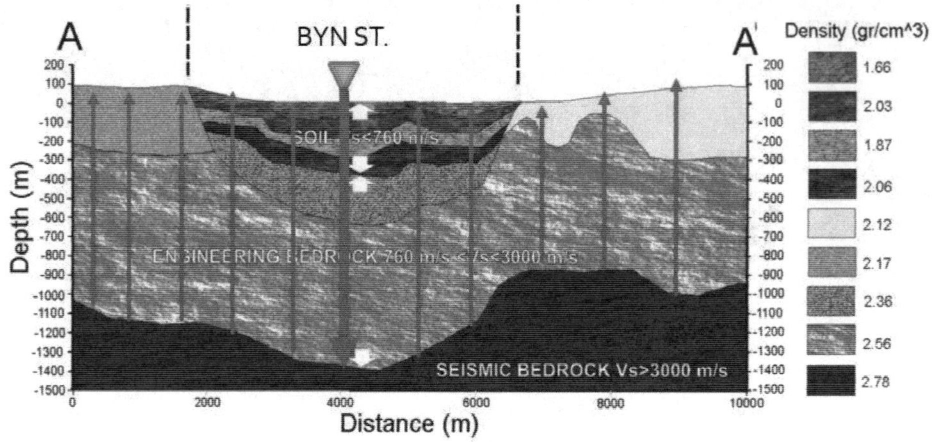

Figure 3. Illustration of the modelled A-A profile.

Using the determined soil model and recorded earthquake data, earthquake data was transferred to the seismic bedrock interface. Deconvolotion of the selected earthquake record is shown in Figure 4.

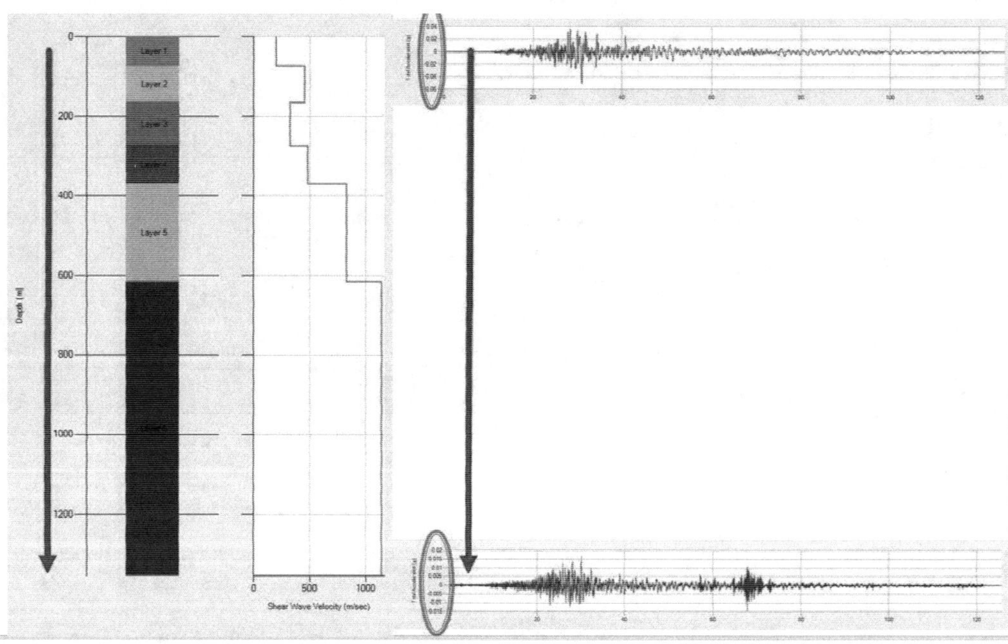

Figure 4. Deconvolotion of the selected earthquake record.

The earthquake data which was then transferred to the seismic bedrock interface was moved upwards from the seismic bedrock interface to the ground surface with a grid of 1000 m along the A-A 'model.

Peak acceleration values of all interfaces were determined during this transport operation and 2D peak acceleration distribution of the earthquake was determined as a result of interpolation of these values (Figure 5).

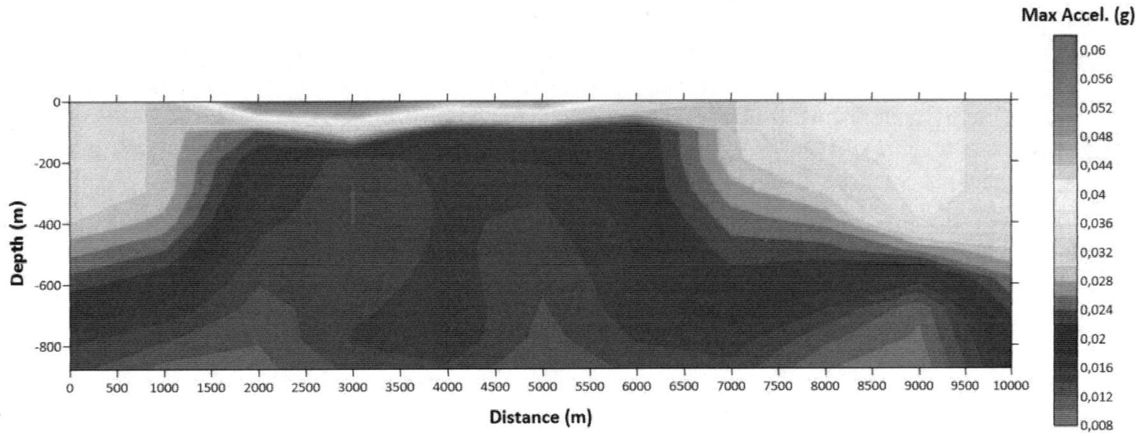

Figure 5. Peak acceleration distribution of the model A-A ((Mw = 6.2 Midilli earthquake at 12:28:37 GMT on 12.06.2017)

4. Results and Discussions

Compared with the predominant period values, it is seen that it is generally compatible. The sections with high pre dominant period values were characterized by low gravity values. When the residual Bouguer gravity anomaly map was examined, the gravity values ranged from -7 to 11 mGal. These values are between -7 and -3 in the middle parts of the study area where the ground thickness is expected to be deeper than the surrounding area. A partial increase in the values is observed as is approach from the sea. The gravity values increased as expected in the northern and southern parts of the study area, where the topography increased and the soil thickness decreased.

The study area is generally represented by a 3-layer model. Accordingly, the first layer was modeled as soil, the second layer was modeled as an engineering bedrock and the last layer as seismic bedrock. In the middle of the study area where the soil thickness is greater, the maximum acceleration values increases. In areas where S wave velocities are higher than 760 m / sec, acceleration values are observed to be between 0.008-0.0024g. In addition, there is an increase in acceleration values on both sides of the area due to the basin effect.

In the ESPAC study at the northernmost of the area, The S wave velocity is calculated as 3000 m / s after 1200 m depth and this unit probably points to the Menderes Massif. This unit, which corresponds to the seismic bedrock, could not be reached in other ESPAC studies.

Acknowledgements

This work was realized within the scope of Mrs. Yaprak İpek's PhD thesis in Dokuz Eylül University. Also, the authors thanks to Turkish Republic Ministry of Interior Disaster and Emergengy Management Presidency for earthquake data.

References

[1] Akgün M, Gönenç T, Pamukçu O, Özyalın Ş and Özdağ Ö C 2013 Integrated geophysical methods for the determination of engineering bedrock: İzmir New City Center. Turkish J Geophys 18, 67–80 (in Turkish).

[2] Akgün M, Gönenç T, Tunçel A and Pamukçu O 2013 A multi-approach geophysical estimation of soil dynamic properties in settlements: A case study in Güzelbahce-İzmir (Western Anatolia). Journal of Geophysics and Engineering, 10(4) 045001.

[3] Erdoğan B 1990 Tectonic relations between Izmir-Ankara zone and Karaburun belt. *Maden Tetkik*

ve Arama Dergisi, 110(110),pp. 1-15.

[4] Özbek D 1981 Geology of Altındağ Village (İzmir) and engineering geology of Altındağ quarries. Graduation Project, Ege University Faculty of Earth Sciences, Department of Geological Engineering, İzmir.

[5] Özer S and İrtem O 1982 "Işıklar-Altındağ (Bornova-İzmir) alanı Üst Kretase kireçtaşlarının jeolojik konumu, stratigrafisi ve fasiyes özellikleri." *Türkiye Jeoloji Kırımu Bülteni* 25;pp. 41-47.

[6] Pamuk E, Akgün M, Özdağ Ö C and Gönenç T 2017 2D soil and engineering-seismic bedrock modeling of eastern part of Izmir inner bay/Turkey. *Journal of Applied Geophysics*, 137, 104–117.

IOP Conf. Series: Materials Science and Engineering **800** (2020) 012026 doi:10.1088/1757-899X/800/1/012026

A comparative study of critical failure surface determination in slope stability assessment

A Iravanian[1*] and A Shlash[1]
[1]Near East University, Civil Engineering Department, Lefkosa, N. Cyprus, 99138, Mersin 10, Turkey

*E-mail: anoosheh.iravanian@neu.edu.tr

Abstract. Slope stability analysis and design of manmade slopes has always been a challenging geotechnical problem. Different methods have been developed and created to examine two and three dimensional slope models based on various methods of assumption and assessment. Throughout this study two limit equilibrium, and finite element methods has been used to examine the stability of the slope and to determine the critical slip surface of failure by using PLAXIS, FLAC/Slope and SLOPE/W software programs respectively. Different values of shear strength parameters were chosen to investigate the efficiency of these methods. Various values of unit weight, cohesion, internal friction angle, and their impact on the safety factor were examined. Comparing the results for factor of safety indicated that finite element calculation used in PLAXIS gives more conservative results by 0.5% compared to SLOPE/W program. FLAC/Slope was observed as the most complicated software and finite difference method generally gave the lowest value for factor of safety in comparison with the other methods. Impact of internal friction angle, unit weight, and cohesion on length of failure arc was also evaluated, and no significant relation between length of failure arc (La) and safety factor was detected.

1. Introduction

There are many problems encountered by the engineer, student or executor, so it should be ready to set the groundwork for predicting such problems and devising appropriate solutions in case of occurrence. The goal of any natural or manmade slope's slope stability analysis is to determine the critical failure surface with the minimum value of safety factor. With the development in technology software packages use the finite difference and finite element methods have increased in popularity as they head to possess a wide range of characteristics [4]. At present, the problem of slope stability is one of the main issues facing workers and engineers in soil field; there is still a big distinction between the scientists about the best way to calculate the stability of the slopes [5]. The main objectives of this study are using various slope stability analysis software programs to evaluate the impacts of soil strength parameters (cohesion, unit weight, internal friction angle) and slope geometry parameters (Height of the slope, angle of the horizental studied slope, Alpha, and the vertical angle of the slope, Beta) as shown in Figure 10 on factor of safety and critical slip surface, in addition to compare the results of the different analysis software programs of slope stability in this study. Over 100 types of soil with different strength parameters have been used in this study to be modeled and analyzed.

IOP Conf. Series: Materials Science and Engineering **800** (2020) 012026 doi:10.1088/1757-899X/800/1/012026

2. Factor of safety, FOS

Simply the ratio of resisting force of deriving force, F = S(Ꞇ)$^{-1}$ [6], or ratio of shear strength to that required to keep the slope stable in case only stable slope (without any structure on it). [7] and [8] found that factor of safety which computed by using 3D analyzes will always be equal or higher than to factor of safety computed by using 2D analyzes. There are various methods for formulating FOS, usually each of the techniques of analysis has its own formulas for FOS, but the most popular formula assumes that the FOS is constant and can be divided into two types: Moment equilibrium and Force equilibrium [9].

3. Impact of geometry and soil strength parameters on factor of safety

To evaluate the influence of strength parameters of soil and geometry parameters on the safety factor, FOS versus these parameters was offered and drawn in the following figures. To study the influence of cohesion C, internal friction angle (Phi) φ and unit weight (UW) γ on factor of safety FOS, the values of three parameters diverge from 15 to 32 kPa, degree, and kN/m^3 respectively were chosen while the other two parameters kept constant at 15.

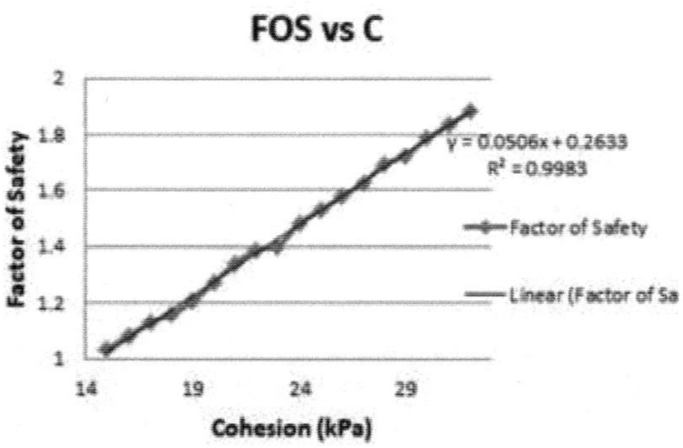

Figure 1. Cohesion effect on FOS.

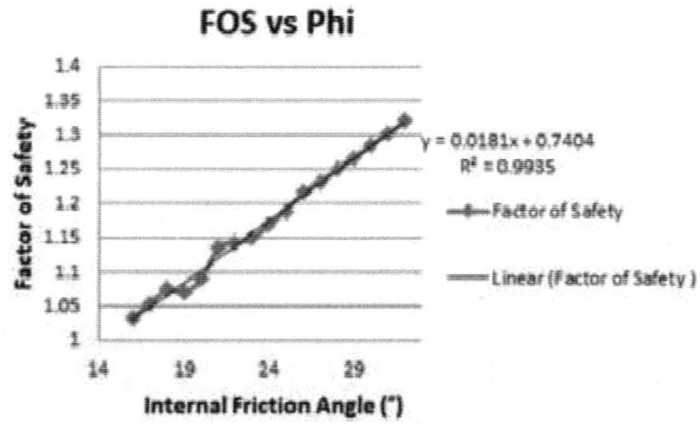

Figure 2. Internal friction angle impact on the FOS.

The cohesion effect on FOS has been shown in figure 1 while Figure 2 the relationship between Phi and FOS. Increasing the value of cohesion and Phi as a resistant forces, as expected, increased the value of FOS which is nearly linear with R^2 factor of 0.99.

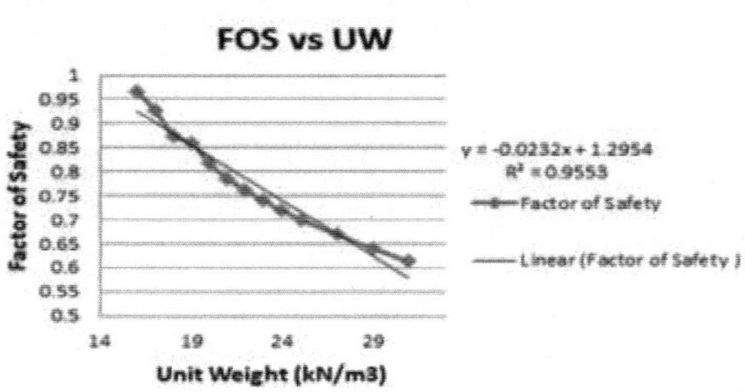

Figure 3. Effect of UW on FOS.

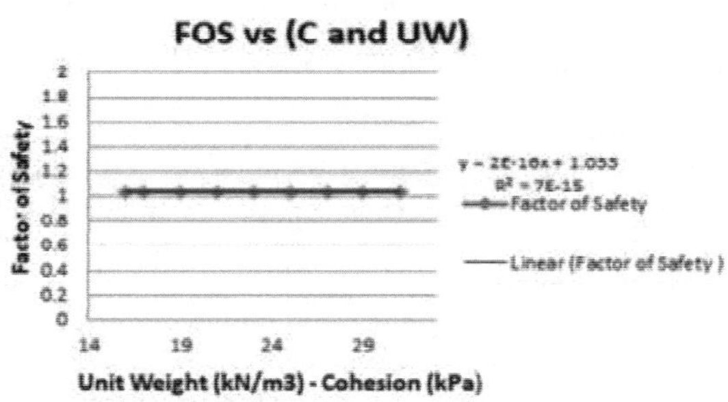

Figure 4. The combined effect UW and cohesion the on FOS.

As can be shown from the Figure 3, the soil's unit weight is inversely proportional to the safety factor because of the raise in unit weight which is the main reason of the derivative forces. In Figure 4 Cohesion and soil unit weight were raised together here, but the ratio stayed constant. The outputs dictate that the potential surface of the slip is affected by the combination of c and φ, the function of which is defined as λ equivalent to

$$\lambda = C / (\gamma * h * \tan\varphi) \quad (1)$$

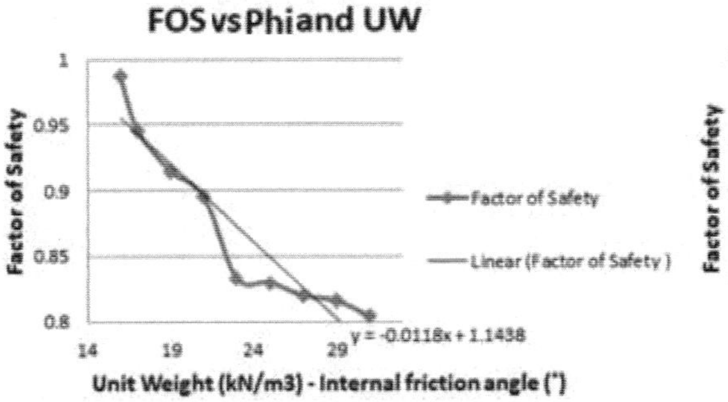

Figure 5. The combined effect of Phi and UW on FOS.

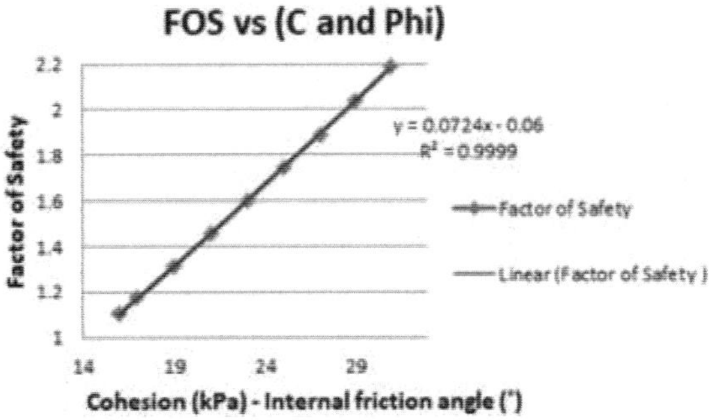

Figure 6. The effect of Phi and C on FOS.

As shown in Figure 5, decrease in the value of safety factor has achieved by raising phi value so γ value. This is due to the failure surface movement to the top, thus reducing the length of the arc of failure and so reducing the impact of resisting forces. Since the surface of potential failure is expected to be influenced by both of C and phi values, Figure 6 shows the relationship between the safety factor and C, φ so because both of these parameters of shear strength are holding (resisting) forces, raising these two values results in an raise in the value of the safety factor. FOS is nearly linear with R^2 factor of 0.9999.

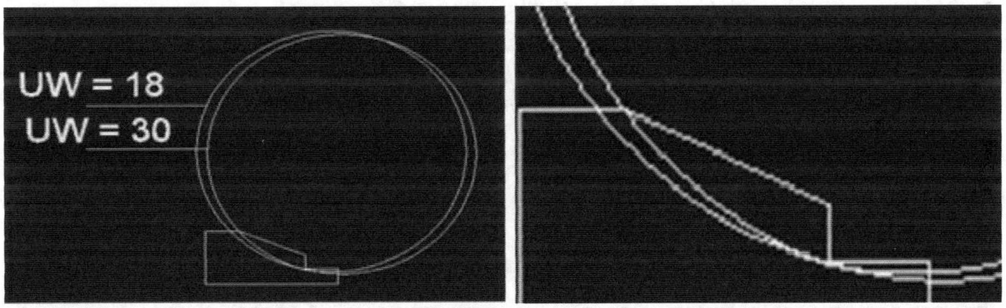

(a) (b)

Figure 7. (a) Unit weight impact on slip surface. (b) cropped part of (a).

Figure 7 (a) shows the impact of unit weight on failure slip surface (whereas Figure 7 (b) is zoomed variant of (a)). The tests follow a practical rule, by maximizing the soil's unit weight, the failure surface is moved to the right, leading in a smaller volume of soil failure and thus decreasing the slip surface length. Less resistant force is activated due to the smaller surface for holding factors (friction and cohesion). For these reasons, a lower value for FOS is accomplished. Two angles of the slope α and β used for this to test the influence of geometry on the safety factor. The results presented in the figures below.

IOP Conf. Series: Materials Science and Engineering **800** (2020) 012026 doi:10.1088/1757-899X/800/1/012026

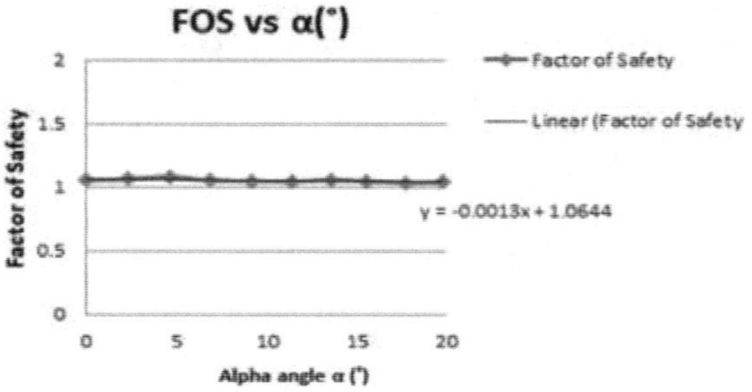

Figure 8. Alpha angle impact on factor of safety.

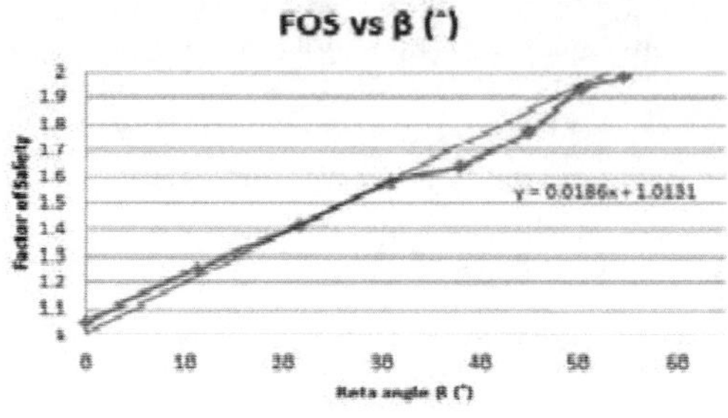

Figure 9. Beta angle impact on FOS.

Figure 8 represents that no noticeable variation is observed in the safety factor except the value of FOS between angles 4.6° and 6.8°, it decreased unpretentiously after the angle 6.8° although it was increasing slightly before that value of Alpha angle. These decrease because it is possible by increase the Alpha angle as if adding an additional overhead load on the surface of the slope. While this rise in the surface of failure produces more resistant force, it simultaneously produces an raise in derivative force (weight of the surface of failure). The safety factor therefore remains constant, for angles more than 6.8°, the rise in derivative force approach the value of the resisting force also from this angle value onwards, and the derivative force becomes greater than the resisting force, and therefore a decrease could be seen in safety factor's value. Figure 9 represents the safety factor significantly increases by increasing Beta angle. The reason for this behavior is that only the failure arc length increase as resisting force and the weight of the failure shape as driving force stays constant by increasing the beta angle. For that by increasing the length of the arc result an increase in the resisting force and thus the safety factor.

4. Impact of geometry and soil strength parameters on the slip surface

Based on what was debated in the past part, it is easy to estimate that a correlation should exist between the strength parameters of soil and the geometry of the slope and the surface of failure; the following analyzed models will be studied in order to examine this condition. Many models were generated using SLOPE/W software in this step. The output in this aspect will become the safety factor and slip circle center coordinates as well as the circular failure surface radius. The circles have been drawn using AutoCAD software to determine failure arc's length and locate entry and exit points in the slope area. Figure 10 reflects the general shape of the slope geometry that going to be used in the next models.

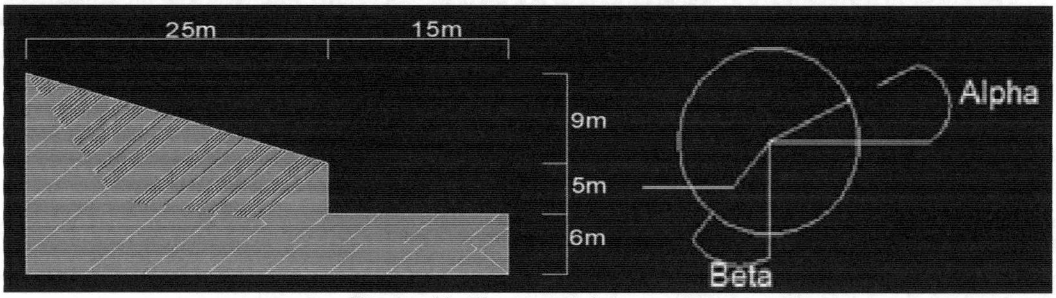

Figure 10. Slope model geometry.

To analyze the impact of any soil parameter on the slip surface, the length of the failure arc was selected to be studied as a quantitative variable. To demonstrate this impact, the following figures will be provided.

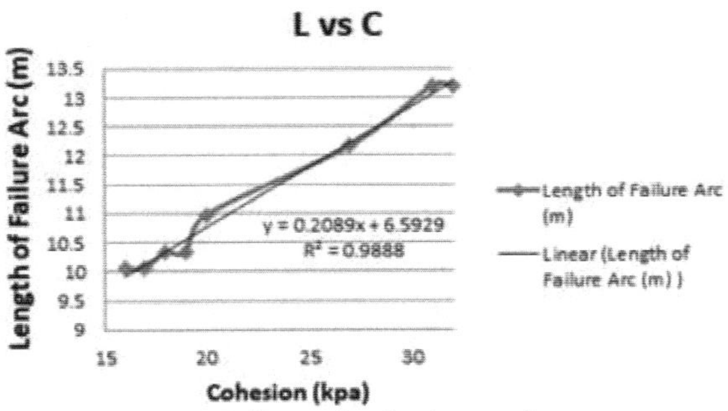

Figure 11. Effect of cohesion on La.

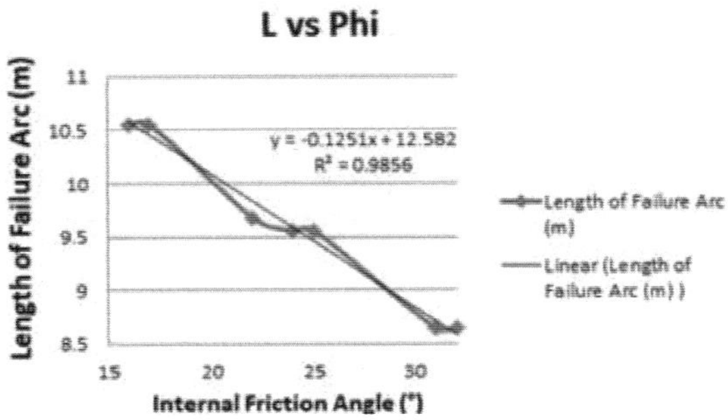

Figure 12. Effect of internal friction angle on La.

Figure 11 represents that the length of failure surface will raise by raising the value of cohesion. The cause is that in the instance of the constant failure surface's location as the cohesion factor raise, the hold forces become greater in addition to the FOS. The deriving forces must raise that can be accomplished by increase the area of slope failure, in order Finding the smallest value of FOS (the initial objective for slope stability analyses). This results in a larger arc failure length and so a greater value of FOS. The impact of internal friction angle on failure surface length is shown in Figure 12. In the previous section, referring to the same explanation, it could be anticipated that the length of the arc must be in a direct correlation with φ, but as can be seen in Figure 12, La and φ are related reversely. This reverse relationship is consistent with the paper [1] which noted that "when the distribution of slope geometry and unit weight in a homogeneous soil slope is given, the slip surface location for specific slice technique is only concerned with the ratio C (tan $\varphi)^{-1}$ of the slope. This paper indicates that slip surface positioning and therefore La is in the inverse relation to the friction angle.

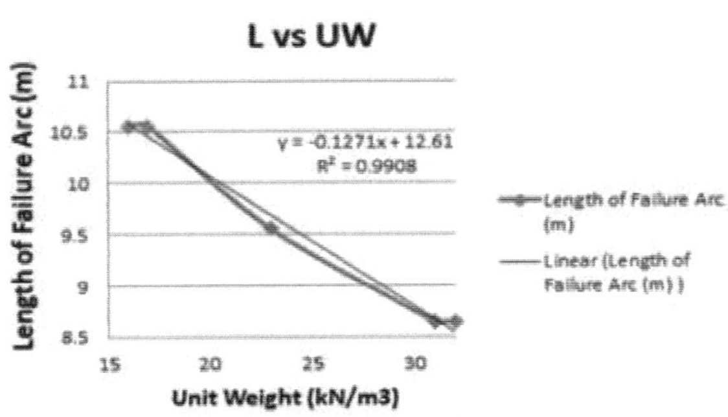

Figure 13. Effect of UW on La.

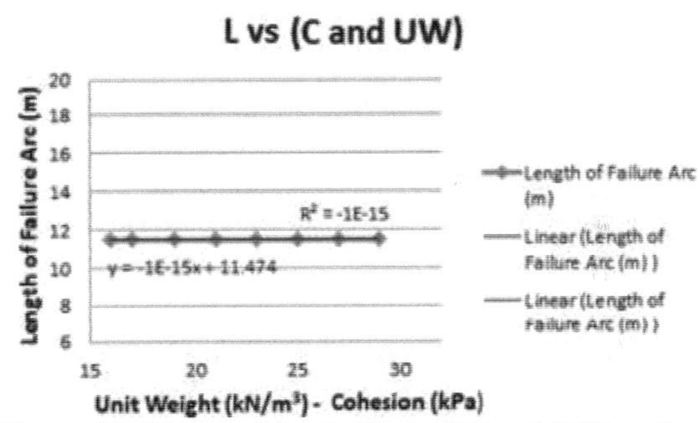

Figure14. The combined effect of C and UW on La.

As shown in Fig 13 by raising the value of soil unit weight, the weight of the failed shape rises, resulting a small safety factor. And by considering lambda the surface of the slip shifts towards the slope's face whereas by reducing La, the impacts of the friction angle and cohesion as the forces of resistance decrease, thus achieving a smaller safety factor. Constant unit weight over C ratio, which results in a constant λ. As noted in the study [10], this means the same shape of failure and therefore a constant value of La as shown in Figure 14.

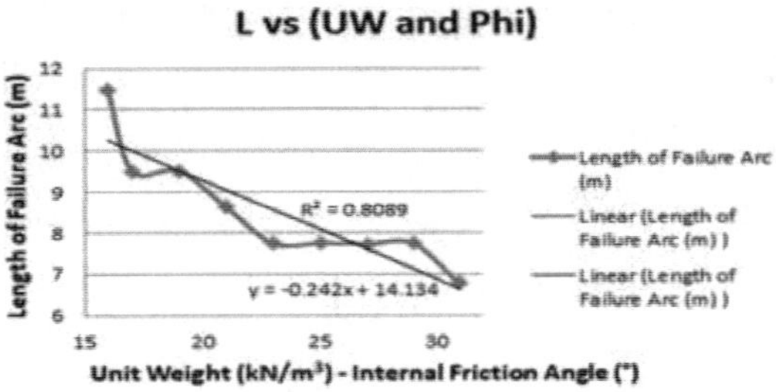

Figure 15. The effect of Phi and UW on La.

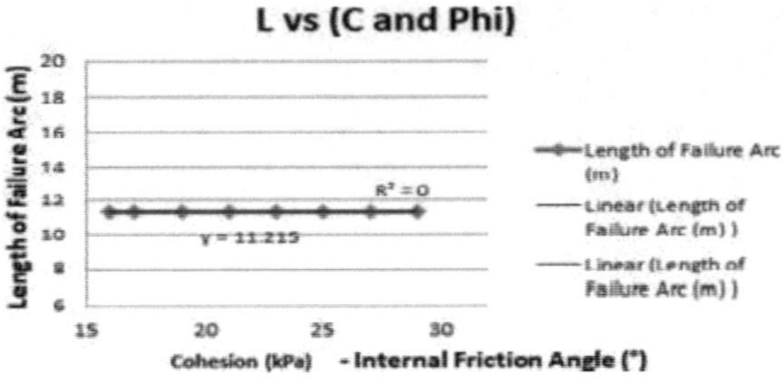

Figure 16. The effect of C and Phi on La.

The results shown in Figure 15, to show the variation impact for both of the strength factors. It can be seen that there will be a reduction in the length of the surface failure by increasing the value of γ and tanφ. It is consists when talking about the λ value, by maximizing this value, λ declines; smaller value of λ imply a surface of failure nearer to the surface of the slope and thus a smaller failure arc length. In Figure 16, it can be seen that La will remain proportionally constant at value (11.2 m) when both of the factors cohesion and friction angle remain constant between the values (16-32). Since constant C and tan φ results in a constant lambda, and constant lambda imply a constant surface of failure, La also stays constant.

The length of the failure arc as a numerical value was estimated and drawn to observe the effect of the slope geometry on the surface of the failure as shown in Figures 17 and 18.. Slope geometry has been shown to have a proportional relation with the stability of the slope also the properties of soil strength [11]. In the last models series, strength parameters of soil remained unchanged at the following values, whereas angles ranged from 0 ° to 63.4° for (Beta) angle while (Alpha) angle ranged from 0 ° to 19.8° in slope geometry (shown in Figure 4.4). Cohesion = 15 kPa, angle of friction = 15°, unit weight = 15 kN/m^3.

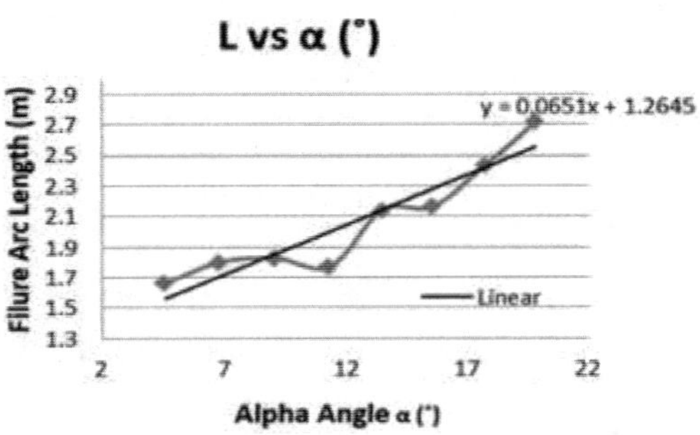

Figure 17. Alpha angle impact on La.

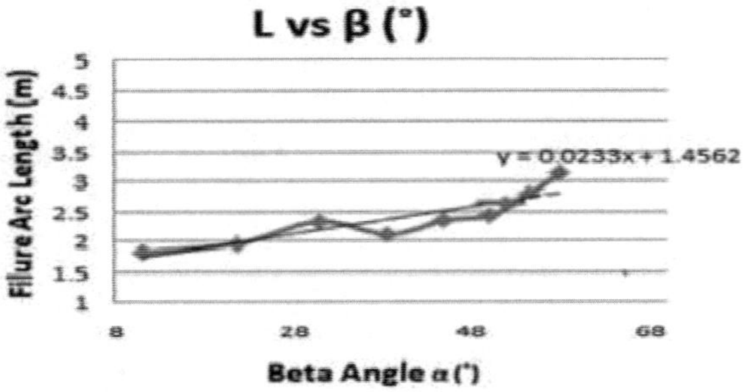

Figure 18. Beta angle impact on La.

Results of analysis of the models indicate that the positioning of the failure surface is not significantly different by increasing the Alpha angle. The cause for increasing the length of the failure arc is only the slope surface motion and therefore the failure arc extension. By increasing Beta angle value and the remaining parameters in the formula (1) won't change and therefore don't affect the value of lambda), the failure surface depth doesn't vary. But, for the other side, an raise in the value of Beta angle can shift the surface of the slope to the left and this is going to cause the arc to be protracted as shown in Figure 18.

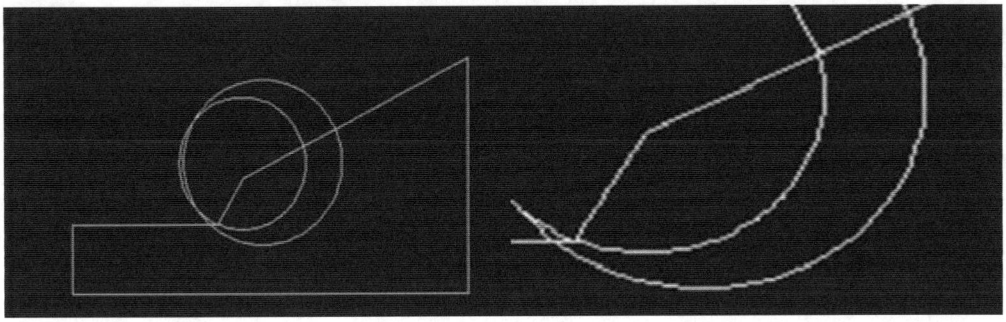

(a) (b)

Figure 19. (a) Impact of Alpha angle on length of failure Arc, (b) zoomed fig of (a).

5. Reanalyzing models by PLAXIS and comparison of results

The studied models were reanalyzed using PLAXIS software program to validate the outputs of the SLOPE/W program acquired in part 3. Table 1 shows the results of these analyzes.

Table 1. Models for the factor of safety analysis – PLAXIS.

Model Number	Unit Weight [kN/m^3]	Phi [°]	Cohesion [kPa]	FOS
2	15	15	16	1.051
6	15	15	20	1.254
10	15	15	24	1.391
16	15	15	30	1.663
19	15	16	15	1.047
23	15	20	15	1.073
29	15	26	15	1.237
33	15	30	15	1.316
37	17	15	15	0.867
41	21	15	15	0.752
45	25	15	15	0.689
48	31	15	15	0.594
49	16	15	16	1.024
52	21	15	21	1.024
54	25	15	25	1.024
57	31	15	31	1.024
58	16	16	15	0.954
61	21	21	15	0.900
63	25	25	15	0.838
66	31	31	15	0.792
67	15	16	16	1.048
70	15	21	21	1.341
72	15	25	25	1.617
75	15	31	31	2.046

Table 2 summarizes the distinction between the FOS acquired from both programs (PLAXIS and SLOPE/W) and compares these results using the following formula.
Difference = FOS(SLOPEW) − FOS(PLAXIS) {FOS (PLAXIS)}$^{-1}$*100

Table 2. The difference of FOS between SLOPE/W and PLAXIS.

Model Number	SLOPE/W FOS	PLAXIS FOS	Difference%
2	1.081	1.051	0.029
6	1.272	1.254	0.014
10	1.483	1.391	0.066
16	1.787	1.663	0.075
19	1.033	1.047	-0.013
23	1.089	1.073	0.015
29	1.215	1.237	-0.018
33	1.284	1.316	-0.02
37	0.927	0.867	0.030
41	0.786	0.752	0.017
45	0.701	0.689	0.006
48	0.613	0.594	0.0095
49	1.033	1.024	0.0045
52	1.033	1.024	0.0045
54	1.033	1.024	0.0045
57	1.033	1.024	0.0045
58	0.988	0.954	0.017
61	0.895	0.9	-0.0025
63	0.83	0.838	-0.004
66	0.804	0.792	0.01515
67	1.103	1.048	0.0275
70	1.456	1.341	0.0575
72	1.745	1.617	0.064
75	2.19	2.046	0.072

Table 3. Impact of slope's geometry on the slope's slip surface – PLAXIS.

Model Number	α [°]	β [°]	Slope Height [m]	Factor of Safety
1	19.8	0	20	1.013
2	17.75	0	19	1.029
4	13.5	0	17	1.061
6	9.1	0	15	1.064
8	4.6	0	13	1.057
10	0	0	11	1.048
11	0	11.3	11	1.216
12	0	21.8	11	1.365
14	0	38	11	1.609
16	0	50.2	11	1.877
20	0	63.4	11	2.274

Table 4. The difference of FOS between PLAXIS and SLOPE/W for slope geometry.

Model No	SLOPE/W FOS	PLAXIS FOS	Difference (%)
1	1.037	1.013	0.012
2	1.033	1.029	0.002
4	1.058	1.061	-0.0015
6	1.048	1.064	-0.008
8	1.079	1.057	0.011
10	1.053	1.048	0.0025
11	1.252	1.216	0.018
12	1.419	1.365	0.027
14	1.642	1.609	0.0165
16	1.928	1.877	0.0255
20	2.302	2.274	0.014

It can be seen in Table 3 and 4 that PLAXIS is a more conservative designing program. On average, PLAXIS gives less than 1% lower FOS (except few models) which gave FOS greater than SLOPE/W software that will make PLAXIS more conservative and therefore safer to design and analyze more significant slopes. By comparison, giving SLOPE/W a higher FOS makes it more beneficial to analyze and design less important slopes.

6. Reanalyzing the past models by using FLAC/Slope

To inspect the outputs from PLAXIS and SLOPE/W, a sample of the models were chosen up to their properties, and these models were reanalyzed by using FLAC/SLOPE program as shown in Tables 5 and 6. Considering that FLAC is not software with a complete LEM, the results may show a distinction between three software.

Table 5. Reanalyze models by using FLAC software (shear strength parameters models).

Model Number	Unit Weight [kN/m^3]	Phi [°]	Cohesion [kPa]	FOS
6	15	15	20	1.029
11	15	15	25	1.221
23	15	20	15	0.607
28	15	25	15	0.932
45	25	15	15	1.049
54	25	15	25	0.842
63	25	25	15	0.764
72	15	25	25	1.436

Table 6. Reanalyze models by using FLAC software (slope geometry models).

Model No	α (°)	β (°)	Factor of Safety
6	9.1	0	0.869
15	0	45	1.827

Table 7. The difference between the software packages (shear strength parameters models).

Model no	Factor of safety of			Difference between FLAC and	
	FLAC	SLOPE/W	PLAXIS	SLOPE/W (%)	PLAXIS (%)
6	1.029	1.272	1.254	0.236152	0.218659
23	0.607	1.089	1.073	0.794069	0.76771
45	1.049	0.701	0.689	-0.33174	-0.34318
54	0.842	1.053	1.024	0.250594	0.216152
63	0.764	0.83	0.838	0.086387	0.096859
72	1.436	1.745	1.617	0.215181	0.126045

Table 8. The difference between the three software packages (slope geometry models).

Model no	FACTOR OF SAFETY OF			Difference between FLAC and	
	FLAC	SLOPE/W	PLAXIS	SLOPE/W (%)	PLAXIS (%)
6	0.869	1.048	1.064	0.205984	0.195
15	1.827	1.774	1.774	-0.02901	-0.053

As shown in Tables 7 and 8, the average distinction between FLAC/SLOPE and other two programs is less than 1 percent, which is appropriate. In addition, it is noticeable that FOSs obtained from FLAC are lower than the other two programs in 75% of the models.

7. Conclusion and recommendations

cohesion (c) and friction angle (φ), as resistance forces are directly related to the safety factor while unit weight as deriving force is inversely related to the safety factor. An increase in the cohesion value results to higher value for length of failure arc (La). An increase in the friction angle value results to reduce the length of failure arc (La) value. Increasing the unit weight value of soil leads to increase in the of the length of failure arc (La) value. An increase in the value of Alpha angle to a specific value doesn't have any noticeable influence on FOS. while, an increase in the value of the Beta angle affects directly on the factor of safety. The greater value of the Alpha angle, leads to a greater value of La. However, varying the value of Beta angle doesn't affect noticeably on the value of La. PLAXIS is more conservative software for slope stability studies, and comparing to SLOPE/W which it gives about 0.5% lower value for factor of safety. FLAC/Slope is the most complicated software and usually gave out the lowest value for factor of safety in comparison with PLAXIS and SLOPE/W. There is no any significant relation between length of failure arc (La) and safety factor. At the end of this study I recommend analysis and modeling wider range in parameters of soil strength. Including the level of water content and consideration of unsaturated soils and the effect of water pressure and pore air. Including more slope geometry variables as overall slope's angle and the slope's length. Conduct a state study to verify the validation of the equations acquired to locate the critical surface of the failure.

1. Cohesion (c) and friction angle (φ), as resistance forces are directly related to the safety factor while unit weight as deriving force is inversely related to the safety factor.
2. An increase in the cohesion value results to higher value for length of failure arc (La).
3. An increase in the friction angle value results to reduce the length of failure arc (La) value.
4. Increasing the unit weight value of soil leads to increase in the length of failure arc (La) value.

5. An increase in the value of Alpha angle to a specific value doesn't have any noticeable influence on FOS. While, an increase in the value of the Beta angle affects directly on the factor of safety.
6. The greater value of the Alpha angle, leads to a greater value of La. However, varying the value of Beta angle doesn't affect noticeably on the value of La.
7. PLAXIS is more conservative software for slope stability studies, and comparing to SLOPE/W which it gives about 0.5% lower value for factor of safety.
8. FLAC/Slope is the most complicated software and usually gave out the lowest value for factor of safety in comparison with PLAXIS and SLOPE/W.
9. There is no significant relation between length of failure arc (La) and safety factor FOS.

8. Recommendations

The following recommendations analysis can be performed for other studies in connection with this thesis study:

1. Analysis and modeling wider range in parameters of soil strength.
2. Including the level of water content and consideration of unsaturated soils and the effect of water pressure.
3. Including more slope geometry variables as the angle of the slope itself and the length of the slope.
4. Conduct a state study to verify the validation of the equations acquired to locate the critical surface of the failure.

References

[1] Jiang JC, Yamagami T (2006) Charts for estimating strength parameters from slips in homogeneous slopes, Computers and Geotechnics, 33(6–7): 294-304, ISSN 0266-352X
[2] Shiguo X, Liping Y, Zhiqiang C (2011) A method combining numerical analysis and limit equilibrium theory to determine potential slip surfaces in soil slopes. Journal of Mountain Science, 8(5): 718–727
[3] Naderi MA (2013) Determination of the Critical Slip Surface in Slope Stability Analysis, Master Thesis Eastern Mediterranean University
[4] Hammouri, N. A., Husein Malkawi, A. I., & Yamin, M. M. A. (2008). Stability analysis of slopes using the finite element method and limiting equilibrium approach. *Bulletin of Engineering Geology and the Environment.*
[5] Oghli, Z. (2011). *Analyzing slop stability by using shear strength reduction method.* Aleppo University.
[6] Duncan, J. M., Wright, S. G., & Brandon, T. L. (2014). *Soil strength and slope stability.* Canada: John Wiley & Sons, Inc. Retrieved from.
[7] Mitchell, J. K., Chang, M., Seed, R. B., Mitchell, J. K. ;, Chang, M. ;, Cahill, E. G., & Cahill, J. R. (1993). The Kettleman Hills Landfill Failure: A Retrospective View of the Failure Investigations and Lessons Learned. *International Conference on Case Histories in Geotechnical Engineering*, 1–15.
[8] Duncan, J. M. (1992). State-of-the-art: Static stability and deformation analysis. In *Proc. of Specialty Conf. Stability and Performance of Slopes and Embankments-II* (Vol. 1, pp. 222-266). ASCE.
[9] Cheng, Y. M., Lansivaara, T., & Siu, J. (2008). Impact of convergence on slope stability analysis and design. *Computers and Geotechnics*, *35*(1), 105–113. https://doi.org/10.1016/j.compgeo.2007.02.011.
[10] Lin, H., & Cao, P. (2011). Potential slip surfaces of slope with strength parameters. In *Advanced Materials Research* (Vol. 243, pp. 3315-3318). Trans Tech Publications.
[11] Namdar, A. (2011). Geometry in Slope Modeling and Design. e -Journal of Science & Technology (e-JST)(22), 9-21.

Swell and compressibility characteristics of expansive clays in north Cyprus

Anoosheh Iravanian[1] **and Amr Abdeh**[2]

[1] Civil Engineering Department, Near East University, Lefkosa, N. Cyprus, 99138, Mersin 10, Turkey
[2] Total Advance Engineering Services Limited, Abuja, Nigeria

e-mail: anoosheh.iravanian@neu.edu.tr

Abstract. Expansive clays in semi-arid regions are known as problematic soils especially for low weight civil structures. Volume change is a critical issue, therefore determining expansive clays and quantifying their expandability, retractility and strength is a major step to be considered in geotechnical engineering. This research provides a study on the behaviour of expansive clays done under different geotechnical laboratory experiments on four different types of expansive clayey soils brought from various areas in Northern Cyprus, aiming to prepare a soil-properties map in these regions. Fundamental assessments were performed for determining soil index properties. Swell and consolidation behaviours were determined using one-dimensional oedometer apparatus. The results showed that the rate of the expansiveness of sample the bentonitic clay samples were the highest for the predicted ultimate swell with expandability index categorized as very high. Also, these samples showed the largest compressibility upon the consolidation test results. The results obtained were used to construct a relationship between compression and plasticity Index.

1. Introduction

Expansive clays are high swelling soils, they are very reactive due to their high plasticity. Clays with high plasticity are composed of fine-grained particles which are prone to a huge volume change whenever water content differs. According to [15], clays with a plasticity index larger than 35 percent are highly plastic. Clay moisture content decreases and increases depending on the environmental conditions surrounding it resulting in shrinkage and swell, this change in moisture is regarded as the main reason for the change in volume. Also, the mineralogy, soil structure, specific surface and stress history all contribute to the volume change [20]. The factors that mainly affect the soil volumetric change potentials are clay mineral type, the ratio of voids and moisture content of a certain soil [8], [13], [17], [19]. Expansive clays moisture content changes when wetted or dried which leads to void ratio and volume change of the soil.

Expansive soils behavior was also studied by other researchers like [12], [18], [21], [22], [26]. Clays with swell potentials are found in semi-arid regions of tropical and climate temperature zones worldwide [9]. Considerable infrastructure damages had been reported caused by high plastic clays due to their shrink and swell behavior. It is a challenging issue for engineers to design substructures on expansive clays, in order to bypass that, the volume change, swell and shrinkage, characteristics of expansive clays must be considered before engineering structures are built. Determining type of clays and evaluating their swell potential before construction can assist in reducing future possible damages.

The geology and climate of Cyprus made expansive clay formations in some parts of the Island [25]. The majority of swelling clays in Cyprus occur in a geological unit of Neogene. The greatest amount of damages caused by swelling clays are contained in stratigraphic sequence ranging in age from Miocene to Quaternary. Therefore, the island's geological and geotechnical evolution contributed to the swelling clay formations. The geological location of Cyprus is at the triple junction of Africa, Eurasia and Arabian plates. Through the complex tectonical and sedimentary process, the triple junction intersection zone occurred. Complex geotectonic activities were found in Cyprus during Late Cenozoic [10]. In this work volume change and mechanical behavior of four typical expansive clay samples of Cyprus is studied. The obtained information could be used as the basic step for engineers and researchers in stabilizing and foundation design works in this area.

2. Experimental work

2.1. Materials

Cyprus is covered with marly and clayey formations bearing montmorillonite clays to bentonitic group of clays. Kythrea formation, Mamonia complex, Nicosia formation, Alluvial soils and Bentonitic clays are the most common soils of Cyprus. Clay deposits consisting of bentonite are mostly found at Lefke (Lefka) and Yiğitler (Arsos) [7]. Swelling clay formations are mostly found at Nicosia, Famagusta, Kyrenia, Kalecik, Çamlibel and Değırmenlık figure 1.

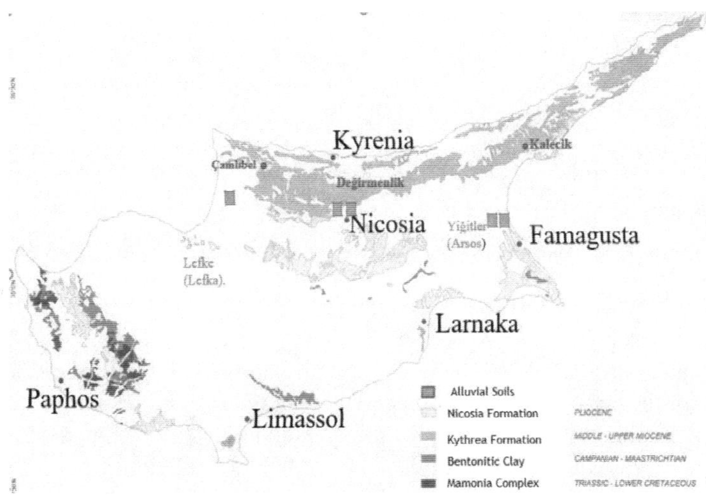

Figure 1: Swelling clay settlements in Cyprus [10]

The laboratory work was planned to understand the properties related to expansive clay mechanical behaviour. Four natural expansive soil samples were taken from different sites, which are located in the South of Taskent village (sample T1), North of Haspolat village (sample T2 and T3) and South of Yigitler village (sample T4) within and around Nicosia, North Cyprus. Theses clays are widely known for their expansiveness and are responsible for the large amount of damage to civil structures resulting from swelling and shrinkage cycles. Different laboratory experiments were applied for obtaining the characteristics of the clays in accordance to [1, 2, 3, 4, 5, 6]. The physical properties of sample T1, T2, T3 and T4 are presented in Table 1.

After determining the basic physical properties, free-swell and consolidation tests were carried out on the obtained soil samples compacted at their optimum moisture content (OMC) and maximum dry density (MDD). The maximum dry density and optimum water content of all samples were measured by standard Proctor compaction test [4]; the obtained results are represented in Table 2.

Table 1. Physical properties of obtained soils

Samples Name	T1	T2	T3	T4
Silt Content (%)	52	40	32.5	25
Clay Content (%)	48	60	67.5	75
Liquid Limit (%)	64	72	115	132
Plastic Limit (%)	33	28	32	40
Plasticity Index (%)	31	44	83	92
Specific Gravity (Gs)	2.55	2.56	2.55	2.48
USCS classification	CH	CH	CH	CH

The main purpose of doing the free swell test is to obtain the maximum ultimate swell vlaue of the obtained compacted soil samples. Also, the consolidation test was performed for finding the properties of compression for all samples at an applied pressure ranging between 6.9-3530 kPa.

Table 2. Compaction test results of investigated samples

Sample Name	Maximum dry density (g/cm^3)	Optimum water content (%)
T1	1.60	21.5
T2	1.62	22.0
T3	1.67	19.0
T4	1.30	39.4

Consolidation and free swell tests were carried out in a one-dimensional oedometer apparatus. The soil samples were compacted into a metal ring of 20 mm height and an inner diameter of 50 mm. The height of the compacted soils was kept at 14 mm in order to allow free swell in the remaining 6 mm of the consolidation ring.

3. Results and discussions

Full swell is to be achieved before knowing the maximum ultimate swell; therefore, the samples were left to swell until no further change in samples height was seen. The response of free swell in percent swell ($\triangle H/H_0*100$) versus time in minutes for the obtained samples is shown below in figure 2, where the percent swell is represented as axial strain (%) and time in logarithmic (min). The overall swell of each sample consists of three stages which are the initial swell stage, primary swell stage and secondary swell stage. The initial stage starts and ends in the first few minutes, while the main part of the whole swell is the primary stage and finally the secondary stage which is the part that builds up progressively from the primary stage taking the most time before completion [25].

Also, Figure 4 represents the results of all four samples as it shows their overall swelling behavior from the start to maximum swell measured. It is clear that T1 has the lowest swelling potentials with a primary swell of 3.45%, whilst T2 and T3 exhibited a higher swell with a primary swell of 9.53% and 7.49% respectively. The highest swell was for T4 with a primary swell of 17.4% which makes it the most expansive among all samples. The potential expansion of the soils is classified by their Expansion Index (EI), calculated by equation (1) shown in Table 3 [1].

$$EI = \frac{\Delta H}{H_0} x\ 1000 \quad (1)$$

Where

EI = Expansion index
ΔH = Final dial reading (mm) – Initial dial reading (mm)

Table 3. Samples classification of potential expansion according to their EI [1]

Samples	Expansion Index, EI	Potential Expansion
T1	38	Low
T2	113	High
T3	133	Very High
T4	189	Very High

The curves shown in Figure 2 flows in three trends; initial escalation in axial strain with time then a leap indicated by a curve ending up with a linear line and finally a slight increase indicated by a horizontal or inclined finishing. The initial, primary and secondary time were extracted by applying the tangent-intersection method. In this method, the S-shaped axial strain-time curve is composed of two non-linear parts at the secondary (upper phase) and initial (lower phase) swelling stages, in addition to that a part which is linearly inclined at the primary (middle phase) swelling stage [23]. The initial and primary swell time for all sample are shown in Table 4.

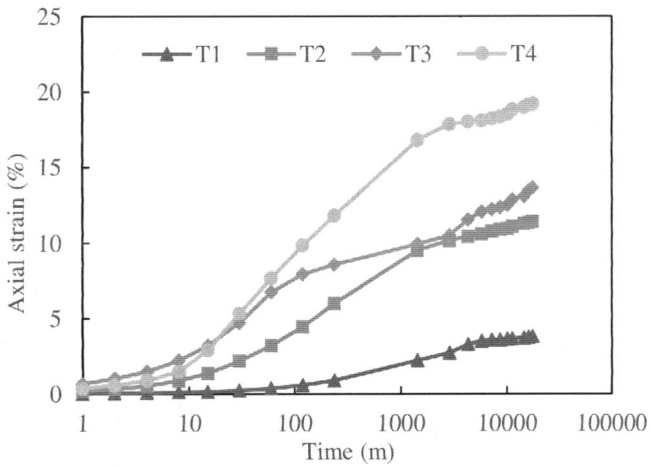

Figure 2. Percent swell of sample T1, T2, T3 and T4 versus logarithmic time

The mechanism of different swelling phases is due to surface hydration of particles during the initial swell caused by the non-swelling fractions within the voids; primary swelling occurs when voids cannot bear any more water causing it to develop faster whereas secondary swelling occurs due to swelling of active minerals [11].

Table 4. Swelling time of sample T1, T2, T3 and T4

Samples	Initial swelling time (min)	Primary swelling time (min)
T1	90	5200
T2	18.5	1440
T3	7.8	91
T4	8	1500

The value of ultimate swell cannot be reached in the laboratories using the normal practical methods because by theory an infinite time is required to get to the ultimate swell. Through the straight lines fitted in figure 3 which shows the time/swell versus time, the ultimate swell can be predicted using equation 2 as proposed by [24]. The values obtained from plotted graphs and equations are presented in Table 5 including the highest R^2 of the straight line.

$$u_{max} = lim_{t\to\infty}\left(\frac{1}{\frac{x}{t}+y}\right) = \frac{1}{y} \quad (2)$$

Where

U_{max} = Ultimate swell
X = represents the ordinates of a line
Y = represents the slope of a line.
t = time

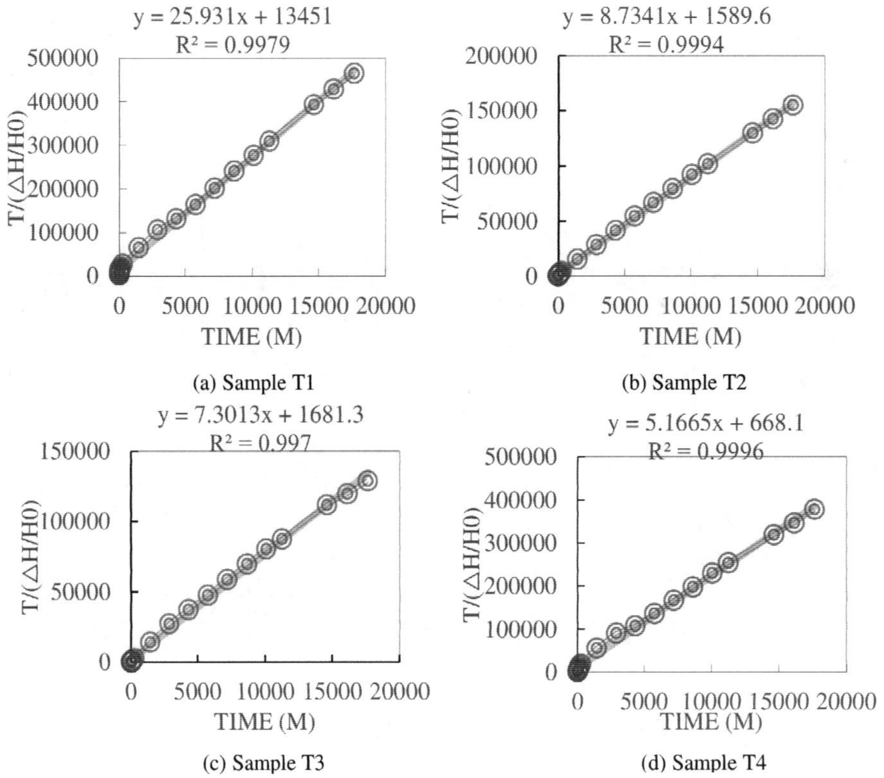

(a) Sample T1 (b) Sample T2

(c) Sample T3 (d) Sample T4

Figure 3. The relationship of time/swell vs time of samples T1, T2, T3 and T4

Table 5. Ultimate swell values prediction and swell properties of tested samples

Samples	**Initial swell (%)**	**Primary swell (%)**	**Max swell measured (%)**	**Hyperbolic constant, y**	**Hyperbolic constant, x**	**Ultimate swell (%)**	**R^2**
T1	0.5	3.45	3.79	25.9310	13451	3.85	0.9979
T2	1.6	9.53	11.39	8.7341	1589.6	11.44	0.9994
T3	2.2	7.49	13.66	7.3013	1681.3	13.69	0.9970
T4	1.4	17.4	19.20	5.1665	668.1	19.36	0.9996

The compression index (C_c) obtained from the test results illustrated in figure 6 which represents void ratio versus logarithmic pressure (kPa) clearly shows that all the samples have decreased in volume well enough, that describes their mechanical behaviour as highly prone to volume change making it undesired for construction. The test results are expressed as consolidation parameters which are compression index (C_c) responsible for compressibility indication of soils, rebound index (C_r) known as swell index after unloading and pre-consolidation pressure (kPa). The pre-consolidation pressure is the pressure where a rapid fall in stiffness of soil occurs and is shown by a concave curve which indicates the maximum effective past pressure [14]. The consolidation parameters obtained from the investigated sample results are all tabulated in Table 6. All the samples show a curve which is concaved before reaching the point where virgin compression line is extended. Also, as the pressure exceeds the pre-consolidation pressure, a continuous decrease in compressibility is observed with the increase in effective stress.

The main difference between all four clays is that T1 showed the least compression index (C_c) whereas T2 and T3 are almost at the average and higher than T1 and finally T4 showing the highest compression index (C_c) among all the samples.

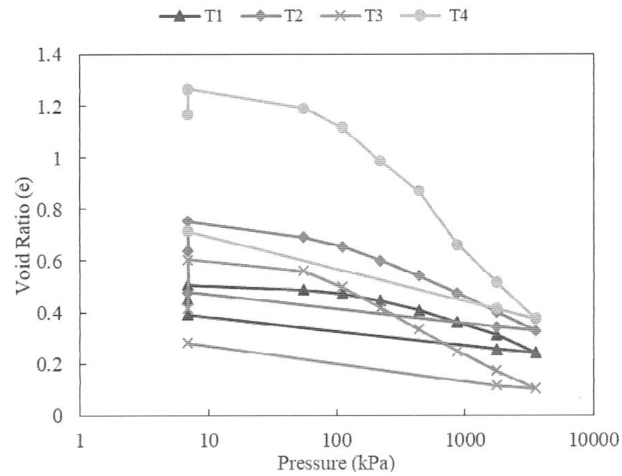

Figure 4. Tested samples consolidation results

Table 6. Consolidation parameter

Samples	**T1**	**T2**	**T3**	**T4**
Swell (%)	3.79	11.39	13.66	19.20
Compression Index (Cc)	0.166	0.199	0.282	0.399
Pre-consolidation P (kPa)	146	115	110	102
Rebound Index (Cr)	0.086	0.042	0.080	0.170

A correlation was made between the compression index (C_c) and plasticity Index (PI) in order to understand the physical and mechanical behaviour of the obtained clays. An important element in civil engineering is the behaviour of soil. Soil properties such as strength, compressibility and plasticity have a great influence on the design during construction. Since index properties such as moisture content and Atterberg limits are basic in soils tests, it will be a wise step to use them for understanding clays behavior [16]. Figure 7 shows the correlation between the plasticity index and the compression index. It was observed that the compression index increased with increasing plasticity index, the compressibility index of soil could be predicted from the relation as, C_c= 0.0033PI+0.0551, by R^2 of 0.8825. The ability of a material to undergo a large amount of deformation is termed as its plasticity; clay soil exerts this property at a large degree especially with an increasing liquid limit. That explains why soils having a high liquid limit, contains high compression index.

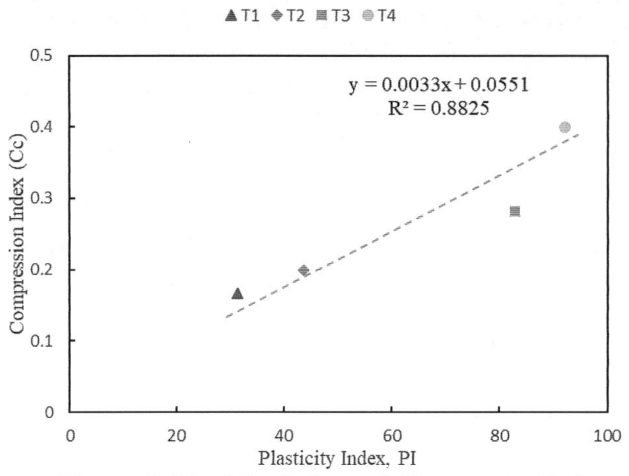

Figure 5: Plasticity Index and Compression Index

4. Conclusions

The results obtained from the hydrometer test showed that all the obtained samples were composed of silts and clays which makes them highly prone to expansion when wet.

- The liquid limit and plasticity index obtained during the Atterberg test showed that all the samples were above 50% LL and 25% PI. Therefore, their volume change was categorized as high to very high [15]. Also, the clays were all beyond the A-line and were categorized as clay with high plasticity (CH) according to Unified Soil Classification System.
- The maximum dry density obtained for sample T1, T2, T3 and T4 were 1.60, 1.62, 1.67, 1.30 g/cm^3 respectively with optimum moisture contents of 21, 22, 19, 39 % respectively.
- The one-dimensional oedometer swell test showed that T4 had the maximum ultimate swell 19.36%, T2 and T3 had a close maximum ultimate swell value of 11.44 and 13.69% respectively, while T1 had the lowest with an ultimate maximum swell value of 3.85% and therefore all of the samples are considered as highly expansive.
- The consolidation test showed that T1 had the highest pre-consolidation pressure, T2 and T3 had a close pre-consolidation pressure while T4 showed the lowest pre-consolidation pressure. The correlation between the compression index and plasticity index showed that the compression index increases with increasing plasticity index in a linear manner.

References

[1] American Society for Testing and Materials. (2011). *Standard test method for Expansion index of soils* (D4829-11). ASTM International.

[2] American Society for Testing and Materials. (2010). *Test Methods for Liquid Limit, Plastic Limit, and Plasticity Index of Soils* (D4318). ASTM International. https://doi.org/10.1520/D4318-10E01

[3] American Society for Testing and Materials. (2011b). *Test Methods for One-Dimensional Consolidation Properties of Soils Using Incremental Loading* (D2435/D2435M). ASTM International. https://doi.org/10.1520/D2435

[4] American Society for Testing and Materials. (2012). *Test Methods for Laboratory Compaction Characteristics of Soil Using Standard Effort (12 400 ft-lbf/ft3 (600 kN-m/m3))* (D698). ASTM International. https://doi.org/10.1520/D0698-12E01

[5] American Society for Testing and Materials. (2014a). *Test Methods for One-Dimensional Swell or Collapse of Soils* (D4546). ASTM International.

[6] American Society for Testing and Materials. (2014b). *Test Methods for Specific Gravity of Soil Solids by Water Pycnometer* (D854). ASTM International.

[7] Atalar, C., & Kilic, R. (2006). Geotchnical properties of Cyprus clays. *International Association for Engineering Geology and the Environment, Formerly International Association for Engineering Geology*, (419), 9.

[8] Bell, F. G. (2000). *Engineering Properties of Soils and Rocks* (3rd ed.). Oxford: Butterwerth-Heinemann Ltd.

[9] Chen, F. H. (1988). *Foundations on Expansive Soils* (2nd edition). Amsterdam ; New York : New York, NY, U.S.A: Elsevier Science.

[10] Constantinou, G., Petrides, G., Kyrou, K., & Chrysostomou, C. (2002). *Swelling Clays: Continuous Threat to the Built Environment of Cyprus* (ETEK). Nicosia, Cyprus: ETEK-Cyprus Scientific and Technical Chamber.

[11] Elsharief, A. M., & Sufian, M. (2018). Time rate of swelling of compacted highly plastic clay soil from Sudan. *MATEC Web of Conferences*, *149*, 02032. https://doi.org/10.1051/matecconf/201814902032

[12] Estabragh, A. R., Parsaei, B., & Javadi, A. A. (2015). Laboratory investigation of the effect of cyclic wetting and drying on the behaviour of an expansive soil. *Soils and Foundations*, *55*(2), 304–314. https://doi.org/10.1016/j.sandf.2015.02.007

[13] Ferber, V., Auriol, J.-C., Cui, Y.-J., & Magnan, J.-P. (2009). On the swelling potential of compacted high plasticity clays. *Engineering Geology*, *104*(3–4), 200–210. https://doi.org/10.1016/j.enggeo.2008.10.008

[14] Ho, M.-H., Chan, C.-M., & Bakar, I. (2010). One Dimensional Compressibility Characteristics of Clay Stabilised with Cement-Rubber Chips, *1*(2), 14.

[15] Holtz, R. D., & Kovacs, W. D. (2010). *An Introduction to Geotechnical Engineering* (2 edition). Upper Saddle River, NJ: Pearson.

[16] Jain, V. K., & Dixit, M. (2015). Correlation of Plasticity Index and Compression Index of Soil. *International Journal of Innovations in Engineering and Technology*, *5*(3), 8.

[17] Jones, D. Earl, & Jones, K. A. (1987). Treating Expansive Soils. *Civil Engineering—ASCE*, *57*(8), 62–65.

[18] Jotisankasa, A., Coop, M., & Ridley, A. (2009). The mechanical behaviour of an unsaturated compacted silty clay. *Géotechnique*, *59*(5), 415–428.

[19] Mitchell, A. R., & van Genuchten, M. T. (1992). Shrinkage of Bare and Cultivated Soil. *Soil Science Society of America Journal*, *56*(4), 1036–1042.

[20] Pusch, R., & Yong, R. N. (2006). *Microstructure of Smectite Clays and Engineering Performance*. CRC Press. https://doi.org/10.1201/9781482265675

[21] Sharma, R. S. (1998). *Mechanical behaviour of unsaturated highly expansive clays*. University of Oxford, Oxford-England.

[22] Sivakumar, V., Tan, W. C., Murray, E. J., & McKinley, J. D. (2006). Wetting, drying and compression characteristics of compacted clay. *Géotechnique*, *56*(1), 57–62. https://doi.org/10.1680/geot.2006.56.1.57

[23] Soltani, A., Taheri, A., Khatibi, M., & Estabragh, A. R. (2017). Swelling Potential of a Stabilized Expansive Soil: A Comparative Experimental Study. *Geotechnical and Geological Engineering*, *35*(4), 1717–1744. https://doi.org/10.1007/s10706-017-0204-1

[24] Komine, H. and Ogata, N. (1994). Experimental study on Swelling Characteristics of Compacted bentonite. Canadian Geotechnical Journal, 31, 478-490.

[25] Sridharan, A., & Gurtug, Y. (2004). Swelling behaviour of compacted fine-grained soils. *Engineering Geology*, *72*(1–2), 9–18.

[26] Wheeler, S. J., Sharma, R. S., & Buisson, M. S. R. (2003). Coupling of hydraulic hysteresis and stress–strain behaviour in unsaturated soils. *Géotechnique*, *53*(1), 41–54. https://doi.org/10.1680/geot.2003.53.1.41

Surface water resource and effect of weather parameters in estimating the annual rainfall: A case study in Lebanon

Y Kassem[1,2*], H Gökçekuş[2], J Aljamal[2]

[1]Department of Mechanical Engineering, Engineering Faculty, Near East University, 99138 Nicosia (via Mersin 10, Turkey), Cyprus
[2]Department of Civil Engineering, Civil and Environment Engineering Faculty, Near East University, 99138 Nicosia (via Mersin 10, Turkey), Cyprus

*E-mail: yousseuf.kassem@neu.edu.tr

Abstract. The quality and quantity of freshwater resources are continually decreasing in the world. The objective of this paper is to review the literature on the water resource with a focus on the surface water, quality of surface water in terms of physical and chemical properties in different locations in Lebanon. Moreover, one of the most important sources influencing the surface water is rainfall. Forecasting rainfall is one of the most essential issues in the hydrological cycle. It is very challenging because is still not possible to develop an ideal model given the uncertainty and unexpected variation. In the present study, prediction models using artificial neural networks (ANN) and multiple linear regressions (MLR) are developed to estimate the annual rainfall as a function of weather parameters and geographical coordinates. The annual data used in this study are recorded in 1942 locations in Lebanon. The latitude, longitude, and altitude of the location, global solar radiation, average temperature, wind speed, and relative humidity are used as the input variables and annual rainfall is estimated as the output variable. The measured values are compared versus those predicted by the ANN and MLR models by evaluating R-squared and Root mean squared error.

1. Introduction

Water availability and use depend on several factors including increased population, energy demand, and related environmental problems [1,2]. Climate change significantly affects the environment and natural resources [2]. Air temperature and precipitation (rainfall or snow) are the major parameters of climate that influence human activities such as urban water resources [3] and agricultural production [4, 5]. Precipitation is one of the most important factors in the Earth's water cycle, affecting several human activities, like agriculture, with significant impacts on the economy [6,7].

Lebanon is a small Mediterranean country (surface area 10,452 km^2, and average width 45 km) located in South-West Asia, between N latitude 34°42′ and 33°3′and E longitudes 35°6′ and 36°37′ [8]. Lebanon’s physiography is unique, dominated by two mountain ranges which run parallel to the sea (NNE-SSW) and are separated by the Bekaa valley. Lebanon has mild, dry summers and cold, wet winters. The heaviest rainfall occurs between November and April, with relatively minimal precipitation, if any, between July and August [9].Lebanon is a Middle Eastern country that is fortunate to have significant water resources, unlike its neighbours. However, rain is mainly concentrated in the winter months. While water is abundant in winter, significant water shortages are still experienced around the country for the rest of the year. Besides, water quality in many areas is questionable.

Recently, various models are used to estimate the monthly or annual rainfall such as the mathematical model, Neuro-Fuzzy Inference System (ANFIS), and Artificial Neural Network (ANN).

The review aims to assess the water resources in Lebanon focused on surface water and physicochemical of the surface water. One of the primary sources surface water is rainfall, thus in the present study, four meteorological parameters including average temperature (Tav), global solar radiation (GSR); relative humidity (RH) and wind speed (WS) are selected for predicting the annual rainfall. Also, geographical coordinates in terms of latitude, longitudes, and elevations are utilized as input variables for the model. For this purpose, Artificial Neural Networks (ANN) and Multiple Linear Regressions (MLR) are developed and tested for predicting annual rainfall in Lebanon. The present study aims to estimate the rainfall at any location in Lebanon where there are no measurements.

2. Surface water resources in Lebanon

Generally, Lebanon's water resources are classified into groundwater and surface water. The sources of surface water resources in Lebanon are from 16 river systems, as shown in Figure 1. The main water resources in Lebanon are groundwater (51%), and surface water (49%). The surface water resources are principally sourced from rivers (46%) and surface storage like dams and lakes (3%) [10].

Lebanon has 40 rivers, 16 of which are permanent (Figure 1). The combined annual flow of rivers is estimated to be about 3,900 million cubic meters, where the majority of the flow (75%) occurs between January and May [11]. Surface water resources are mainly sourced from 16 river mainstream discharges in Lebanon, particularly: El Kabir, Ostuene, El Bared, Abou Ali, El Kjaouz, Ibrahim, EL Kalb, Beirut, Damour, EL Awali, El Zahrani, El Assi, Al Qasmieh, Litani, Wazzani and Hasbani (Figure 1). The Kusba and Abu Samra comprise the Abou Ali river systems. Al-Janin and Al Khodaira constitute the Ibrahim river systems. In addition, El Yamouneh, Qaraoun heights, Qaraoun El Khardali, and El Khardali Sea form the Litani river systems. Table 1 presents the short time-series discharge data from 1971 to 1975 and from 2005-2009 for the 16 perennial rivers in Lebanon. According to the State of the Environment Report of Lebanon, El Assi and Hasbani are the only two rivers that do not discharge into the Mediterranean Sea. Additionally, the highest river flows in Lebanon are associated with Nahr el Litani, Nahr Ibrahim, and Nahr el Assi. Rivers are mainly replenished from springs that are fed from melting snow. Over 2,000 regular springs feed into different streams in Lebanon, creating an aggregate of 1,150-1,200 Mcm/yr of water that is not completely misused [12]. Most of the surface water resources come from springs, where 637 Mcm/yr is currently used [13]. The surface water supply also comes from storage dams, namely the Qaroun dam and the Shibroh Dam, which currently provide about 45 Mcm/yr of water [13]. Table 2 shows the water resources available annually in Lebanon. In addition, exploited resources according to water source type in Mm^3/y are listed in Table 3.

Water resources in Lebanon are heavily polluted, as domestic and industrial wastewater is largely untreated, andintolerable agricultural practices exacerbate the situation. The main rivers in Lebanon have very high levels of bacterial contamination due to the flow of raw sewage, which poses a real threat to public health [16].

Random disposal of solid waste also results in water contamination due to leakage of chemicals. tamination of surface and groundwater from industrial discharges is Also, chemical con common. Untreated wastewater containing heavy metals is often disposed [17], while leaks from underground gasoline tanks and uncontrolled pumping of oil and petroleum by-o products are als common.

Coastal waters in Lebanon are also heavily polluted by wastewater flowing from domestic sewage discharges, industrial wastewater discharges, coastal agricultural waste, and huge wasted dumps (in Tripoli, Burj hammoud, Beirut, Saida, and Tyre).Oil spills especially the large leakage that occurred after the 2006 war with Israel, and coastal power stations (in Beddawi, Zouk, Jiyeh, and Al-zahrani) are the other sources of pollution.. Several studies non have on the coastal and marine waters in Leba found very high levels of pollutants, particularly in areas close to the three main cities of Tripoli,

Beirut, and Sidon, including heavy metals, which are toxic to both marine and human populations [18].

The intensive use of fertilizers and pesticides in agricultural practices, especially during the dry seasons, has led to the filtering of nitrates inside the soil and contamination of groundwater systems with high concentrations found in coastal plain[19], and the Bekaa Valley[20]. As farmers rely on wells for irrigation, health and environmental concerns are increasing. In addition, it is reported that raw sewage is used for irrigation in many areas, including Akkar and Bekaa and farmers throughout the cou ntry are likely to resort to this practice when freshwater is not available[21].

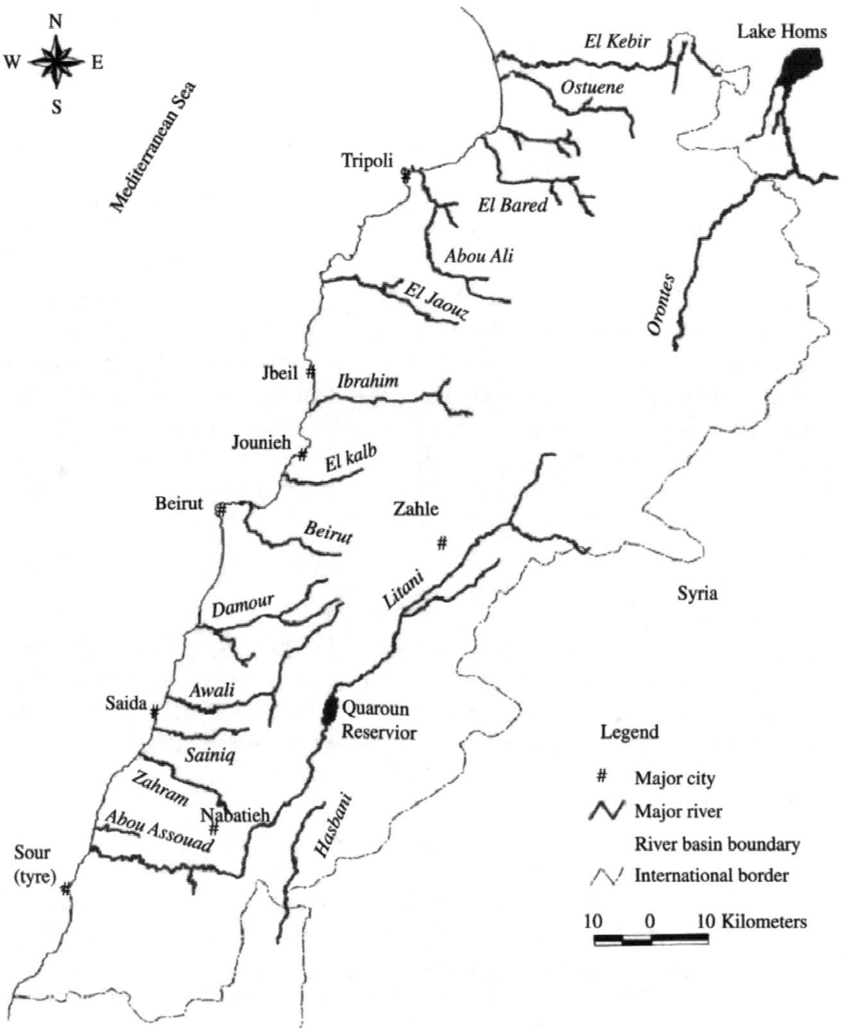

Figure 1. 16 perennial rivers of Lebanon [14].

Table 1. Flow data for the 16 perennial rivers of Lebanon (1971-1975 and 2005-2009) [15].

River name	River length (Km)	Annual average volume		Average flow		Maximum flow		Minimum flow	
		(71-75) Mm^3	(05-09) Mm^3	(71-75) m^3/s	(05-09) m^3/s	(71-75) m^3/s	(05-09) m^3/s	(71-75) m^3/s	(05-09) m^3/s
El Kabir	58	259.2	283.9	9.1	91.3	48.5	190.8	1.5	1.4
Ostuene	44	-	47.0	-	1.6	-	6.9	-	0.0
El Bared	24	132.8	120.1	4.2	3.8	24.0	18.9	0.2	0.5
Abou Ali	45	148.6	206.6	4.6	6.6	25.2	32.5	0.6	1.1
El Kjaouz	38	32.3	44.6	1.0	1.4	11.4	17.9	0.0	0.0
Ibrahim	30	208.6	329.2	6.6	10.5	65.5	79.1	0.1	0.3
EL Kalb	38	154.1	189.3	4.9	6.1	29.3	67.0	0.2	0.0
Beirut	42	47.9	81.8	1.5	2.6	25.1	49.9	0.0	0.0
Damour	38	-	166.9	-	5.4	-	51.0	-	0.1
EL Awali	48	393.7	252.9	12.5	8.1	51.7	32.2	1.9	1.6
El Zahrani	25	19.2	17.5	0.6	0.6	10.6	4.5	0.0	0.0
El Assi	46	326.4	275.5	11.0	8.7	13.8	12.4	8.8	6.0
Al Qasmieh	-	151.7	131.3	4.8	4.2	47.6	46.6	0.8	0.0
Litani	170	-	167.8	-	5.4	-	43.6	-	0.0
Wazzani	-	-	71.9	-	2.3	-	19.5	-	0.5
Hasbani	21	38.4	28.7	1.2	0.9	9.9	14.9	0.0	0.0

Table 2. Annual available resources in Mm^3

Source	Ministry of Energy and Water 2012	Ministry of Environment / ECODIT 2002,3	Ministry of Energy and Water 2010
Precipitation	8,600	8,600	8,200
Evapotranspiration	-4,500	-4,300	-4,100
Losses	**-1,400**	**-1,700**	**-1,333**
Rivers to neighbors	-700	-670	-648
Groundwater	-700	-1,030	-685
Total renewable resources	**2,700**	**0**	**2,767**
Surface water	2,200		2,200
Groundwater	500		567
Net exploitable resources	2,700	2,000	**2,767**

Table 3. Exploited resources according to source type in Mm^3/y

Source	Beirut and Mount-Lebanon Water Establishment	North Lebanon Water Establishment	South Lebanon Water Establishment	Bekaa Water Establishment	Total
Surface water (springs)	174	175	82	206	637
Groundwater (Wells)	198	163	141	193	695
Storage (dams and lakes)	15	-	20	10	45
Total	**387**	**338**	**243**	**409**	**1377**

In recent years, scientific researchers have been examined the quality of surface water in different locations in Lebanon. For instant, Najjar et al. [22] measured twenty-three the physicochemical parameters of Ibrahim River. The results indicated that the quality water was considered as medium to good water quality with an average value of 69.0± 1.9. Haydar et al. [23] measured the physicochemical parameters including pH, T°, TDS, EC, Na+, Ca2+, Mg2+, Clí, SO2í4, NH3+, NOí3, PO2í4, K+ and Heavy metals of Upper Litani River Basin for different seasons. The results demonstrated that the degree of pollution depends on the location and the season. Also, they concluded that the presence of pollution in terms of mineral and anthropogenic came from municipal wastewater and agricultural purposes discharged into the river. Kabbara et al. [24] evaluated the water quality in the coastal area of Tripoli using Landsat 7 ETM+ data. They found that the coastal area of Tripoli is described as moderate eutrophic conditions with fluvial and wastewater runoff sources. Massoud [25] evaluated the water quality index of the small Mediterranean river in Southern Lebanon (Damour River). The results show that the index of water quality is classified as good and anthropogenic activities are mainly source of pollution affect the river. Saadeh et al. [26] assessed the groundwater quality of the Upper Litani River Basin with nearly 300,000 persons depends on it for their daily domestic needs. The authors found that Total Dissolved Solids (TDS) of the river was above the normal range of 100-500 mg/L in the summer season, which indicated that the river received runoff from all other tributaries including the heavily polluted Berdawni. Daou et al. [27] measured the physicochemical and microbiological parameters of the surface water samples taken from the Arka River located in the Akkar District, north of Lebanon. The analysis showed that the most polluted sources in the river come from flatland and the surrounding villages. Therefore, an effective surface water quality management system could be established enabling the proper use of water for irrigation purposes. Fadel and Slim [28, 29] analyzed the physicochemical parameters in terms of pH, electrical conductivity, TDS, turbidity, alkalinity, Ca, Mg, TH, Cl−, SO2 4−, NH3, NO− 3, PO3 4−, Fe, Al, Na, Zn, Cr, Cu and As of Qaraaoun reservoir and the results are compared with water standards. The results showed that suboptimal quality would probably be remediated through customary water treatment processes. Korfali et al. [30] evaluated the water quality of the Qaraaoun reservoir that considered as having three water quality zones and measured the parameters (pH, Eh, DO and temperature) of 15 samples collected from different site. The authors found that the sediment data showed higher metal contents where the river entered the reservoir which matched higher concentrations of water parameters at the influx site. Korfali and Davies [31] investigate the variation of the total metal content (Fe, Mn, Zn, Cu, Pb, and Cd) in bed sediments and water of River Nahr-Ibrahim. The results indicated that the decreasing in water pH caused by the decrease in precipitation rate, lowering the level of water and the dilution of industrial discharges. Korfali and Jurdi [32] examined the quality of domestic water in Beirut city. The results showed deterioration patterns in domestic water quality. Korfali and Jurdi [33] measured the physicochemical and bacteriological parameters of domestic water collected from three household water sources. They found a high frequency of water-borne diseases in the collected water. Korfali and Jurdi [34] determined metal speciation sediment chemical fractions and metal speciation in reservoir water. The measured data indicated that the highest percentages of total metal content in sediment fractions were for Fe in residual followed by reducible, Cr and Ni in residual and is reducible, Cu in organic followed by exchangeable, Zn in residual and inorganic, Pb in organic and carbonate, Cd was mainly in carbonate. Kouzayha et al. [35] analyzed chemically the water quality of drinking water collected from the major cities in Lebanon. The authors found that high pesticide ecotoxicological risk determined by diazinon, chlorpyrifos, fenpropathrin and bifenthrin insecticides in many surface waters. Semerjian [36] analyzed physicochemical and bacterial water quality parameters of domestic bottled water collected from shops and supermarkets throughout Lebanon. The results showed negative growth for fecal Coliforms and positive results for total coliforms. Houri et al. [37] studied the chemical and microbiological properties of Lebanese perennial coastal rivers during the dry season. They found that the most polluted rivers in most categories were Abu Ali and Antelias.

Table 4 summarizes the physical parameters for selected rivers in the dry season (July, August, and September). It is found that pH values for all rivers with exception Damour river are within the limit range of limit drinking water. In addition, it is observed that the lowest value of saturation was recorded in Abu Ali and Antelias rivers and the highest value was obtained in the Damour river. Increase algal growth produced oxygen from photosynthesis led to an increase in the level of dissolved oxygen. In the dry season, the highest average TDS (Total dissolved solids) was found in the Awali river (1863 g/s) and the lowest was found in the Damour river (44 g/s) as shown in Table 5.

Table 4. Characteristics of selected rivers in the dry season [37].

Month	July								
River	**pH**	**Temperature [°C]**	**Dissolved oxygen [mg/l]**	**Saturation [%]**	**Biological oxygen demand [mg/l]**	**Conductivity [μS/cm]**	**Flow [m³/s]**	**Total coliform**	**E. coli**
Kabir	8.32	28.4	5.7	64	-	550	5	230	10
Bared	8.42	24.9	6.90	78	-	456	2.3	350	90
Abu Ali	8.37	22.0	5.55	63	-	498	7.2	12000	5000
Ibrahim	8.34	21.5	7.05	80	-	338	2.5	4900	50
Antelies	8.01	22.0	5.84	65	-	638	1.32	105000	3400
Damour	8.98	30.5	7.05	79	-	370	0.2	470	6
Awali	8.18	20.7	6.00	66	-	462	4.8	140	0
Kassmieh	8.20	29.4	6.15	68	-	576	0.3	40	9
Month	**August**								
River	**pH**	**T [°C]**	**DO [mg/l]**	**% sat**	**BOD [mg/l]**	**Cond [μS/cm]**	**Flow [m³/s]**	**Tsniotsl colliform**	**E. coli**
Kabir	8.30	28.5	6.75	79	14.4	548	4.3	900	20
Bared	8.35	25.0	6.97	82	28.2	470	2	610	17
Abu Ali	8.06	23.5	4.40	51	39.3	604	3.3	26500	3000
Ibrahim	8.41	24.5	6.94	81	62.8	328	2	3500	200
Antelies	8.06	22.5	5.10	60	53.2	569	1	28000	6000
Damour	8.60	25.0	8.31	93	21.3	421	0.1	490	15
Awali	8.17	21.5	7.24	80	33.4	410	9.1	710	1
Kassmieh	8.06	28.4	8.55	94	22.5	608	0.2	80	0
Month	**September**								
River	**pH**	**T [°C]**	**DO [mg/l]**	**% sat**	**BOD [mg/l]**	**Cond [μS/cm]**	**Flow [m³/s]**	**Totsl colliform**	**E. coli**
Kabir	8.26	24.4	5.64	68	12.8	551	6	77	10
Bared	8.45	24.0	7.28	89	20.9	469	3.4	70	20
Abu Ali	8.28	21.5	6.15	75	14.7	561	2.3	1300	200
Ibrahim	8.38	19.5	6.70	83	25.2	356	1.4	60	3
Antelies	7.97	21.0	4.33	54	23.2	625	0.6	11000	5000
Damour	9.09	29.0	10.37	100	24.9	384	0.4	110	4
Awali	8.15	22.5	7.40	82	27	372	9.1	40	1
Kassmieh	7.73	25.0	6.84	76	28.8	590	0.3	290	10

Table 5. Pollutant loading from active coastal rivers in the dry season [37].

River	BOD [g/s]	Nitrate [g/s]	Phosphate [g/s]	Sulfate [g/s]	TDS [g/s]	E.Coli 10-5 [CFU/sec]	Coliform 10-5 [CFU/sec]
Kabir	69.4	13.6	1.1	86.7	1360	68	2052
Bared	63	5.3	0.4	66.3	578	109	881
Abu Ali	115.1	11	1.6	76.8	1146	11662	56604
Ibrahim	86.5	1.1	0.3	17.4	322	166	5546
Antelies	37.3	2	1.4	27.5	290	4693	46928
Damour	6.8	0.4	0.03	8.3	44	2	83
Awali	281	34.9	1.3	191	1863	6.2	2759
Kassmieh	5.4	1	0.04	5.1	76.5	2	36
Total	664.5	69.3	6.17	479.1	5679.5	16708.2	114889

3. Material and method

Climate change affects water resources through changes in temperature, rainfall, runoff, and groundwater recharge. Furthermore, the changes in air temperature and rainfall could affect river flows, and hence, the mobility and dilution of contaminants [38]. Consequently, this study aims to study is to

IOP Conf. Series: Materials Science and Engineering **800** (2020) 012028 doi:10.1088/1757-899X/800/1/012028

predict the annual rainfall using an Artificial Neural Network (ANN) and Multi-linear regression (MLR) approaches in Lebanon.

3.1. Study region and data

In the current study, the used data consist of the annual rainfall, average temperature, relative humidity, global solar radiation and wind speed. The data are recorded in 1942 locations in Lebanon. The geographical area of the selected location is bounded by the latitude of 34.691° and 33.060° and by the longitudes of 36.575° and 35.179° (see Figure 2), whose elevations range from 1 to 2300 m above mean sea level.

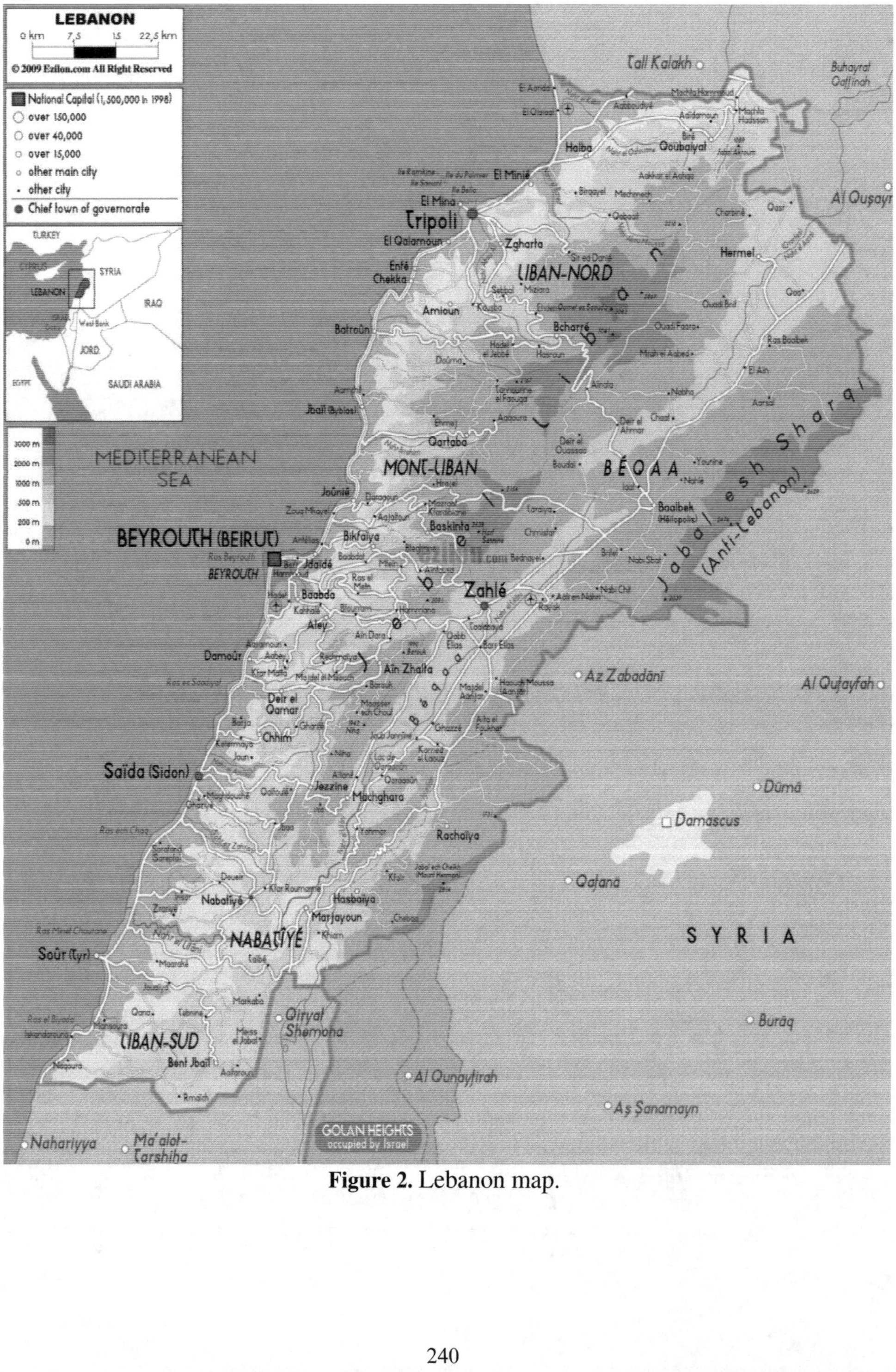

Figure 2. Lebanon map.

3.2. Artificial neural networks

The most widespread technique used in calculating outputs of many systems is the artificial neural networks (ANN) model. A large number of academicians in many different fields have used ANN in their studies [39-42]. The artificial neural networks (ANN) model, also known as the black-box model, is composed of interconnected processing units called artificial neurons or nodes [43]. Generally, the multilayer feed-forward neural network is widely used for solving engineering problems. It consists of three layers, namely the input layer(s), Hidden layer(s) and output layer(s). Besides, the number of these layers depends on the nature of the problem.

In this study, the feed forward architecture with the three layers is used. TRAINLM is used as a training function that updates the weight and bias values of the neuron connections according to Levenberg-Marquardt (LM) optimization. The back-propagation algorithm is used as a learning algorithm and it is a gradient descent algorithm. The activation function for the neurons can be linear or non-linear. The logistic-sigmoid (logsig) and tangent-sigmoid (tansig) were used as an activation function whose output lies between 0 and 1. By trial and error, the optimum number of nodes in the hidden layers, the most suitable transfer function and the number of neurons are determined. To obtain the best performance results, various ANN models are designed.

In this research, a conventional data division technique was used to divide the data, whereby the sets were divided on an arbitrary basis and the statistical properties of the respective data sets were not considered [44]. Approximately 80% of the data was used for training, while the remaining 20% was reserved for testing. The training data was used to train the ANN models with the LM algorithm. The testing data do not affect training and provide an independent measure of network performance during and after training. Moreover, normalization of the data is required for improving the performance of the ANN model. The minimum (min) and maximum (max) values of the inputs and output parameters are shown in Table 6. In general, the number of hidden layers and the number of neurons are the most factors that can affect the performance of the ANN model. Figure 3 shows the structure of the ANN model used in this study. In this study, the number of epochs and performance goal were 100,000 and 0.001, respectively. In addition, the number of the hidden layers varied between 1 and 10, while the number of neurons varied between 5-50 neurons.

Table 6. Limit values for the input and output variables

Parameters	Type	Parameter description	Abbreviation	Minimum	maximum	Unit
Input 1	Meteorological parameters	Annual average temperature	Tav	12.49	23.56	°C
Input 2		Annual relative humidity	RH	55.33	76.50	%
Input 3		Annual global solar radiation	GSR	184.93	230.62	kWh/m^2
Input 4		Annual wind speed	WS	1.74	2.75	m/s
Input 5	Geographical coordinates	Latitude	La	33.69	33.06	°
Input 6		Longitudes	Lo	35.18	36.58	°
Input 7		Elevations	El	1	2300	m
Output 1		Annual rainfall	R	90	1650	mm

3.3. Multiple linear regressions (MLR)

Multiple linear regressions are described as the relationship dependent (y) and independent variables (x). It can be expressed as

$$y_i = \beta_0 + \beta_1 x_1 + \cdots + \beta_i x_i \quad i = 1,2 \ldots n \tag{1}$$

where y_i denotes the dependent variable (rainfall) and x_i where $i=1,2,..,n$, denotes the explanatory or independent variables and β is called the intercept. Minitab 17 was used for the regression and testing of the data.

4. Results and discussions

The annual values of rainfall and the other parameters (Tav, RH, GSR, WS) at 1942 locations for 2016 are recorded and La, Lo and El of each location are used as explanatory input variables. First, the data are randomly split into 80% training data (1554 data) and 20% testing data (388). The result of the ANN model is compared with the MLR model. The statistics of the data are summarized in Table 7.Different statistical measures, including the mean, standard deviation (SD),coefficient of variation(CV), minimum, maximum, skewness and kurtosis are calculated for each variable. It is found that the annual mean rainfall varied from 90 mm to 165 mm. The maximum and minimum annual rainfall recorded in choueifat and El Khraïbé locations, respectively. The CV values are high, ranging from 0.76 to 68.32. During the investigation period, the Skewness value for rainfall is negative, which indicates that all distributions are right-skewed. In addition, Lebanon has a maximum and minimum average temperature of 23.561°C and 12.487°C, respectively. Moreover, it is found that the annual mean wind speed in the country was 2.04m/s. additionally, the global solar radiation and relative humidity values ranged between 184.93-230.62 kWh/m^2 and 55.326-76.501%, respectively. Consequently, it can be established that this country has considerable solar potential. Generally, the mean and standard deviation values suggest that there is good consistency in the meteorological parameter behaviour.

Table 7. Statistical parameters of the meteorological data.

Variable	Mean	SD	CV	Minimum	Maximum	Skewness	Kurtosis
La	33.923	0.361	1.07	33.06	34.691	0.07	-0.47
Lo	35.737	0.271	0.76	35.18	36.575	0.68	-0.14
El	647.8	442.6	68.32	1	2300	0.45	-0.41
TAV	16.488	1.894	11.49	12.487	23.561	1.52	1.59
RH	71.55	3.619	5.06	55.326	76.501	-1.91	5.02
GSR	200.7	11.54	5.75	184.93	230.62	1.11	0.14
WS	2.0449	0.2937	14.36	1.7427	2.7431	1.08	-0.2
R	812.13	223.84	27.56	90	1650	-0.63	0.32

The La, Lo, El, Tav, RH, GSR, WS are utilized in the input layer as input data for the feed forward architecture with a back propagation algorithm (Figure 3). The annual rainfall is the outcome of the output layer. In this study, various ANN configurations were designed and the number of hidden layers and neurons was determined by trial and error for obtaining the best performance results. The logistic-sigmoid and the tangent-sigmoid functions are tried in the hidden layer and output layer. The best performance of the network was obtained by training the developed ANN architecture several times until the MSE showed the minimum value. The same-trained network was tested with the new datasets to check the performance of the network. Table 8 shows the best number of hidden layers and neurons, training rule, activation function, epochs, R-squared and mean squared error (MSE) that were chosen for each ANN model. Also, the R-squared and RMSE for testing data are tabulated in Table 8. It is found that the best ANN model has two hidden layers with 5 neurons and having TANSIG as the activation function.

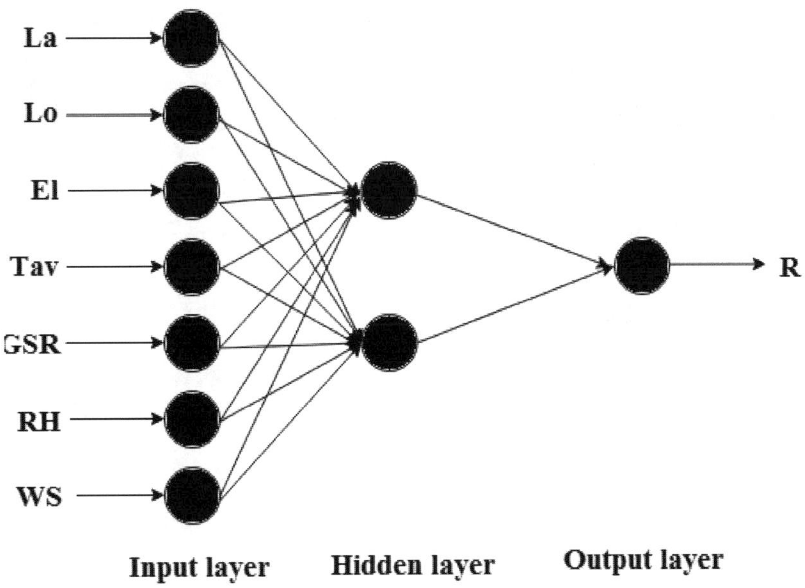

Figure 3. The structure of the ANN for estimating annual rainfall

Table 8. Evaluation of the networks and Statistical tools' performance of the ANN models.

Transfer function	Hidden layer	Neurons	Epoch	MSE	R^2-training	R^2-testing	RMSE (training)	RMSE (testing)
	1	5	85	0.00458	0.7665	0.7458	0.0686	0.0756
	1	10	23	0.00462	0.7700	0.7422	0.0681	0.0761
TANSIG	**2**	**5**	**118**	**0.00333**	**0.7882**	**0.7436**	**0.0653**	**0.0759**
	2	10	188	0.0047	0.7813	0.7314	0.0664	0.0777
	2	15	18	0.00448	0.7667	0.7324	0.0685	0.0776
	1	5	28	0.00778	0.7115	0.6908	0.0811	0.0881
	1	10	110	0.00782	0.7150	0.6872	0.0806	0.0886
Logsig	**2**	**5**	45	0.00653	0.7332	0.6886	0.0778	0.0884
	2	10	73	0.0079	0.7263	0.6764	0.0789	0.0902
	2	15	10	0.00768	0.7117	0.6774	0.0810	0.0901

The MLR model is formed for the data used for training and MLR equations are checked with the testing data. The MLR equation is given below. The Minitab 17 software was applied to the model for the regression analysis of the actual data. Eq. (1) is the model equation that correlates the response (rainfall) to the independent variables in terms of La, Lo, El, Tav, RH, GSR, WS.

$$R = 0.476 + 0.798 \cdot La - 0.976 \cdot Lo + 0.166 \cdot El - 0.128 \cdot Tav - 0.058 \cdot RH + 0.38 \cdot GSR - 0.048 \cdot WS \quad (1)$$

Figures 4 and 5 show the comparisons of the actual and predicted data of training and testing phases, respectively. It is found that the ANN model makes better predicts than the MLR model.

IOP Conf. Series: Materials Science and Engineering **800** (2020) 012028 doi:10.1088/1757-899X/800/1/012028

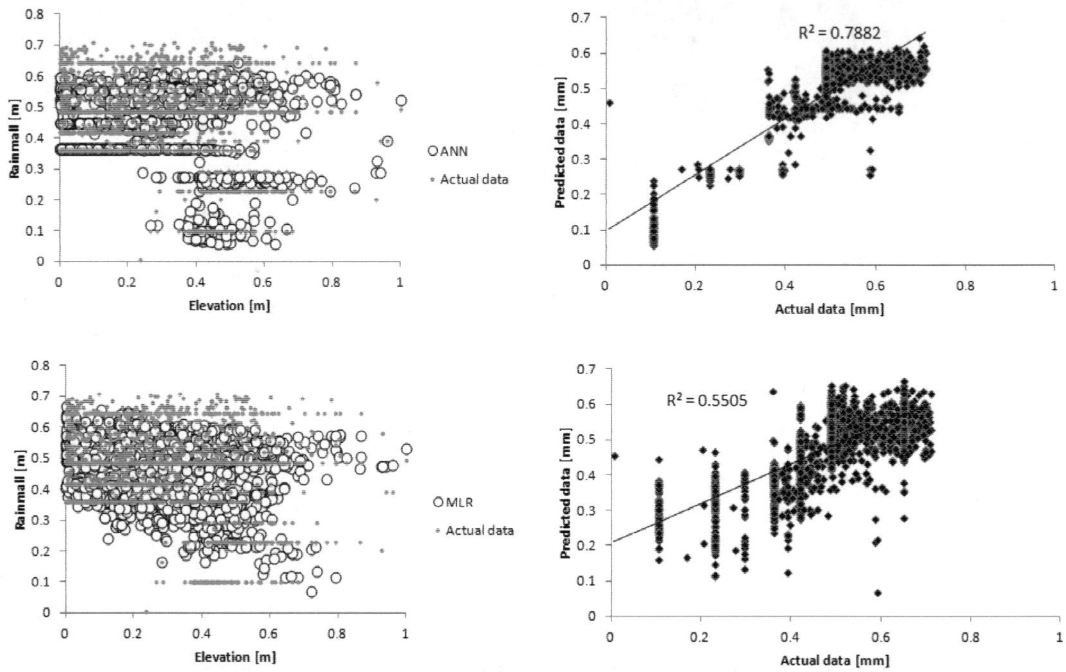

Figure 4. Comparisons of the predicted and actual values of the annual rainfall for training phase.

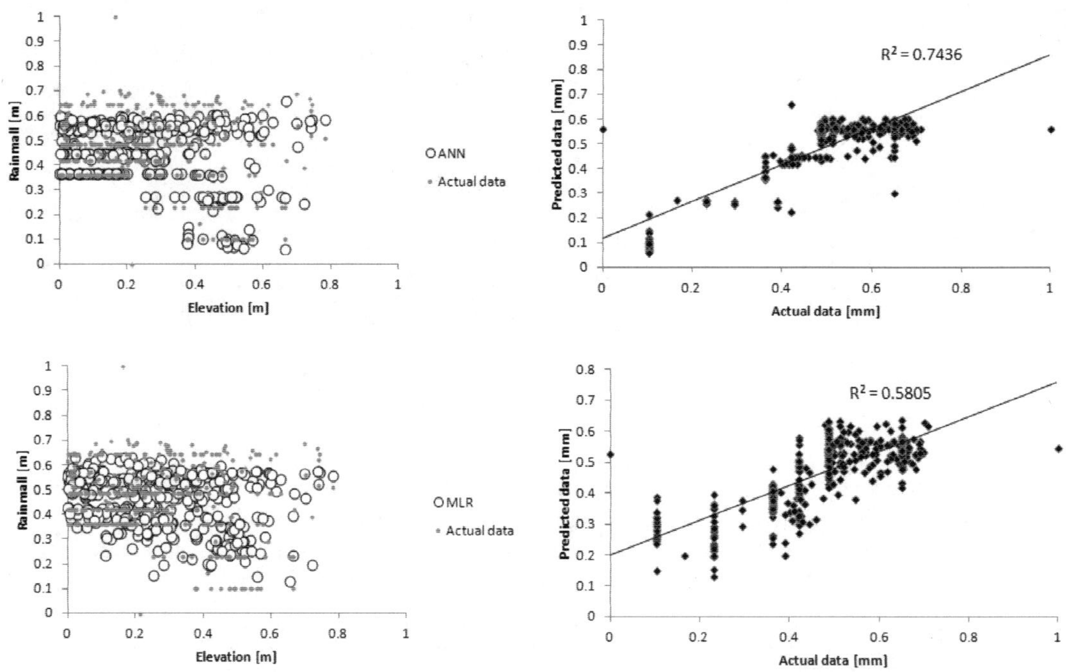

Figure 5. Comparisons of the predicted and actual values of the annual rainfall for testing phase.

In addition, it is found that the scatter diagrams, the noises of the predicted values around the best-fit lines are wider for MLR models. Furthermore, the R^2 for annual rainfall estimated by the ANN and MLR for test phases are found to be 0.7436 and 0.5805, respectively (see Table 9). It is concluded that the ANN model is a capable model to define a non-linear relationship as output variable and weather parameters and geographical coordinates of any location in Lebanon as input variables without needing a priori information and without having to make preliminary assumptions.

Table 9. Statistical tools' performance of the ANN and MLR models.

Model	R^2-training	R^2-testing	RMSE-training	RMSE-testing
ANN	0.7882	0.7436	0.0653	0.0759
MLR	0.5505	0.5805	0.0950	0.0971

5. Conclusions

The paper has highlighted that water resources in Lebanon focused on surface water and the quality of the water. Since the main surface water sources are rainfall thus, annual climate data in terms of rainfall, average temperature relative humidity, global solar radiation, and wind speed were analyzed statistically to show the effect of climate parameters on the rainfall. This study has shown the power of ANN to evaluate the most influencing input parameters in the prediction of monthly rainfall. ANN model using a back-propagation algorithm was developed. Out of the ANN and MLR models, the ANN model has given the best prediction with the highest R-squared and minimum RMSE. These models can be used for determining the level of groundwater based on the amount of rainfall and rainfall distribution at any site in Lebanon. Therefore, it can be used for the assessment of trends in groundwater levels across Lebanon.

References

[1] M. Song, R. Wang, and X. Zeng, Water resources utilization efficiency and influence factors under environmental restrictions. Journal of Cleaner Production, 184, (2018) 611-621.
[2] S. Kundu, D. Khare, and A. Mondal, Future changes in rainfall, temperature and reference evapotranspiration in central India by least square support vector machine. Geoscience Frontiers, 8(3) (2017) 583-596.
[3] N. Seino, T. Aoyagi, and H. Tsuguti, Numerical simulation of urban impact on precipitation in Tokyo: How do urban temperature rise affect precipitation? Urban Climate, 23 (2018) 8-35.
[4] Y. Kang, S. Khan, and X. Ma, Climate change impacts on crop yield, crop water productivity, and food security – A review. Progress in Natural Science, 19(12) (2009) 1665-1674
[5] T. Iizumi, and N. Ramankutty, How do weather and climate influence cropping area and intensity? Global Food Security, 4 (2015) 46-50.
[6] C. P. McMullen, and J. R. Jabbour, Climate change science compendium 2009. Nairobi: United Nations Environment Programme, (2009).
[7] OECD (Organization for Economic Co-Operation and Development), The Land-Water-Energy nexus biophysical and economic consequences. Paris: OECD Publishing, (2017).
[8] ECODIT, National Action Plan for the Reduction of Pollution into the Mediterranean Sea from Land-Based Sources, Technical report for UNEP and Ministry of Environment, Lebanon, (2015).
[9] K. J. Sene, T. J. Marsh, and A. Hachache, An assessment of the difficulties in quantifying the surface water resources of Lebanon. Hydrological Sciences Journal, 44(1) (1999) 79-96.
[10] P. Schuler, and A. Margane, Water Balance for the Groundwater Contribution Zone of Jeita Spring using WEAP Including Water Resources Management Options & Scenarios, (2013).
[11] F. Comair, Water Management and Hydrodiplomacy in Lebanon. Notre Dame University Press, Zouk Mikayel, (2009).
[12] Fanack Water, Retrieved November 5, 2018, from https://water.fanack.com/ar/lebanon/water-resources-in-lebanon/

[13] World Bank, Republic of Lebanon - Water sector: Public expenditure review. Retrieved November 5, 2018, from http://documents.worldbank.org/curated/en/965931468265767738/Republic-of-Lebanon-Water-sector-public-expenditure-review

[14] El-Fadel, M. (2001). Water resources management in Lebanon: institutional capacity and policy options. Water Policy, 3(5), 425–448. doi: 10.1016/s1366-7017(01)00079-4

[15] Ministry of Environment, Lebanon's Initial National Communication to the United Nations Framework Convention on Climate Change. Retrieved November 5, 2018, from http://climatechange.moe.gov.lb/national-communications

[16] Nehme, N., Haydar, C., Koubaissy, B., Fakih, M., Awad, S., Toufaily, J., … Hamieh, T. (2014). The Distribution of Heavy Metals in the Lower River Basin, Lebanon. Physics Procedia, 55, 456–463. doi: 10.1016/j.phpro.2014.07.066

[17] Raven, P. H., Berg, L. R., & Johnson, G. B. (2001). Environment. Fort Worth: Harcourt College

[18] Fallah, R. E., Olama, Z., & Holail, H. (2016). Marine Quality Assessment of Northern Lebanese *Coast: Microbiological and Chemical Characteristics and their Impact on the Marine* Ecosystem. International Journal of Current Microbiology and Applied Sciences, 5(1), *.389-376*

[19] Saad, Z., Kazpard, V., Kazpard, K., & Nabhan, P. (2003). Natural and anthropogenic influence on the quality of Ibrahim river water, Lebanon. Journal European D'Hydrologie, 34(1), 85-99.

[20] Darwish, T., Atallah, T., Francis, R., Saab, C., Jomaa, I., Shaaban, A., . . . Zdruli, P. (2011). *soil and groundwater contamination with nitrate: A case study Observations on from* Lebanon-East Mediterranean. Agricultural Water Management, 99(1), 74-84.

[21] Farajalla, N., El Hajj, R., Kerkezian, S., Farhat, Z. and Matta, M. (2015). The Way Forward to Safeguard Water in Lebanon: National Water Integrity Risk Assessment. American University of Beirut, Lebanon

[22] Najjar, P. E., Kassouf, A., Probst, A., Probst, J.-L., Ouaini, N., Daou, C., & Azzi, D. E. (2019). High-frequency monitoring of surface water quality at the outlet of the Ibrahim River (Lebanon): A multivariate assessment. Ecological Indicators, 104, 13–23. doi: 10.1016/j.ecolind.2019.04.061

[23] Haydar, C. M., Nehme, N., Awad, S., Koubaissy, B., Fakih, M., Yaacoub, A., … Hamieh, T. (2014). Water Quality of the upper Litani River Basin, Lebanon. Physics Procedia, 55, 279–284. doi: 10.1016/j.phpro.2014.07.040

[24] Kabbara, N., Benkhelil, J., Awad, M., & Barale, V. (2008). Monitoring water quality in the coastal area of Tripoli (Lebanon) using high-resolution satellite data. ISPRS Journal of Photogrammetry and Remote Sensing, 63(5), 488–495. doi: 10.1016/j.isprsjprs.2008.01.004

[25] Massoud, M. A. (2011). Assessment of water quality along a recreational section of the Damour River in Lebanon using the water quality index. Environmental Monitoring and Assessment, 184(7), 4151–4160. doi: 10.1007/s10661-011-2251-z

[26] Saadeh, M., Semerjian, L., & Amacha, N. (2012). Physicochemical Evaluation of the Upper Litani River Watershed, Lebanon. The Scientific World Journal, 2012, 1–8. doi: 10.1100/2012/462467

[27] Daou, C., Nabbout, R., & Kassouf, A. (2016). Spatial and temporal assessment of surface water quality in the Arka River, Akkar, Lebanon. Environmental Monitoring and Assessment, 188(12). doi: 10.1007/s10661-016-5686-4

[28] Fadel, A., & Slim, K. (2018). Evaluation of the Physicochemical and Environmental Status of Qaraaoun Reservoir. Water Science and Technology Library The Litani River, Lebanon: An Assessment and Current Challenges, 71–86. doi: 10.1007/978-3-319-76300-2_5

[29] Shaban, A., & Hamzé, M. (2019). Litani River, Lebanon: an assessment and current challenges. S.l.: Springer.

[30] Korfali, S. I., Jurdi, M., & Davies, B. E. (2006). Variation of Metals in Bed Sediments of Qaraaoun Reservoir, Lebanon. Environmental Monitoring and Assessment, 115(1-3), 307–

319. doi: 10.1007/s10661-006-6556-2

[31] Korfali, S., & Davies, B. (2003). A comparison of metals in sediments and water in the river Nahr-Ibrahim, Lebanon: 1996 and 1999. Environmental Geochemistry and Health, 25, 41–50.

[32] Korfali, S. I., & Jurdi, M. (2007). Assessment of domestic water quality: case study, Beirut, Lebanon. Environmental Monitoring and Assessment, 135(1-3), 241–251. doi: 10.1007/s10661-007-9646-x

[33] Korfali, S. I., & Jurdi, M. (2008). Provision of safe domestic water for the promotion and protection of public health: a case study of the city of Beirut, Lebanon. Environmental Geochemistry and Health, 31(2), 283–295. doi: 10.1007/s10653-008-9218-1

[34] Korfali, S. I., & Jurdi, M. S. (2010). Speciation of metals in bed sediments and water of Qaraaoun Reservoir, Lebanon. Environmental Monitoring and Assessment, 178(1-4), 563–579. doi: 10.1007/s10661-010-1713-z

[35] Kouzayha, A., Ashi, A. A., Akoum, R. A., Iskandarani, M. A., Budzinski, H., & Jaber, F. (2013). Occurrence of Pesticide Residues in Lebanon's Water Resources. Bulletin of Environmental Contamination and Toxicology, 91(5), 503–509. doi: 10.1007/s00128-013-1071-y

[36] Semerjian, L. A. (2010). Quality assessment of various bottled waters marketed in Lebanon. Environmental Monitoring and Assessment, 172(1-4), 275–285. doi: 10.1007/s10661-010-1333-7

[37] Houri, A., & Jeblawi, S. W. E. (2007). Water quality assessment of Lebanese coastal rivers during dry season and pollution load into the Mediterranean Sea. Journal of Water and Health, 5(4), 615–623. doi: 10.2166/wh.2007.047

[38] Whitehead, P. G., Wilby, R. L., Battarbee, R. W., Kernan, M., & Wade, A. J. (2009). A review of the potential impacts of climate change on surface water quality. Hydrological Sciences Journal, 54(1), 101–123. doi: 10.1623/hysj.54.1.101

[39] Kalogirou, S. A. (2001). Artificial neural networks in renewable energy systems applications: a review. Renewable and Sustainable Energy Reviews, 5(4), 373–401. doi: 10.1016/s1364-0321(01)00006-5

[40] Budcema, M., & Sacco, P. L. (2000). Feedforward networks in financial predictions: the future that modifies the present. Expert Systems, 17(3), 149–170. doi: 10.1111/1468-0394.00137

[41] Cobaner, M., Unal, B., & Kisi, O. (2009). Suspended sediment concentration estimation by an adaptive neuro-fuzzy and neural network approaches using hydro-meteorological data. Journal of Hydrology, 367(1-2), 52–61. doi: 10.1016/j.jhydrol.2008.12.024

[42] Parlak, A., Islamoglu, Y., Yasar, H., & Egrisogut, A. (2006). Application of artificial neural network to predict specific fuel consumption and exhaust temperature for a Diesel engine. Applied Thermal Engineering, 26(8-9), 824–828. doi: 10.1016/j.applthermaleng.2005.10.006

[43] Nourani, V., Mousavi, S., Dabrowska, D., & Sadikoglu, F. (2017). Conjunction of radial basis function interpolator and artificial intelligence models for time-space modeling of contaminant transport in porous media. Journal of Hydrology, 548, 569–587. doi: 10.1016/j.jhydrol.2017.03.036

[44] Bowden, G. J., Maier, H. R., & Dandy, G. C. (2002). Optimal division of data for neural network models in water resources applications. Water Resources Research, 38(2). doi: 10.1029/2001wr000266

Acoustic and structural design of a highway noise barrier

S Kesten[1], Ö Umut[2] and B Ayva[3]
[1] Assistant Professor at IŞIK University Engineering Faculty, Department of Civil Engineering
[2] Assistant Professor at IŞIK University Engineering Faculty, Department of Civil Engineering
[3] Teaching Assistant at IŞIK University Engineering Faculty, Department of Civil Engineering

E-mail: sercan.kesten@isikun.edu.tr

Abstract. Highway noise barriers are designed to mitigate the effects of traffic noise along the highway. Noise barriers primarily block the direct path of the sound between the source on the highway and the receiver exposed to the sound. Noise mitigation capability, aesthetics, cost and constructability and structural capacity against wind and earthquake loads are critical parameters to be considered in noise barrier design. The height and length of a noise barrier are determined by acoustical considerations, aesthetics and cost. In addition, the structural design of the noise barrier should meet the established standards. The aim of this study is to evaluate the noise mitigation capabilities, aesthetic features and costs of acoustic designs for the residential buildings located near the urban highway and evaluate the structural performances against wind and earthquake. Through the interviews, the designs higher than 10m are not found acceptable by the residents and also due to the aesthetic concerns the upper part of the noise barrier is set to be constructed as glass. Therefore, the noise reduction capabilities and construction cost and the alternatives that complies with the constraints are determined and the corresponding costs are calculated accordingly. Noise simulation software is used to identify the effectiveness of selected noise barriers and structural analysis and design software is used to evaluate the structural performance. The results show that designed precast lightweight concrete and glass composite noise barrier meet the required noise reduction goal along with a resilient structural performance and affordable cost.

1. Introduction

Despite the recognition of the importance to reduce environmental noise has had a lower priority than that the other environmental problems such as air and water pollution, noise pollution is one of the phenomena that should be addressed in many cities in the world. The adverse impacts of transportation-related noise in general and road traffic noise in particular on the society have gained recognition in the last 40 years in the world where more than 55% of the population has been started to live in urban cities. [1,2]

There are a number of methods for controlling road traffic noise and noise barriers are arguably the most simple, effective and commonly used method among alternatives. Turkey and many other countries have established their national noise barrier standards such as TS EN 1793-1,2,3, EN 1794-1,2,3 and ZTV-LSW 06 [3,4,5]. Noise barriers are commonly used in urban expressways in Japan. In figure 1 noise barriers installed on Tomei Expressway are presented.

Highway noise barriers are designed to mitigate the effects of traffic noise along the highway. Noise barriers primarily block the direct path of the sound between the source on the highway and the receiver exposed to the sound. Noise mitigation capability, aesthetics, cost and constructability and structural capacity against wind and earthquake loads are critical parameters to be considered in noise barrier design. The height and length of a noise barrier are determined by acoustical considerations, aesthetics and cost. Therefore, the need for a noise barrier is primarily designed to meet the acoustic standards; however, the final design is generally shaped by the non-acoustic factors as well.

The aim of this paper is to present a road traffic noise calculation method, road traffic noise modelling and calibration and design of a noise barrier by considering acoustic, aesthetic and structural standards. Section 2 highlights the theoretical concepts of SRM II highway noise level calculation method, in Section 3 road traffic noise modelling of high-rise residential building along the highway is given, in Section 4 noise barrier acoustic and structural design are presented and the corresponding costs are calculated, in the last section the conclusions are drown and the possible areas of further research is discussed.

Figure 1. A noise barrier along Tomei Expressway around Yokohama [Maps Google Image]

2. SRM II highway noise level calculation method

Several acoustic emission and propagation models derived by national standards can be applicable for transportation noise pollution. Generally, models are based on the solution of boundary integral equation of the wave equation in two dimensions and the results are transformed to derive a pseudo-three-dimensional (3-D) solution. [6] Most of highway noise level calculation models are regional and specific for countries such as Denmark's NORD2000 [9] [7], Germany's DIN18005-1 [8] and SCHALL 03 [9], France's NMPB [10] and Dutch's RMR96-SRMII [11]. Since each model is specific to the transport infrastructure of the country in which it is developed, various calibration and correction coefficients are needed when used in another country. When noise mapping became mandatory throughout Europe after the European Directive on the Assessment and Management of Environmental Noise, 2002/49/EC [12], studies on this subject gained momentum. Because as a controversial decision, countries not having country-specific noise prediction model are recommended by the same European Directive to use Dutch SRM II Model. The Republic of Turkey doesn't have its own noise calculation model for transportation noise. However, Dutch SRM II noise calculation model is widely used according to directive of 2002/49/EC and TR/2004/IB/RN/02 in The Republic of Turkey [13].

RMR96 SRM II noise prediction model work on the octave bands such as median frequencies of 63, 125, 250, 500, 1000, 2000, 4000 and 8000 Hz and two different equations for each road and railway traffics. Road traffic prediction model in SRM II examines road vehicles in three different groups such as light vehicles (lv), medium vehicles (mv) and heavy vehicles (hv) [14]. The corresponding equation for the road traffic noise in RMR96 SRM II is given below:

$$L_{Ei,m} = 10log\left(\frac{Q_m}{v_m}\right) + \alpha_{i,m} + \beta_{i,m}log\left(\frac{v_m}{v_{0,m}}\right) + C_{surface_{i,m}} + C_{H,m} \quad (1)$$

Where i and m are index for octave band and vehicle category respectively. Q, v, and v_0 are mean flow of vehicle per hour, mean speed in km/h and reference speed respectively. This model also contains correction constants such as $C_{surface}$ for road surface type and C_H for inclination of road. Alpha and beta are parameters which determine the sound power level for vehicle categories. The constants are tabulated in the table 1 below.

Table 1. SRM II Prediction Model Constants for Road Traffic Noise

Octave frequency [Hz]	alpha lv	alpha mv	alpha hv	beta lv	beta mv	beta hv
62	74.5	79.9	84.1	-.05	-0.2	9.8
125	84.5	91.1	91.4	24.6	16.6	11.4
250	89.9	97.1	97.7	27.6	2.5	2.6
500	94.0	100.5	104.8	26.1	26.6	23.2
1000	101.1	103.3	106.5	26.8	22.3	20.8
2000	99.0	100.4	102.4	22.5	16.6	15.0
4000	90.9	93.9	95.6	22.2	16.2	12.4
8000	81.0	85.6	87.0	11.7	-1.9	-3.1

Directive on the Assessment and Management of Environmental Noise demands on cities to produce noise maps [12]. During the producing process, several proprietary software like IMMI [15], SoundPlan [16], CadnaA [17], MITHRA-SIG [18] etc. are used. These software are based on standard methods like SRM II, XPS 31-133 etc. and need several inputs such as transportation networks, buildings, distribution of population and topography.

3. Road traffic noise modelling of high-rise residential building along the highway

3.1. Noise resources and standard limits

According to the Turkish regulation on the assessment and management of environmental noise was first published in 2005, and later amended in June 2010 and 2015, the noise levels are varied with respect to the time of day for the commercial buildings and dense residential areas with noise sensitive locations. The limits are given in Table 2, where the night noise limit is set as 58 dBA [19].

Table 2. Noise Limits

Locations	Existing Roads L_{day} (dBA)	Existing Roads $L_{evening}$ (dBA)	Existing Roads L_{night} (dBA)
Commercial Buildings and Dense Residential Areas with Noise Sensitive Locations	68	63	58

With the increase in urbanization rate the need for high-rise building are also increased. As a result of this, the value of the land is also raised which pushed the contractors to use the construction areas all the way to the borders. Therefore, most of the high-rise buildings in big cities in Turkey are facing highways.

One of the prominent examples of this type of high-noise impacted area is selected as the case study. The residential high-rise complex is located in Istanbul at Beylikdüzü and Esenyurt region. The complex is near a busy arterial that connects Ambarlı Port to a major highway. The volume of traffic and the ratio of heavy vehicles is considerably high on the road. The aerial and the 3D view of the residential complex is given in Figure 2 and Figure 3.

Figure 2. The location of case study

Figure 3. 3D view of residential high-rise complex

3.2. Noise Level Measurements

Noise level measurements were made at noon on 02.09.2019. As the measurement points, the side of the highway, the garden of the residential building on the highway front and the roof of a 4th storey of building on the same front were selected. The results of the measurements are given in figure 4 and noise level statistics are tabulated in table 3. High noise levels are detected on the façade of the building. The legal limit for daylight hours is 68 dB (A) and the measured value in the site yard is above the limit values that is around 71 dB (A).

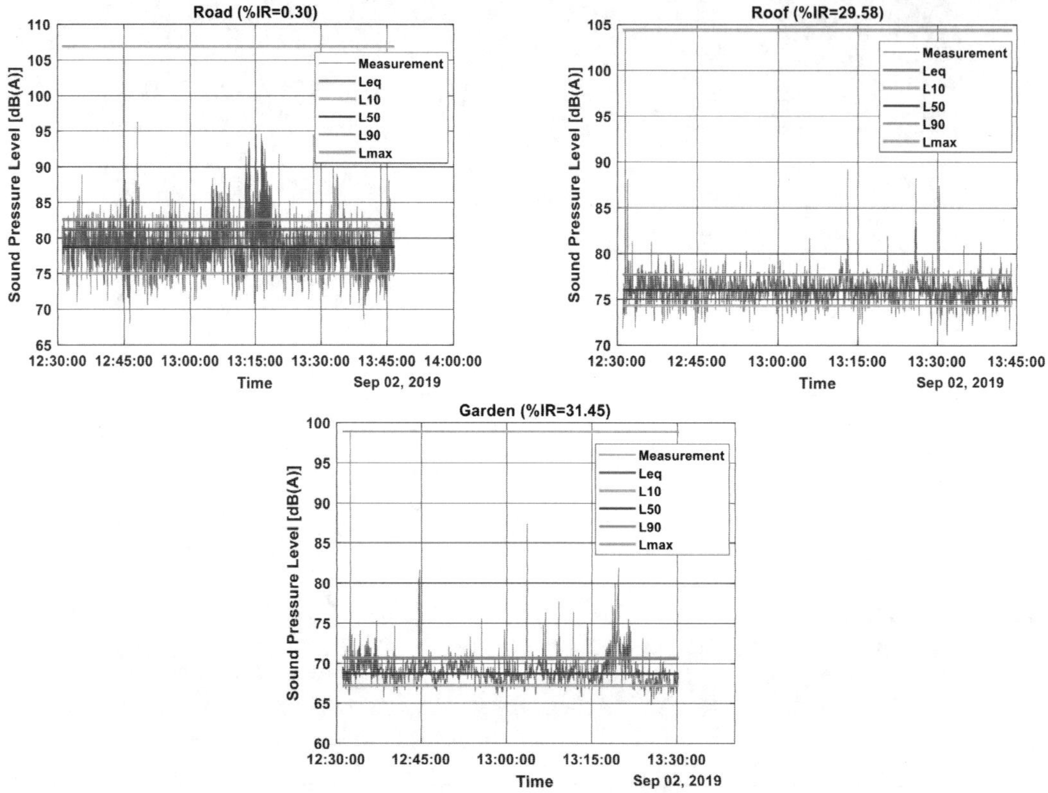

Figure 4. Noise levels measured on site

Table 3. Noise level statistics

	L_{eq} [dB(A)]	L_{10} [dB(A)]	L_{50} [dB(A)]	L_{90} [dB(A)]	IR [%]
Garden	70.61	67.20	68.70	70.00	31.45
Roof (h=12m)	77.67	74.30	76.00	77.70	29.58
Near road	81.21	75.00	78.80	82.60	0.30

The noise model of study area is modelled on IMMI 2018 software, and the topography model was created through NASA's SRTM data. The project area has been brought to its current elevations through the project files obtained from the contractor of the residential complex. It was found that the garden elevation on the façade of the HEP Istanbul housing estate was at 31.2 m. The height of the designed noise barriers is defined at 31.2 m elevation. For the defined heavy vehicle traffic, it is assumed that Ambarlı Port will operate with 75% capacity and handle 3.65 million TEU cargo annually, thus creating a heavy vehicle traffic of 10.000 vehicles per day [20]. It is assumed that the traffic of this heavy vehicle will increase to 830 truck/hour and the average cruising speed will be 65 km/h. Even though the speed limit indicates 50km/h for heavy vehicles it is observed that the heavy vehicle are not complying the speed limits on the road. The measurements were used for the calibration of the noise model and vehicle volume data were used as the calibration variable.

3.3. Modelling and calibration of road traffic noise and exposure levels

Noise levels obtained by noise simulation model and the field measurements are presented in table 4. In this respect, automobile traffic was used as a calibration variable. The measured L_{eq} value and the calculated L_{eq} values are significantly close to each other and the logarithmic relative error is about 1%. Considering the fact that, uncertainty level in noise modelling is widely accepted as 2.5 dBA, it can be considered that the model established in IMMI 2018 software can mimic the actual situation with a high precision. It should be noted that noise levels are affected by environmental variables such as wind and temperature, the results could also vary for traffic volumes and conditions.

Table 4. L_{eq} levels for noise model and site measurements

	L_{eq} [dBA]$_{simulation}$	L_{eq} [dBA]$_{measurement}$
Garden	70.25	70.61
Roof (h=12m)	77.90	77.67
Near road	81.56	81.21

The building blocks are modelled as 3D solids, the receiver points are placed in each floor on the facade and the noise is defined as line source. Figure 5 shows the building blocks and line noise sources. As can be seen from figure 6, the noise levels on the facade of the buildings in the residential complex are above the limits set in Turkish regulation on the assessment and management of environmental noise. The width of the noise resource and the high traffic volume on the road narrow the noise mitigation solution set. Therefore, a simple and cost-effective noise barrier design is proposed for mitigating the impacts of highway traffic noise.

Figure 5. Modelled building blocks in 3D

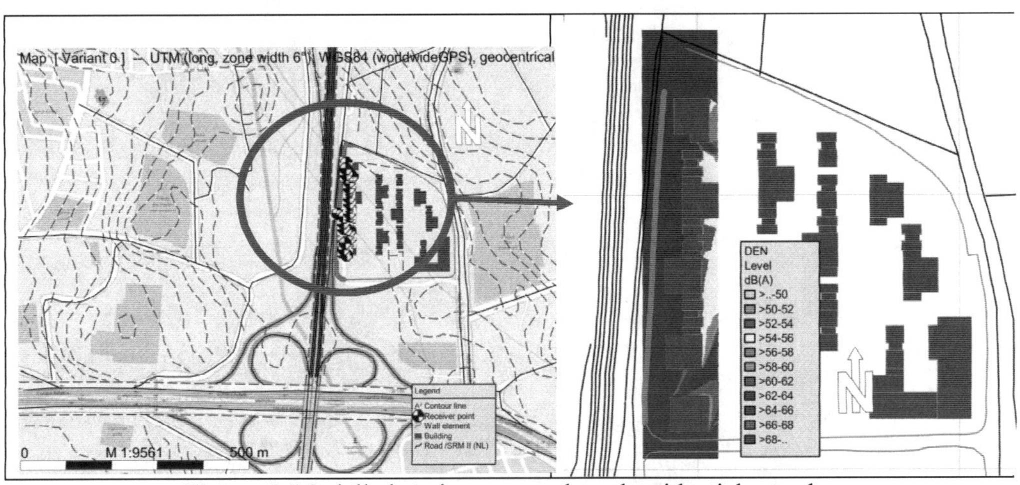

Figure 6. Modelled study area, roads and residential complex

4. Noise barrier design and costs

The acoustical performance of a noise barrier can be measured with the ratio of path length difference to the acoustic wavelength. The diffracted path when a barrier is installed in between noise resource and a receiver is presented in Figure 7 schematically.

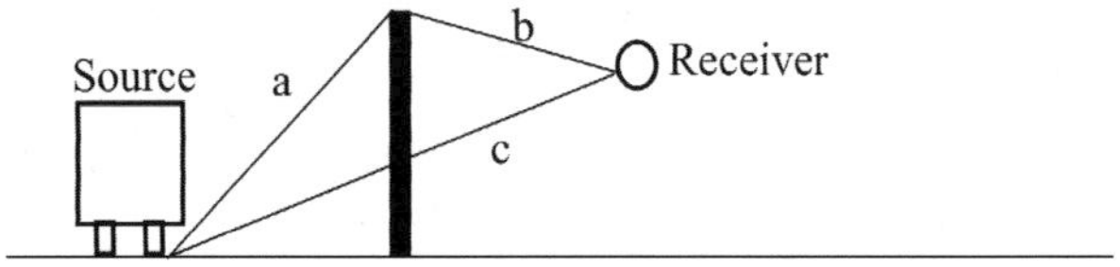

Figure 7. Path length difference

The path difference can be defined as the difference between the direct sound path from source to receiver when there is no noise barrier and the diffracted path when a barrier is installed in between.

$$\delta = a + b - c \tag{2}$$

A non-dimensional variable namely Fresnel number, N, is defined as the ratio of the path length difference, δ, over the wavelength of sound in air, λ. Fresnel number has been widely used in simple models as a performance indicator of a noise barrier.

$$N = \frac{2\,\delta}{\lambda} \tag{3}$$

4.1. Geometric design and acoustic performance

The acoustical performance of a noise barrier designs are evaluated in noise simulation environment. The initial designs higher than 10m are not found acceptable by the residents and also due to the aesthetic concerns the upper part of the noise barrier is set to be constructed as glass. The conceptual design of alternative noise barriers are given in Figure 8.

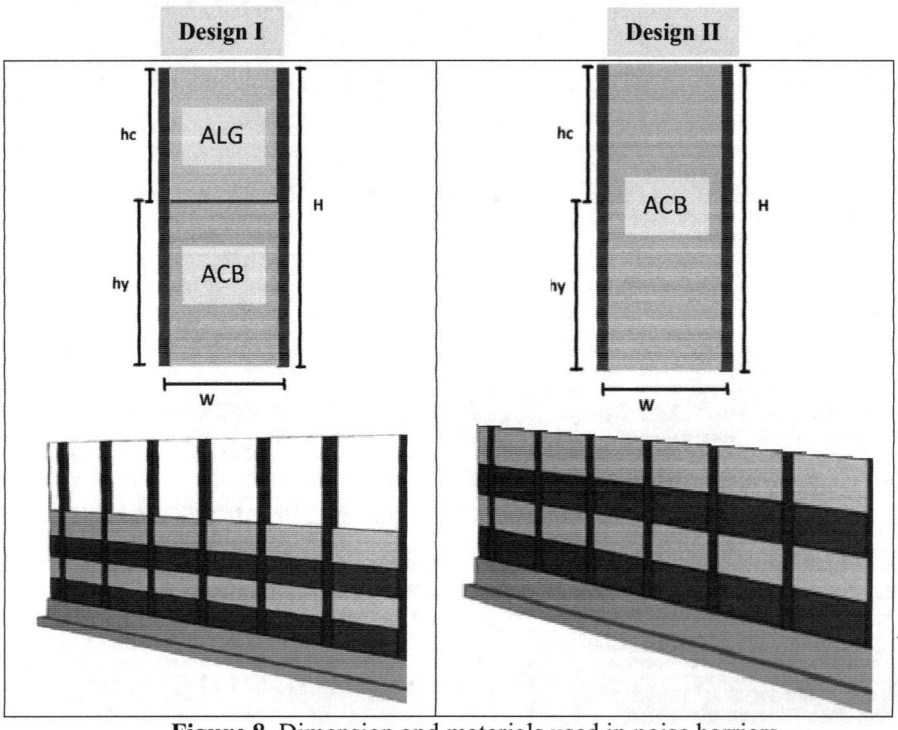

Figure 8. Dimension and materials used in noise barriers

The unit width is 3m (W) and the total height is 7.4m (H) for both designs. Design I is a composed of a 4.2m (h_y) aerated concrete blocks (ACB) and 3.2m (h_g) of acoustic laminated glass (ALG). ACB's internal structure is comprised of 60%-70% air, and AAC absorbs much more sound than conventional concrete and other materials, effectively reducing sound wave transmission. ALG is produced by joining two or more glass plates under heat and pressure with the help of special acoustic polyvinyl butyral (PVB) layers that provide sound insulation. The thickness of the ACB is 12mm and the unit weight is 30 kg/m^2.

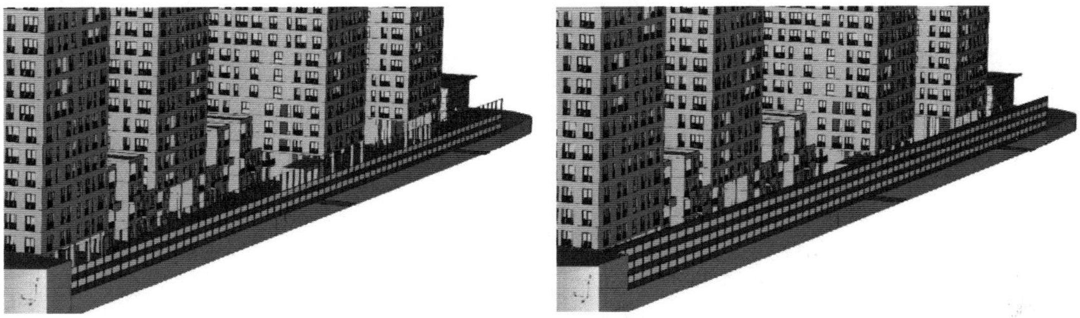

Figure 9. 3D view of Design I and Design II

The acoustical performance of noise barrier designs is evaluated in a noise simulation environment. Both designs provide a reliable highway noise mitigation and the acoustic performance of the designs is statistically indifferent. The receiver points read 79.89 dBA on average for all four blocks' 6th-meter height on the façade of buildings without a noise barrier (NB). In Design I, the noise barriers manage to reduce this average traffic noise level to 66.90 dBA with an average 12.99 dBA reduction where the daytime noise level is limited with 68 dBA. The summary of noise performance of designs are given in Table 5.

Table 5. Noise level comparison with measurement and designs

	Without NB	Design I	Design II
Block 1	79,78	68,56	68,90
Block 2	80,46	66,42	66,76
Block 3	79,12	64,89	65,18
Block 4	80,21	67,76	68,00
Average	79,89	66,90	67,21

The noise maps are given in Figure 10 present the effectiveness of designed noise barriers for the residential complex located near a highway.

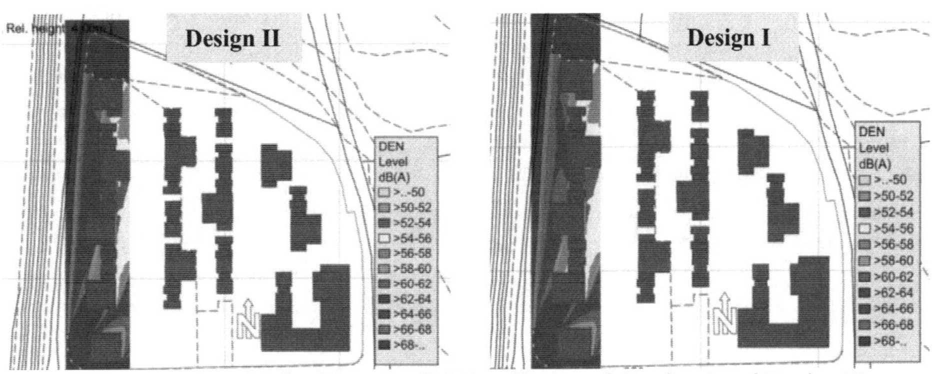

Figure 10. L_{DEN} Noise Maps in dB(A) at H=4m of Design I and Design II

IOP Conf. Series: Materials Science and Engineering **800** (2020) 012029 doi:10.1088/1757-899X/800/1/012029

4.2. Structural design

Steel framing systems are becoming a more preferred structural system due to their ductility and high performance in meeting the horizontal loads such as earthquake and wind. The specifications and regulations prepared for the steel structures create a solid base for the reliable design and practical implementation. The noise barriers discussed in this study is planned to be constructed in an area with both high seismic activity and high wind speeds. Therefore, steel frame is selected as the structural system due to its capability for resisting high seismic and wind loads, along with the additional advantages in terms of precision and construction time.

Considering the earthquake and wind effects, the selection of the appropriate structural system is of great importance. In order to determine the effects of the earthquake on the steel structured noise barrier, Revised Probabilistic Seismic Hazard Map of Turkey [21] is used. Probabilistic Seismic Hazard Map of Turkey is a contour map based on geographic coordinates. The 10% probability of exceedance in 50 years and it is corresponding return period of 475 years of rare earthquake ground motion of a (DD2) is selected form table 6 and the earthquake parameters are determined. S_S is the short period map spectral acceleration coefficient, S_1 is the map spectral acceleration coefficient for 1.0 second period, S_{DS} is the short term design spectral acceleration coefficient, S_{D1} is the design spectral acceleration coefficient for 1.0 second period, PGA is the maximum ground acceleration [g] value.

Table 6. Earthquake parameters

Earthquake Ground Motion Level	DD-2
Local Soil Type	ZC
S_S	1.133
S_1	0.308
S_{DS}	1.36
S_{D1}	0.462
PGA	0.464

The Eurocode-1 [22] regulation was used to determine the responses to the noise barrier due to wind. The basic wind speed, which is one of the most important parameters in wind calculations, should be determined correctly. The fundamental wind speed, which is of great importance in structural calculations, is the wind speed which corresponds to the possibility of exceeding the average 10-minute wind speeds measured in any direction at an open area at 10m height at least once every 50 years [24]. According to Istanbul High Buildings Wind Regulation [24], the basic wind speed for Istanbul is defined as V_b = 25 m / s. If sufficient wind records are available, the actual wind records obtained in the area where the structure will be constructed can be used for the analysis. Wind speed records, depending on the type of land on which the recording is taken (uneven areas, sea shorelines, forest areas, villages, city centres, etc.), surface roughness coefficient, topography coefficient expressing the height of the place where the high structure is located and structures around the high structure. They show different characteristics depending on various factors. In order to obtain basic wind speeds, wind speed data obtained from the Meteorological Data Archive System of the General Directorate of State Meteorological Affairs was used for the region where the noise barrier will be constructed. As a result, the basic wind speed is 28 m/s is taken for the analysis.

The steel frame systems proposed in the study are designed in accordance with the Design, Calculation and Construction Principles of Steel Structures (2016) [25], ASCE 7 (2016) [26], AISC 360 (2016) [27] and AISC 341 (2016) [28]. The structure system model used for the analysis is shown in Figure 11.

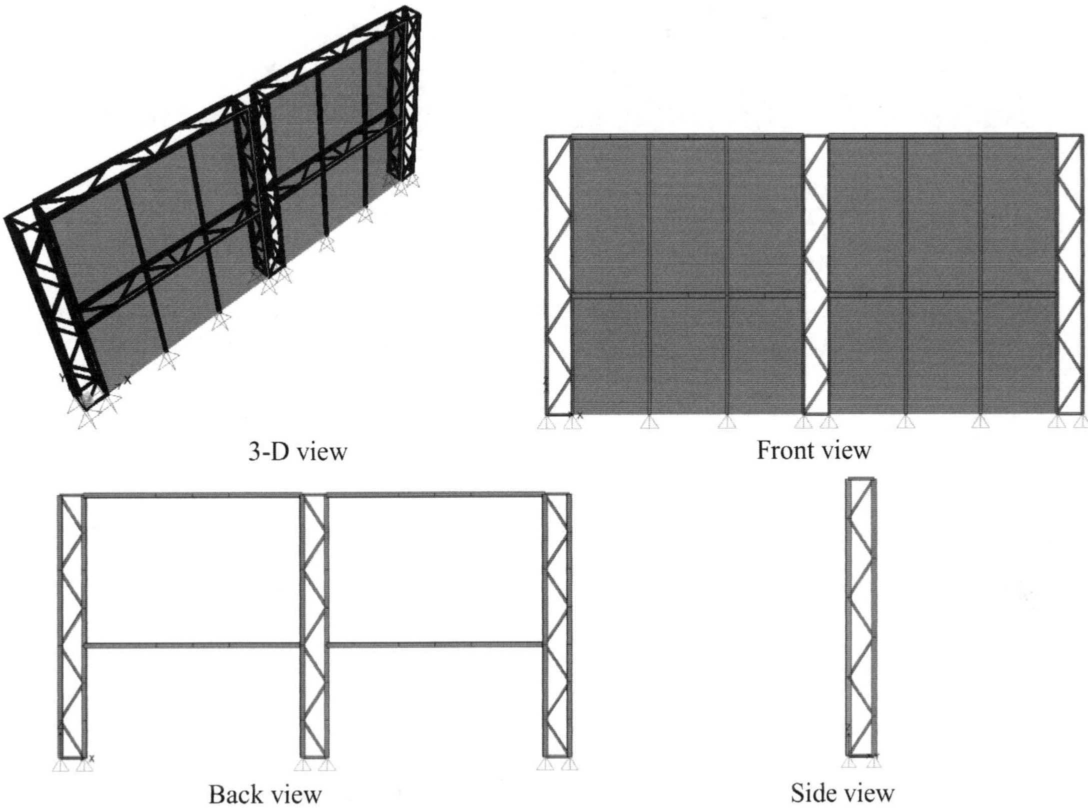

Figure 11. Structural system model

As it is mentioned in the geometric and acoustical design, two different solutions have been proposed to ensure the sound insulation of the noise barrier, which is planned to be 220 meters in length and 7.4 meters in height. It is assumed that the columns of all frames are supported with a simple connection to the foundation. In the structural system, the columns are 1m*1m 3D truss, beams 2D truss, the distance between the columns is 9 meters and the distance between the posts is 3 meters. S275 quality IPE profiles are used in the columns, S235 quality box section profiles are used in beams, trusses and mullions. The system is modelled with SAP2000 finite element program [29]. Linear analysis of the systems was carried out in accordance with the Design, Calculation and Construction Principles of Steel Structures (2016) and AISC 360 (2016) by using load resistance coefficients and design method (LRFD). The profiles selected to meet the requirements of the regulations are given in Table 7. The required amount of steel for both systems was obtained as 45 kg/m^2.

Table 7. Steel members used in the structure

Column	IPE 220
Upright	200-120-6
Column shear	100-100-4
Beam	200-200-6
Beam Shear	100-100-4

4.3. Approximate cost of designs

In order to calculate the total cost of noise barriers, the unit material costs and the total amount of construction is determined. The cost of the steel framework, the cost of ACB and the cost ALG is determined for each design. Table 8 presents the total costs of the corresponding designs.

Table 8. The costs of noise barrier designs

		Design I		Design II	
	Unit Material Cost	**Quantity**	**Total Cost**	**Quantity**	**Total Cost**
ACB	115 TL/m^2	924 m^2	106,260 TL	1,628 m^2	187,220 TL
ALG	190 TL/m^2	704 m^2	133,760 TL	-	-
Steel	4 TL/kg	72,260 kg	284,040 TL	72,260 kg	284,040 TL
Concrete	450 TL/m^3	110 m^3	49,500 TL	110 m^3	49,500 TL
		TOTAL	**573,560 TL**	TOTAL	**520,760 TL**

The unit cost of ALG is higher than ACB however, the acoustic performance of ALG+ACB is relatively better. Hence, Design I has a higher cost with a slightly better acoustic performance. In addition to that Design I is considered as more acceptable by the residents which makes the design to be preferable.

The average cost of the composite type noise barrier, Design I, is around 350 TL/m^2. This value decreases to 320 TL/m^2 in Design II where only ACB is used. Reinforced concrete noise barriers unit cost is given as 270 TL/m^2 by Calis et al. in 2018 [30]. Similarly, the unit cost of the 2A group of sheet piles and retaining walls in 2019 is given as 450 T TL/m^2. When all these unit prices are evaluated, the total cost for the composite ACB and ALG design, Design I, is expected to be around the 500 TL/m^2 for 2019 prices.

5. Conclusion

In this paper, the noise mitigation capabilities, aesthetic features and costs of acoustic designs for the residential buildings located near the urban highway is evaluated and also structural performances against wind and earthquake are examined. Design I presented a better acoustic performance along with a lower total cost. There is no established noise barrier geometric and material optimization method for highway traffic noise mitigation. In order to find the best design, the designer should consider not only the noise level standard limits, but also the cost, aesthetic and the structural standards. However, an acoustic optimization method is planned to be developed to generate an optimum dimension of a noise barrier.

According to the results, a complete glass noise barrier design can be examined further. However, due to the transparent nature the thermal and aesthetic consequences along with the resilience against the traffic accidents of a full glass design should also be researched in detail.

References

[1] Commission of the European Communities, Green Paper on "Future Noise Policy", COM (96)540, Brussels, November 1996

[2] The European Parliament and the Council of the European Union, Directive 2002/49/EC, Assessment and management of Noise, 25 June 2002

[3] TS EN 1793-1,2,3 "Road traffic noise reducing devices — Test method for determining the acoustic performance Part 1: Intrinsic characteristics of sound absorption", EN 1793-1, (1997). "Road traffic noise reducing devices, Test method for determining the acoustic performance Intrinsic characteristics of airborne sound insulation", EN 1793-2, (1997). Road traffic noise reducing devices, Test method for determining the acoustic performance, Part 1: Normalized traffic noise spectrum", EN 1793-3, (1997).

[4] EN 1794-1 and 2 "Road traffic noise reducing devices, Non-acoustic performance, Part 1: Mechanical performances and stability requirements", EN 1794-1, (1998) and "Road traffic noise reducing devices, Non-acoustic performance, Part 2: General safety and environmental requirements", EN 1794-2, (1998)

[5] ZTV-LSW 06, Forschungsgesellschaft für Straßen und Verkehrswesen: Zusätzliche Technische Vertragsbedingungen und Richtlinien für die Ausführung von Lärmschutzwänden an Straßen, Ausgabe 2006 (ZTV-Lsw 06), Köln

[6] Martin, S. J., and D. C. Hothersall. "Numerical modelling of median road traffic noise barriers." Journal of Sound and Vibration 251.4 (2002): 671-681.

[7] Jonasson, H., Storeheier, S. (2001) Nord 2000. New Nordic Prediction Method for Road Traffic Noise; Borås, Sweden.

[8] Noise Abatement in Town Planning—Part 1: Fundamentals and Directions for Planning; DIN 18005-1. 2002. [online] Available at: https://www.beuth.de/en/standard/din-18005- 1/46725515 [Accessed 1 Oct. 2019].

[9] Gesetze-im-internet.de. (2019). Berechnung des Beurteilungspegels für Schienenwege (Schall 03). Anlage 2 16. BImSchV - Einzelnorm. [online] Available at: https://www.gesetze-im-internet.de/ bimschv_16/ anlage_2.html [Accessed 1 Oct. 2019].

[10] Dutilleux, G.; Defrance, J.; Écotiere, D.; Gauvreau, B.; Bérengier, M.; Besnard, F.; Le Duc, E. (2010). NMPB-routes-2008: The revision of the french method for road traffic noise prediction. Acta Acust. United Acust. 96, 452–462

[11] Reken-en Meetvoorschrift Railverkeerslawaai'96, (1996) Ministerie Volkshuisvesting, Ruimtelijke Ordening en Milieubeheer

[12] Directive on the Assessment and Management of Environmental Noise, 2002/49/EC from 25.06.2002.

[13] Çevresel gürültü değerlendirmesi ve yönetimi ile ilgili ab direktiflerinin uyumlaştirilmasi ve uygulamamalar avrupa birliği eşleştirme projesi TR/2004/IB/EN/02

[14] AR-INTERIM-CM (CONTRACT: B4-3040/2001/329750/MAR/C1), Adaptation and revision of the interim noise computation methods for the purpose of strategic noise mapping, WP 3.1.2: Road traffic noise - Noise emission: databases

[15] Woelfel.de. (2019). Wölfel - IMMI. [online] Available at: https://www.woelfel.de/en/ products/ immi/ [Accessed 2 Oct. 2019].

[16] SoundPLAN GmbH, L. (2019). Home - SoundPLAN GmbH. [online] Soundplan.eu. Available at: https://www.soundplan.eu/english/ [Accessed 2 Oct. 2019].

[17] Software, C. (2019). CadnaA – State-of-the-art Noise Prediction Software. [online] DataKustik GmbH. Available at: https://www.datakustik. com/products/cadnaa/cadnaa/ [Accessed 2 Oct. 2019]. [18] Geomod. (2019). MithraSIG - Acoustic simulation software software for exterior area. [online] Available at: https://www.geomod.fr/en/land-geomatic/mithrasig/ [Accessed 2 Oct. 2019].

[19] Turkish regulation on the assessment and management of environmental noise. (2010). [Online], Available at:https://www.mevzuat.gov.tr/Metin.Aspx?MevzuatKod=7.5.14012&MevzuatIliski=0&source Xml Search %C3%A7evresel [Accessed 2 Oct 2019].

[20] Kapasite ve Elleçleme Bilgileri. (2019). [online] Available at: http://www.mardas.com.tr/tr-TR/kapasite-ve-ellecleme-bilgileri/4/39/ [Accessed 4 Sep. 2019]

[21] Revised Probabilistic Seismic Hazard Map of Turkey [online] Available at: https://tdth.afad.gov.tr/, 2018

[22] Eurocode 1: Actions on Structures. Part 1-4: Wind actions. ENV-1991-1-4:2005, C.E.N., European Committee for Standardization, 2005

[23] Türkiye Bina Deprem Yönetmeliği, Afet ve Acil Durum Yönetimi Başkanlığı, Ankara, 2018

[24] İstanbul Yüksek Binalar Rüzgâr Yönetmeliği (İYBRY), Deprem Mühendisliği Anabilim Dalı Kandilli Rasathanesi ve Deprem Araştırma Enstitüsü Boğaziçi Üniversitesi Çengelköy, İstanbul, 2009

[25] Çevre ve Şehircilik Bakanlığı. Çelik Yapıların Tasarım, Hesap ve Yapım Esasları, , Ankara, Türkiye, 2016

[26] American Society of Civil Engineers (ASCE), "Minimum Design Loads for Buildings and Other Structures", American Society of Civil Engineers, Reston, VA, 2016

IOP Conf. Series: Materials Science and Engineering **800** (2020) 012029 doi:10.1088/1757-899X/800/1/012029

[27] ANSI AISC 360-16, “Specification for Structural Steel Buildings”, American Institute of Steel Construction, Chicago, IL, 2016
[28] ANSI/AISC 341-106, “Seismic Provisions for Steel Structural Buildings”, American Institute of Steel Construction, Chicago, IL, 2016
[29] SAP2000, Computers and Structures Inc., Berkeley, California, U.S.A
[30] Çaliş M and Öztürk Z. "Beton Gürültü Bariyeri İçin Perlit Kaplama ve Maliyeti." Teknik Dergi 29.3 (2018): 8433-8457

Adaption of construction management aspects and building economic principles to prevent cost overruns in modern timber construction

J Koppelhuber[1*], M W Bok[1]

[1] pm holzbau – Sporgasse 11, 511a, 8010 Graz, Austria

*Email: koppelhuber@pmholzbau.at

Abstract. The timber construction industry is subject to continuous development and steady modernisation worldwide. This leads to a compelling necessity regarding highly industrialised and standardised processes – both technological and economical – along the value chain of today's timber construction. Within this field innovative materials and products (such as cross laminated timber – CLT) in combination with a deeply rooted systematic approach are used to create large span and multi-storey timber structures. The results are displayed in an increased use of (partially) automated machines and production plants, novel (semi-)prefabricated components as well as complex software solutions. However – next to numerous technical researches – numerous aspects within construction management and building economics require intensive investigation. Since the influencing factors from economical point of view have been criminally neglected over the past decades it becomes apparent that aspects such as cost planning and estimation, invitation and call for tender as well as cost calculation of timber construction projects demand comprehensive research. Therefore, an outstanding and all-encompassing investigation has been carried out over the past 6 years in Austria at the University of Technology in Graz [1] to identify, develop and implement building economic principles and relating tools focussing exclusively on timber construction. Therefore, the emphasis was set on the development and practical implementation of timber-specific solutions in order to avoid cost overruns and hidden costs during design, construction and use. Possible approaches to solve this lack of economical and operational instruments in timber construction are discussed in this paper.

1. Introduction

The evolution of timber constructions in the 21st century is primarily driven by technological advancements on product level and its production methods. Meanwhile, the principles of building economics regarding cost planning, cost calculation and invitation and call for tenders have not been adapted adequately to this rapid technological development. Practical tools specially adapted for the timber construction sector have received insufficient attention resp. have been neglected entirely for a long time. This results in a flagrant information deficit, based on the combination of a lack of reliable data and a largely missing processing of the few existing findings and parameters for the construction companies, developers, architects as well as other parties involved along the construction and design process.

A sole transfer of universal principles and values from conventional concrete and masonry construction into timber construction is only possible to a limited extent. Because of company-significant demands and industry-specific requirements as well as the fact of a completely contrary production and building process, the development and establishment of independent procedures and implementation tools is necessary.

However, the timber construction industry is – due to the latest developments in recent years as described below – on a continuous path to liberate itself from its infancy regarding building economics and to carry out a holistic professionalization from a contractual perspective. In addition to numerous other issues such as the ongoing debates concerning the overall design process and the long-term strategic challenge to find a way out of the so called "price war" within the construction industry, the topics of cost planning, tendering and cost calculation need to be solved in the near future.

In the following passages the status quo (SQ) for economic aspects within the timber construction industry regarding especially these three thematic fields is presented and feasible solution approaches (FSA) are discussed. Current developments which have already begun or are still pending as well as their potential long-term impact on the industries are subsequently described.

2. Aspect I – budgeting / cost planning

Austrian Standards International which publishes all Austrian Standards as public national institutions divides the planning of construction costs (as part of the quality, cost and time triangle) in six subdivisions following the Austrian Standard OENORM B 1801-1. Within the cost sequences specific cost management tasks need to be carried out for each stage of the construction and by becoming more precisely the further the project is developed. These stages of budgeting are depicted in the following figure.

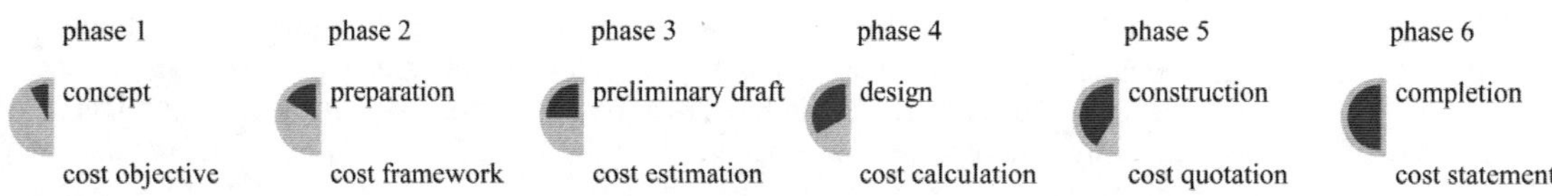

Figure 1. Budgeting in six stages of cost planning referring to the OENORM B 1801-1[2]

The cost objective and the cost framework of a project, which represent the first two stages within the cost planning procedure, as well as their consistent compliance are decisive in the pursuit of a positive project completion. The responsibility for an accurate cost planning and monitoring lies in the hands of the contracting authority resp. the planning representative o the client. Different interests of the builder, the authorized designers or the local construction supervision often result in an opportunistic behaviour towards the building contracts. The cost calculation within the cost planning process (conducted by the contracting authority in the design phase) must be distinguished to the calculation of the construction costs (conducted by the contractor during the tender process) descripted in chapter 4.

2.1. Status quo in the timber construction industry

Timber construction companies increasingly offer in many cases a high added value on the market in comparison to general contractors or even as design and build-contractors. This is especially necessary when large-scale multi-storey construction projects are realized primarily with timber. Therefore, timber construction must prove itself technically but above all financially to the competitors building with concrete. A comparable and sound cost planning is therefore without any alternative.

Due to insufficient or non-existing cost indicators within the budgeting cost estimations (phase 3 preliminary draft), cost calculations (phase 4 design) and cost quotation (phase 5 construction) for timber constructions is often inadequate, less resilient and generated on the wrong basis. The missing data leads directly to the circumstance that a neutral and conclusive cost comparison between timber structures and conventional structures is rarely possible in practice. At this stage the costs of a timber construction are mainly estimated from literature sources generated outside the industry mainly not timber-specific and not reflecting the current technical innovations at the product level. A temporary solution is often a superficial cost planning based on internal empirical values and rough estimations. Only in a few cases adequate cost parameters on a neutral and technical accurate basis are available. The result is a strongly distorted and unjustified budgeting in which timber structures are often portrayed as financially disadvantageous.

Furthermore, most construction projects are designed to be realized with reinforced-concrete, bricks or other conventional building materials. Timber construction companies therefore frequently have to offer their value proposition as alternative, which consequently provokes additional costs as the design is not timber specific and their inadequate. This focus on a certain construction method in early project phases complicates a valid cost comparison on a neutral basis.

An additional complication is the unequal quality expectations regarding different construction methods. Experience have shown that a higher level of quality standards is more often required in timber structures than compared to conventional concrete structures. The average timber construction provides a higher equipment standard than other building methods (e.g. wood-aluminium windows and doors instead of PVC-products, parquet instead of laminate, shingles instead of thin plaster, wood-fiber-insulations instead of EPS-insulation, etc.). Often cost planning based on empirical data does not exclude the cost of interior components and design elements, which contributes to the common misconception regarding the economic efficiency of timber constructions.

STATUS QUO (SQ) – BUDGETING / COST PLANNING

- insufficient cost indicators and parameters on a neutral on a technical accurate basis
- construction method requires a specific design process and cost estimation
- cost comparison of structures is not on an equal equipment standard

Figure 2. SQ budgeting – cost planning phases in timber construction

2.2. Feasible solution approaches for the timber construction industry

Following these facts, it is of considerable importance to pursue the goal of providing the key decision-makers in the cost planning process with comprehensive and (company-)neutral cost indicators, figures and parameters which are in line with the current state of the art.

Furthermore, a cost comparison between different building materials and/or construction methods must be based exclusively on the shell construction to eliminate unequal demands regarding interior elements and equipment features from the equation. The wall, ceiling and rooftop elements as well as other structural components distinguish certain construction methods from each other, while must finishing elements of the expansion as well as the equipment are not construction specific. All elements such as windows, doors, facades, roof coverings, floors, building services (heating, air conditioning, ventilation, sanitary provisions) and electrical wires are therefore not to be considered in the original cost comparison of the building material / construction system / construction method. These will only be included in the overall calculation at the time of the final cost calculation, which enables a neutral cost comparison based on an equivalent with respect to analogousness referring to statics and building-physics (same static function, deflection, heat transition coefficient, sound insulation and fire protection etc.). Without these measurements timber construction methods cannot be assessed on a neutral base and fair arrangement. The current conditions deprive developers and architects of a real alternative to conventional construction methods and unfairly hinders the progression of environmentally friendly construction methods.

Numerous construction surveys following the systematic of REFA have been undertaken by the author within his former function as scientific researcher at the Institute of Construction Management and Economics (BBW) at University of Technology in Graz (Austria). They have shown that timber construction can be identified as cost neutral in identical circumstances [3][4][5]. This applies to timber frame construction, solid timber construction (CLT) as well as timber-concrete-composite constructions (TCC). Additional soft facts, such as lower wall thicknesses (and therefore a greater floor space), shortened construction times (earlier building utilization or rental/sale income) and less CO_2-emissions (avoidance of potential carbon taxes) are additional aspects, which will also have to be considered financially in the future.

**FEASIBLE SOLUTION APPROACHES (FSA) –
ASPECT II BUDGETING / COST PLANNING**

- collection and aggregation of accurate cost indicators and parameters in a neutral and up-to-date database
- definition of component properties (equal static and building physic) instead of specific structures
- promotion of a cost estimation and cost comparison based exclusively on the shell construction

Figure 3. FSA – budgeting / cost planning in timber construction

3. Aspect II – invitation and call for tender

In accordance with the fundamental principle "a good project requires a good tender" the question occurs how to translate a sound design into a complete tender dossier to provide the best possible basis for a contractual framework which prevents cost overruns, construction time extensions and legal disputes. The underlying tendering process plays an essential role in achieving a smooth transition from design to construction in the traditional design-bid-build project delivery method (figure 4) and helps to avoid the previously mentioned undesired monetary, temporal and legal impacts on the overall project.

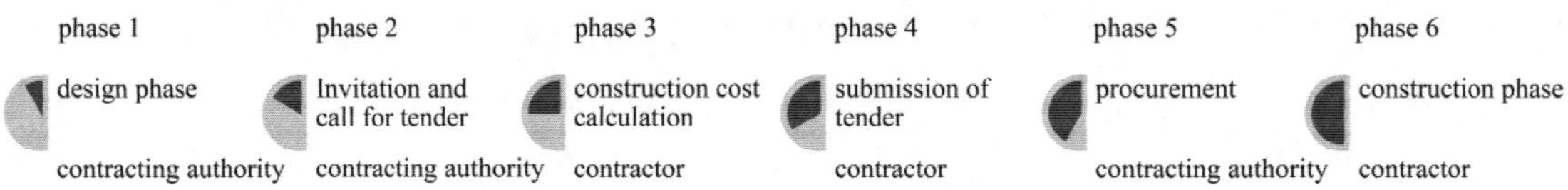

Figure 4. Six stages of the tender process in the design-bid-build project delivery method

Especially in private / commercial tenders an additional negotiation stage between the submission of tender and the procurement procedure is quite common. Furthermore, a complete, transparent and resilient tendering process is highly important between contractor and subcontractor as well es between contracting authority (client or authorized architect) and bidders (design and build contractors or general contractors). The previously stated budgeting procedure shown as cost planning sequence (especially the cost calculation by the contracting authority) sets out a precise presentation of the technical specifications, components and services to be produced as well as the expected circumstances for the construction works. The latter must be recognizable in quality and quantity to allow a precise construction cost calculation by the bidder.

In order to take the aforementioned facts into account the contracting authorities in Austria have relied on the federal government´s standardised specifications in form of standard texts for tenders for the entire construction (LB-HB) for the past decades, which covers approximately 80% of all construction works carried out in the general building sector. The descriptions are in a textually precise and comprehensive manner and pursue the principle that building services are to be clearly, completely and neutrally describable.

3.1. Status quo in the timber construction industry

Due to the rapid product innovations, the continuously adaption of new detail solutions and modern connection methods as well as the ever-increasing performance levels in timber construction the aged carpenter specific group of technical specifications included in the LB-HB had to be updated significantly to adapt to the state of the art. The newly released version of the LB-HB 021 (officially released by the Austrian government in December 2018 [6]) includes an up to date timber specific description and extensively recreated group of technical specifications (the so-called LG 36) for the entire timber construction industry. This allows a comprehensive and detailed technically adequate tender in timber construction due to a completely different structure sectioned in components (wall, ceiling, roof, etc.) as well as in vertical and horizontal layers (support structure, insulation, interior clothing, facade, etc.). General and particular technical specifications in a standardised way enable both the contractor and the contracting authority to tender as well as calculate according to a clear scheme and provide a structure similar to the widespread tender procedure of the conventional construction methods. It also saves time and costs, because it allows the bidder to provide a detailed cost calculation of the individual items quickly and effectively.

While the reworked standardised specifications for timber construction in Austria is a first step into the right direction other countries with a lower market share of timber buildings lack comparable procedures for a timber-adequate tender process which leads to an unlevel playing fields on the international level. Cross-border standards (e.g. Eurocodes) which represent the state of the art in timber construction in a technical way are currently not available for an international standardised tender referring to contractual topics.

While the rapid technological progress in the timber industry is accountable for its sharp increased market share it aggravates the development of timber-specific solutions. National and international standards and regulations as well as implementation tools in the timber building industry must be updated continuously to keep up with the technological advancements. Due to well-filled order books timber construction companies often do not have the time, interest and/or motivation to deal with advancements in the field of building economics and construction management. This results in a loss of opportunities, time and money plus a risk of legal consequences as well as an exclusion during the tender process because of insufficient tender documents or inadequate formulations.

STATUS QUO (SQ) – TENDER PROCESS AND PRINCIPLES

- standardised specifications for timber construction are adjusting to the traditional procurement
- international and national standards and regulations often do not represent the state of the art
- limited usage of building economic and construction management tools by timber companies

Figure 5. SQ – tender process and principles in timber construction

3.2. Feasible solution approaches for the timber construction industry

In order to achieve a comprehensive and nevertheless detailed tender incorporating timber construction methods the use of the previously described standard specification for building construction is vital to allow little to no possibility of additional cost claims and construction time extensions. A neutral tender regarding building material (with technical objectives regarding static and building physics as suggested in chapter 2.2) as well as a timber-specific tender must describe the technical specifications, components and services adequately with enough detail to allow a precise construction cost calculation by the contractor.

Due to the ascent of timber construction companies from a regional niche to global players international standardisation of products, services and technical components as well as a timber-compatible tendering process are overdue – if not worldwide at least pan-European. This development is imminent to provide sustainable and innovative construction methods like timber with an equal chance to succeed.

The usage of such standardised tender concepts and economical tools within the tender process in all construction projects also means to deviate from the mostly contradictory, company-specific tender texts and preliminary remarks that have been widespread in the timber industry for decades. Thus, a product- and company-neutral tender in timber construction is possible, which in turn allows a solid, detailed and comprehensible construction cost calculation as well as a specific cost determination without a high risk premium due to insufficient clarity resulting from inadequate design and non-standardised contractual specifications.

IOP Conf. Series: Materials Science and Engineering **800** (2020) 012030 doi:10.1088/1757-899X/800/1/012030

Ultimately, this step towards standardised tendering in timber construction also forms the basis for a continuous compliance with the previously cost framework that lasts until the project is completed and does not result in a contractual framework far from reality – which is often recognizable in modern construction industry. Although extensive use of standardised specifications for building construction requires a trial period of familiarization for the contracting authorities it also offers the certainty of ensuring unambiguous comparability of all offers and reduces the contract risk by eliminating inadequate formulations and contradictions in the tender. In the end, this also effects the price of the bidder, because the contractor also experiences a minimization of the contract risk as a result of a standardised call for tender, which becomes part of the construction contract.

FEASIBLE SOLUTION APPROACHES (FSA)
ASPECT II – TENDER PROCESS AND PRINCIPLES

- widespread use of standardised technical and contractual specifications for (timber) construction
- adjustment and harmonisation of international and national standards and regulations
- promotion of implementation tools and standards through spread of knowledge

Figure 6. FSA in tender process in timber construction

4. Aspect III – cost calculation of construction

The calculation of construction costs resp. prices by the bidder / contractor is clearly regulated in Austria by standardized cost calculation form sheets (K-sheets) following the Austrian Standard OENORM B 2061 [7]. This cost determination is based on the tender documents which has a position regarding the building material as well as the construction method and is nationwide uniformly executed with calculation forms depicted in figure 7.

K3	K4	K5	K6	K7
average wage price calculation	material price calculation	price calculation of products and services	price calculation of constructio machinery	actual price calculation

Figure 7. Cost calculation forms for the determination of construction prices acc. to Austrian Standard

Associated auxiliary sheets support the calculation of construction costs resp. prices. The national standard provides additional information relating to the correct structure of the cost calculation and deals with clear presentability as basis for a review procedure regarding the appropriateness of the construction cost submitted in the tender.

This calculatory framework is used in the award procedure as a fundamental criterion in the comparison of tenders and is mandatory in public construction projects in accordance with the federal procurement law (Bundesvergabegesetz BVergG [8]). A detailed construction cost calculation according to this scheme is also recommended for private and commercial construction projects to ensure the formal correctness and appropriateness of the submitted tender.

4.1. Status quo in the timber construction industry

Nowadays timber is much more than just a construction material for single-family houses and agricultural buildings. Timber frame constructions as well as solid timber (and hybrid) constructions are, due to its elegance in construction, architectural diversity and ecological benefits, increasingly used for multi-storey housing and office buildings as well as large-scale public buildings such as educational institutes, sport facilities and administration offices. To enable timber construction companies to act as general contractors a complete and comprehensive cost calculation according to the national code as well as pan-European normative requirements is inevitable. This leads to the fact that all detailed cost calculations are carried out and fully disclosed in the case of deepened cost resp. price appropriateness reviews.

Therefore, it is no longer sufficient in today's timber industry to calculate solely to the level of unit prices for individual advertised positions, but to deal in detail with the connection details, processes and the expected circumstances on site and in relation to other trades including all cost factors as well as a transparent and resilient documentation of the cost determination process. This form of a detailed cost calculation – as it is custom in conventional mineral construction methods for decades – is finding its way also into timber construction, since without it an excretion of the tender process is the inevitable consequence.

Thus, it is up to the timber construction companies to carry out a sound and valid calculation, which is verifiable and plausible for the contracting authority or the authorized person responsible for the deeper tender evaluation. As such the cost calculation of timber constructions do not give a cause for an exclusion in the procurement process.

The detailed cost calculation must be divided (as depicted in figure 7) into wages, materials, products and external services as well as construction machinery and equipment. Since industrialized timber constructions have a high share of material costs in comparison to the labor or wage costs, the economic calculation tools also need to be adapted to this specific situation. However, due to varying suppliers (semi-finished parts and prefabricated structural elements), the entrepreneur itself can have little to almost no influence on the production processes of construction components, which makes a detailed and all-encompassing cost calculation even more complicated.

On the other hand, the assembly works on a construction site are found to be relatively consistent due to manageable assembly crews of only three to five skilled workers (mostly carpenters). However, they do differ in reliance to the degree of prefabrication, the complexity of the components and the quality of the work done by other subcontractors (e.g. in case of a reinforced concrete building core).

These circumstances lead to the fact that the cost resp. price determination of timber construction works must be conducted more accurately and a complete renunciation from an exclusive assessment from the gut must be enforced to ensure a realistic assessment of the overall construction costs. This development is highly depended on auxiliary and timber-specific cost calculation tools to encourage the use of standardized cost calculation forms and a substantiated cost database which represents the current state of the art on the product level.

STATUS QUO (SQ) – COST CALCULATION

- insufficient cost calculations restrict the market access for timber construction companies
- high degree of prefabrication limits the possibility of holistic cost calculations
- calculation guidelines and tools depicting the timber constructions state of the art are missing

Figure 8. SQ – cost calculation in timber construction

4.2. Feasible solution approaches for the timber construction industry

Although the experience of established entrepreneurs is theoretically a good foundation for an initial construction cost resp. price assessment. However, they are hardly able to withstand any appropriateness test and are not transparent for external reviewers in any way. Consequently, it is necessary to implement a standard costing (pricing) index and calculative scheme in the timber construction industry in which each process is specified according to the same schemes as in other trades. This leads to a comparable and comprehensible cost calculation in which only individual cost approaches as well as the quantities must be adapted between various construction projects. The calculation process itself remains the same.

In the past years standardized cost calculation approaches were rigorously collected and published in the course of construction process (acc. to REFA analysis – see chapter 2). The objective of such studies is to generate a well-founded database for the cost calculation of timber construction companies. In the process, a large amount of data can be collected and scientifically substantiated. Cost indices are derived from observations and can be used in individual calculations. The result of several research projects is currently being implemented in a specific cost calculation software (especially for Austria) – which follows the structure and positions of the timber-specific technical specifications in the new LB-HB.

Ultimately, there is the possibility in timber construction – following the example of various other trades in the past – to conduct a sound and valid cost determination. This facilitates the market access of timber construction companies regarding large-scale public projects and encourages cost optimizations throughout a detailed cost calculation. It is inevitable for an overall cost reduction to examine every cost center. If these sub-processes are also calculated and a full-cost calculation is carried out at the beginning of every project, it is possible for the entrepreneur to be competitive and ensure a necessary monetary surplus. Ecological attributes are the drivers for the development of timber constructions, but ultimately the price of a system decides on its selection.

FEASIBLE SOLUTION APPROACHES (FSA) –
ASPECT III – COST CALCULATION

- implementation of a standard costing index and calculative scheme for timber constructions
- conduction of empirical studies to derive state of the art cost parameters
- cost optimization of timber construction based on full-cost calculations

Figure 9. FSA – cost calculation in timber construction

5. Current developments and long-term impact

In addition to these three essential facts – cost planning, tendering and cost calculation – another topic is directly related to the developments inside timber construction: Building design as advance input and building contract law as legal consequence. Above all, the topic of planning resp. design in timber construction has been the focus of numerous considerations and (research) projects for several years. Due to the increasing volumes and complexities in timber construction as well as the often-ambiguous situation between the various architects and specialist designers, it seems to be necessary to define unambiguous performance profiles or detailed descriptions of the scope of work for all parties concerned. These would ensure clarity regarding the responsibilities and areas of competence of architects, structural engineers, building physicists and designers of technical services as well as timber construction companies and enable smooth cooperation between all of them along with the client and/or investor. Ultimately, it is about the question of who plans what, when and with which level of detail and how these services are adequately remunerated according to the actual effort within each construction project. The timber-specific advancement of design services is of considerable importance, because early completion of construction drafts is highly important to enable prefabrication and short construction times.

In order to guarantee clear and adequate responsibilities, to define interfaces and to specify the timber-specific contents of the individual planning stages, a comprehensive performance profile is currently being prepared at the Graz University of Technology (BBW – Prof. H. Lechner) together with the author [9]. This is done in accordance with the Austrian service phases of LM.VM.2014 (also written by Prof. H. Lechner) as well as the German HOAI (also written by Prof. H. Lechner) and serves as a fundamental basis for a clear distribution of competencies of the designers specifically in timber construction and for a fair payment solution for everybody involved.

Numerous other issues regarding construction management, building economics and building contract law are currently being considered in timber construction, as they form the basis together for making timber construction more successful and sustainable in the economic sense. These include topics of organization (process engineering, work preparation, production planning, logistics etc.), of economic optimization and cost calculation (further calculation approaches and databases, comparative evaluation of procedures, cost analyzes based on completed buildings, etc.) as well as areas of timber construction contracting and personnel development. Ultimately, this is going to lead to a restructuring process across the borders of the timber construction industry as well as beyond the product and technological level. New strategic innovations and business models are going to challenge the prevailing corporate strategies and push the whole construction industry into a higher productivity and greater sustainability. Due to the technical achievements in recent years and the supplementary economic tools, construction data and innovations regarding corporate management the use of timber construction will accelerate even further soon.

This development is sustained, inter alia, through a systematic and modular approach to timber construction. With consistent specification from a technical point of view and extensive implementation in the construction companies as well as planning departments, timber system construction is going to allow the widespread use of timber in a way which has seemed impossible up to now. This is possible, because the higher productivity resulting from the increased usage of modular timber systems can radically reduce costs and thus the comparison of construction methods and especially structural materials is tilting more and more in the favor of the construction material timber.

At this point it must be noted, that the presented scenario can only be realized if the timber industry ignores the prevailing lethargy of the construction industry and the blatant deficits regarding labor productivity as well as the degree of digitization are solved. Only this development enables the possible for timber to succeed over other construction materials – in addition to the ecological benefits – in terms of costs in the long term. The goal must be to reduce construction costs drastically, especially in the area of affordable housing. At the same time, it is important to consider the global ecological developments and to assume responsibility for the numerous environmental and ecological-social problems in the construction industry. Timber construction can play a leading role in the evolution of construction from an ecological as well as an economic point of view if there is a consistent implementation of building management approaches and strategic tools. A technologically outstanding product alone is not enough to survive in the business world of the 21st century – not even in the construction industry [10].

6. Conclusion

The liberation of its economical infancy has begun in the timber industry a few years ago and is now leading to a continuous change in the industry towards a holistic construction process management. This includes all construction and management related areas, which – in addition to all the technical achievements – provide the basis for a continuous development of timber construction. The construction material timber and thus the underlying companies can develop out of a niche market into global players and initiate a new age of building. Numerous measures were and are set out now in order to create the armament, which makes it possible for timber construction companies to realize a lot more than just lighthouse projects and to be dismissed as nice to have in an ecological point of view. Regardless of the construction method – timber frame constructions, solid timber constructions and timber hybrid constructions – the target must be to conquer the mass market. This is especially true regarding multi-storey housing, office buildings, public buildings such as schools, libraries and governmental buildings as well es industrial buildings. The presented developments in the fields of construction management and building economics in the form of standardized processes and tools for cost planning, tendering and cost calculation ultimately serve the big picture and are necessary steps to further professionalize and legitimize timber construction. They are the indispensable foundation of the successful implementation of the building material timber.

References

[1] Koppelhuber J 2018 *Bauprozessmanagement im industriellen Holzbau – Ableitung eines Bauprozessmodells zur Prozess- und Bauablaufoptimierung im Holzsystembau (Building Process Management in Industrial Timber Construction Defining a Model for the Optimisation of Building Processes and Construction Sequences in Timber System Construction)* Dissertation, TU Graz (Graz: University of Technology)

[2] Austrian Standard Institute 2015 *OENORM B 1801-1 Bauprojekt- und Objektmanagement Teil*

1: Objekterrichtung (Project and object management in construction — Part 1: Object construction) (Vienna: Austrian Standard International)

[3] Heck D Koppelhuber J 2014 *Tagungsband 1. Forum Holzbau trifft Bauwirtschaft – Kosten und Kalkulation im Holzbau 27.03.*2014, TU Graz (Graz: University of Technology)

[4] Heck D Koppelhuber J 2015 *Tagungsband 2. Forum Holzbau trifft Bauwirtschaft – Bauwirtschaftliche Optimierungsansätze im Holzbau 29.04.2015*, TU Graz (Graz: University of Technology)

[5] Heck D Koppelhuber J Wall J 2017 *Tagungsband 3. Forum Holzbau trifft Bauwirtschaft – Bauwirtschaftliche Ansätze in der Ausschreibung, Kalkulation und Kooperation im Holzbau 12.05.2017*, TU Graz (Graz: University of Technology)

[6] BMDW Bundesministerium für Digitalisierung und Wirtschaftsstandort *Standardisierte Leistungsbeschreibung Kennung: HB Version 021 (Ausgabe 31.12.2018)* https://www.bmdw.gv.at/ (Vienna: BMDW Bundesministerium für Digitalisierung und Wirtschaftsstandort)

[7] Austrian Standard Institute 1999 *OENORM B2061 Preisermittlung für Bauleistungen 1999 (Determination of price in building and construction – General principles)* (Vienna: Austrian Standard International)

[8] Bundeskanzleramt Österreich 2018 *Bundesgesetz über die Vergabe von Aufträgen (Bundesvergabegesetz 2018 – BVergG 2018) – BGBl. I Nr. 100/2018 idgF.* (Vienna: Bundeskanzleramt Österreich)

[9] Lechner H Koppelhuber J release planned for 2020 *Planung und Bauaufsicht für Holzbauten – Leistungen der Holzbauplanung (PREFAB)* Fachbücher Planung und Bauen Institut für Baubetrieb und Bauwirtschaft TU Graz (Graz: University of Technology)

[10] Bok M 2019 *Geschäftsmodelle im industriellen Holzbau – Ableitung strategischer und technologischer Parameter anhand einer visionsorientierten Geschäftsmodell-Systematik (Business Models for Industrialized Timber Construction – Derivation of Strategic and Technological Parameters on the Basis of a vision-oriented Systematic)* Masterthesis Institut für Baubetrieb und Bauwirtschaft TU Graz (Graz: University of Technology)

IOP Conf. Series: Materials Science and Engineering **800** (2020) 012031 doi:10.1088/1757-899X/800/1/012031

Building services in timber construction – Investigation on difficulties and reliefs during the installation procedures in multi-storey timber residential buildings

D Koppelhuber*, A Magg

Graz University of Technology, Institute of Construction Management and Economics, Lessingstraße 25, 8010 Graz, Austria

*Email: daniela.koppelhuber@tugraz.at

Abstract. Multi-storey timber construction has undergone a strong positive development in Austria in recent years. However, while there are constant advancements in materials, construction technologies, production stages and assembly methods in timber construction itself, the design as well as the installation procedures of building services in multi-storey timber buildings lags far behind these technological progresses. The principles of planning and installation of building equipment are generally similar to those within solid mineral constructions that could be identified in the holistic research project TGA-Timber. A comprehensive investigation of the installation procedure on 3 multi-storey timber-construction projects has been carried on site and during prefabrication within the past 2 years in Austria at Graz University of Technology. The emphasis was set on the identification of disorders and delays caused by installation procedure of HVAC-building equipment in timber construction. The results of this research release the large potential of optimisation of piping runs as well as time- und cost savings within the installations process of building equipment as part of an overall economical optimisation. This publication is an excerpt of the results generated for these construction site analysis and deals in particular with difficulties and disturbances that have occurred.

1. Introduction

The percentage of multi-storey timber constructions, in particular housing projects, has undergone a strong positive development in Austria in recent years. Between 1998 and 2018 the percentage of timber residential buildings increased from 14 % up to 24 % measured on living area [1]. However, while there are constant advancements in materials, construction and manufacturing technologies and prefabrications and assembly methods in timber construction itself, the design as well as the installation procedures of building services in multi-storey timber buildings lags far behind those technological progresses. However, due to the advancement of higher energy efficiency of buildings and denser building skins building services become more and more important.

In order to demonstrate how the installation work in timber construction sites is currently performed and which challenges and problems arise, workflow-analysis were carried out on three timber construction sites for the research project TGA-Timber [2]. This research project under the leadership of Holzforschung Austria, aimed, among other resarch objectives, at an economic analysis of the current state of installation work. Part of this analysis was the survey of the current state of the installation

IOP Conf. Series: Materials Science and Engineering **800** (2020) 012031 doi:10.1088/1757-899X/800/1/012031

process and a comparison of different levels of prefabrication. The analysis identified, that the principles of planning and installation work of building services are generally similar to those within solid mineral constructions.

2. Analysis of building systems in multi-storey timber constructions

2.1. Initial situation

Hausladen [3] defined already in 2008 six design guidelines for the prefabrication of buildings equipment in multi-storey timber projects as part of a research project. On the one hand, these guidelines meet the demands on a high degree of prefabrication and a high degree of flexibility, on the other hand, they also offer increased protection against moisture leakage. In addition to the principle of the central vertical and horizontal piping runs is a separation of the building services equipment from the structure, the permanent accessibility and thus also control and maintenance a major requirement of buildings services in timber constructions.

2.2. Research objectives

But, in which way do the buildings services engineers involve Hausladen´s guidelines and how do HVAC installers handle these new challenges which arise at timber construction sites? Some answers could be given through the workflow-analysis for the research project TGA-Timber in the years 2017 and 2018 coupled with the investigation of the MEP plans and tenders. Additional effort values have been examined in detail within the analysis which show the differences of efficiency of installation procedures of the varying prefabrication levels.

2.3. Analysis of the plans and tenders

Before starting the workflow-analysis the HVAC-Plans and tenders had to be investigated to get an overview of the underlying documents for the installers. The investigations revealed, that the plans for building services for the two on-site-constructions-projects didn´t differ to those of conventional solid brick or concrete constructions. Hausladen´s six design guidelines were not considered at all, on the contrary, there were no timber-specific peculiarities of the piping runs recognizable. In some cases, the layout of the pipes was shown on the plans only in one floor, the piping runs of the other floors was incumbent on the executing HVAC-technicians. This fact resulted in deviations between the plan layout and the execution with varying floor plans or window arrangements. In contrast, the building services plans for the turnkey project showed exactly the pipelining in walls and at floors, which resulted in no scope for the installers.

A similar picture as in the planning provided the analysis of the specifications for the HVAC tenders for the construction site assembly. Regardless of the timber construction, these descriptions were thus adopted unchanged from the standardized specification for building services LB-HT (current version 012) or from previous projects. An adaptation of the position wordings to the requirements of timber construction, or at least a revision of the wording, was completely absent in all three analysed tenders. The additional research objective therefore arose as to how inaccuracies in the planning and tendering process affect the assembly work.

3. Data collection on site

To answer the preceding questions three multi-storey housing projects for the workflow-analysis at the construction-site were found. Two of these buildings were on-site-construction projects with partly prefabricated walls and ceilings, the third project was a turnkey-prefabricated project with about 120 almost similar apartment-units. Due to the differences of these projects promising results were finally achieved.

3.1. Aim and methodology

For the workflow-analysis the methodology according to REFA [4] which is commonly used for solid brick and concrete building systems was used. Specific characteristics of the installation procedure of HVAC-building equipment were implemented consequently into the analysis.

The main aim of these data collections was the derivation and analysis of disturbances resulting from the new building systems of multi-storey timber construction. While the HVAC installation process in mineral solid construction can be regarded as standard for the involved planners and technicians, there are new details, assembly methods and interfaces with other crafts, which planners and installers have not met yet. In the workflow-analysis special attention had therefore been paid to timber specific design details.

3.2. Presentation of the investigated projects

The investigated projects were multi-storey residential buildings for urban areas. The two on-site projects consist of two to four timber-storeys on a concrete basement with 38 (Project 1) respectively 56 apartments (Project 3). The walls are prefabricated 2d timber frames and CLT-elements supplemented with prefabricated CLT-ceilings. The buildings services equipment was assembled completely at the construction-site. On the other hand, the third investigated project (Project 2) exists of almost 120 turnkey prefabricated room-units with an integrated bathroom which were finished in a production hall. They are constructed of CLT-ceilings and -walls and timber frame walls. The HVAC-installations were partly prefabricated in a separated workshop-area within the production hall and then assembled in the timber-module, partly the installations procedure was done completely in the timber-module without prefabrication.

3.3. Data recording

The data survey for those buildings was done by the workflow-analysis following the REFA-technique. The HVAC installers of the three constructions sites were observed over a period of 66.25 hours between August 2017 and February 2018. Especially the multi-moment-recording-system MMR with a five-minute observation interval showed good research results, in particular for the timber-construction related interruptions of the installation process. Accordingly, a total of 2107 data were recorded, which corresponds to an observation period of about 176 hours of installations time at all three construction sites. The number of concurrent working installers varied between two and six. In addition, discussions with HVAC planners and site managers for the timber construction and the buildings services supplemented the results of the workflow-analysis.

4. Evaluation of collected data

Due to the comprehensive data collection, the required results listed above could be evaluated and derived. In the first step, the entire workflow was broken down into the three categories of activities, interruptions and non-identifiable activities. The collections of the individual areas of activity was done by means of multi-moment-recording. Based on this, the occurred dysfunctions were filtered, their causes identified, and a time loss evaluation carried out.

In the analysed buildings two different production and assembly methods were inspected. One was the in-situ HVAC-installation process assembled on prefabricated CLT-ceilings, CLT-walls and plasterboard-walls and the other was a turnkey prefabricated room-unit including the building services. Aim was to gain realistic time-values for disruptions and to find out who is responsible for these interruptions.

4.1. Analysis of the study

The analysis of the workflow according to REFA provides for a subdivision of the data into the main categories activity and interruption. Additional non-identifiable activities, e.g. the absence of the technician or the observer, are also going to be recorded. Subsequently, the interrupts were broken down into interrupts caused by dysfunction (e.g., waiting time), recreation and personal needs (e.g., telephone call). The data were recorded by means of multi-moment-recording, with the activities of all subjects recorded at intervals of five minutes.

Table 1 summarizes these three main categories of the three analysed building projects. In project 3 the percentage of interruptions was significantly smaller (9 %) compared to the other two construction projects with 14 % and 26 %. But more important, at 2 %, project 3 had the lowest incidence of dysfunctions during the period of investigations, whereas in project 1 about 14 % of the investigated activities fell into the category of interruptions due to dysfunctions.

Table 1. Overview of the recorded categories of activities according to REFA for all three investigated projects [5]

	unit	*Project 1 On-site-construction 38 apartments*	*Project 2 120 turnkey room-units*	*Project 3 On-site-construction 56 apartments*
Number of collected data MMR	[n]	740	1017	350
Corresponding time	[h]	61.67	84.75	29.17
Activities	[%]	73	73	90
Interruptions	[%]	26	14	9
Non-identifiable activities	[%]	1	13	1
Interruptions doe to dysfunctions	[%]	14	6	2

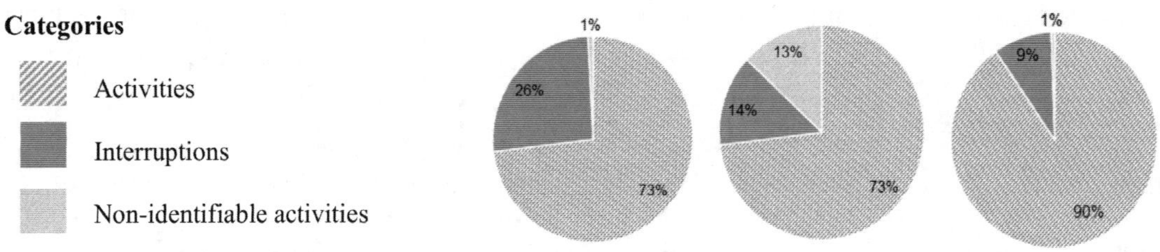

In the next step, on the one hand, for project 1 and 2 the delays due dysfunctions were assigned to the spheres of causation, on the other hand, timber-building-specific disturbances were identified, as shown in Table 2.

4.2. Results disorders and delays

First it was identified whether the disturbances were attributable to timber construction or whether they would also occur in a solid construction such as brick or concrete. Timber-specific dysfunctions are those that arise on the one hand by special new execution details, on the other hand, malfunctions due to improper tools and additional interfaces with other trades. General disturbances are e.g. planning errors (missing information) but also waiting times caused by inadequate working layout or correction of execution errors.

4.3. Results field of responsibility

Table 2 shows the results of the field of responsibility of the disturbances in the individual spheres of planning (architectural and buildings services), construction monitoring and site management, working layout and prefabrication (considered for project 2).

Table 2. Share of timber-constructions related dysfunctions and field of responsibility of dysfunctions [5]

	unit	*Project 1 On-site-construction 38 apartments*	*Project 2 120 turnkey room-units*
General interruption	[%]	62	8
Timber related dysfunctions	[%]	38	92
Caused by planning	[%]	27	79
Caused by working layout	[%]	1	0
Caused by construction monitoring and site management	[%]	10	3
Caused by assembly line production	[%]	0	10
Proportion of interruptions caused by disturbances	[%]	14	6

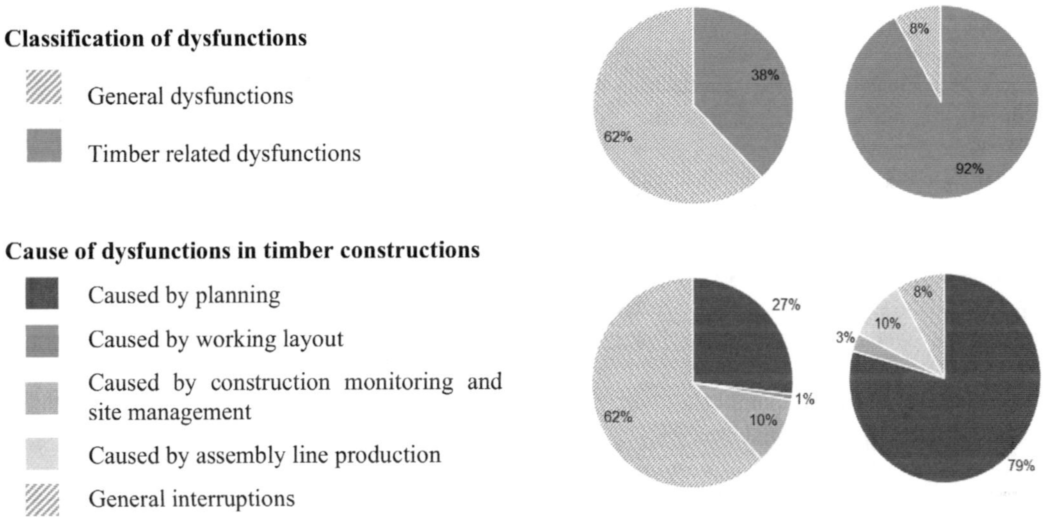

Causes of the high number of disturbances in the area of planning were e.g. extensive drilling in aluminium and steel profiles within plasterboard walls, as the pipelines in internal walls were not planned. That means, that the dimensions of the walls and aluminium profiles were too small and no installation space considered. Also, boreholes through CLT-walls, which are very complicated to perform without suitable tools, have not been considered separately in the plans.

Figure 1 shows the most problematic detail points for planners and installers in Project 1: The main challenge was the waterproofing of CLT-ceilings which is required in the Austrian Standard ÖNORM B 2320:2017 – Wooden residential houses – Technical requirements [6]. So far, little attention had been paid to this new detail. In some cases, planners did not take account of the waterproofing in plans and tenders, which led to negligence in execution. If the properties and schedule of the waterproofing process are not fixed, disturbances in the assembly process and associated additional costs are the consequence. In connection with the waterproofing of CLT-ceilings, new piping runs on the floors must also be taken into account or even better, the pipelines are completely installed in internal walls or as wall surface installations. These details require more space in internal walls, which must early be considered.

Another challenging new timber-specific detail was the covering of openings in CLT-ceilings to gain an appropriate fire protection in installation shafts. The enumerated detail points are just a few excerpts from the list of challenges and problems that had arisen due to incomplete plans, tenders and the lack of training of the installers.

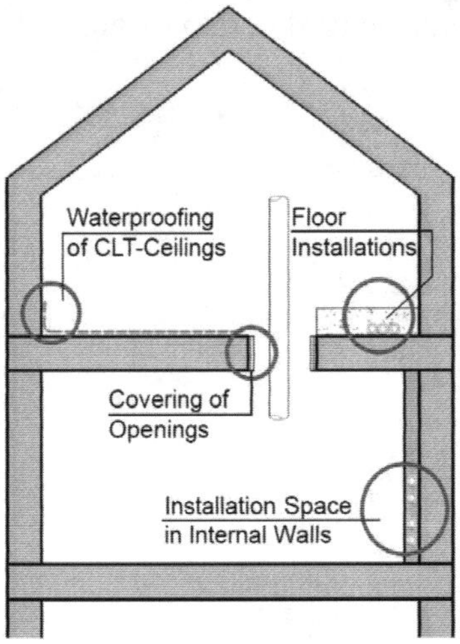

Figure 1. Problematic detail points for the installation process of building services in multi-storey timber buildings [7]

Project 2, which was entire prefabricated in the production hall, has the largest percentage of timber-specific disturbances with 92 %. One reason is that turnkey 3d-modules needs an exact planning and coordination of the assembly processes, so most of the disruptions are due to timber-specific services. Table 2 shows that 79 % of the dysfunctions of Project 2 are caused by planners. During the workflow-analysis a subsequent rescheduling led to the modification of 14 already equipped installation shafts. This fact shows, that changes of plans have an even more negative effect on the production flow caused by the high production speed of similar 3d-units through a production line.

Another special feature could be observed in project 2 in the course of the assembly line prefabrication: Due to the exact scheduling of the prefabrication and a person-specific work distribution waiting times occurred during the assembly process again and again. In this context, a great advantage of the prefabrication in the production hall with integrated workshop for all trades was recognizable: The proportion of disturbances caused by waiting times could be reduces by around 20 %. The reason was that installers could carry out module-independent prefabrication in the workshop, e.g. cutting, bending and joining of pipes. On the one hand, numerous of module-independent work outside the "installation location" were possible through the workshop area in the production hall, on the other hand, the installers were trained to bridge waiting times with necessary preparatory work at the workshop. There – in contrast to the assembly work in the modules, where a clear division of labour was found – the pending preparatory work was completed as needed.

These results of the construction process analysis illustrate the optimisation potential of the work processes with a high degree of prefabrication in a production hall. Considering appropriate workshop areas in on-site construction projects and training of contractors, that improvement in workflow could also be considered in conventional site assembly.

5. Output and conclusion

The presented abstract of results of the research project TGA-Timber with included workflow-analysis at three construction-sites illustrate the large number of challenges in planning and assembling of building services in multi-storey timber construction. Many years of experience in mineral solid construction are of little use to the planners, site managers and contractors in these new construction techniques, as early exact plan specifications, additional interface coordination with other trades and modified details must be considered. Failure to take these parameters into account will result in both time loss and economic consequences. For example, the initially small proportion of disturbances in Project 1 in the workflow-analysis recording period of 578 minutes added up, accumulated on the entire construction period to a loss of time of almost 60 hours. In addition, improper detailing leads to deterioration in quality and thus to the risk of water leakage in timber construction. Finally, there is the requirement of clients after a regular check, done manually or by means of technical devices (sensors) in finished timber structures in order to detect moisture damages early on here.

Business models that allow qualified HVAC companies to intervene in the planning, execution and facility management of objects would partly close the detected knowledge gap, which was verified through the investigations of plans, tenders and workflow-analysis. However, this would require a comprehensive training of planners and installers in order to carry out planning, tendering and construction management for timber-construction-adequate piping runs and to be able to offer additional consulting services for companies with little timber-construction experience. Another important aim for multi-storey timber constructions should be the promotion of prefabrication including the buildings services, in order to transfer the time savings of prefabrication in timber construction to the HVAC installation process. In addition, continuous quality controls are easier to carry out in the production hall and assembly processes through prefabrication workshops are easier to optimize.

The results of this research release the large potential of optimisation of piping runs as well as time- und cost savings within the installations process of building services as part of an overall economical optimisation. They also can make a significant contribution to the sensitisation of planners and contractors, in order to guarantee an economic and technically correct execution of building services installations in timber construction projects.

References

[1] Teischinger A et al 2019 *Holzbauanteil in Österreich – Statischische erhebung aller Hozbauvorhaben in den Jahren 1998-2008-2018* (Vienna: University of Natural Resources and Life Sciences) p 5

[2] Nusser B et al 2019 *TGA-Timber – Technische Gebäudeausstattung im mehrgeschoßigen Holzbau – Endbericht* (Vienna: Holzforschung Austria)

[3] Hausladen G et al 2008 *Holzbau der Zukunft – Teilprojekt 12. Modulare, vorgefertigte Installaionen in mehrgeschoßigen Holzbauwerken* In *Reihe Holzbauforschung Holzbauforschung – Band 7/12* (Stuttgart: Fraunhofer IBR Verlag) p P 12-5

[4] REFA – Verband für Arbeitsstudien und Betriebsorganisation e.V. 1984 *Methodenlehre des Arbeitsstudiums – Teil 2 – Datenermittlung* (München: Carl Hanser Verlag)

[5] Koppelhuber D 2019 *Zeit- und Kostenfallen vermeiden – Wirtschaftliche Betrachtungen zur TGA-Planung im Holzbau* In *Bauphysik Forum 2019, Tagungsband* (Vienna: Holzforschung Austria) pp 82-94

[6] Austrian Standards Institute *2017 ÖNORM B 2320:2017 – Wooden residential houses – Technical requirements* (Vienna: Austrian Standards Institute) p 14

[7] Magg A. 2018 *Gebäudetechnik im mehrgeschoßigen Holzbau: Eine baubegleitende Analyse des Ablaufs und der konstruktiven Ausführung haustechnischer Installationen. Masterthesis* (Graz: University of Technology) p 128

Acknowledgements

This paper is the result of investigations for the research project “TGA-Timber”, financially supported by the Austrian Research Promotion Agency FFG.

Moringa Oleifera seeds and Chitosan as alternatives to conventional coagulation agents

T Kučera[1*], L Hofmanová[1]

[1]Brno University of Technology, Faculty of Civil Engineering, Institute of Municipal Water Management, Žižkova 17, Brno, 602 00, Czech Republic.

*Email: tomas.kucera@vut.cz

Abstract. This article discusses the possibilities of using natural polymers in the water treatment process to replace conventional chemical-based reagents. Mainly part of article presented the results of a laboratory-scale research focused on the use of alternative coagulants in water treatment. Attention was particularly paid to natural cation coagulants derived from Moringa oleifera seeds and chitin. These two coagulant agents were used in laboratory research for treating real raw water. Several series of laboratory jar tests were performed to identify the effectiveness of the selected coagulants in removing turbidity from water.

1. Introduction

The introduction of a coagulation process is often required for treating water from surface sources to decrease turbidity and color, as well as the volume of pathogenic organisms. The process may also be optimized for the removal of natural organic matter (NOM) and heavy metals [6]. Various coagulant agents are used for the process and the selection of a suitable coagulant depends upon the type of substance removed, as well as the overall water treatment process [8]. Coagulation may be combined with other physical, chemical or biological processes.

Coagulants are primarily divided into the organic and inorganic. Inorganic coagulants are metal salts, or possibly in a hydrolyzed polymer form. Organic coagulants are available in many types of substances and create long chains. Coagulants may further be divided based on their charge into cation and anion and, according to their origin, divided into synthetic or natural.

Salts of aluminum and iron appear to be most commercially successful. However, in recent years, polymer coagulants are also used [25], most often as auxiliary agents used to increase the efficiency of water treatment on the one hand and decrease costs on the other.

Various coagulants display differing efficiency in destabilizing individual pollutants [18]. Therefore, the properties of the raw water must be known in order to establish the type and volume of the necessary coagulant. The degree of concentration also affects the potency of their coagulation.

Choice of coagulants primarily depends upon the composition of the water treated, including its pH, temperature, oxidation-reduction potential, volume of substances suspended in the water, etc. The choice of an optimal destabilization agent must always be made based on laboratory tests, ideally by testing under partial operating conditions. The size of floccules is the major physical parameter affecting the treatment process. Another important property is the resistance of these floccules to decomposition into smaller parts. Smaller parts decrease the effectiveness of the subsequent separation, as they settle more slowly than larger particles of similar density [18].

2. Inorganic coagulants, their properties and use

2.1. Metal coagulants

Metal salts are probably the most frequently used coagulants in practical water treatment. These are most often trivalent compounds of aluminum and iron [8]. The effectiveness of their coagulation is affected by dosage, the method of homogenization, pH and the properties of the substance removed [29]. Water temperature is also an essential factor for their function [9], particularly in the case of aluminum-based coagulants. Satisfactory coagulation effectiveness will not be achieved at low temperatures during the colder part of the year, even by increased doses of the agent.

Inorganic coagulants are typically dosed in higher concentrations, resulting in the production of a larger volume of sediment. A certain amount of the metal cation of the respective salt remains in such treated water. Metal coagulant residues alone pose certain health and/or environmental risks.

The disadvantages of inorganic salts, along with their sensitivity to seasonal changes in the physical properties of raw water (pH and temperature) as well as their minimal effectiveness in removing fine particles, brought a demand for different types of coagulants. Effectiveness is improved by polymerization of inorganic coagulants and their use in water treatment [29].

Examples of metal coagulants used in water treatment are iron(III) sulfate, aluminum sulfate, aluminum chloride, iron(III) chloride, as well as titanium disulfide and zirconyl chloride octahydrate [11].

2.1.1. Pre-hydrolyzed metal coagulants

To minimize the disadvantages of inorganic coagulants, they may be hydrolyzed, creating an inorganic polymer with improved coagulation properties. But generally, we could say that they are much more effective than the monomers of salts from which they are produced. Among their most frequently mentioned benefit is a higher effectiveness at lower dosage. Because a smaller amount of the coagulant is used during the water treatment, a far lower volume of sludge is produced [12], nor do they leave as many residual cations in the treated water. Unlike their monomers, they function within large ranges of pH and are less sensitive to temperature [21]. These groups of coagulants include the frequently used polyaluminium chloride (PAC), as well as polymers of iron(III) sulfate, polyaluminium silicate-chloride and polyaluminium silicate sulfate.

3. Organic polymers as coagulation agents

Natural or synthetic macromolecular water-soluble substances are primarily used as destabilizing or auxiliary aggregation agents (flocculants). The use of polymers enables faster floccule sedimentation (the formation of floccules with stronger, denser structure) and improved filterability of the formed suspension. Further, polymer flocculants decrease the content of organic substances, including turbidity as well as residual coagulant, thus improving the ultimate quality of the treated water. The use of these substances also enables a decrease in dosage of inorganic coagulants.

Unlike metal coagulants, organic polymers function within a larger pH range, while having no effect on the pH of the treated water. Their use produces formations that are large, compact, thick and display good sedimentation properties. Unlike metal coagulants, organic polymers leave no metal cations in the treated water. High effectiveness is achieved even at relatively small doses of the coagulant agent, reducing the production of sludge [21]. The addition of an appropriate polymer to an inorganic coagulant may significantly improve its effectiveness in removing impurities, because it forms larger thicker floccules with better sedimentation properties [4].

Organic polymers are also effective in removing small-size particles. They have a double effect on coagulation – aside from charge neutralization, they also have a bridging effect [7]. Bridging is specifically most notable in long molecules with large molecular weight. During the reaction, the polymer is bound to the polluting particle in several places and the loops or loose ends of the polymer float in space, ready to bind to another particle. Among other factors, the effectiveness of coagulation is determined by the dosage of the agent. Should the dose be too low, a certain ratio of impurities remain in the water. To the contrary, should the dose be too high, the pollutants are dispersed again and the polymers subsequently plug the filters [3].

Organic polymers are currently used primarily as auxiliary coagulation and flocculation agents [15], although they may also be used as primary coagulants. Wider use of organic polymers is prevented by their higher cost. Currently, they are also a center of focus regarding possible health risks [30].

Organic polymers may primarily be divided into either **synthetic or natural**. Depending upon their overall charge they further divide into cation, anion, and charge neutral. Synthetic organic polymers (such as cation polyacrylamides) are usually very efficient coagulants, but unlike natural polymers, they tend to be more costly, are not biodegradable and show certain dangerous properties (toxicity, carcinogenic properties). Natural organic polymers are somewhat less effective, but are biodegradable, not toxic and do not produce any secondary pollution [21].

4. Natural polymers

Over time, there has been an effort to replace classic coagulants with natural non-toxic and environmentally-friendly substances. The reason may be the high cost or limited availability of commonly used coagulants (aluminum sulfate, iron(III) sulfate, synthetic polymers etc.), the requirement of highly effective doses of these agents, production of an notable volume of sludge [12] and the undesirable toxicity of the residual coagulants that remain in the treated water.

The use of seeds of certain plants (for example *Lens esculenta, Tamarindus indica, Cyamoppsis psoraloides)* presents a cost-effective, practical and suitable solution for water treatment in developing countries and beyond. A certain preparation is necessary in order to use these seeds, including drying, grinding and dilution with water.

4.1. Cation polyelectrolytes

Strong cation polyelectrolytes have a positive charge regardless of the pH of the solution. To the contrary, weak cation polyelectrolytes have a positive charge only in an acidic environment, as covered [4]. Natural coagulants were used long before chemical salts. But in the end they were pushed to the background, as the use of salts appeared to be an easier and better solution to treating drinking water. In this regard, with the increasing necessity to treat water in developing countries, natural polymers are returning to the center of attention [23]. Natural polymers are easily available, cost-effective and easy to handle.

Among the common natural cation polymers are the **polysaccharides**. The main advantage of polysaccharides is that they are biodegradable, although they are the least effective and must be used in high concentrations. A combination of synthetic polymers onto polysaccharides forms a branched polymer with improved effectiveness [4].

The most frequently used natural cation polymer is **chitosan**. Chitosan is not water-soluble and must therefore be dissolved in a solution of a carboxylic acid, such as acetic acid [22].

Among other natural cation polymers are both starch and tannins [23]. Tannins include many secondary metabolites of plant origins. They are found in bark, leaves or seeds of plants. A very promising cation polymer among the tannins is an extract obtained from the seeds of the ***Moringa oleifera* tree**, containing a cation protein. The main advantage of this extract is that it is not toxic for either humans or animals, while serving as a rather effective coagulant. Further, it does not cause problems with corrosion and produces less sludge. Compared with aluminum or iron salt, the extract shows a lesser effect in removing turbidity. Treatment of water with this extract increases the content of organic substances (particularly orthophosphates and nitrates) in the treated water. These residue substances may worsen the organoleptic properties of water, such as color, taste and odor [4, 16, 19].

Other options for water treatment use extracts from the seeds of the European horse-chestnut (*Aesculus hippocastanum*), English oak (*Quercus robur*), Australian oak (*Quercus cerris*), Northern red oak (*Quercus rubra*) and sweet chestnut (*Castanea sativa*). The concentration of the obtained cation proteins are approximately ten-times less than in the case of *Moringa oleifera*, yet, despite this, good coagulation properties were confirmed [5, 24].

Organic cation polymers may be used both as auxiliary coagulation-flocculation agents and as primary coagulants. Coagulation properties of cation polymers used as primary coagulants may additionally be improved by the addition of insoluble solid substances such as clay or metal oxides (iron, aluminum and manganese oxides). For example, chitosan alone has a low efficiency in removing turbidity and color, but when mixed with bentonite, its effectiveness improves synergically. Solid particles function as sorbents for NOM and, simultaneously, as a core for condensing a NOM-polymer complex.

4.1.1. Moringa oleifera tree seeds

Moringa oleifera is an oil-producing tree originally from Northern India (Himalayas), currently spread across tropical regions worldwide. The plant is notable for its high resistance to drought. *Moringa* seeds are composed of approximately 31% protein, 18% saccharides and 37% fats. Powder from ground seeds contains soluble proteins capable of flocculating turbidity contained in water. The solubility of these proteins is increased with the content of salts in water. Sodium chloride (NaCl) is thus sometimes used during laboratory preparations of the coagulation agent. The coagulation effect of the extract prepared with the use of salt shows greater effectiveness than agents prepared using only tap or distilled water. Studies prove that, in the case of *Moringa oleifera*, adsorption, charge neutralization and formation of polymer bridges between particles are the predominant mechanisms for removing turbidity [13, 26].

The processes of preparing a coagulation agent from *Moringa oleifera* seeds may differ. Nonetheless, the polyelectrolyte used for water treatment is in most cases obtained by grinding dried seeds from the tree and mixing them with water. The proteins with predominantly positive charge on their surface are thus transferred into the water solution. A typical dose for preparation ranges from 10 to 50 grams of ground seeds per liter of water. The suspension of seeds with water is shaken intensively to allow the development of molecules, followed by filtration. The dosage of coagulation agent ranges from 75-200 $mg \cdot l^{-1}$ depending on the properties (particularly turbidity) of treated water [16].

In case cooling of the suspension is possible, its effect is extended by up to one week. Otherwise it is necessary to prepare a fresh suspension daily. The effect of *Moringa oleifera* extract in removing turbidity from water decreases by extending the storage period of the prepared coagulation agent [26].

4.1.2. Chitosan

Chitosan is a substance insoluble in most organic solvents. However, it dissolves in both organic and inorganic acids (such as acetic acid, formic acid, etc.). This polysaccharide is produced through an alkaline deacetylation of chitin, which is found in the outer shells of crustaceans, insects and gastropods, as well as within the cellular walls of yeast and certain fungi. Chitin and chitosan may be produced from the waste products of processing marine animals [28]. Chitosan produced for the purposes of water treatment is typically prepared using acetic acid where it is a subject of hydrolysis.

Chitosan is a cation polyelectrolyte used for the reduction of water turbidity [20], sorption of metal irons, as well as removal of organic pollution. Destabilization primarily takes place through the mechanism of charge neutralization. Chitosan is a biodegradable substance and its use in water treatment is therefore more environmentally friendly than commonly used coagulants. The frequent use of chitosan is increasing in recent years. This substance is used for example in Norway, where it was selected as a coagulant in several water treatment facilities, used either exclusively or in a combination with metal-based coagulants [20].

Chitosan has many advantages in comparison with traditional coagulants (flocculants). Among them are effectiveness in low doses, success in the reduction of chemical oxygen demand (COD), the ability to remove metal ions, the formation of large (voluminous) floccules with higher sedimentation rates, biological degradability, major effect in the removal of algae and suspended particles, anti-bacterial effects and zero formation of secondary pollution [1]. Chitosan is capable of binding both fats and metals, including arsenic, molybdenum, cadmium, chromium, lead and cobalt [14].

Many studies have been conducted regarding the use of chitosan as a single coagulant or in combination with other agents in removing turbidity [2, 20].

Chitosan, as a positively charged polymer with long chains, is able to coagulate negatively charged and finely dispersed colloidal particles through adsorption, charge neutralization, hydrophobic effect or through the formation of polymer bridges. Considerable savings in terms of the volume of agents used may be achieved by using such natural coagulants in combination with the more classic forms, resulting in additional cost reductions for sludge removal, as it is produced in smaller quantities [1].

A coagulation jar test was [20] performed using raw water from the Evans Lake in California, comparing the effects of chitosan, aluminum sulfate and iron(III) chloride. Upon performing the coagulation, sedimentation and filtration using filtration paper, the resulting values of turbidity reduction were 98.8% (aluminum sulfate), 96.9% (chitosan) and 98.9% (iron(III) chloride). The removal of coliform bacteria was most effective with iron(III) chloride and least effective using chitosan (upon optimal doses of coagulants and pH values). Chitosan was the most effective of all agents used in removing color and also caused only minor water softening (by 22 %). In the case of chitosan, excessive doses caused re-stabilization of the dispersion. Optimum pH must be kept in mind when treating raw water into drinking water.

5. Methods

Our laboratory research focused on establishing the effect of selected natural polymer coagulants originating from the seeds of the *Moringa oleifera* tree and chitosan. The effectiveness of both these agents was compared to a conventional coagulant – aluminum sulfate (50% solution).

5.1. Raw water

Raw surface water sampled from the Svratka River was used in the laboratory test. The research used a volume of approximately 30 liters. At the time of testing, the raw water had a temperature of 18.2° C, turbidity 8.92 (NTU) and a pH value of 7.85. The pH value was adjusted using sulfuric acid to a value of 7.5. This value was stipulated with regard to previous tests with aluminum sulfate.

Our research was performed with a standard jar test using a blending column with a paddle blender. Turbidity was used as the indicator of effectiveness. Aside from turbidity, the parameters of water temperature and pH were also monitored. Process parameters were identically maintained for all three agents with the particular goal of verifying whether other conditions of an actual application process would require adjustment upon changing the coagulant.

Fast blending was first performed for a duration of 2 minutes at the rate of 150 rpm, followed by slow blending at 40 rpm for 20 minutes. Subsequently, a 60-minute undisturbed sedimentation completed the process.

5.2. Moringa oleifera seeds

A coagulant was obtained by processing 9 pieces of *Moringa oleifera* seeds and used in a series of tests. The white rounded cores under the peel were pulverized in a mortar and pestle to powder form. The amount obtained was transferred to a porcelain bowl and weighed. The weight of the agent amounted to 1.262 g.

A coagulant was prepared by mixing the powder from the seeds with 306 ml of drinking water (the additional 6 ml were added due to loss of water in the subsequent filtration by being absorbed by the cloth). The suspension was intensively shaken for the duration of 5 minutes and subsequently filtered through a clean cotton cloth to separate larger particles. A dose ranging from 2.5 to 42.5 ml of the solution was added to the individual vessels, representing 20.6 to 350.9 mg of powder.

5.3. Chitosan

The effect of chitosan in removing turbidity was examined last. The agent was used in the form of 25 g of white powder – chitosan originating from shrimp shells. Chitosan was dissolved in acetic acid (99%), that was subsequently mixed with 32.6 ml of drinking water and used in the form of a solution at the concentration CH_3COOH 1 mol·dm^{-3}. Subsequently, 0.1 g of chitosan was added to the acid solution. This formed a liquid of a rather thick consistency that was further diluted with drinking water to a 100 ml volume. This coagulant (100 mg of chitosan in 100 ml of the agent) was dosed to the vessels in an optimization jar test.

6. Results

Our research focused on the comparison of effectiveness in reducing turbidity after applying three coagulants used in water treatment. The selected coagulants were aluminum sulfate commonly used as a conventional coagulant, as well as chitosan and *Moringa oleifera* as alternatives. An optimization jar test was performed with the tested aspect being turbidity. The research was carried out in the laboratory format as a series of jar tests.

6.1. Aluminum sulfate

Aluminum sulfate was selected as representative of conventional agents used in water treatment. The purpose of its use in this research study is primarily as a comparison with the other two agents that are not as frequently used and practically never in the Czech Republic.

The authors of this article have extensive previous experience with this coagulant in relation to raw water used in this research. As is apparent from previous jar tests, the optimum pH for turbidity removal is 7.5, the rate for the first phase of mixing is 150 rpm (for the duration of 2 minutes) and the rate for the second phase of coagulation is 40 rpm (for 20 minutes). The same values were also used for the comparison tests with the other agents. The results are apparent from Table 1 and Figure 1.

Table 1. Turbidity removal – aluminum sulfate

Sample	Dosage			Turbidity	Turbidity removal
	Aluminum sulfate		Al	@ 60 min	
	[ml]	[mg]	[mg]	[NTU]	[%]
1	0.050	65.5	2.6	3.56	60.09
2	0.075	98.3	3.9	2.13	76.12
3	0.100	131.0	5.2	0.94	89.46
4	0.115	150.7	6.0	0.46	94.84
5	0.130	170.3	6.8	0.65	92.71
6	0.175	229.3	9.2	0.85	90.47
7	0.220	288.2	11.5	1.00	88.79
8	0.265	347.2	13.9	1.26	85.87

The lowest measured value of turbidity after sixty minutes of sedimentation was 0.46 NTU in case of 0.115 ml dose of aluminum sulfate. The decrease of NTU by 8.46 shows removal of turbidity at 94.84 %. The worst result was recorded at doses of 0.050 ml – the value of the turbidity 3.56 ZF and turbidity removal at only 60.09 %.

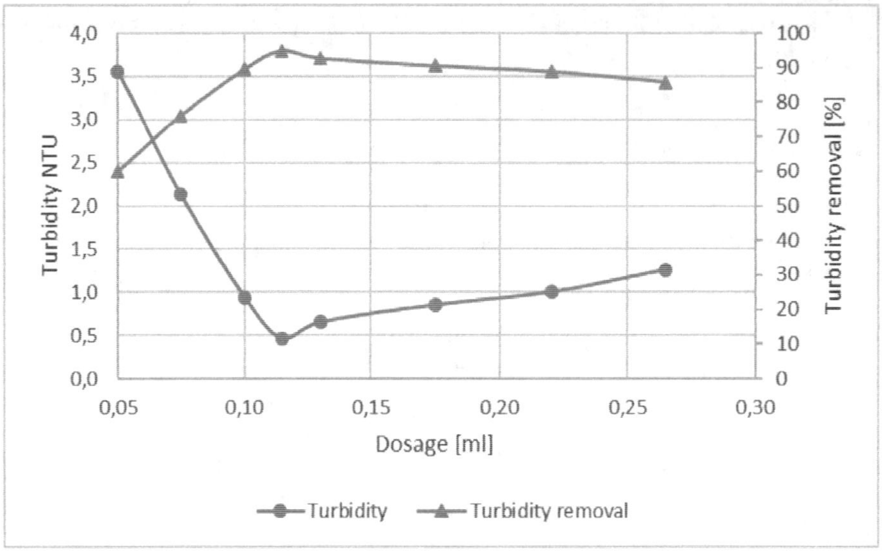

Figure 1. Turbidity removal – aluminum sulfate

A large number of discernable floccules formed when using aluminum sulfate, even in the early stage of fast blending. The most aggregates were formed in the water to which most of the agent was dosed. Upon 60 minutes of sedimentation the floccules settled at the bottom, where they formed sludge of a light brown color. The optimal dose of aluminum sulfate 150.7 $mg \cdot l^{-1}$ managed to reduce the turbidity by 94.84 % to a value of 0.46 NTU.

6.2. Moringa oleifera

The coagulant obtained from *Moringa oleifera* seeds by the method described above was used to treat water of identical quality as above and also within a laboratory-scale test. The same parameters were set for the coagulation test. The formation of small light-colored floccules in the treated water was recorded as early as the slow blending stage.

The amount of these small observed clusters grew with increased amounts of the coagulation agent. However, the floccules remained suspended in the volume of the vessels throughout the entire phase and were only partly sedimented. Only a portion of the floccules settled on the bottom, along with sediment. Despite this, lower levels of turbidity were measured than prior to water treatment. Results are shown in Figure 2.

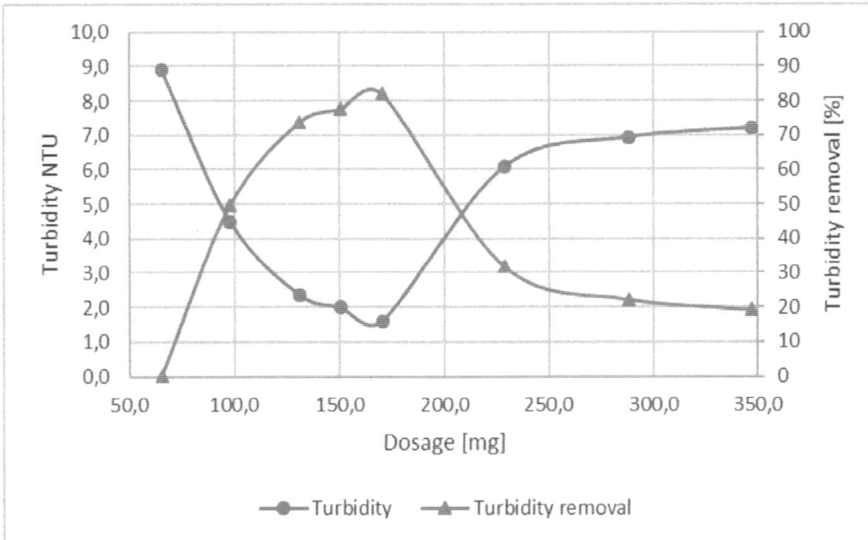

Figure 2. Turbidity removal – Moringa oleifera

As is apparent from the above results, the most effective dose of *Moringa oleifera* for turbidity removal was 164.6 mg. The initial turbidity value was reduced to 1.60 NTU, representing the removal of 82.04% of turbidity. The lowest value of turbidity removal was measured at the dose of 20.6 mg of *Moringa oleifera*. Here, the turbidity value only decreased by 0.01 NTU.

When using *Moringa oleifera*, the floccules formed during coagulation were of pale color. They were visible to the naked eye only in the second phase of blending. The sedimentation of the aggregates took place very slowly. Even after one hour of sedimentation, there were no apparent changes. The optimum dose was 164.6 $mg \cdot l^{-1}$, which is the largest in comparison with both aluminum sulfate and chitosan. Better results of turbidity removal could apparently be achieved by prolonging the sedimentation period or using filtration.

6.3. Chitosan

The next coagulation tests were using chitosan as an agent – also under the same conditions. The formation of floccules in raw water was visible from approximately the middle of the slow blending phase. The smallaggregates had a light brown color and settled more quickly than the aggregates using *Moringa oleifera*. Although most floccules settled after 60 minutes of sedimentation, a portion of the floccules remained suspended. As regards turbidity removal, the resulting values are shown in Figure 3.

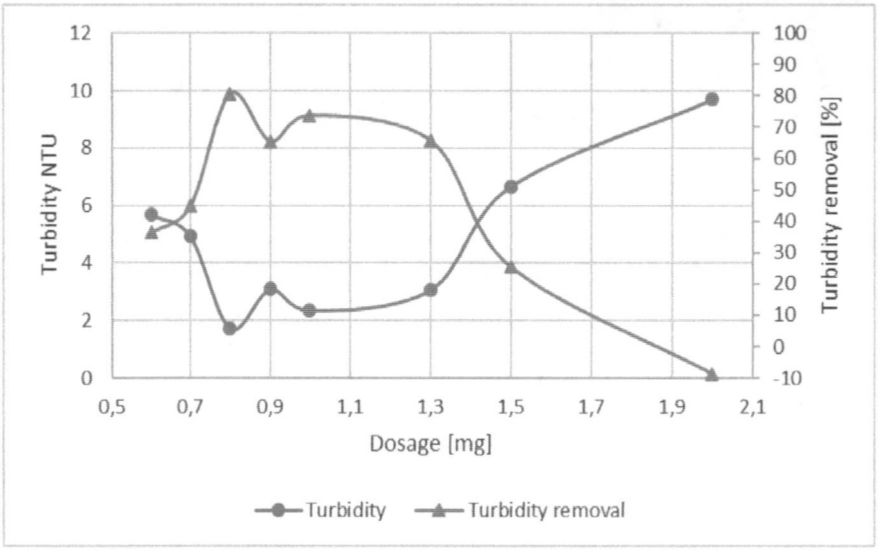

Figure 3. Turbidity removal – chitosan

In the case of chitosan, the optimal dose was low – the mentioned 0.8 mg·l^{-1}. At this dose, the remaining turbidity measured had a value of 1.73 ZF, representing 80.66% less than measured in raw water. In the jar test, the effect of chitosan was only visible during the second half of slow blending. Very tiny light brown floccules were formed in the vessels and their amount was in direct proportion to the coagulant dose. The aggregates showed much better sedimentation properties than in the case of *Moringa oleifera*.

7. Conclusions and Discussion

The laboratory experiments described evaluated the removal of turbidity from a water environment using a coagulation process with three different coagulation agents. As is apparent from Figures 1-3, the course of coagulation and flocculation during the optimization jar test, as well as the dependency of the agent dosage on turbidity removal, were different for each agent.

The most effective coagulant in terms of turbidity removal under the given conditions is without doubt aluminum sulfate. But in case of chitosan, the much lower optimum dosage compared to the other two agent is noteworthy (Table 2).

Table 2. Comparison of the results upon using three different coagulants

Coagulant	Optimal Dosage	Turbidity	Turbidity removal
		@ 60 min	
	[mg]	[NTU]	[%]
Aluminum sulfate	150.7	0.46	94.84
Moringa oleifera	164.6	1.60	82.04
Chitosan	0.8	1.73	80.66

The best results in removing turbidity from raw water were achieved by aluminum sulfate. However, the use of aluminum sulfate in water treatments from secondary products had related health risks. The results of the experiment using *Moringa oleifera* are less impressive, although removing turbidity by 82% indicates a very good overall effectiveness of this agent. Better results could be obtained through different preparation of the coagulant (the use of NaCl, distilled water). In the case of chitosan, a slightly better result was expected. However, the reduction of turbidity by 80.66% indicates very good properties of this polyelectrolyte in removing turbidity from raw water. Chitosan efficiency may be supported by using it in combination with traditional coagulants.

Studies researching this version prove that the application of certain amounts of substances with natural origins reduces the necessary optimal amounts of commonly used coagulants. Water treatment using a combination of both substances is more environmentally friendly. Lesser amounts of harmful residual products are formed during the course of such process (residual Fe and Al) and the amounts of necessary agent are reduced, resulting in savings as the volume of the formed sludge is reduced.

The authors of this study are aware that the results in case of *Moringa oleifera* seeds and chitosan might even be better, as the jar test conditions were optimized for aluminum sulfate. In the case of a possible replacement of an existing coagulant (aluminum sulfate) by an alternative in a real-application situation, it would be necessary to also adjust the parameters of the coagulation process, such as blending rates and the period of placement in blending and separation basins.

An interesting criterion for discussions of possible replacements of conventional coagulants for natural polymers may also be their cost. The prices of all three coagulants are subject to fluctuations depending on the location of use. Our research did not establish the price of the optimum dosage, as the natural polymers used for the purpose of our study were obtained under price-inefficient conditions at low quantities.

Acknowledgements
The article paper was drawn up within project No. LO1408 "*AdMaS UP - Advanced Building Materials, Structures and Technologies*" supported by the Ministry of Education, Youth and Sports as part of the targeted support programme "National Programme for Sustainability I".

References
[1] Bhalkaran, S.; Wilson, L.D. Investigation of Self-Assembly Processes for Chitosan-Based Coagulant-Flocculant Systems: A Mini-Review. Int. J. Mol. Sci. 2016, 17, 1662.
[2] Bina, B., Mehdinejad, M.H., Nikaeen, M., Movahedian Attar, H. Effectiveness of chitosan as natural coagulant aid in treating turbid waters (2009) Iranian Journal of Environmental Health Science and Engineering, 6 (4), pp. 247-252.
[3] Bolto, B.A. Soluble polymers in water purification (1995) Progress in Polymer Science, 20 (6), pp. 987-1041. DOI: 10.1016/0079-6700(95)00010-D
[4] Bolto, B., Gregory, J. Organic polyelectrolytes in water treatment (2007) Water Research, 41 (11), pp. 2301-2324. DOI: 10.1016/j.watres.2007.03.012
[5] Camacho, F.P., Sousa, V.S., Bergamasco, R., Ribau Teixeira, M. The use of Moringa oleifera as a natural coagulant in surface water treatment (2017) Chemical Engineering Journal, 313, pp. 226-237. DOI: 10.1016/j.cej.2016.12.031
[6] Charerntanyarak, L. Heavy metals removal by chemical coagulation and precipitation (1999) Water Science and Technology, 39 (10-11), pp. 135-138. DOI: 10.1016/S0273-1223(99)00304-2.
[7] Ebeling, J.M., Rishel, K.L., Sibrell, P.L. Screening and evaluation of polymers as flocculation aids for the treatment of aquacultural effluents (2005) Aquacultural Engineering, 33 (4), pp. 235-249. DOI: 10.1016/j.aquaeng.2005.02.001
[8] Edzwald J. K. 1993 Coagulation in drinking water treatment: Particles, organics and coagulants. *Water Science and Technology,* **27**(11), 21-35.
[9] Fitzpatrick, C.S.B., Fradin, E., Gregory, J. Temperature effects on flocculation, using different coagulants (2004) Water Science and Technology, 50 (12), pp. 171-175.
[10] Gao, B.Y., Hahn, H.H., Hoffmann, E. Evaluation of aluminum-silicate polymer composite as a coagulant for water treatment (2002) Water Research, 36 (14), pp. 3573-3581. DOI: 10.1016/S0043-1354(02)00054-4
[11] Jarvis, P., Sharp, E., Pidou, M., Molinder, R., Parsons, S.A., Jefferson, B. Comparison of

coagulation performance and floc properties using a novel zirconium coagulant against traditional ferric and alum coagulants (2012) Water Research, 46 (13), pp. 4179-4187. DOI: 10.1016/j.watres.2012.04.043

[12] Keeley, J., Jarvis, P., Judd, S.J. Coagulant recovery from water treatment residuals: A review of applicable technologies (2014) Critical Reviews in Environmental Science and Technology, 44 (24), pp. 2675-2719. DOI: 10.1080/10643389.2013.829766

[13] Lea, M. Bioremediation of turbid surface water using seed extract from the Moringa oleifera Lam. (Drumstick) tree (2014) Current Protocols in Microbiology, (SUPPL.33), art. no. 1G.2, DOI: 10.1002/9780471729259.mc01g02s33

[14] Li, S., Lian, Y., Mai, J., Tan, J., Hu, S., Xia, W., Duan, W. Removal of hexavalent chromium and turbidity in drinking water by Chitosan (2012) Advanced Materials Research, 433-440, pp. 793-797. DOI: 10.4028/www.scientific.net/AMR.433-440.793

[15] Matilainen, A., Vepsäläinen, M., Sillanpää, M. Natural organic matter removal by coagulation during drinking water treatment: A review (2010) Advances in Colloid and Interface Science, 159 (2), pp. 189-197. DOI: 10.1016/j.cis.2010.06.007

[16] Ndabigengesere, A., Subba Narasiah, K. Quality of water treated by coagulation using Moringa oleifera seeds (1998) Water Research, 32 (3), pp. 781-791. DOI: 10.1016/S0043-1354(97)00295-9

[17] Niquette, P., Monette, F., Azzouz, A., Hausler, R. Impacts of substituting aluminum-based coagulants in drinking water treatment (2004) Water Quality Research Journal of Canada, 39 (3), pp. 303-310.

[18] Pivokonský, M. Tvorba suspenze při úpravě vody: teorie a praxe. Líbeznice: Medim, 2011, 218 s. ISBN 9788087140185.

[19] Pritchard, M., Craven, T., Mkandawire, T., Edmondson, A.S., O'Neill, J.G. A comparison between Moringa oleifera and chemical coagulants in the purification of drinking water - An alternative sustainable solution for developing countries (2010) Physics and Chemistry of the Earth, 35 (13-14), pp. 798-805. DOI: 10.1016/j.pce.2010.07.014

[20] Pontius, Frederick W. Chitosan as a Drinking Water Treatment Coagulant. *American Journal of Civil Engineering* [online]. 2016, **4**(5), 205- [cit. 2018-05-29]. DOI: 10.11648/j.ajce.20160405.11. ISSN 2330-8729.

[21] Renault, F., Sancey, B., Badot, P.-M., Crini, G. Chitosan for coagulation/flocculation processes - An eco-friendly approach (2009) European Polymer Journal, 45 (5), pp. 1337-1348. DOI: 10.1016/j.eurpolymj.2008.12.027

[22] Rizzo, L., Di Gennaro, A., Gallo, M., Belgiorno, V. Coagulation/chlorination of surface water: A comparison between chitosan and metal salts (2008) Separation and Purification Technology, 62 (1), pp. 79-85. DOI: 10.1016/j.seppur.2007.12.020

[23] Sánchez-Martín, J., González-Velasco, M., Beltrán-Heredia, J. Surface water treatment with tannin-based coagulants from Quebracho (Schinopsis balansae) (2010) Chemical Engineering Journal, 165 (3), pp. 851-858. DOI: 10.1016/j.cej.2010.10.030

[24] Šćiban, M., Klašnja, M., Antov, M., Škrbić, B. Removal of water turbidity by natural coagulants obtained from chestnut and acorn (2009) Bioresource Technology, 100 (24), pp. 6639-6643. DOI: 10.1016/j.biortech.2009.06.047.

[25] Tomaszewska, M., Mozia, S., Morawski, A.W. 2004, „Removal of organic matter by coagulation enhanced with adsorption on PAC", Desalination, Vol. 161, pp. 79-87.

[26] Yarahmadi, M & Hoseini, Mohammad & Bina, Bijan & Mahmoudian, M.H. & Naimabadi, Abolfazl & Shsavani, A. (2009). Application of Moringa Oleifera Seed extracted and polyaluminum chloride in water treatment. World Applied Sciences Journal 7 (8), pp. 962-967.

[27] Ying, F. Investigation on flocculation process of composite poly-Si-Fe coagulant (2011) Desalination and Water Treatment, 30 (1-3), pp. 122-133. DOI: 10.5004/dwt.2011.1932

[28] Younes, I., Rinaudo, M. Chitin and chitosan preparation from marine sources. Structure,

properties and applications (2015) Marine Drugs, 13 (3), pp. 1133-1174. DOI: 10.3390/md13031133

[29] Yu, W., Gregory, J., Campos, L.C., Graham, N. Dependence of floc properties on coagulant type, dosing mode and nature of particles (2015) Water Research, 68, pp. 119-126. DOI: 10.1016/j.watres.2014.09.045

[30] Zhao, Y.X., Gao, B.Y., Shon, H.K., Wang, Y., Kim, J.-H., Yue, Q.Y. The effect of second coagulant dose on the regrowth of flocs formed by charge neutralization and sweep coagulation using titanium tetrachloride (TiCl 4) (2011) Journal of Hazardous Materials, 198, pp. 70-77. DOI: 10.1016/j.jhazmat.2011.10.015

[31] Zhou Z., Yang Y., Li X., Liu Y. & Su, Z 2013 Fluorescence characteristics of dissolved organic matter in coagulation process of recycling filter backwash water. *Huagong Xuebao/CIESC Journal,* **64**(5), 1825-1832. DOI: 10.3969/j.issn.0438-1157.2013.05.044.

IOP Conf. Series: Materials Science and Engineering **800** (2020) 012033 doi:10.1088/1757-899X/800/1/012033

Modular timber structures

D Kuda[1*], M Petříčková[1]

[1] Department of Engineering, Faculty of Architecture - Brno University of technology, Poříčí 273/5, 639 00 Brno, Czech Republic

*Email: xakuda@stud.fa.vutbr.cz

Abstract. Related to sustainability movement and minimizing the carbon footprint, timber structures are becoming more attractive. Wood, as main structural material, offers many benefits relate mostly to economic and ecological aspects, compared to other materials as steel or concrete. On the other hand, physical characteristics of wood complicate the usage of a timber for high-rise or large-span structures. It brings a new challenge for architects and engineers to deliver feasible solution for usability of timber, despite its features. One of the possible solutions could be implementation of CLT (Cross-Laminate Timber) panels in structural systems developed earlier for buildings made of prefabricated concrete slabs. SOM in cooperation with Oregon State University are currently testing composite slabs made of CLT and thin concrete layer reinforcing the wood and protecting it from fire. Although the system solution looks promising, and could bring the result, slabs limit using of the space in layout. On the other hand, frame structures would be much more efficient. This article comes up with an idea of modular frame structure, which could help to solve the problem. The scheme is based on "gridshell" type systems, where rods form a more efficient shell for dealing with stress forces.

1. Introduction

Currently, glue laminated timber is commonly used in standard buildings. For high rise structures (HRS) and long span structures (LSS) wood, as a main structural material, is used only sporadically. These structures are mostly experimental buildings and the project that usually requires specialized software which is developed exactly for designed structure. On case studies related to finished and experimental buildings, characterized by load bearing timber structure, is based on a development of universal structural prototype, which can be used for timber high rise buildings. The system is supposed to be universal and should be able to meet various architectonic and space requirements.

2. Case studies

Excellent example of load bearing frame timber structure can be seen in Tamedia Office Building in Zurich, Switzerland - 'Figure 1'.

Figure 1. Tamedia Office Building.

Structural principle of Tamedia is actually pretty simple. It is a six-storey high frame structure made of a glue-laminated timber - 'Figure 2'. Wooden beams and pillar in the whole building are not covered and are protected from the weather by a transparent façade, which makes the whole interior structure visible from outside. An important element is using of joints – architect Shigeru Ban designed them according to traditional Japanese carpentry techniques called "Miya-daiku" and "Sukiya-dayiku". These joints are made only of wood - 'Figure 3', which eliminates need of other material for the structure.

Figure 2. Tamedia – structure.

Figure 3. Tamedia – joint.

For a better distribution of stress in frame structure it is possible to use a system called gridshell. This structure is characterized by (usually double) curved plane made of rod type elements. Grid is usually orthogonal, but also irregular schemes exist. We can illustrate the principle on Trondheim gridshell - 'Figure 4', an experimental structure builds by students of Norwegian University of Science and Technology.

Figure 4. Trondheim gridshell – model.

The structure is a result of large analysis of existing gridshells and following development of the optimal form (curve) of the shell. There were used parametrical software for the form finding phase and special computing add-on software for optimization of the parameters. The final structure is based on orthogonal planar grid. The final shell was reached by deforming the structure by rising the centre of the grid and fixing its foundations. The structure is then actually pre-stressed therefore, more effective in diverting the load.

Orthogonal gridshell systems are suitable for regular forms of shell. For free (organic) forms, like Pompidou Metz - 'Figure 5', it is better to use more sophisticated systems. The entire building is wrapped in a fabric-clad hexagonal layered gridshell. The architect, Shigeru Ban, found inspiration in woven canework of traditional Chinese hat. Each of three layers of laminated timber goes in different direction, forming a hexagonal net. Timber blocks are added between the layers, which increasing the depth of structure and thus the structural performance of the system [1] - 'Figure 6'. The final approximately 90 metres wide 80 metres high structure is then very rigid and stable. The shell was divided into elements, which were prefabricated using CNC machines and assembled on site.

Figure 5. Centre Pompidou Metz.

Figure 6. Centre Pompidou Metz – structure.

Potential of usage of frame structures seem to be limited due to physical-mechanical characteristic of wood. Pillars and beams have to be very massive to bear the loads from upper storey and possible bending moments caused by wind. As an attractive solution seem to be implementation of slabs in the structure. CLT (Cross-Laminate Timber) panels can create a "box" system, which is much more stable. One of these systems was firstly presented and patented in Czech Republic in 1964 by prof. Ing. Václav Rojík, DrSc. The structure was firstly developed for reinforced concrete buildings, but has never been used. Research based on the original system using CLT panels is currently underway [2]. According to parametrical models, the high of twenty or more storeys can be reached with using only 150 mm thick panels.

3. Possibilities for high-rise buildings

The question is, if we can use the knowledge from existing structures to develop a frame system for high rise buildings. Static and material disadvantage of wooden rod elements were already mentioned in this article. Nevertheless, the solution mentioned above is relevant only for the specific case and cannot be applied generally. The system is usually developed for specific use regard to architectonic design. For general use of the results of the study it is necessary to develop a universal timber structure system. This system should provide sufficient freedom in design and possibilities in usage of space according to functions (i.e. housing, administration, etc.).

The key attribute for designing a universal system is modularity. Module is an integral part of structure, represented by a single construction component, a segment of structure or the whole part of the building, i.e. a functional unit, story, etc. By repeating the basic module we should reach many different structures. Module should be also simple for prefabrication, should be easy to connect to another module, transport, etc. Problem obviously comes with using the same universal module in different part of the structures or building, where different requirements for use of space or typically for HRS and LSS in different types of stress are identified. Because the frame structure creates a free space (according to Le Corbusier´s "Plan Libre") it is ease to design different layout in to the same modular structure. Complications come with variable requirements related to structural stability.

Generally, three principles can be used as a basic solution approach. Modules can be designed for the maximum of stress forces found in the structure, which leads to a waste of material in less loaded modules. Unnecessary material in upper storeys causes additional load for lower parts of the structure, so these parts must be designed for higher load than would be necessary.

In a nutshell, it is possible to design special modules for different parts of the building. Nevertheless, as there are always different specifics and requirements, it will inevitably lead to lose of universality in the system due to specification of its parts. Another possibility is the application of reinforcing parts, like monolithic concrete core or steel parts, which also lowering the universality of the system - 'Figure 7'.

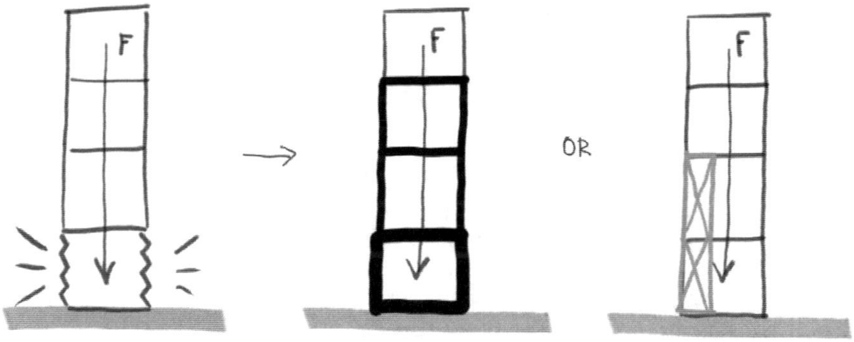

Figure 7. Stability.

This article presents another possible way of designing the structure. If there are variable requirements, the module needs to be also variable. To maintain the universality of the system, the module must be adaptable according to clear principle. This principle should be able to react to as many different requirements, as possible. The promising approach could be the fractality principle.

IOP Conf. Series: Materials Science and Engineering **800** (2020) 012033 doi:10.1088/1757-899X/800/1/012033

4. Fractals

Mathematical definition says that fractal is a subset of a Euclidean space for which the Hausdorff dimension strictly exceeds the topological dimension. Better definition is "A fractal is a shape made of parts similar to the whole in some way" [3]. The principle could be easily explained by the following picture of "Sieprinski Carpet" - 'Figure 8'.

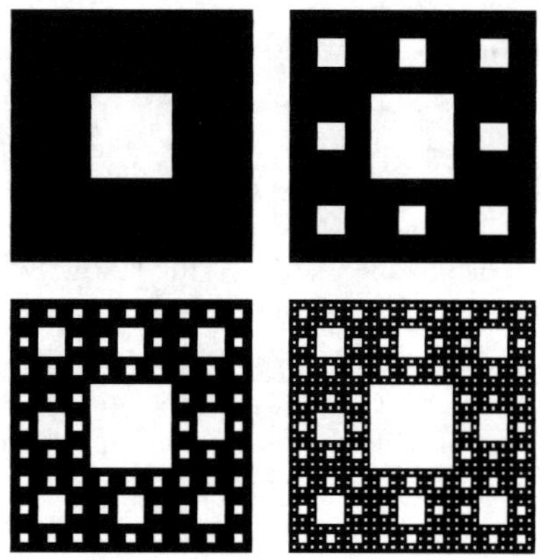

Figure 8. Sieprinski Carpet.

5. Fractal modular system

The idea of the system is the replacing the standard module by a predetermined number of sub-modules according to structural and space requirements. The easiest way of application is to define a module by the standard height of one storey - 'Figure 9'.

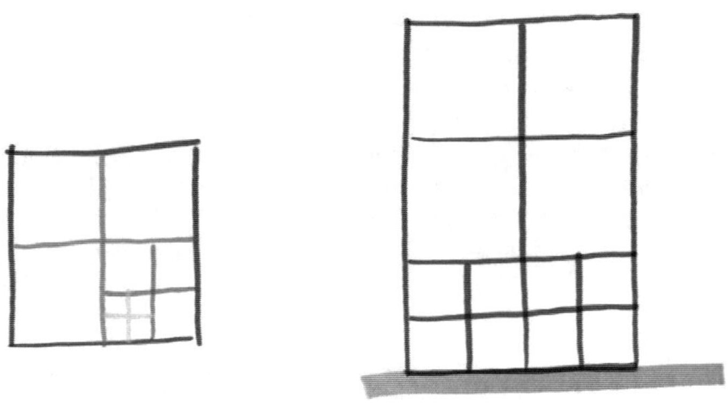

Figure 9. Basic principle.

6. General theory of structures for high rise structures

The idea works well for vertical load, but rectangular frame is not very effective in resisting to horizontal stress. Especially in high-rise timber structures, which are very light and also height (for standard housing unit can be the weight of timber load bearing structure set to about fifteen tons, compared to concrete, which is generally between 30 to 60 tons per unit), bending moments and twisting caused by wind are very dangerous. The module, on its own, should be stable to improve the aspects of the whole structure. The most stable geometric shape is triangle, so the final system should include it.

7. Geometry

The geometry of the system described in this article is based on a square, which diagonal has a dimension of one story – in this example its three meters. Modules connected together are creating one grid, so the structure is based on load bearing facade, as same as in professor Rojík´s system. Floor slabs or grid-slabs can be attached to joints of modules that repeat every half-storey height. Each module can be divided into sub-modules, according to structural requirements. Standard module height is suitable for housing (according to Czech legislation), one and half module is perfect for administration. Other possibilities can be found on the following picture - 'Figure 10'.

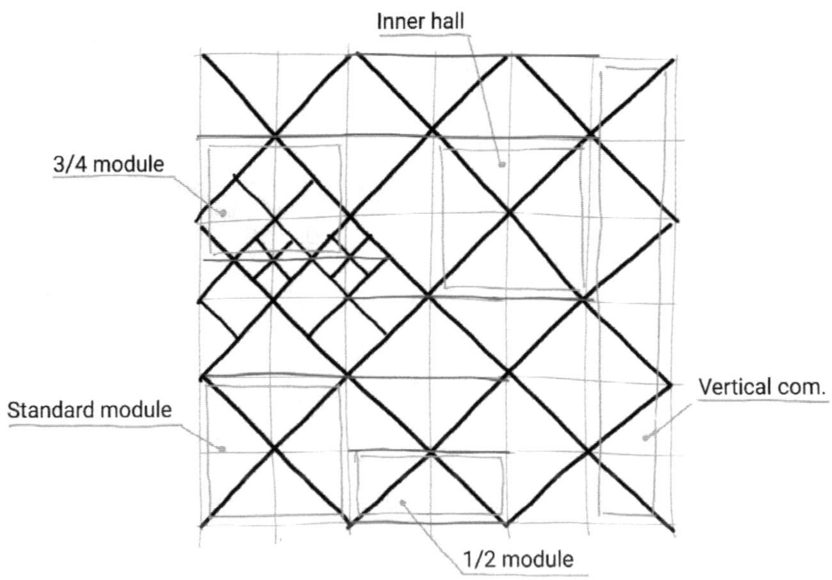

Figure 10. Geometry.

8. Application

In design phase of the building, the system is used as free shape load bearing façade. In the layout, the shape of the structure is free. It may be orthogonal, but also free curved shapes are possible. It is preferred to make multiple layers forming inner atriums or caverns. The final shape of layout could significantly improve the stability of the whole building. It is possible to create tilted facades, forward or receding stories. There are only several conditions which needs to be always followed as to keep the module planar and to keep it in the consistent dimensions (twisting and scaling needs to be avoided). The free shape of the structure makes the load bearing facade to distribute loads similarly to gridshell structures.

The dimensions of section of laminated timber beams are assumed to be 300/300 mm. Clearly the most important part is the joints. It is possible to use steel segments, but the general aim of this research is to maximize the use of timber. Similarly, as in structure of Tamedia, it is possible to use traditional carpentry techniques to connect the modules. It is possible to connect modules together in main connections by using the traditional joints. Also, additional connections in sub–modules can be created by the same principle. All timber part should by prefabricated on CNC, but also it would be possible to make a joint in-situ.

Floor slabs are also formed by the same grid modules, only horizontal orientation is considered. If needed, modules can be reinforced by dividing into submodules or adding vertical modules. As the result, the finale structure should be very rigid. As the protection from the weather condition a double transparent façade anchored to modular structure can be used.

IOP Conf. Series: Materials Science and Engineering **800** (2020) 012033 doi:10.1088/1757-899X/800/1/012033

On the next pictures there is presented a twenty-storey high building which is designed by using the modular system. On the left side -'Figure 11.'- there is the load bearing structure, right side - 'Figure 12.' – there is the same building viewed in x-ref mode.

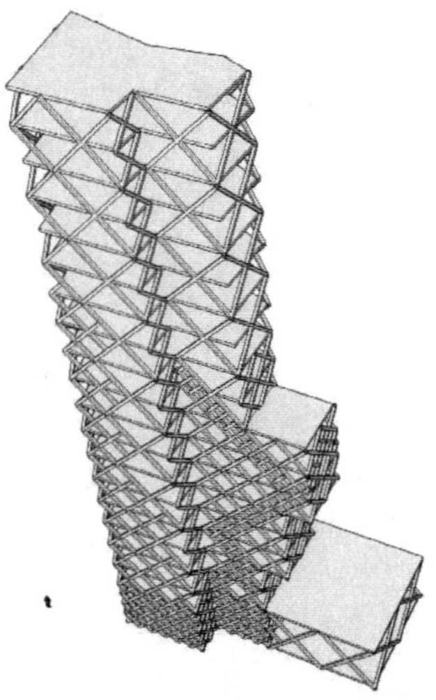

Figure 11. Application.

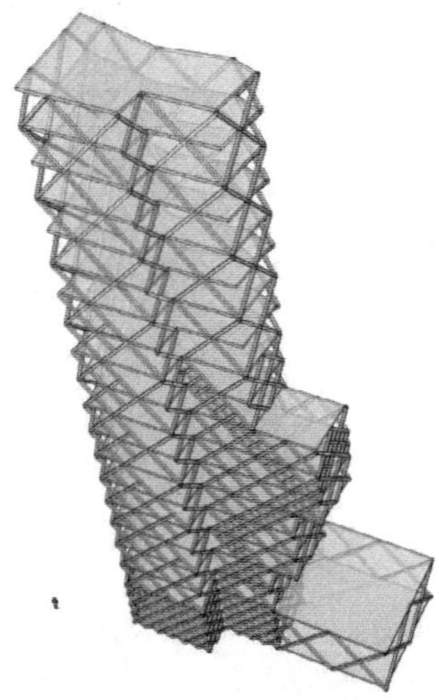

Figure 12. Application – x-ref

9. Conclusion

The aim of the research is to find a modular timber structure for high-rise buildings. The modularity of the structure makes the system very universal. By using the same principle for different types of buildings and repeating the same principle, a great amount of energy can be saved in developing the software and can be used for deeper development of the structure.

Although, there are still many parts of the system that need to be developed. This article shows the general idea and the possible approach. Next possibility could be layering of the structure, as demonstrated on Pompidou Metz roof. Combination of structural layers could simplify the joints of each module, could bring material savings and improve mechanical performance of the system. Effect of replacing the one joint in specific spot between modules by multiple joints on different spots of every layer is now in testing phase.

References

[1] Silver P, Mclean W, Evans P, 2013 *Structural Engineering for Architects: A Handbook* Laurence King Publishing 194 ISBN 978-178067-055-3
[2] Rojík V, 2019 *Časopis Stavebnictví* Statické problémy vícepodlažních budov ze dřeva 14 40-44
[3] Jens Feder 2013 *Fractals* Springer Science & Business Media 11 ISBN 978-1-4899-2124-6

An evaluation of level of knowledge, behaviour and attitude towards environmental footprint in secondary school students in Libya

I F G Laama[1] , D Özdal[1*] , T Onay[2]

[1]European University of Lefke, Faculty of Engineering, Department of Environmental Sciences, Lefke-Northern Cyprus, TR-10, Mersin, Turkey

[2]Boğaziçi University, Institute of Environmental Sciences, Bebek, 34342, İstanbul, Turkey

*Email: devrimozdal@eul.edu.tr

Abstract

The indications of the impact of activities of people on environment are called footprints. Footprints are targeting various concerns related with environment and they quantify resource appropriation or waste generation, or both. Knowledge on protecting environment is one of the important factor that may affect environment adversely, because knowledge determines attitudes and behaviours of someone to environment. This study aimed to assess knowledge, attitudes and behaviours of Libyan secondary schools students towards environmental footprint. A quantitative research employing a questionnaire including questions on demographic characteristics of students and 35 questions on student's knowledge, attitudes and behaviours of students towards environment was carried out. Data were collected from 400 students in the last quarter of 2018 in four secondary schools located in Tripoli. Data were analysed using Statistical Package for Social Sciences, SPSS vs. 21, employing both descriptive and inferential statistical techniques. Majority of students (91.8%) were observed as having high knowledge on environmental footprint. A higher proportion of students were having moderate levels of attitudes (76.0%) and behaviours (65.8%) towards environment. When the impact of demographic characteristics on level of knowledge, attitudes and behaviours was studied using t-test, it was found that students whose fathers were graduated from master or PhD had significantly greater knowledge level compared to others. It is important to implement programmes for improving knowledge of different parts of population by both empowering school curriculum and involving public and private sectors for involving in policies and programmes for both improving knowledge of different population groups and encouraging them to safe behaviours towards environment.

1. Introduction

Traces of the effect of activities by populations in the environment are defined as environmental footprint. There are two main types of footprint, namely environmental footprint and carbon emissions related footprint. Environmental footprints have a very wide range, because it includes many areas of human activities in the environment. Environmental footprint involves the arithmetic of assessing the human utilization of environmental endowments in relation to the effects on the immediate environment; the concept assesses or weighs the nature endowments utilization with regards to natural supply of such endowments; these could be applicable to single person, households, people, settlements, academic populations of various categories, municipal areas, states and countries across the world. A number of environmental concerns may therefore be related to footprints and to reduce these concerns appropriate resource allocation and improvement of waste generation are required [1,2]. Environmental footprint, in addition, includes the indications of consumption of populations on the entire world, bio-diversity endowments on the biosphere, hydrosphere, and atmospheric environment. Government, non-governmental organizations and formal business organizations must consider the people consumption amounts in the way that if people consumption exceeds the environmental carriage capacity rate, the environmental problems will increase [3]. Since in the today's world there are huge amount of activities of industries, high technology advancement, increase in the number of population and the expansion of cities on wide areas are increasing the requirements of research on environmental footprint and viable sustainable life [4].

The level of knowledge on how to protect environment is one of the important factor that can affect environment negatively. The reason for this is because knowledge is an important factor that determines people's attitudes and behaviours against environment. Person related institutional factors and socioeconomic status; for example, age, ethnicity, gender, education status and income level of people, can impact the knowledge and attitudes of people towards environmental concerns [5]. It is also discussed that knowledge on protecting environment is also important in raising awareness on environment. It was stated that awareness on the environment is about know people realize and get to know information about the environment. Supportive efforts must be given to encourage people for contributing to finding solutions to environmental positions. It was further shown in the evidence that the awareness on the environment significantly changes based on the sex, age, education, household income, source of environmental information [6,7].

Proving knowledge on how to behave environmentally friendly can help changing attitudes of people towards environment and attain behaviours that are towards protecting environment. Reduced or lack of knowledge on the environment and importance of protecting it is an important concern and can be overcome by providing education or training on importance of protecting environment [8,9,10]. There are further emphases on the importance of enhancing knowledge of young populations regarding environment. It is important to assess the knowledge level of young people on environment and implement tailored environmental programmes to increase awareness of young towards environmental problems and assist them in relating the knowledge they gained to their actions and attitudes. There are wide evidence on to control ecological conditions, interactive teaching methods must be used for enhancing knowledge on protecting environment in young populations and children [11,12,13].

The main aim of this study was to evaluate knowledge, behaviour and attitude of the Libyan secondary school students towards environmental footprint and sustainable life based on their demographic characteristics, such as gender differences, age, nationality, educational background of the father, educational background of mother and income level of the family.

2. Methodology

2.1. Study Setting and Participants

This research aimed to evaluate knowledge, attitude and behaviour of students regarding environmental footprint and sustainable life in four selected secondary schools in Tripoli Metropolitan City, Libya. In Libya students aged between 16 to 18 years are given education in schools called secondary schools. Students studying in secondary schools were targeted and based on the number of total populations of schools, the representative population sample size was determined as 400 to attain 95% confidence level and 5 confidence interval in the results obtained. As the number of target population size was similar in each school, 100 students were involved in each school. The response rate to the questionnaires circulated in each school were 100%.

2.2. Study tool
The survey tool compromising questions on demographic information and a sum of thirty-five questions, including questions addressing knowledge, attitude and behaviour on environmental footprint and sustainable life was used. The questionnaire was adapted from the research of Meyer, 2009 and it was translated in to Libyan Language using a back to back translation method. There were 24 questions assessing knowledge; 17 questions assessing attitude and 13 questions assessing behaviours towards environmental footprint. The responses to questions were employing using five Likert scales to provide uncompromising options for the respondents in all the questions that are stipulated in the questionnaire (Strongly agree 1; Agree 2; Neutral 3; Disagree 4, and Strongly disagree 5). The questionnaire was approved for the level of understanding by three expertise in the area and for improved validity reverse questions were involved in the questionnaire.

2.3. Data collection
The collection of data was done in the last quarter of 2018. The required permissions were arranged prior to the visits to schools. The students were involved in an orientation program before the data collection to socially motivate and educate them to enable them express their selves in accordance with the objectives and research questions of the particular study. The study methodology and the tool to be used was approved by the Ethical Committee of the European University of Lefke.

2.4. Data analysis techniques
The data collected during the course of this study were processed and analysed using the Statistical Package for Social Sciences, SPSS v21. The validity of the statistical tool was first evaluated by using the responses from 15 students and Cronbach Alpha value was formulated for assessing the validity. The Cronbach Alpha value was 0.87, which was greater than 0.7 that is used as a threshold value for determining how valid the tool is. After the data were collected from all targeted students, the data were entered in to the statistical package and assessed for any missing data and data were organized before the analysis. During the data processing, the data were assessed against normality using Skewness and Kurtosis tests and frequency distribution charts. Whilst scores on attitude and behaviour towards environmental footprint were normally distributed (Skewness and Kurtosis scores close to 0), knowledge scores were not distributed normally in the study sample. Data analyses were done using descriptive statistics, including frequency tests, distributions graphics, Student's T-test. Inferential statistical technique, ANOVA (Analysis of Variance) was used for testing the statistical difference in level of knowledge, attitude and behaviour by different demographic characteristics. Since knowledge scores were not normally distributed, the reciprocal of knowledge scores were taken, which distributed the data normally. P-values of less than and equal to 0.05 (≤ 0.05) were accepted as statistically significant.

3. Results

3.1. Characteristics of study population

The study was performed on overall 400 secondary school students, most of which were female (83.3%) and aged 16 years (40.8%) (Table 1). Majority of students were Libyan (96.8%) and approximately half of the students have low family income earning 500 to 1000$ a month.

Table 1. Characteristics of students studying in four secondary schools in Tripoli Metropolitan City, Libya

Characteristics	Frequency	Percent
Gender		
Male	67	16.8
Female	333	83.3
Age		
16	163	40.8
17	102	66.3
18	135	33.8
Nationality		
Libya	387	96.8
Other	13	3.3
Education status of father		
Has not been graduated from primary school	20	5.0
Primary school	50	12.5
Secondary school	98	24.5
University/Faculty	194	48.5
Master/PhD	38	9.5
Education status of father		
Has not been graduated from primary school	13	3.3
Primary school	43	10.8
Secondary school	79	19.8
University/Faculty	185	46.3
Master/PhD	80	20.0
Montly income of the family		
500-1000$	188	47.0
1001-2000$	75	18.8
2001-3000$	34	8.5
> 3000$	103	25.8
Total	400	100

3.2. Level and ranking of knowledge, attitude and behaviours towards environmental footprint:
The scores from the questions on each dimensions, namely knowledge, attitude and behaviours towards protecting environment were calculated by summing up the scores from the marks in likert scales. Questions having positive and negative responses were determined and they were scored based on the state of the responses: negative responses were marked inversely. The average score was 39.5 (38.9-40.2) for knowledge; 34.1 (33.5-34.7) for behaviour and 43.3 (42.6-44.0) for attitude. The level of three dimensions were ranked based on the ranking scale shown in Figure 1. Since Likert scale was ranked from 1 to 5, from positive to negative responses (Strongly agree to Strongly disagree), smaller scores of dimensions indicate better scores for knowledge, attitude and behaviours. For example; a score of knowledge between 24 and 48 indicate good knowledge towards environmental footprint and was shown as rank 1 for knowledge.

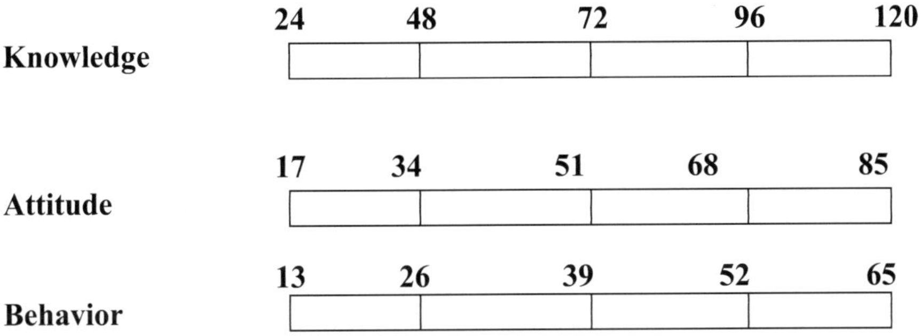

Knowledge	Rank 1 = 24 - 48	Rank 2 = 49 - 72	Rank 3= 73 - 96	Rank 4= 97-120
Attitude	**Rank 1 = 17 - 34**	**Rank 2 = 35 - 51**	**Rank 3= 52 - 67**	**Rank 4= 68-85**
Behaviour	**Rank 1 = 13 - 26**	**Rank 2 = 47 - 39**	**Rank 3= 40 - 72**	**Rank 4= 53-65**

Figure 1. Linear Scales Rating for Knowledge, Attitude and Behaviour towards environmental footprint

The frequency distributions for level of knowledge, attitude and behaviour towards environmental footprint and sustainable life are given in the Table 2. In the Libyan secondary school students, it was shown that the level knowledge on environmental footprint and sustainable life was high, with most of the students having high level of knowledge (91.8%) and moderate level of attitude (76.0%) and behaviour (65.8%) regarding environmental footprint (Figure 2a to c).

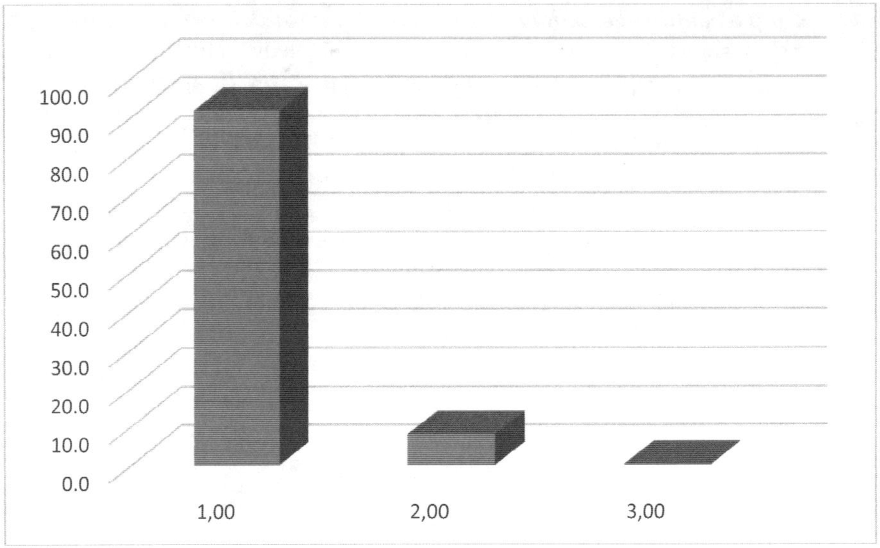

Figure 2a. The ranking of level of knowledge in students studying in secondary schools in Libya

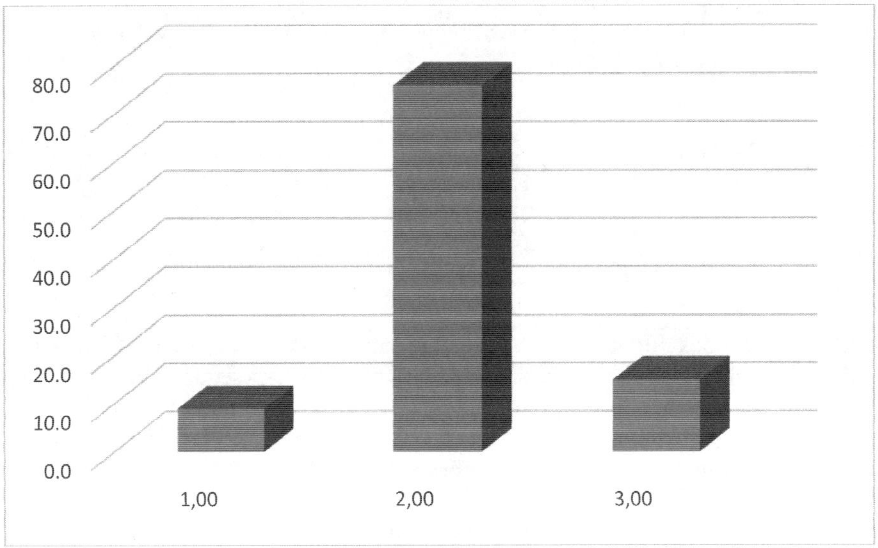

Figure 2b. The ranking of level of attitude in students studying in secondary schools in Libya

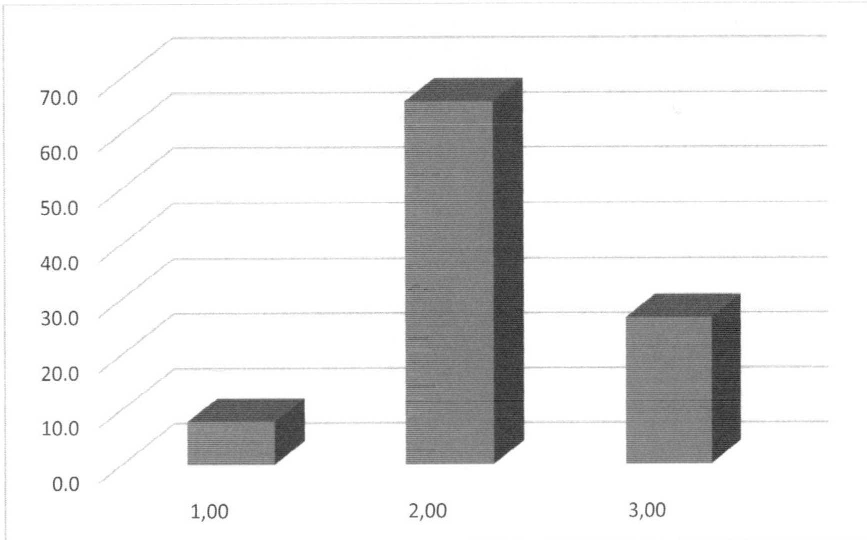

Figure 2c. The ranking of level of behaviour in students studying in secondary schools in Libya

3.3. The relationship between demographic characteristics of students and knowledge, attitude and behaviour on environmental footprint

The level of knowledge in secondary school students was not significantly but greater in those aged 17 years compared to those aged 16 years (mean score of 40.578 ± 7.712 compared to 38.871 ± 6.915, p-value= 0.063). Students whose fathers were master or PhD graduates were shown as having higher knowledge on environment compared to those having fathers graduated from secondary school only (41.105±7.082 in those having fathers graduated from a masters or PhD degree compared to 39.061 ±6.192 in students whose fathers were secondary school graduates, p-value: 0.05). Students having better income level were shown as having lower knowledge compared to those having lower income (36.824 ± 6.293 in those having 2001-3000$ monthly income compared to 39.840 ± 6.498 in those having 500-1000$ monthly income, p-value= 0.013). The behaviour on environment was shown to be better in Libyan students compared to others (34.199±6.124 in Libyan students compared to 30.769 ± 6.760, p-value: 0.048 in students from other origins). There was no difference in the mean level of attitude with different demographic characteristics.

Inferential statistics using ANOVA test were performed to test the differences in the level of knowledge, attitudes and behaviours among the students with different demographic characteristics (Table 3). There was no statistically significant difference in the level of knowledge and any demographic characteristics, except the level of income of the family of students. Those having lower income level were shown as having better knowledge than those having higher income levels (p-value = 0.05) Although there was no difference shown in attitude of students towards environmental footprint between any demographic group, inferential test have shown significant difference. Those from other nationalities had lower level of attitudes towards environment compared to Libyan students (p-value < 0.05).

IOP Conf. Series: Materials Science and Engineering **800** (2020) 012034 doi:10.1088/1757-899X/800/1/012034

Table 2. The differences in knowledge, attitudes and behaviours of students between different population groups using Student's t-test

	Knowledge			**Attitude**			**Behaviour**		
Characteristics	X (±sd)	t	p-value	X (sd)	t	p-value	X (sd)	t	p-value
Gender									
Male	40.269	0.000	0.000	43.507 (8.598)			34.328 (7.450)	0.000	0.000
Female	39.423	0.928	0.354	43.261 (6.792)	0.258	0.796	34.039 (5.887)	0.350	0.727
Age									
16*	38.871 (6.915)	0.000	0.000	42.509 (7.213)			33.601 (6.131)	0.000	0.000
17	40.578 (7.712)	-1.870	0.063	43.863 (7.461)	-1.467	0.144	34.588 (6.545)	0.350	0.727
18	39.659 (5.576)	-1.068	0.287	43.837 (6.677)	-1.068	0.287	34.296 (5.914)	-0.990	0.323
Nationality									
Libya*	39.641 (6.672)	0.000	0.000	43.416 (7.032)			34.199 (6.124)	0.000	0.000
Other	37.539 (8.491)	1.107	0.269	39.923 (8.930)	1.746	0.082	30.769 (6.760)	1.980	0.048
Education status of father									
Has not been graduated from primary school	37.846 (5.535)	-1.100	0.274	40.000 (8.631)	-1.614	0.110	32.923 (7.621)	-0.772	0.442
Primary school	39.302 (6.232)	-0.737	0.463	44.186 (6.929)	0.334	0.739	34.000 (5.774)	-0.409	0.683
Secondary school*	40.317 (7.762)	0.000	0.000	40.317 (7.762)	0.000	0.000	34.494 (6.662)	0.000	0.000
University/Faculty	40.317 (7.762)	-0.262	0.793	43.573 (7.143)	-0.152	0.879	34.405 (6.243)	-0.103	0.879
Master/PhD	38.113 (6.201)	-1.979	0.050	42.325 (6.340)	-1.264	0.208	33.188 (5.429)	-1.356	0.177
Education status of mother									
Has not been graduated from primary school	40.050 (8.212)	0.614	0.541	43.700 (7.794)	0.240	0.811	34.150 (6.900)	-0.108	0.914

Primary school	40.640 (6.546)	1.439	0.152	43.340 (7.935)	0.064	0.949	33.400 (6.575)	-0.811	0.419
Secondary school*	39.061 (6.192)	0.000	0.000	43.255 (7.515)	0.000	0.000	34.327 (6.678)	0.000	0.000
University/Faculty	39.206 (6.808)	0.177	0.860	43.217 (6.230)	- 0.045	0.946	34.124 (5.888)	-0.267	0.790
Master/PhD	41.105 (7.082)	1.658	0.100	43.605 (7.332)	0.245	0.806	34.158 (5.768)	-0.139	0.890
Montly income of the family									
500-1000$	39.840 (6.498)	0.000	0.000	43.617 (7.083)	0.000	0.000	34.090 (6.182)	0.000	0.000
1001-2000$	40.560 (7.795)	- 0.765	0.445	43.080 (7.555)	0.545	0.586	33.867 (6.204)	0.265	0.791
2001-3000$	36.824 (6.293)	2.503	0.013	42.441 (7.093)	0.258	0.797	33.794 (6.114)	0.258	0.797
> 3000$	39.272 (6.303)	0.721	0.471	43.175 (6.920)	0.514	0.608	34.340 (6.207)	0.721	0.471

Table 3. ANOVA test results assessing the differences in demographic characteristics and level of knowledge, attitude and behaviour regarding environmental footprint

	Knowledge		Attitudes		Behaviours	
Characteristics	Mean Squares	p-value	Mean Squares	p-value	Mean Squares	p-value
Gender						
Intergroup	39.004	0.354	3.381	0.796	4.669	0.727
Intragroup	45.389	0.354	50.741	0.796	38.109	0.727
Age						
Intergroup	91.202	0.131	86.584	0.181	35.002	0.399
Intragroup	45.137	0.131	50.441	0.181	38.040	0.399
Nationality						
Intergroup	55.592	0.269	153.454	0.082	147.950	0.048
Intragroup	45.348	0.269	50.364	0.082	37.749	0.048
Education status of father						
Intergroup	75.493	0.155	69.799	0.239	28.621	0.558
Intragroup	45.068	0.155	50.428	0.239	38.120	0.558
Education status of mother						
Intergroup	50.616	0.348	2.092	0.997	7.438	0.941
Intragroup	45.320	0.348	51.114	0.997	38.335	0.941
Montly income of the family						
Intergroup	117.625	0.050	16.405	0.809	4.381	0.952
Intragroup	44.826	0.050	50.882	0.809	38.280	0.952

4. Discussion

4.1. Main findings of the study

This study aiming to evaluate the knowledge, attitude and behaviour of secondary school students was conducted on 400 secondary school students studying in four secondary schools in Tripoli Metropolitan City, Libya. Since male students are predominantly studying in the schools in Libya, the majority of students in the study sample were males (83.3%). A high proportion of students were Libyan and most of them were aged 17 years (66.3%). Whilst 91.8% of students were reported as having high knowledge level, 76.0% of them had moderate level of attitude and 65.8% of students had moderate level of behaviour on environmental footprint and sustainable life. The analysis on the correlation between demographic characteristics and level of knowledge, attitude and behaviour towards environmental footprint showed that students having lower income level had better level of knowledge compared to those who have better socioeconomic status. In only ANOVA test results, students from Libyan origin were shown as having better attitudes towards environment compared to others.

4.2. Explanation of results and comparison with previous evidence

Libyan students between the age of 16 to 18 years were predominantly observed as having good knowledge on environmental footprint and sustainable life. This may indicate that the information on protecting environment and sustainable life in secondary school curriculums are strong. It was stated previously that the education and learning on environmental concerns have been enhanced in todays world and this enhancement in the knowledge on environmental concerns are expected to improve actions and attitudes that would help for solving environmental problems [14]. The ranking of level of attitudes towards environment have shown that the level of attitudes towards environment in the majority of students was moderate. In opposite to the evidence on the effectiveness of education on the environmental concerns enhancing attitudes towards environment [15,16], our results suggest that although students gain adequate knowledge on environmental risks and protecting environment, there would be gaps in applying the knowledge into attitudes towards environment. Behaviours towards environmental footprints were at moderate level, indicating that the good level of knowledge may not be reflected fully to adjust the unfriendly situations in the environment. Although students may gain adequate knowledge on the situations and actions dangerous for environment, they may not be encouraged to apply their knowledge into practices to behave friendly in the environment [17].

There are supporting evidence that socio-economic status, personal and psychological characteristics can lead to variations in the ecological attitudes of individuals in the environment [18,19]. These variations in attitudes are also reflected to behaviours, showing variations in behaviours of people towards matters related to environmental issues [20]. Empirical evidence is suggesting that unfavourable behaviours in the environment are highly observed in the males, younger people, those poorly educated, socioeconomically deprived people, and people living in rural regions [21]. It has also been reported earlier that morality and spiritual faith are also the factors that may be associated with the attitudes and behaviours towards environmental issues and environmental sustainability [22]. In opposite to the given evidence, the present study did not show any disparities in attitudes in those with different demographic characteristics. Knowledge on environmental issues was reported as related with the income level of families of the students. Evidence supports that those with better income earning tend to have better life styles, greater opportunities of having better education and more influence on the other people [23].

Studies have therefore shown that rich people have more concerns relative to environmental issues compared to other population groups [24]. Our findings, however, showed results that are opposing this evidence, with better knowledge on environmental footprint in those having lower income levels. This difference in the knowledge level between socioeconomic groups are probably due to the limitations in the data because the number of those having greater income level were low, probably leading to results, which are opposite to the empirical evidence.

5. Conclusions

Environmental footprint is the sign of the activities of human on environment. Since knowledge on the ways of protecting environment against environmental hazards is an important determinant of attitudes and behaviours towards environment, it is crucial to research the level of knowledge, attitudes and behaviours towards the environment and the factors that may influence these. The current research aimed at looking the level of knowledge towards environmental footprint in a young population using a sample from a secondary school in Libya. The study further evaluated the impact of level of knowledge on the attitudes and behaviours in environment in young people. It was overall shown that although level of knowledge was high in the study population, the level of behaviours and attitudes were lower, with most of the subjects having moderate levels of behaviours and attitudes towards environment. This may indicate that although level of knowledge on protecting environment is high, young people must be supported in their practices in the environment towards protecting it against environmental hazards.

There were no critical differences observed in knowledge, attitudes and behaviours towards environment in the different student groups in the study sample. Further research is required to assess any impact of demographic issues that influence the particular three dimensions crucial in reducing environmental footprints.

References

[1] Bowers C A 2011 Let them eat data: How computers affect education, cultural diversity, and the prospects of ecological sustainability, *University of Georgia Press.*

[2] Wackernagel M and Rees W 1998 Our ecological footprint: reducing human impact on the earth (Vol. 9). *New Society Publishers.*

[3] Wiedmann T and Barrett J 2010 A review of the ecological footprint indicator—perceptions and methods. *Sustainability* **2 (6)** 1645-1693.

[4] Südaş H D and Özeltürkay E Y 2015 Analyzing the Thoughts of Ecological Footprints of University Students: A Preliminary Research on Turkish Students. *Procedia-Social and Behavioral Sciences* **175** 176-184.

[5] Kollmuss A and Agyeman J 2002 Mind the gap: why do people act environmentally and what are the barriers to pro-environmental behaviour ? *Environmental education research* **8 (3)** 239-260.

[6] Mamady K 2016 Factors influencing attitude, safety behavior, and knowledge regarding household waste management in Guinea: A cross-sectional study. *Journal of Environmental and Public Health.*

[7] Garcia A P Iniesta-arandia I Garcia D Montes C and Martin B 2015 Factors influencing local ecological knowledge maintenance in Mediterranean watersheds : Insights for environmental policies *Journal of the Human Environment* **44** 285-296.

[8] Birhanu A 2014 Environmental Knowledge , Attitude and Participatory Behavior towards Land Degradation in Injibara Secondary and Preparatory School , Northwestern Ethiopia *Journal of Environment and Earth Science* **4 (17)** 89–96.

[9]Anne K Lieflander 2015 Effectiveness of environmental education on water : connectedness to nature , environmental attitudes and environmental knowledge *Environmental Education Research* **21 (1)** 145-146.

[10] Sumargo B 2018 Comparing Better Environmental Knowledge Based on Education and Income Using the Odds Ratio *Procedia Computer Science* **135** 105–112.

[11] Trumper R 2010 How Do Learners in Developed and Developing Countries Relate to Environmental Issues? *Science Education International* **21 (4)** 217-240.

[12] Harring N and Jagers S C 2018 Why do people accept environmental policies? The prospects of higher education and changes in norms, beliefs and policy preferences *Environmental Education Research* **24 (6)** 791-806.

[13] Asunta T 2003 Knowledge of environmental issues: where pupils acquire information and how it affects their attitudes, opinions, and laboratory behaviour (No. 221) *The University of Jyväskylä.*

[14] Kennedy E H Beckley T M Mcfarlane B L and Nadeau S 2009 Why we don't" walk the talk": Understanding the environmental values/behaviour gap in Canada *Human Ecology Review* **16 (2)** 151-160.

[15] Roberts B W and Bogg T 2004 A longitudinal study of the relationships between conscientiousness and the socialenvironmental factors and substance- use behaviours that influence health *Journal of Personality* **72 (2)** 325-354.

[16] Schutzer K A and Graves B S 2004 Barriers and motivations to exercise in older adults *Preventive medicine* **39 (5)** 1056-1061.

[17] Carver C S and Scheier M F 2012 Attention and self-regulation: A control-theory approach to human behaviour. *Springer Science & Business Media.*

[18] Boyes E *et al.* 2014 An international study of the propensity of students to limit their use of private transport in light of their understanding of the causes of global warming *International Research in Geographical and Environmental Education* **23 (2)** 142-165.

[19] Redondo I and Puelles M 2017 The connection between environmental attitude-behaviour gap and other individual inconsistencies: a call for strengthening selfcontrol *International Research in Geographical and Environmental Education* **26 (2)** 107-120.

[20] Ambusaidi A Boyes E Stanisstreet M and Taylor N 2012 Omani students‘ views about global warming: Beliefs about actions and willingness to act *International Research in Geographical and Environmental Education* **21 (1)** 21-39.

[21] Kollmuss A and Agyeman J 2002 Mind the gap: why do people act environmentally and what are the barriers to pro-environmental behaviour? *Environmental education research* **8 (3)** 239-260.

[22] Albayrak T Moutinho L and Herstein R 2011 The influence of scepticism on green purchase behaviour *International Journal of Business and Social Science* **2 (13)** 189-197.

[23] Mayer F S and Frantz C M 2004 The connectedness to nature scale: A measure of individuals‘ feeling in community with nature *Journal of environmental psychology* **24 (4)** 503-515.

[24] Fairbrother M 2012 Rich People, Poor People and Environmental Concern: Evidence across Nations and Time *European Sociological Review* **29** 910-922.

IOP Conf. Series: Materials Science and Engineering **800** (2020) 012035 doi:10.1088/1757-899X/800/1/012035

Service quality factors in the construction sector: A literature review

M F B Landy[1,2], S Sousa[2] and F Romero[2*]

[1]Department of Applied Chemistry and Production Systems, Faculty of Chemical Sciences, University of Cuenca, Cuenca, Ecuador
[2]ALGORITMI Research Centre, University of Minho, Guimarães, Portugal.

[*]Email: fromero@dps.uminho.pt

Abstract. This paper consists of an exploratory literature review whose main research objectives were the identification of the service quality factors that are considered more important in the construction sector, the service quality models that are used to measure quality, and how are they related to success factors. The research methodology is documental and based on a review of articles obtained from six major scientific databases. The description of the main results follows. In all cases, the traditional models of service quality were used as guidelines to explain and adapt to specific contexts. The typical parameters used in these models are reliability, responsiveness, communication, credibility, assurance, empathy with the client and construction quality. In some cases, additional dimensions were added. Success factors seem to be intrinsically related to project management, communication skills, professional skills, quality of the final product, design, aesthetic and innovation, where the three latter ones represent parameters that have gained recently preponderance. A holistic, flexible and adaptable attitude seems to be relevant to face dynamic and turbulent conditions and changing customer's requirements and expectations. Overall, the results indicate a generalised conservative approach which characterises this sector.

1. Introduction

The Construction Sector (CS) is a complex sector and satisfying customers' expectations represents a challenge for organizations within this sector. Service Quality (SQ) is related to the satisfaction of clients. Identifying Success Factors (SF) of SQ represents an important approach that aims to improve the quality of the organizations within the CS.

The International Standard Industrial Classification and the Central Product Classification [1], consider that the CS has two parts: first, construction that covers the physical outputs of construction activities (e.g., buildings or civil engineering works); and second, construction services that cover services provided in constructing these objects [1]. Construction is also a process of delivering physical outputs to the client through a temporary production system that consist of elements shared with other projects [2]. Therefore, the CS can offer goods and services at the same time [3]. Both dimensions are generally present in all construction projects and are dynamically and non-linearly related, which justifies why the CS can be considered and treated as a complex system [4].

The aim of SQ in an organization is "to evaluate the satisfaction of consumers through the analysis of the gap between the expectations and the perceptions of the consumers in a specific time after receiving a service" [5, 6]. SQ is dependent on technical quality and functional quality but corporate image plays an important role [7]. Dimensions have been defined to measure SQ, such as reliability, responsiveness, assurance, empathy, and tangibles [8, 9]. Each SQ dimension has several attributes and factors that may influence with different levels of intensity customer satisfaction. Some of these attributes and factors can be considered as SF. The identification of SF in construction services and their relationships, represents the basis for developing a process approach in construction organizations [10].

The goal of this document is to present a comprehensive literature review related to the evolution of SQ factors within the CS. Identifying the SQ factors that are considered more important in the CS, the SQ models that are used to measure quality, and how are they related to SF. The remaining paper is organized as follows:

- The research methodology is addressed in the following section (Section 2).
- A presentation of the main concepts in the beginning of Section 3 was developed to explain SQ and SF (Subsection 3.1).
- An identification of traditional and new factors presented in the literature was developed through an analysis of the research of important authors (Section 3.2) with an explanation of dimensions for future study (Figure 1).
- The paper ends with a discussion and conclusion section.

2. Research methodology

This work is an exploratory study and the literature review was made by analysing documents from high quality journals, conferences and institutes, addressing several topics of the CS, with a focus on the concept of SQ. The search was executed using six general scientific databases: Scopus (www.scopus.com), Web-of-Science (www.webofknowledge.com), IEEE Xplore Digital Library (ieeexplore.ieee.org), Emerald insight (www.emeraldinsight.com), Science Direct (www.sciencedirect.com), and Microsoft Academic (academic.microsoft.com). The list of keywords used in this search included combinations of the keywords "Service Quality" with "Construction Industry" and/or "Construction Sector". The search results were analysed manually and systematically and reduced to 33 relevant documents related to this review aim, and 17 more as complements that provided valuable insights and new research directions.

3. Literature review

3.1. A short presentation of the main concepts

3.1.1. Service Quality. Service Quality (SQ) is related to the satisfaction of the clients in the short term, and it is related with the factors that change continuously. SQ has no singular definition. According to [5] to assess SQ in an organization is "to evaluate the satisfaction of consumers through the analysis of the gap between expectations and perceptions in a specific time after receiving a service". SQ depends on technical quality and functional quality with an important role of the corporate image [7]. The technical quality interprets what are the expectations, the functional quality finds the way in which the organization interprets the characteristics and translates them to technical specifications of work, and the corporate image deals with the perceptions of the consumers about the organization of the service that depends on: technical and functional quality, price, external communications, physical location appearance of the site, and the competence and behaviour of firms' employees and the associated service [11].

Some propositions are presented that provide different understandings of SQ, which is viewed from different angles: SQ "is the consequence not only of the performance of service, but also of the interaction between customer and firms" [12]; SQ "is an important determinant of customer satisfaction which in turn influences customers' loyalty" [13]. Therefore, SQ within CS organizations contribute to develop a new way of understanding them.

3.1.2. Success Factors. Success Factors (SF) are mechanisms that allow organizations to survive, adapt and prosper in a competitive market. SF of SQ represents an important approach that aims to improve the quality of the organizations. Five dimensions or factors can be defined to measure SQ [8, 9]: (i) reliability, as the ability to perform the service in a careful and reliable manner, (ii) responsiveness, willingness to help clients and provide fast service, (iii) assurance, knowledge and attentions shown by employees and their abilities to generate credibility and confidence, (iv) empathy, personalized attention with kindness and courtesy, (v) tangibles elements, aspect of physical facilities, equipment, personnel and communication materials.

SQ processes developed with customers helps to create an organization culture oriented to generate customer satisfaction all the time. SF influence the decision making process, the improvement of plans and the formulation and implementation of strategy. Therefore, SF within the SQ process help to improve the organization competitiveness and the adaptation to the market.

3.2. Service Quality Factors in the Construction Sector: exploring the literature

3.2.1. Traditional Factors. In general, SQ traditional models are widely used by firms as guidelines to explain and adapt to specific contexts. Thus, in this section, we explore the literature in terms of the use and the chronological evolution of traditional factors (the presentation of authors follows roughly a chronological perspective).

Segura [14] and Asahara [15], as all the authors identified in this section, identified in their studies the use of the traditional factors to measure SQ, in some form or another: reliability, responsiveness, competence–construction, access, courtesy, communication, credibility, confidentiality, understanding the customer, and tangibles. The literature identified these traditional factors in several studies of sub- sectors of the CS, or in studies of the construction process. For instance, Holm & Bröchner [16] studied SQ in housing refurbishment and craftsmen and suggest that "service delivery in conversion projects appears to be the key to improved reputation for the contractors, where contractors meeting the users' expectations and craftsmen behaviour are central elements" (p. 543); and "satisfaction with the service is dependent on the interaction between the service provider and the user" (p. 536). Segura [14] also proposed some propositions about SQ and customer satisfaction related to reputation, and to the capacity of the firm to provide value added to the costumer by adding and mixing tangible and intangible dimensions, or product and service, to the overall product. Asahara [15] emphasized the phases in the construction process, such as planning and design, contact, and construction phases and some SF were identified. From the user perspective, SF included: project management and satisfaction; incidents; information; company background; and landlord information incident. From the service provider perspective, SF included: project management and satisfaction; details; and company reputation and experience. Holm [17] analysed SF into housing refurbishment in two ways: (i) Opinions from workers (such as satisfaction and communication, quality control and information, behaviour and quality of work; (ii) Opinions from tenants such as: landlord behaviour and dust exposure; quality, expectations, recommendation and trustworthiness; workers' behaviour; and interaction and voting for contractor. In all of these studies, research concluded that SQ constitutes an important aspect within CS and it is necessary to incorporate a continuous improvement process to generate customers' satisfaction.

Al-Momani [18] and Hoxley [19] identified SQ SF in the construction process and UK construction professional services, respectively. The attributes suggested by [18] were related to project management, contractual issues and reliability issues. Hoxley [22] also analysed SQ in the public sector by professionals' services in relation to the determinants that the UK government suggests. In all cases SQ SF oriented to the analysis of the professional's quality within CS proved to be very important.

Maloney [3] considers that the construction provides a service that consists of three elements: service product; service environment; and service delivery, where firms often have several roles. Organizations in the CS can be analysed as firms that provide products and services at the same time in a complex and interdependent way. For instance, Love et al. [20] analysed SF in architecture and engineering through the supply chain, arguing that the management of the supply chain is very important in terms of providing reliability. Siu, Bridge & Skitmore [21] introduced a SQ reflection in building maintenance from the perspective of Mechanical and Engineering Services (M&Es), which includes electrical firms, plumbing firms or air-conditioning firms, and identified important discrepancies in expectations and perceptions between service users and service providers. Concerning the interactions inside the supply chain, Segura [14] proposed SF that were grouped in the following categories: contractor-customer relationship, the contractor's project management skills, the contractor's safety performance, prepared/skilled workforce, cost and general satisfaction. The overall conclusion of these studies was that a permanent feedback process seems to be necessary to improve SQ and organizations need a flexible and adaptable structure to respond to frequent changes in requirements and expectations and to manage the effective provision of these demands through their networks of contractors or partners.

Arditi & Lee [23] and Ling & Chong [24] argued that customers have high expectations of SQ from D&B contractors but Cheng, Proverbs, & Oduoza [26] argued that those expectations are frequently not met and that, although there are many causes underlying that discrepancy, the main ones are attributable to project management deficiencies and inferior quality. Ling, Ibbs, & Hoo [25], Kärnä, Sorvala, & Junnonen [27] and Kärnä, Junnonen, & Sorvala [28] corroborate this by identifying project management but also communication skills as key SF that contribute to the successful performance of projects by national or international construction firms. Ismail, Othman, & Amat [29], Hui & Zheng [30], Sunindijo, Hadikusumo, & Phangchunun [31] and Durdyev et al. [32], identified the dimensions of "empathy", "assurance" and "reliability" as the most significant ones in terms of its effects on customer satisfaction and on the success of CS projects and firms.

The studies referred above seem to suggest that, besides the factors related to professional skills and tangible aspects of construction quality, as pointed earlier in this section, SQ requires a continuous communication process to assess the effectivity of the actions. A holistic vision seems necessary to understand the design and later the service process to develop SQ SF. SQ needs an understanding of the environment changes alongside the improvement of the communication process. The studies also suggest that in many instances firms adopt a rather conservative approach to these issues.

As a way of synthesis, Table 1 shows a summary of the most important aspects identified by representative authors, as explained above, classified by SQ Dimension, emphasizing the use of traditional models.

Table 1. Success factors by SQ Dimension - traditional

SQ Dimension	Most important SF	Selected representative authors
Reliability	Total quality of work output. Reputation and experience Delivery times	[4], [15–22]
Responsiveness	Professionals skills Workers behaviour Technological tools Incidents resolution Work disruptions	[14], [20–24], [26], [28], [29], [31], [34], [39], [41], [44], [45]
Assurance	Competence Credibility Confidentiality	[3], [20], [23], [26], [28], [40], [41], [43]
Empathy	Access Courtesy Communication Understanding the customer Interaction with customers	[3], [23], [26], [28], [33], [39–41]
Tangibles	Technological tools	[3], [26], [33], [39]–[41]

3.2.2. New Factors. In this section we explore some of the new SQ SF that are proposed in the literature, which are different from the traditional ones that were explored in the last section, and which reflect additional perceptions of other stakeholders.

Forsythe [33] developed a model to understand the SQ in the sector of Housing Construction (HC). He adapted the gap model to suit the design and construction process by merging it with SQ in construction projects and he incorporated some new factors such as: aesthetic (construction) workmanship, technical (construction) workmanship, and design detailing by workers onsite. Thus, a different vision to understand SQ is incorporated by this author, which differs from more traditional perspectives [34].

Bjeković & Kubicki [35] argued that it is necessary a deep understanding of business activities to generate service innovation processes. They argued that SQ is the "umbrella concept abstracting some non-functional aspects of service" (p.514) suggesting a direct connection between service innovation and SQ. In addition, they identified quality categories and factors that affect SQ such as: (i) business stability: business domain adequacy, effect on collaborative practices, reputation within the sector; (ii) stability: reliability, availability, accuracy; (iii) performance: time behaviour, resource utilization; (iv) security: confidentiality, integrity, non-repudiation, accountability, authenticity; (v) usability: understandability, learnability, ease of use, user error protection; and (vi) regulatory and interoperability: supported standards, interoperability. These identified SF are related to knowledge and other intangibles dimensions not previously considered in traditional models. Similarly, McClements, Odeyinka, & Eadie [36] developed a construction consultant pre-selection factor model in relation with the intangible attributes related to trust.

Forsythe [37] investigated SQ as an input to pre-purchase (pre-contract) expectations in the made-to-order HC, and discovered that "customers tend to be highly involved in the design process and also observe the construction as a live act with a vested interest in the speed, cost and quality of construction" (p.590) arguing that the needs recognition, in relation to design, but also to other aspects, is important in managing customers' satisfaction [38]. Tan et al. [39] identified SF within the building maintenance business in Hong Kong. SF identified were maintenance service, organization and project management, certification, people, relationship, marketing, but also, and similarly to above mentioned authors, innovation and sustainability.

Forsythe [40] developed a SQ model called BUILDSERV that is a combination of the traditional dimensions and new variables in a specific context in Australia. SF defined were reliability, responsiveness and communications, empathy, assurance, tangibles, care in execution of work, and work output. Also, an important contribution was the features oriented to generate positive or negative incidents during the construction process. They were (i) situational context, (ii) service provider involvement, (iii) sources of exposure, (iv) core content, and (v) customer awareness of involvement. These features emphasise the need to consider dynamic characteristics of the environment and their influence in the SQ process.

Forsythe [41] argued that "SQ is likely to be important to customer satisfaction during changing dynamics, unexpected events and different personal relations that occur during construction activities onsite" (p. 323) stressing the importance of dynamic factors to SQ. In this vein, Prakash & Phadtare [42] developed a scale to measure SQ having in mind that "uncertainty and turbulence in a business environment may bring progressive sophistication in design and construction" (p.670), requiring extensive amounts of information exchange. They identified the SF that affect SQ such as design quality, project administration quality, communication quality, relationship quality and dependability quality.

As a way of synthesis, Table 2 shows a summary of the most important aspects and representative authors by SQ Dimension, emphasising the surge of new factors that influence SQ.

Table 2. Success factors by SQ Dimension - new

SQ Dimension	Most important SF	Selected representative authors
Quality Aesthetic	Aesthetic (construction) workmanship Technical (construction) workmanship	[28], [33], [39]–[41], [44]
Design	Flexibility Adaptability	[20], [23], [25], [42], [43], [46]
Care in execution of work	Social responsibility	[40], [41]
Innovation	Sustainability Social responsibility	[39], [42]

Complementing the synthesis and the presentation of the main identified SQ dimensions and the respective indication of a set of the most important factors, Figure 1 presents a map of the literature that was reviewed, integrating two perspectives: a chronological one and a conceptual one in relation to the study dimensions proposed. Traditional and new SF can signal new trends and challenges for firms related to (i) *performance*; by offering increasingly better value to stakeholders through organizational sustainability and learning, and the dynamic capabilities of employees; (ii) *service loyalty*; striving to maintain permanent customer satisfaction and their perception of reliability; (iii) *servitization* as a process by which manufacturers move from a product-oriented business model to a service-oriented business model; (iv) *communication and feedback*; to reduce uncertainty and ambiguity within the construction process and (v) *environment changes*; dynamic and turbulent environment are complex, and it is necessary a flexible and adaptable business model in firms within the CS [47], [48].

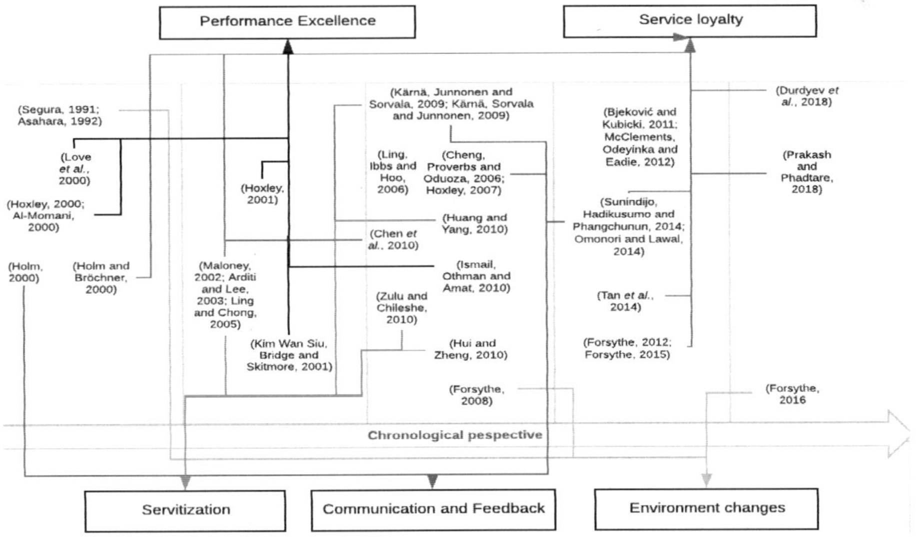

Figure 1. Literature map

4. Discussion and conclusions

CS is a complex sector and satisfying customers' expectations represents a great challenge. Understanding SF of SQ from the customers' perspective represents an important approach and for a construction organization is important to know, in the best possible way, the customers' expectations [30]. Important implications are implicit in this analysis, including the improvement of quality, adaptability, flexibility and organizational learning oriented to customer satisfaction, customer loyalty [32] and to achieve a good reputation. SQ within CS organizations contributes to develop a new way of understanding them [49], and SF within SQ processes helps to improve the organization competitiveness and the adaptation to the market. It is in the interest of CS organizations to develop a SQ culture within their value chain to generate customer's satisfaction in the medium and long term.

Additionally, in the literature some propositions were presented about SQ such as (i) SQ constitutes an important aspect within CS and it is necessary to incorporate a continuous improvement process to generate customers satisfaction; (ii) SQ is dependent on the supply chain management; (iii) SQ represents a factor of competitive advantage and generates customer's loyalty; (iv) SQ helps to understand quality requirements. Therefore, SQ represents an improvement mechanism that needs a permanent feedback process and/or a continuous communication process to assess the effectiveness of the actions.

In this work, the results showed that traditional dimensions of SQ are still pervasive, with customers considering "reliability as the most important determinant in project SQ" [24] and the other dimensions, such as responsiveness, empathy, assurance and tangibles, can be regarded as important complements to generate customer's satisfaction in different contexts and circumstances. The new factors that were identified, such as quality aesthetic, design, care in execution of work and innovation represent an important trend that needs deeper study. In addition, traditional and new factors can contribute to study dimensions such as: performance excellence, service loyalty, environment changes, communication and feedback and servitization (Figure 1). They can be considered for future studies.

An organization within CS also needs a flexible and adaptable structure to respond to the great challenge of "customer's satisfaction" especially in more dynamic and turbulent environments. Thus, the understanding of drivers of customer satisfaction, which may be related to new factors, is the key element of SQ [24] and SQ culture developed in organizations could help to adapt their structure and processes to respond to the environment [50]. It is important to consider customer satisfaction to increase the market of customers and the future behavioural intention [24] and the features of the dynamic environment and their influence in the SQ process [24]. Therefore, a holistic, flexible and adaptable attitude seems to be relevant to face understand and face those conditions and changing customer's requirements and expectations.

5. Acknowledgments

This work has been supported by FCT - Fundação para a Ciência e Tecnologia within the Project Scope: UID/CEC/00319/2019 and SENESCYT - Secretaría de Educación Superior, Ciencia, Tecnología e Innovación – Ecuador. Department of Applied Chemistry and Production System - Faculty of Chemistry Sciences - University of Cuenca, Cuenca – Ecuador.

References

[1] United Nations 2015 *Central Product Classification (CPC) Version 2.1*, **4**, March. New York
[2] Bertelsen S and Emmitt S 2005 The Client as a Complex System, *13th Int. Gr. Lean Constr. Conf. Proc.*
[3] Maloney W F 2002 Construction product/service and customer satisfaction, *J. Constr. Eng. Manag.*, **128**, 522–9 [4] Bertelsen S 2003 Construction as a complex system *Proc. IGLC*, **11**, 143–68
[5] Parasuraman A, Zeithaml V and Berry L 1985 A Conceptual Model of Service Quality and Its Implications for Future Research *J. Mark.* **49**, 41–50
[6] Parasuraman A, Zeithaml V and Berry L 1988 SERVQUAL: A Multiple-Item scale for Measuring Consumer Perceptions of Service Quality *J. Retail.* **64** 12-40
[7] Grönroos C 1984 A Service Quality Model and its Marketing Implications *Eur. J. Mark.* **18** 36–44

[8] Parasuraman A, Berry L and Zeithaml V 1991 Refinement and reassessment of the SERVQUAL scale *J. Retail.* **67** 420-50

[9] Parasuraman A, Zeithaml V and Berry L 1994 Alternative scales for measuring service quality: A comparative assessment based on psychometric and diagnostic criteria *J. Retail.* **70** 201–30

[10] Czajkowska A and Kadłubek M 2015 Management of Factors Affecting Quality of Processes in Construction Enterprises," *POLISH J. Manag. Stud. Czajkowska A* **11** 28–38

[11] Seth N, Deshmukh S and Vrat P 2005 *Service quality models: a review* **22** 913-49

[12] Rew D, Jung J and Cha W 2018 Service productivity vs service quality: a zero-sum game? *Int. J. Qual. Reliab. Manag.* **35** 1835–52

[13] Izogo E and Ogba I 2015 Service quality, customer satisfaction and loyalty in automobile repair services sector *Int. J. Qual. Reliab. Manag.* **32** 250–69

[14] Segura C 1991 Customer service quality in the construction industry *Massachusetts Institute of Technology.*

[15] Asahara T 1992 Customer Service quality management in the construction industry a comparative study of the US and Japan *Massachusetts Institute of Technology*

[16] Holm M and Brochner J 2000 Satisfying the ultimate customer in housing refurbishment" *Doktorsavhandlingar vid Chalmers Tek. Hogsk.*, no. 1613, 205–12

[17] Holm M 2000 Service quality and product quality in housing refurbishment *Int. J. Qual. Reliab. Manag.* **17** 527–40.

[18] Al-Momani A 2000 Examining service quality within construction processes *Technovation* **20** 643–51

[19] Hoxley M 2000 Measuring UK construction professional service quality: The what, how, when and who *Int. J. Qual. Reliab. Manag.* **17** 511–26

[20] Love P, Smith J, Treloar G and Li H 2000 Some empirical observations of service quality in construction," *Eng. Constr.Archit. Manag.* **7** 191–201

[21] Siu G, Bridge A and Skitmore M 2001 Assessing the service quality of building maintenance providers: Mechanical and engineering services *Constr. Manag. Econ.* **19** 719–26

[22] Hoxley M, 2001 Purchasing UK public sector property and construction professional services: competition vs quality *Eur. J. Purch. Supply Manag.* **7** 133–9

[23] Arditi D and Lee D 2003 Assessing the corporate service quality performance of design-build contractors using quality function deployment *Constr. Manag. Econ.* **21** 175–85

[24] Ling F and Chong C 2005 Design-and-build contractors' service quality in public projects in Singapore *Build. Environ. 40 815–23*

[25] Ling F, Ibbs C and Hoo W 2006 Determinants of international architectural, engineering, and construction firms' project success in China *J. Constr. Eng. Manag.* **132** 206–1

[26] Cheng J, Proverbs D G and Oduoza C F 2006 The satisfaction levels of UK construction clients based on the performance of consultants *Eng. Constr. Archit. Manag.* **13** 567–583

[27] Kärnä S, Sorvala V-M and Junnonen J-M 2009 Classifying and clustering construction projects by customer satisfaction *Facilities* **27** 387–398

[28] Kärnä S, Junnonen J M and Sorvala V-M 2009 Modelling structure of customer satisfaction with construction *J. Facil Manag.* **7** 111–12

[29] Ismail S, Othman M and Amat S 2010 A study on service quality of refurbishment contractors towards clients satisfaction *CSSR 2010 - 2010 Int. Conf. Sci. Soc. Res.* 1047–52

[30] Hui E and Zheng X 2010 Measuring customer satisfaction of FM service in housing sector *Facilities* **28** 306–20.

[31] Sunindijo R Y, Hadikusumo B H W and Phangchunun T 2014 Modelling service quality in the construction industry *Int. J. Bus. Perform. Manag.* **15** 262–276

[32] Durdyev S, Ihtiyar A , Banaitis A and Thurnell D 2018 The construction client satisfaction model: a PLS-SEM approach *J. Civ. Eng. Manag.* **24** 31–42

[33] Forsythe P 2008 Modelling customer perceived service quality in housing construction *Eng. Constr. Archit. Manag.* **15** 485–96

[34] Zulu S and Chileshe N 2010 Service quality of building maintenance contractors in Zambia: A pilot study," *Int. J. Constr. Manag.* **10** 63–81

[35] Bjeković M and Kubicki S 2011 Service quality description - A business perspective in *2011 Federated Conference on Computer Science and Information Systems, FedCSIS 2011* 513–20

[36] McClements S, Odeyinka H and Eadie R 2012 Pre-selection of construction consultants based on attributes of trust," in *Ass. of Researchers in Constr. Manag., ARCOM 2012 - Proceedings of the 28th Annual Conference* **1** 717–26 [37] Forsythe P 2012 Profiling customer perceived service quality expectations in made-to-order housing construction in Australia *Eng. Constr. Archit. Manag.* **19** 587–609

[38] Omonori A and Lawal A 2014 Understanding Customers' Satisfaction in Construction Industry in Nigeria," *J. Econ. Sustain. Dev.* **5** 115–20

[39] Tan Y, Shen L, Langston C, Lu W and Yam M 2014 Critical success factors for building maintenance business: A Hong Kong case study *Facilities* **32** 208–25

[40] Forsythe P. 2015 Monitoring Customer Perceived Service Quality and Satisfaction during the Construction Process *Constr. Econ. Build.***15** 19–42

[41] Forsythe P 2016 Construction service quality and satisfaction for a targeted housing customer *Eng. Constr. Archit. Manag.* **23** 323–348

[42] Prakash A and Phadtare M 2018 Service quality for architects: scale development and validation *Eng. Constr. Archit. Manag.* **25** 670–686

[43] Al-Momani A 2000 Examining service quality within construction processes *Technovation* **20** 643–51

[44] Huang R B and Yang X. 2010 Framework for evaluating the service quality of project preparation consultants *2010 Int. Conf. Manag. Serv. Sci. MASS 2010* 1–5, 2010.

[45] Durdyev S, Ihtiyar A, Banaitis A and Thurnell D 2018 The construction client satisfaction model: A pls-sem approach *J. Civ. Eng. Manag.* **24** 31–42

[46] Holm M and Bröchner J 2000 Office conversions: The effects of craftsman-user interaction *Facilities* **18** 535–45. [47] Millar C C J M, Groth O and Mahon J F 2018 Management Innovation in a VUCA World: Challenges and Recommendations *Calif. Manage. Rev.* **61** 5–14

[48] Barbosa F, Romero F C and Cunha J Innovation 2018 *Sustainability, and Organizational Change in a Social Portuguese Organization: A Strategic Management Perspective Maintaining Sustainable Accounting Systems in Small Business* eds L C Carvalho and E Truant (Hershey: IGI Global) chapter 9 pp 179–202

[49] Beltagui A and Candi A 2018 Revisiting service quality through the lens of experience-centric services *Int. J. Oper. Prod. Manag.* **38** 915–932

[50] Pekkola S, Saunila M and Rantanen H 2016 Performance measurement system implementation in a turbulent operating environment *Int. J. Product. Perform. Manag.* **65** 947–958

IOP Conf. Series: Materials Science and Engineering **800** (2020) 012036 doi:10.1088/1757-899X/800/1/012036

The information exchange in the construction sector in the age of BIM: a multiple case study from Italy

A D Leto

Politecnico di Milano, piazza Leonardo da Vinci 32, 20133, Milano, Italy

E-mail: alessiodomenico.leto@polimi.it

Abstract. In recent years, due to the request of public clients, Building Information Modelling (BIM) projects are more vastly deployed in construction sector. Furthermore, BIM seems to be the digital methodology with the potential of enhancing the productivity in the construction sector. However, due to the large number of actors involved in a project, the internal information flow of the same actors and the information flow among the stakeholders need to be refined.
This article, therefore, by focusing on the Italian contractors' perspective aims to examine the contractors' internal information flow and the information exchange with the stakeholders.
13 interviews were conducted with several individuals in 3 Italian construction contractors. The informants were selected from different levels of the enterprises, spanning from the owners to the foremen. Combined with site visits, these interviews provide an insight of the contractors' internal information flow and an overview of the information exchange among the contractors and their stakeholders.
The article highlights the current criticalities in the information exchange between the contractors and other actors of the projects. Furthermore, the study points out the opportunities of the digitalisation of the internal and external information flow that could improve the information exchange and the productivity in the Italian construction sector.

1. Introduction

Global competition, technological changes and advances in information technology are pushing organizations of different sectors to improve their processes to deliver high quality products in a short time. Correspondingly, the construction industry has been facing several changes related to technology and innovation in order to improve the productivity, that is on a significantly lower level than other industries [1]. However, this digitalization process is not easy since it requires investments in digital technology, in new equipment, in staff training and especially needs a radical change in the information exchange among the project stakeholders. One of the methodologies which could be well-matched for the optimization of information exchange is Building Information Modelling (BIM). The peculiarities of BIM concern the information management across the all construction process (design, construction, operations and maintenance, decommissioning). With the implementation of the European Directive 2014/24/EU, EU Member States may require in the coming years the use of specific electronic tools for public tenders. Consequently, the expected large-scale application of BIM will imply a strong change within the construction enterprises due to the EU Directive recommendations and to the rapid evolution of the market in this era of digital transformation.

By focusing on the perspective of the Italian contractors, this short paper wants to answer the following research questions.

RQ1: how the Italian contractors exchange the information internally?

RQ2: how the Italian contractors exchange the information with the stakeholders?

RQ3: how the digital innovation could modify the internal and external information flow of the Italian contractors?

The study is structured as follows: Section 2 presents the method used; Section 3 provides the results and the discussion; Section 4 presents the conclusions.

2. Method

Given the RQs defined in the previous section, the research approach of this study can be defined as an interpretative approach. The intention of the researchers was indeed to enter and observe the field without any action on it, opening further examination of the phenomenon observed. Therefore, exploratory case study was chosen as the most suitable research method. By means of exploratory case studies it is possible to observe actual practices, and to understand the nature of phenomena that are not completely known [2]. This methodology is indeed the most pertinent to explain and understand in-depth phenomena within a real-life context on which the researcher has little control [3]. The decision taken for the study was to perform a multiple case study that represents a critical case in testing the research question, since this research is related to 3 small and medium-sized enterprises (SMEs) with at most 12 employees. The SMEs represent almost the totality (over 99%) of the Italian construction enterprises [4].

In this research investigations were performed by using semi-structured interviews. A semi-structured format allows researchers to structure interviews and ensure they follow a logical order. The unit of analysis was the Italian construction contractor. The projects were followed for 7 months and 13 respondents from 3 enterprises were interviewed. The language was Italian. The data collection involved the following three activities:

- 13 individual interviews;
- 2 group interviews;
- 1 meeting observations.

Before to start the interviews, the enterprises owners introduced the interviewer about the position of the respondents inside their organisation. The individual interviews were from 20 to 60 minutes long. Multiple informants, with various functions, spanning from the owner to the foreman were interrogated. The 2 group interviews lasted about 30 minutes and involved:

1. two technical officers in the first group interview:
2. one technical officer and a foreman in the second group interview;

The group interviews aimed to reflect collective understandings of the respondents. The one meeting observation in a technical office permitted to observe directly the carrying out of the work without any intervention on it.

3. Results and discussions

Once finished the data collection, it was possible to analyse data and to interpret the results obtained.

3.1. Current contractor's internal flow

According to enterprises interviewed, folders are the preferred way to store information. In all the enterprises the data of each work order are organized in numbered folders according to an internal procedure. The folders can be accessed by a corporate network. In the enterprises interviewed a poor communication among the employees is found. In two enterprises, during the procurement phase, the technical office makes scant reference to the suppliers estimates already obtained from the sales office in the tender stage and requires new offers from the suppliers. Furthermore, at the end of a group interview the structural engineer, speaking to the electrical engineer of the same enterprise, affirms "we never talked about the communication with the suppliers and we thought that the information problems about the structural and electrical sector were so different. Now we are discovering best practice that we can exchange with each other". In two firms interviewed an Enterprise Resource Planning (ERP) system is used: both the technical officers of the two enterprises affirm that using ERP based on their knowledge is not very good. A technical officer affirms "for the metric calculation is needed to create ad hoc items and it is impossible to link the calculation to the Work Breakdown Structure (WBS). Furthermore, our ERP does not automatically integrate information from other ways such as e-mail, and from various formats such as *docx*, *pdf* or *jpg,* which must be searched and analysed manually". In one enterprise the employees can access data both from the corporate network and from the ERP: sometimes the data are not consistent, but the files duplication is necessary because in the past they encountered a problem with the ERP system and they lost all the data saved on it. In one of the enterprises interviewed the technical office is beginning to implement a BIM system, but the use is still marginal.

3.2. Information exchange with the contractors' stakeholders
This subsection consists of two subsubsections: the first is about the information exchange with the suppliers, the second is about the information exchange with the clients.

3.2.1 Information exchange with the suppliers. According to the enterprises investigated, there is no standard procedure for requesting offers to the suppliers and the relationship with the suppliers is very much founded on personal factors. The technical officer or the project manager base the choice of the suppliers on their own memory and often the contact information of the suppliers is not uploaded into the enterprise database. The enterprises know some suppliers with which exist a long-standing relationship of trust and base principally the own business on them. Nevertheless, a manager affirms that is very important to select carefully the suppliers in the first moments of a project considering also the possibility to contact new ones. To this end, the manager developed a contact list of qualified suppliers that is occasionally updated. There is no standardized procedure about the estimates from the suppliers (for example a form that suppliers must return filled out). The information exchanged is mainly based on e-mail and telephone calls: a technical officer in charge of building orders asserts "on the telephone you can interfere with the human side of the suppliers" and adds "the use of a digital system, as ERP, to store information from the suppliers could be for us a constraint and a waste of time, indeed we must enter manually the information from e-mail, *doc* and *pdf* in the ERP". Another technical officer, instead, uses the telephone calls exclusively to check in real time if the e-mail sent to the supplier is correct. Because the difficult to compare the offers from the different suppliers in different formats, the contractors prefer frequently to consider only the trusted suppliers. However, in cases where a comparison is needed or preferred, it is done differently by the enterprises interviewed and it is done differently within the same enterprises according to the work to be done: in some cases there is an expert who, based on his own knowledge, evaluates the offers and then selects the best; in others cases, for continuous supplies, this process is instead semiautomatic by using a spreadsheet in order to compare mainly the price of the offers. Each approach is obviously consolidated inside the enterprises but it will not be preserved for ever: according to the interviews, the future intention of the managers is to adopt a semiautomatic digital approach to consult the data history of the offers, and to not depend purely on the knowledge of an individual expert. In the case of complex supplies, all the enterprises prefer to meet in person the suppliers.

3.2.2 Information exchange with the client. In the information exchange with the client it is necessary to distinguish the private case from the public one.
Private case. The enterprise receives the request from the customer and makes the offer. Depending on the scale of the intervention, the time required for this phase may be long. According to the respondents there in no problem of communication for small projects. Instead, in case of bigger projects some

problems could exist because the enterprises need to interface with the design team of the client and often there is a hard hesitancy to transmit data.

Public case. The enterprise receives the project documentation and analyses the data received assessing the completeness. The analysis of completeness is done directly by the technical officers based on their experience and, sometimes, by using checklists. Often data are missing or incorrect. Two technical officers assert that there is a strong reluctance to transmit data and is needed to make a lot of requests. Many calls are made and e-mails are exchanged with the public institution in charge of the project. In the case of deficiencies (i.e. drawings, metric calculations, etc.) a technical officer states "our enterprise takes the risk of finding the solution rather than expecting it, in order to carry on the work".

3.3. Implementing a digital information flow

Some respondents affirm that the adoption of a digital flow is not so hard. All the three enterprises interviewed admit that could be strategical to invest on the digital innovation of the enterprise. The owners questioned affirm that implementing a management system could be very useful to improve the internal and the external information flow. Two of the three enterprises have already the tools as ERP and one of them uses BIM, but the employees affirm they do not have the knowledge and the ITC skills to use these instruments in a proper way. Indeed, in this respect a senior technical officer asserts "we have an ERP system but at the moment we are not able to use all its potential". In a group interview the technical officer and the foreman sustain that it is the time to improve and innovate the own enterprise in a digital way. They agree that a 3D model provided by BIM tools could be very useful to avoid misunderstanding among the construction site and the technical office. The foreman interrogated affirms "talking on the telephone in the while you see the 3D model could enables us to focus better on the problem and to avoid wasting time'. Furthermore, the employees have a desire to create automatic registration of transport documents by using Barcode or QR Code: a foreman says "it would lose much less time and probably there would be fewer errors". Furthermore, he claims that this system is already being used successfully in some supply warehouses. Additionally, he states that uploading a *digital site activity log* inclusive of photos in a cloud system might allow real-time control by the technical office.

3.4. Recap: the criticalities of the current flow and the opportunities of a digital flow

According to the contractors' viewpoint, the main criticalities of the current information flow and the main opportunities of a digital information flow are presented in Table 1 and in Table 2.

Table 1. Criticalities of the current information flow.

Criticalities
• Poor internal collaboration and poor dialogue between the employees
• Inability to standardize the selection of the suppliers
• Suppliers submit offers on their own form that the enterprises must decipher
• Many data to be processed and entered manually
• Running-in phase of the management tools in possession, such as ERP and BIM (there is potential that is not yet exploited)
• Data from the client are not completed

IOP Conf. Series: Materials Science and Engineering **800** (2020) 012036 doi:10.1088/1757-899X/800/1/012036

Table 2. Opportunities of a digital information flow.

Opportunities
• Use of a cloud system to instantly upload files even form the construction site
• 3D visualization of the project to avoid misunderstandings between the technical office and the construction site office
• Use of QR code or barcode for automatic registration of the transport documents
• *Digital Site Activities Log* for efficient communication between the construction site office and technical office
• More efficient use of the ERP and BIM systems

4. Conclusions

This study investigated the internal flow of the Italian contractors and the information exchange among the contractors and their stakeholders. 13 interviews were conducted with several individuals in 3 Italian construction enterprises. The informants are selected from different levels of the enterprises, spanning from the owner to the foremen.

The following topics were investigated:

1. internal information flow of the Italian contractors;
2. information exchange of the Italian contractors with the stakeholders;
3. evolution of the internal information flow and evolution of the information exchange among the stakeholders with the spreading of the digital innovation.

According to the respondents there are several criticalities in the current information flow and several opportunities in the adoption of a digital information flow. The current problems in the information exchange could be solved by the adoption of an efficient and effective digital information flow that consequently might improve the construction sector productivity. It should be emphasized that, in this digital transformation age, chances of successful innovation could consist of finding a transitional phase between the traditional modus operandi and the new one. Future works may start by investigating the adoption of a digital information flow by general contractors in order to verify the changes of the internal and external information exchange.

4.1. Acknowledgements

I warmly thank M Mancini and A Pavan for helping me to get in contact with the enterprises.

References

[1] Shehu, Z, Endut, I R, Akintoye, A, & Holt, G D (2014). Cost overrun in the Malaysian construction industry projects: A deeper insight. *International Journal of Project Management*, *32*(8), 1471-1480.

[2] Voss, C, Tsikriktsis, N and Frohlich, M, 2002. Case research in operations management. *International Journal of Operations & Production Management*, 22(2), pp.195–219.

[3] Yin, R K, 2009. Case study research: Design and methods . Thousand Oaks, CA: Sage.

[4] ISTAT. Annuario statistico italiano 2018. (https://www.istat.it, retrieved October 18th, 2019)

IOP Conf. Series: Materials Science and Engineering **800** (2020) 012037 doi:10.1088/1757-899X/800/1/012037

Bio-precipitation of $CaCO_3$ for soil improvement: A Review

A Miftah[1*], H Khodadadi Tirkolaei[2] and H Bilsel[3]

[1]Doctoral Student, Department of Civil Engineering, Cyprus International University, Nicosia, Cyprus.
[2]Assistant Research Professor, Center for Bio-mediated and Bio-inspired Geotechnics (CBBG), Arizona State University, Tempe, AZ, USA.
[3]Associate Professor, Department of Civil Engineering, Cyprus International University, Nicosia, Cyprus.

*Email: ahmadshouib1980@gmail.com

Abstract. The bio-mineralization of calcium carbonates is an innovative and eco-friendly technique for improving soil, which has recently become increasingly popular in the field of geotechnical engineering. Different bio-cementation methods are employed for improving granular soils, including microbially-induced $CaCO_3$ precipitation (MICP) and enzymatic-induced carbonate precipitation (EICP). These methods offer innovativeness, effectiveness and sustainability when applied in geotechnical engineering, and have been suggested as possible solutions for various different geotechnical issues. More research has been conducted into MICP compared to EICP. Nevertheless, urease sourced from plants likely offers many benefits over the use of bacteria urease in bio-cementation processes. The present study reviews the mechanisms, possible areas for implementation, related benefits and drawbacks of the bio-cementations techniques. Additionally, it will concentrate on the techniques used in the precipitation of $CaCO_3$ involving the urease hydrolysis of urea. The study findings will enhance the comprehension of biotechnical mechanisms and ensure that geotechnical specialists are suitably informed with up-to-date knowledge on this subject.

1. Introduction

Bio-cementation methods comprise biochemical processes that occur inside a soil matrix, which produces calcium carbonate precipitation to improve the geotechnical characteristics of soil [1]. $CaCO_3$ causes a reduction in permeability while simultaneously enhancing the stiffness, strength, dilatancy through filling pores, roughening particles and inter-particle bonding. In the field of geotechnical engineering, the bio-mediated and bio-inspired ground improvement methods have been developed to offer different options to conventional methods of grouting, which has many negative impacts on the environment and global climate. In the two most known types of bio- cementation techniques, enzymatic induced $CaCO_3$ precipitation (EICP) and microbial induced $CaCO_3$ precipitation (MICP), the precipitation is induced via hydrolysis of urea (ureolysis).

With regard to the current research into bio-cementation techniques for improving soil, MICP has received the greatest attention. It involves a biological process where the bacteria are harnessed utilising urea hydrolysis to induce calcite precipitation between particle contacts within soil voids. EICP is another method for precipitating the $CaCO_3$, which is based on a urease enzyme sourced from plants. In comparison to MICP, EICP is not impacted by problems related to the growth and activity of microbes within soil, including insufficient oxygen for ureolytic bacteria to become active in deeper soil, and implementation in soil where the bacterial cells are larger than the soil pores [2].

The effective design and application of these methods would enable them to be implemented for various significant geotechnical issues, including fugitive dust management [3], liquefaction mitigation [4, 5], surface water erosion control [6, 7]. Such bio-minerals have the potential for use in engineering applications as they offer sufficient strength along with minimal environmental effects in comparison to traditional techniques. For instance, the manufacture of Portland cement requires extreme amounts of energy and produces a significant volume of carbon dioxide (CO_2) emission in addition to nitrogen oxides and sulphur. Therefore, biological precipitation of $CaCO_3$ has a prospect to be an environmentally- friendly geotechnical approach for ground improvement.

2. Microbially induced carbonate precipitation (MICP)

2.1. Mechanisms

Microbially-Induced Calcite Precipitation (MICP) processes occur frequently in natural environments. Calcite, the most stable form of calcium carbonate, is generated by various different microbial reactions, including microbial denitrification reactions and microbial ureolysis [8, 9]. The MICP method based on urea hydrolysis is largely dependent on the Bacillus pasteurii, which produces a high yield of urease [10]. The fundamental mechanism through which soil is improved in the MICP method is as follows: (i) The urease enzyme is microbially synthesised via metabolic activities, followed by hydrolysis of urea by the urease to generate carbonate and ammonia ions; (ii) the carbonate ions are bound to the calcium to assemble insoluble $CaCO_3$ in a medium rich in calcium [10]; (iii) particles of soil are coated in the resulting precipitated carbonate, which also acts as a filler in the empty spaces between particles and binds the particles together, thus enhancing the mechanical characteristics of the soil. Additionally, hydrolysis of urea generates a net rise in the pH level as a result of ionic equilibrium [11]. This process is formulated in Equations 1 and 2 below:

$$CO\,(NH_2)_2 + 2H_2O \longrightarrow 2NH_4^{+} + CO_3^{2-} \quad (1)$$

$$Ca^{2+} + CO_3^{2-} \longrightarrow CaCO_3 \quad (2)$$

Bacteria have a high propensity to adhere to soil particle surfaces. The formation of calcite surrounding the bacterial cells enables soil pores to be filled and bridged, which subsequently reduces the soil's permeability and enhances its strength.

2.2. Limitations

It has been proposed that the MICP technique can be employed in various different geotechnical engineering fields. However, practical application of this approach presents numerous difficulties and constraints that require further investigation, such as problems associated with cost, environmental concerns, uniform treatment and types of soil in which the method is applicable.

2.2.1. Cost

Producing highly-active ureolytic bacteria as part of the MICP technique is associated with significant costs [12]. However, MICP could be more cost-effective in the long term as it is possible to reuse the bacterial enzyme in following treatment applications (two to three) utilising the same cementation solution [13, 8]. Ivanov and Chu [14] made a comparison of the costs of raw materials when using chemical and microbial techniques for grouting, and determined that the costs with chemical grouting (\$2 – \$7 per m^3 of soil) were less than those for microbial grouting (\$0.5 – \$9 per m^3 of soil). Various studies in the literature have proposed alternative approaches for reducing the costs associated with the practical application of MICP. For example, the utilisation of Ca^{2+} ions dissolved in salt water could be a possible replacement for commercially derived $CaCl_2$ [12]. How-Ji Chen et al. [15] investigated the application of pig urine as a substitute for industrial urea. Additionally, modification of the constituent concentrations was found to increase the chemical efficiency, thus enhancing the yield of [$CaCO_3$]p / [Ca^{2+}] and [Ca^{2+}] / [urea] in each treatment cycle [16].

2.2.2. Soil type

The MICP technique could be constrained to deeper soil environments as the growth and movement of bacteria is limited in subsoil. Furthermore, it potentially may only be applied to soils comprised of minimal amounts of fines as a result of the smaller pore spaces in fine-grained soils. Generally, the abundance of microbes is higher in soils with larger particles [17]. Hence, the application of MICP is restricted to soils of fine to medium-sized sand or larger. [18].

2.2.3. Environmental impact

Microbial-based urease leaves the organisms in the soil. Hence, injecting ureolytic bacteria into soil presents numerous challenges including the necessity to obtain permission and licenses from the relevant authorities and the continual inspection of the microbial ecology to ensure it is safe [19]. The elements and by-products of MICP could lead to significant environmental concerns. For instance, urea hydrolysis produces ammonia, which can be particularly harmful for people in addition to the aquatic eco-system and air, while also increasing the potential for corrosion due to elevated pH levels [20]. Furthermore, when the calcium is derived from calcium chloride, this increases the possibility that groundwater will become contaminated by chloride [21]. Consequently, further studies are required to solve these problems prior to considering this technique for practical applications.

2.3. Potential implementations of MICP

MICP has reportedly been opooused in various areas of geotechnical engineering for reducing permeability, mitigating liquefaction, increasing the strength of granular soil and controlling erosion of coastlines.

2.3.1. Permeability

The MICP technique can lead to a sustained decrease in the permeability of soil. Whiffin et al. [8] investigated the effects of microbial sedimentation of calcite as a technique for improving soils for different purposes-. Subsequent to the injection process, it was found that the soil permeability had decreased approximately by 30%. van Paassen [9] stated that a reduction in the initial permeability of 60% was observed in soils treated with biological methods at around 100 kg/m^3 $CaCO_3$ precipitation, while Ivanov [22] reported a decrease in permeability of 50-99% when a 1 M cementation solution was used. Al Qabany and Soga [23] utilised a 0.5 M cementation solution and observed a 20% decrease in the initial level of permeability at 2% $CaCO_3$ precipitation.

2.3.2. Erosion

MICP technique can be employed in for controlling the erosion of granular soils. Yufeng Gao et al. [24] researched the seepage control of sandy soils stabilised with MICP. Their findings revealed that durable crusts with low permeability could be generated on the surface of the soil using MICP treatment, which led to a 379-fold reduction in the rate of seepage. Ning and Soga [7] studied the durability of MICP treated gravel-sand mixtures against erosion, and concluded that MICP can effectively mitigate the internal erosion of 25%-75% mix of sand and gravel.

2.3.3. Liquefaction mitigation

Various researchers have demonstrated that bio-cementation technologies have considerable promise for the efficient mitigation of liquefaction. Burbank et al. [25] made a comparison between the results of cyclic shear tests conducted on sand treated with bio-cementation and Portland cement revealing that the cyclic stress ratio (CSR) of the soil treated with MICP was greater than the soil treated with Portland cement. Zamani and Montoya [26] performed cyclic direct simple shear tests (DSS) on silty Nevada sand samples subjected to a cyclic stress ratio (CSR)... The findings revealed that the increased resistance to liquefaction was due to a rise in density as well as enhanced load transfer and undrained shear strength from the cementation bonds. Xiao et al. [4] demonstrated that calcareous sands treated with MICP reduced surplus pore pressure and compressive strains, while considerably more cycles were required to initiate liquefaction in comparison to untreated calcareous sands.

3. Enzyme induced carbonate precipitation (EICP)

3.1. Mechanisms

Enzyme-induced carbonate precipitation is another method for precipitating the calcium carbonate ($CaCO_3$). EICP has certain similarities to ureolytic MICP, apart from the fact that the bacteria that produces urease is replaced by agricultural-free urease. Nevertheless, the EIPC is considered to be superior to MICP as it is more efficient and cost effective. Urease can be derived from various species of commonly-found plants, such as the pigeon pea as well as certain beans, squashes, melons and plants that belong to the Pinaceae family [27]. In geotechnical implementations, the most well-known and researched urease enzymes are sourced from the jack bean plant (Canavalia ensiformis) [28-30], soya bean [31, 24], and watermelon seeds [31]. The resulting chemical reaction is the same as presented in Equations (1) and (2) above. As with the MICP technique, the soil particles are coated in the precipitated $CaCO_3$, thus allowing them to bind together and improving the characteristics of the soil based on the amounts of precipitation.

3.2. Limitations

3.2.1. Cost

The enzyme may constitute the costliest ingredient of EICP treatment with 57%-98% of the EICP solution costs attributed to the enzyme [32]. One method of reducing the costs associated with the enzyme is to use crude extract rather than urease that has been purified commercially, or alternatively, the enzyme could be extracted from agricultural waste [21].

3.2.2. Soil type

The enzyme is approximately 12 nm in dimension [16]. Thus, the solubilized urease enzyme gives a distinct advantage over microbial urease due to its small size. The use of plant-derived urease can also reduce the possibility of bio-clogging as a result of $CaCO_3$ precipitation and biofilms formation, which could both limit the applicability of MICP [8]. In addition, it increases the chance of injecting the cementation solution into deeper and finer soils, with no concern about bacterial growth related to MICP.

3.2.3. Environmental impact

Similar to ureolytic MICP, the ammonia by-product is the primary environmental issue related to EICP. The lifespan of the urease enzyme is relatively short and its function and activity are progressively decreased [34]. This property has potential benefits in certain engineering fields as the enzyme will degrade in a natural manner, thus reducing long-term effects on the environment.

3.3. Potential implementations of EICP

Enzyme-induced carbonate precipitation (EICP) has similar application potential to MICP but over a broader selection of soils as a result of the smaller dimension of the urease enzyme and its solubility, to mitigate slope stability, erosion and scour, bearing capacity of shallow foundations, as well as seismic settlement and liquefaction.

3.3.1. Permeability

Nemati and Voordouw [35] utilised enzymatic calcium carbonate for reducing the permeability of unconsolidated porous media and Berea sandstone cores. Yasuhara et al. [5] investigated the hydraulic characteristics of sand specimens that had been treated by utilising carbonate precipitation induced enzymatically via permeability tests. The permeability of the treated specimens was found to be greater than one order of magnitude smaller than untreated samples.

3.4. Erosion and dust control

Knorr [6] investigated the impacts of water and wind erosion on soil treated with EICP employing various kinds of soil, such as Ottawa F60 sand, silty-sand and mine tailings. The findings revealed that EICP has significant promise for use in managing erosion caused by both wind and runoff of surface water. Hamdan and Kavazanjian [3] treated different types of sand, including uniform medium-grained silica sand, native Arizona silty sand, and fine sand-sized mine tailings, using a process in which the sand was sprayed in a series of pans. The effectiveness of the EICP treatment in stabilising the soil against fugitive dust emissions was tested in a wind tunnel environment. It is interesting to note that the results indicated the considerable promise of the EICP technique for mitigating fugitive dust emissions for all soil types. Cuccurullo et al. [36] examined the potential of utilising crude extracts of urease from soybeans to mitigate the effects of water erosion on silty clay. Preparation of the sample involved the compaction of the soil within an oedometric ring. The durability of treated soil was analysed on the basis of a series of immersion tests on treated and untreated specimens. The findings demonstrated that the treated samples showed a 3-fold improvement compared to the untreated specimens with regard to lost mass.

3.5. Liquefaction mitigation

Minson and Okamura [5] investigated the ability of sand cemented utilising EICP to resist liquefaction at distinct saturation levels (30% and 97%) during curing in undrained triaxial equipment at effective confining pressures of 50, 100 and 200 kPa. The results indicated that the capability of the treated soil to resist liquefaction was significantly improved, particularly when the calcite was precipitated at low levels of saturation. Simatupangb et al. [5] additionally performed undrained cyclic shear strength tests on Toyoura and Keisha No. 4 sands, taking into account various testing factors, such as the particle dimensions, percentage of calcite, confining pressure and the saturation level in the precipitation process. The study findings also supported the suitability of the EICP technique for mitigating liquefaction. Nafisi et al. [37] recently examined the effects of EICP and MICP treatments on the shear responses of two conventional sand soils: Ottawa 20–30 and Ottawa 50–70. It concluded that the shear wave velocity rate was increased in the EICP sample in comparison to the MICP sample, thus enhancing the liquefaction resistance capacity.

4. Conclusions

This study reviewed shortly the mechanisms, possible areas for implementation, advantages and disadvantages of the MICP and EICP methods. These techniques are still restricted to limited environmental conditions such as soil types, particle size, and treatment depth. MICP and EICP have been suggested as a promising solution for numerous geotechnical problems including permeability reduction, liquefaction mitigation, strengthening of granular soil, and controlling coastal erosion. Nevertheless, more studies are required to solve environmental problems associated with its by-products while reducing the treatment costs.

References

[1] DeJong J , Mortensen B M, Martinez B and Nelson D 2010 Bio-mediated soil improvement Ecol. Eng. 36 197–210
[2] Khodadadi H, Kavazanjian E and Bilsel H 2017 Mineralogy of calcium carbonate in MICP-treated soil using soaking and injection treatment methods Geotechnical Frontiers 2017 Conf. 195–201
[3] Hamdan N and Kavazanjian E 2016 Enzyme-induced carbonate mineral precipitation for fugitive dust control Geotechnique 66 546–55
[4] Xiao P, Liu H, Xiao Y, Stuedlein A, Liu H, Stuedlein A and Evans T 2018 Liquefaction resistance of bio-cemented calcareous sand Soil Dyn. Earthq. Eng. 107 9–18
[5] Simatupang M, Okamura M, Hayashi H, Yasuhara H 2018 Small-strain shear modulus and liquefaction resistance of sand with carbonate precipitation Soil Dyn. Earthq. Eng. 115 710–8
[6] Knorr B 2014 Enzyme-induced carbonate precipitation for the mitigation of fugitive dust (Master Dissertation, Arizona State University)
[7] Jiang N and Soga K 2019 Erosional behavior of gravel-sand mixtures stabilized by microbially induced calcite precipitation (MICP) So. Found. 59 699–709
[8] Whiffin V, van Paassen L and and Harkes M 2007 Microbial carbonate precipitation as a soil improvement technique Geomicrobiol. J. 24 417–23
[9] van Paassen L, Ghose R, van der Linden T, van der Star W and van Loosdrecht M 2010 Quantifying biomediated ground improvement by ureolysis: Large-scale biogrout experiment J. Geotech. Geoenvironmental Eng. 136 1721–8
[10] Wang Z, Zhang N, Cai G, Jin Y, Ding N and Shen D 2017 Review of ground improvement using microbial induced carbonate precipitation (MICP) Mar. Georesources Geotechnol. 35 1135–46
[11] Mobley H and Hausinger R 1989 Microbial ureases: significance, regulation, and molecular characterization Micro. and Mo.r Bio. R. 53 85-108
[12] Cheng L, Cord-Ruwisch R and Shahin A 2013 Cementation of sand soil by microbially induced calcite precipitation at various degrees of saturation Can. Geotech. J. 50 81–90
[13] Ismail M, Joer H, Randolph M, and Meritt† A 2002 Cementation of porous materials using calcite Géotechnique 52 313–24
[14] Ivanov V and Chu J 2008 Applications of microorganisms to geotechnical engineering for bioclogging and biocementation of soil in situ Rev. Environ. Sci. Biotechnol. 7 139–53
[15] Chen H, Huang Y, Chen C, Maity J and Chen C 2018 Microbial induced calcium carbonate precipitation (micp) using pig urine as an alternative to industrial urea W. and Bio. Valorization 10 2887–2895
[16] Martinez C, Dejong T, Ginn R, Montoya M, Barkouki H, Hunt C, Tanyu B and Major D 2013 Experimental optimization of microbial-induced carbonate precipitation for soil improvement J. Geotech. Geoenvironmental Eng. 139 587–98
[17] Rebata-Landa V and Santamarina C 2006 Mechanical limits to microbial activity in deep sediments Geochemistry, Geophys. Geosystems 7 n/a-n/a
[18] Hamdan N 2015 Applications of enzyme induced carbonate precipitation (EICP) for soil improvement (Doctoral dissertation, Arizona State University)
[19] Akiyama M and Kawasaki S 2012 Microbially mediated sand solidification using calcium phosphate compounds Eng. Geol. 137–138 29–39
[20] Pacheco-Torgal F and Labrincha J 2014 Biotechnologies and bioinspired materials for the construction industry: an overview. Int. J. of Sustainable Eng. 7(3) 235-244.
[21] Khodadadi H, Kavazanjian E, van Paassen L and Dejong J 2017 Bio-Grout Materials: A Review Geotechnical Special Publication (American Society of Civil Engineers (ASCE)) pp 1–12
[22] Ivanov V 2010 *Microbiology of Environmental Engineering Systems Environmental Biotechnology* (Humana Press) pp 19–79
[23] Qabany A and Soga K 2013 Effect of chemical treatment used in MICP on engineering properties

of cemented soils Géotechnique 63 331–9

[24] Gao Y, He J, Tang H, Chub J 2019 Calcium carbonate precipitation catalyzed by soybean urease as an improvement method for fine-grained soil So. Found. 59 5 1631-1637

[25] Burbank M, Weaver T, Lewis R, Williams T, Williams B and Crawford R 2013 Geotechnical tests of sands following bioinduced calcite precipitation catalyzed by indigenous J. Geotech. Geoenvironmental Eng. 139 6 928-936

[26] Zamani A and Montoya M 2018 Undrained monotonic shear response of MICP-treated silty sands J. of Geotech. and Geoenvironmental Eng. 144 6

[27] Das N, Kayastha A and Srivastava P 2002 Purification and characterization of urease from dehusked pigeonpea (Cajanus cajan L.) seeds Phytochemistry 61 5 513–521

[28] Yasuhara H, Neupane D, Hayashi K and Okamura M 2012 Experiments and predictions of physical properties of sand cemented by enzymatically-induced carbonate precipitation Soils Found. 52 539–49

[29] Almajed A, Tirkolaei H and Kavazanjian E 2018 Baseline investigation on enzyme-induced calcium carbonate precipitation J. Geotech. Geoenvironmental Eng. 144 11

[30] Miftah A, Khodadadi H and Bilsel H 2019 Strengthening beach sand by enzyme induced calcium carbonate precipitation 8th Geotechnical Symposium Conf. pp 1–9

[31] Javadi N, Khodadadi H, Hamdan N and Kavazanjian E 2018 EICP treatment of soil by using urease enzyme extracted from watermelon seeds 2018 IFCEE 2018 115-124. pp 115–24

[32] Almajed A A 2017 Enzyme induced carbonate precipitation (EICP) for soil improvement (Doctoral dissertation, Arizona State University)

[33] Robert L and Blakeley B 1984 Jack bean urease: the first nickel enzyme J. Mol. Catal. 23 263–92

[34] Pettit N, Smith A, Freedman R and Burns R 1976 Soil urease: Activity, stability and kinetic properties Soil Biol. Biochem. 8 6 479–84

[35] Nemati M and Voordouw G 2003 Modification of porous media permeability, using calcium carbonate produced enzymatically in situ Enzyme Microb. Technol. 33 635–42

[36] Cuccurullo A, Gallipoli D, Bruno A, Augarde C, Hughes P and La Borderie C 2020 Soil stabilization against water erosion via calcite precipitation by plant-derived urease CNRIG 2019 Conf. 40 753–62

[37] Nafisi A, Safavizadeh S and Montoya B 2019 Influence of microbe and enzyme-induced treatments on cemented sand shear response J. Geotech. Geoenvironmental Eng. 145 9

Assessment of Bedestan (St. Nicholas Church) on the first decade of conservation works

D Mısırlısoy*[1], N Bilsel[1] and I Safkan[1]

[1]European University of Lefke, Lefke, Northern Cyprus, TR-10 Mersin, Turkey

*Email: dmisirlisoy@eul.edu.tr

Abstract: St. Nicholas Church (Bedestan) is one of the most significant monuments located exactly in the centre of the walled city of Nicosia, in the Selimiye Quarter. The significance of the heritage building lies on its temporal transformation. Bedestan initially built by Byzantines in the 5th century, and was later enlarged during the Lusignan, Venetian, Ottoman and British period. Between 2004 and 2009 had gone through conservation works by United Nations Development Programme Partnership for Future (UNDP-PFF) which is funded by European Union. In 2009 the project has won Europa Nostra Award between 140 projects. The project consisted of both architectural and structural strengthening interventions. The aim of the study is to assess current condition of the structure on the first decade of conservation works. The method had three objectives. Firstly, the literature was reviewed in order to understand the historical process of the building. Then, site survey was completed for further understanding of the applied intervention to the structure. Finally, the current condition of the structure was observed after ten years of completed restoration works. Deterioration on used materials as well as durability of strengthening works was assessed and observations were listed herein. **Keywords:** Restoration, adaptive reuse, cultural heritage, structural durability, masonry deterioration.

1.Introduction

Cyprus, being at the crossroads of many civilizations from the antiquities till today embodies diverse traces of these dominations. Historical research reveals that the first inhabitants of Cyprus were people who arrived from coastal regions of Anatolia and Syria through the Neolithic and Bronze ages. Later Hittites, Phoenicians, Assyrian, Egyptians, Persians, and Romans ruled the island. Towards the end of the 10th century the Byzantains gained power. In the following centuries Cyprus was ruled by Lusignans, Venetians, Ottomans and the British, respectively, who left in the island the most historical structures that still stands still today [1]. These diversities of civilizations brought together a multiethnic and multi religious character to the urban and suburban architectural identity of Cyprus. Greek and Turkish Cypriots became the most permanent residents of the island, dividing the islands' community into two distinctive and significant ethnic identities.

The city of Nicosia has successfully preserved its rich cultural heritage, regardless of changes in its social, cultural, economic and political structure throughout its history. In Nicosia, despite the changing physical and social structure and also other key factors, especially after 1974 political interferences, traditional architecture in the historic city centre is well kept as much as possible without losing its original identity.

The Nicosia city walls built in the 15th century during the Venetian period and many immovable cultural assets within the city walls are among the most important cultural heritages of the entire island. However, after 1974, the division of the island and the inner part of the city were divided into two parts, North - South Nicosia, and problems started to be preserved regarding the protection of this cultural heritage. The residents of the city walls started to migrate to new settlements outside the city walls by building newer and larger houses. Settlement of immigrants in these abandoned houses brought with it many problems.

The division of the city is the main cause of its ongoing problems. This situation causes the process of city planning and future foresights related to the city not to be implemented in a healthy manner. There are many attempts to revive Nicosia in both regions of the city. However, the division of the city limits development and creates various problems in order to develop urban plans and conservation strategies, while also neglecting social values.

In the early 2000s, important developments took place for the city of Nicosia. One of the most important of these is the conservation-restoration works carried out with the European Union funds that have brought a breath of fresh air to the region, and the other is the opening of the Lokmacı/Ledra border gate in 2007, which contributed economically, socially, culturally, politically and physically. Restoration works of monumental buildings such as Bedestan, Municipal Market, Grand Bath, and Armenian Monastery and the improvement works in the Arabahmet region created a catalyst effect within the walls and the region started to attract the attention of the locals again. Many domestic investors have started to open new businesses in the region. At the same time, as the effect of this change, an increase is observed in the number of Greek and foreign tourists crossing from Southern Cyprus to Northern Cyprus.

There has been a noticeable increase in renewal projects especially in recent years within the walls that change, evolve and transform depending on urban dynamics. As a result of the financial support provided by different institutions and organizations to domestic investors in order to support their restoration projects, such initiatives are increasing. Some traditional residences have been reused with functions such as cafes, restaurants and shops.

In 2009, the project was awarded the European Union Prize for Cultural Heritage/Europa Nostra Award in the research category amongst a selection of 140 projects from 24 countries [2].

The methodology consists three components. Firstly, the literature was reviewed in order to understand the historical process of the building. Then, the site survey was completed for further understanding of the applied intervention to the structure. Finally, the current condition of the structure was observed after ten years of completed restoration works. Deterioration on used materials as well as durability of strengthening works was assessed and observations were listed herein.

2.Historical framework of Bedestan/St. Nicholas Church

Being the capital city since the 10th century, Nicosia has been the central node where the majority of the historical monuments resides. Enlart [3], in the most comprehensive work dedicated to medieval architectural history of Cyprus, points out one of the very first references to St. Nicholas Church.

St. Nicholas Church has a long and a complex history. The construction of St. Nicholas dates back to 6th century Byzantium period. In later centuries it was expanded and some parts were rebuilt. During the Lusignan period (1192 – 1432) the building was operating as a non-cathedral church [4]. The church was built on top of the remainings of the basilica with added French Gothic architectural elements such as pointed arches, vaults and a dome. The vaults were added to insert further support for the building to withstand pressure and to add longevity by preventing its collapse [5]. Later on, with Venetians coming into power, several structural transformations were made and it was converted into a Greek Orthodox Cathedral, dedicated to Virgin Hodegedria [6].

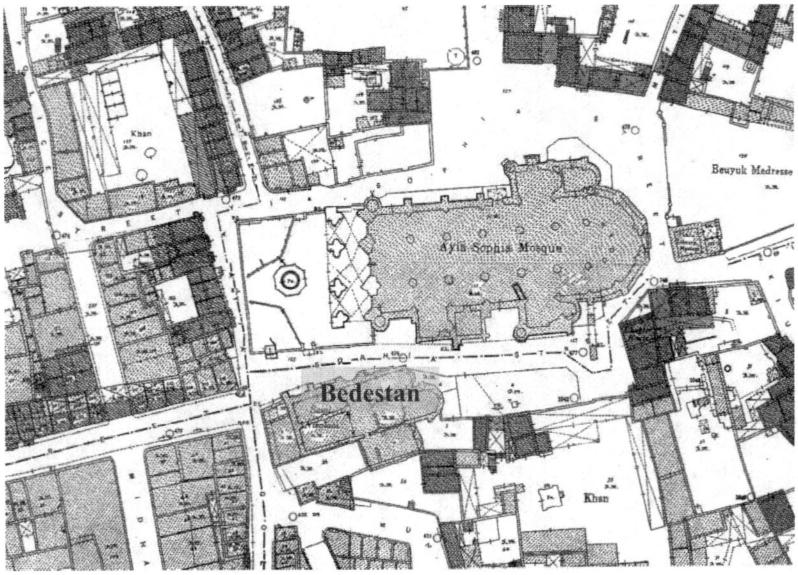

Figure 1. Plan of St. Nicholas Church/Bedestan and St. Sophia Mosque [7].

St. Nicholas church exemplifies the major architectural and structural characteristics of Byzantium church design of the period, embodying an apse, basilica, domed roof and a cross shape plan. As many of the Byzantine churches, St. Nicholas personifies the eclectic approach embodying cross shape layout together with the architectural characteristics of medieval church such as high ceilings with naves [8].

In the Ottoman period (1571-1878), following an Ottoman tradition of converting existing structures (mostly churches) into a new use, the St. Nicholas Church was converted into a closed bazaar, called *Bedestan*. 'Here merchats and people from all over the island gather, flocking through the city gates to sell and buy goods' [9]. The conversion took place in the early years of the Ottoman conquest, and a fountain was built next to the Bedestan to supply water both for the St. Sophia Cathedral (converted into a mosque) and the market place [10]. In one of the earliest mentions of Bedestan, the Dutch traveler Cornelis van Bruyn, who arrived Cyprus in 1683 and stayed there for eight years, writes 'Close by [St. Sophia] is another fine building, almost entire, with a vault resting on two columns. I think it was formerly a church: now it is a bazar or market' [11]. We learn from the writings of the traveller Abbé Mariti who lived in Cyprus between 1760 and 1769 that it was 'a market of all goods', and it was 'the business resort of the chief merchants of Nicosia, Turks, Greeks and Armenians' [12].

From mid-18th century until late 19th century Bedestan was used as a wheat warehouse. According to Bağışkan [4] it was a flour market it 1873. It lost its function at the beginning of the 20th century [13]. Later, in the British period (particularly during early 1930s), it was used as a warehouse [4]. Bedestan was completely abandoned after the British built a nearby closed market (*Bandabuliya*) in 1932.

Since 2003, using EU funds the UNDP/PFF has been undertaking the restoration of Bedestan with the aim of revitalization of the Selimiye quarter. The project was completed in two phases. The first phase (2004-2006) was focused on structural consolidation and anti-seismic improvements. The second phase (2007-

IOP Conf. Series: Materials Science and Engineering **800** (2020) 012038 doi:10.1088/1757-899X/800/1/012038

2009) was focused on conservation and re-use. The inauguration of Bedestan as a cultural centre took place on 9th November, 2009. The total cost of the restoration was 2 million Euros [14].

Figure 2. Bedestan (Authors, 2019)

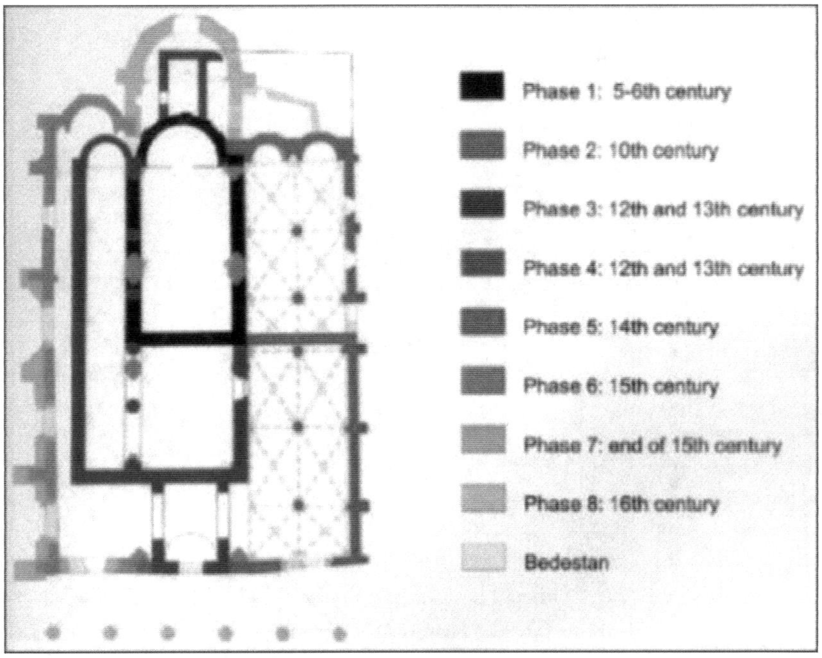

Figure 3. Historical evolution based on a stratigraphic interpretation [15]

3.Restoration project and its interventions

The Bedestan (St. Nicholas Church) is one of the most important cultural heritage sites located within the walled city of Nicosia. Research, investigations and design works for the restoration of the Bedestan (St. Nicholas Church) had started in 2003 by UNDP-PFF that was a EU funded project under the framework of Nicosia Master Plan. The main objective of the conservation of the Bedestan (St. Nicholas Church) was to revitalisation of the Selimiye Quarter, which is the focal point of the Walled City of Nicosia [2].

The Bedestan is located in the center of the walled city and it is one of the most significant monuments of Nicosia. When the restoration works started, the heritage buildings was partly demolished and the restoration aimed to strengthen remaining structural parts. The Bedestan was closed to the public due to dangerous structural conditions before the project starts. Conservation works of the structure includes interventions on the roof, floor and windows. All interventions intended as being reversible. [2].

The project was completed in two phases. First phase includes the structural works, which was performed between 2004 and 2006. And in the second phase the conservation works were completed between 2007-2009. The first phase of the project aimed to improve the Bedestan and involved survey and investigation of the structure. The heritage buildings include different historic layers and had different interventions. In order to understand the involved process of the building detailed analysis was completed. Tomography resistivity analysis was made and stratigraphic interpretation used in defining historical evolution of the structure [15].

The main aim of the restoration was to preserve all historic layers of the heritage building. In this respect, minimum intervention is made in order to preserve the authenticity of Bedestan. Interventions during the conservation works includes also consolidation of elevated structures. In British period, buttresses were constructed to support the elevated southern façade of the structure (Figure 4). In the conservation works, it was decided not to touch the buttresses, since that part of the building could be collapsed.

Figure 4. Southern Façade of Bedestan (Authors, 2019)

The building has 3 entrances built in different periods located side by side as Lusignan, Byzantine and Venetian which is a symbol of showing sympathy to different cultures (Figure 5). Each door reflects architectural characteristics of their period which all ornamentation preserved during restoration.

Figure 5. Doors from Lusignan, Byzantian, Venetian respectively (Authors, 2019)

During restoration every characteristics and detail has been preserved. All interventions to the structure are compatible, legible and reversible. Arches on the south façade are covered with polyurethane which is a transparent material like glass but much lighter. A special hinges hold the windows and the roof which allows expansion due to settlements (Figure 6). Since the original roof was collapsed, a new roof added that comprises contemporary materials and techniques. A lightweight contemporary roof addition covers the main hall of the church. During restoration original roof of the building had been search through all literature but there was not any information regarding to the original condition of the roof. In this respect, it was decided to add a contemporary roof with its form and material that can be easily perceived that it is new and can be removed easily.

Elevated floor was designed to cover the floor of the monument. Wooden elevated floor was selected as the system since it is reversible. Some parts of the floor left opened in order to show original floor to the visitors.

Figure 6. Connection details (Authors, 2019)

Conservation works also ensured the accessibility to the monument for people with reduced mobility. However, interventions to ensure accessibility to the monument do not harm the originality of the heritage building. Ramps have been added with contemporary material and can be removed easily (Figure 7).

Figure 7. Ramp for disabled people, View from main hall (Authors, 2019)

Between 2010 and 2012, the cultural center is supported on the cultural heritage management of Bedestan. A workshop designed to train specialized personnel in the field of maintenance, restoration and conservation of historical buildings was conducted for a team of architecture and engineering students, as well as for the local technical craftsmen assisting the planners and sub-contractors in the drawing of the project and the restoration phase [2].

IOP Conf. Series: Materials Science and Engineering **800** (2020) 012038 doi:10.1088/1757-899X/800/1/012038

4.Assessment of Bedestan on the first decade of conservation works

4.1 New use and its activities
The Bedestan represents a cultural role for the whole city of Nicosia by functioning as a cultural centre. Different activities including concerts, exhibition, seminars, etc. were organized in the venue since 2012. Sufi Dance performances are also performed twice a day in Bedestan.

4.2 Structural durability of the heritage building
Significant structural interventions were made during the conservations works. Due to the excessive soil shrinkage, tensile cracks were initially developed. The conservation project, targeted not only the repair of the damage but, a comprehensive strengthening strategy. First, micro piles were utilised for further soil related deterioration. Then, beam additions on masonry walls provided an improved diaphragm behaviour which increased the lateral resistance of overall structure.

Furthermore, advanced structural intervention techniques were also utilised. Post-tensioning the tilted masonry column and pendulum system for the tower structure [15]. The applied intervention techniques successfully managed the targeted seismic performance of the building. Furthermore, crack monitors were placed on existing crack patterns (Fig. 8) during the conservation works and no such progress was monitored since then.

Rainwater drainage systems were repaired during the conservation works and drainage systems were provided for alteration of rainwater. However, due to the lack of continues maintenance of masonry walls, vegetation re-occurred on several parts (Fig. 8).

Figure 8. Photos of the Bedestan (Current condition) (Authors, 2019)

4.3 Proposals for future use and maintenance
The project contributed to create new jobs during the restoration works, and local expertise in the field of cultural heritage preservation. At the same time, through its re-establishment as a new cultural pole in the heart of the revitalized walled city of Nicosia, it also brought together an awareness towards the preservation of cultural heritage. As of November 2012 the EVKAF administration put in place a ticket system of an entrance fee in as a financial sustainability. The financially improved management can now take position on developing bodies for maintenance purpose.

Although the current state of Bedestan is in a favourable condition regarding to its social cultural and economic inputs, since it has been 10 years post conversation, the maintenance of the building should be scheduled. The ownership of Bedestan belongs to Evkaf, but the organisation does not embody experts in conservation/maintenance works. This makes Bedestan vulnerable against deterioration due to external/internal conditions. It would be beneficial to form a body which will ensure the continues maintenance of the building.

Vegetation on some locations were monitored on building. It is strongly suggested to use anti-vegetation coatings on sandstone to avoid any further damage to the building.

5. Conclusion

Compatibility with the heritage building is one of the most important criteria of the conservation principles. The proposed addition and the historic structure should be compatible in terms of scale and material selection. Generally, additions are successful in terms of conservation principles. All additions on the building are compatible with the heritage building. New additions are differentiated from the old and is distinguishable from the existing building. New additions are not only harmoniously integrated with the context, but it also respects the relation of the building with the surrounding in terms of architectural characteristics. Historic buildings could have different historic layers. Aim of the conservation does not mean to remove all historic layers and preserve one of them. Restoration work respects the building's history and cultural significance. All additions are designed with steel and wood, which is a demountable and removable structure.

Apart from above mentioned physical interventions made to the building, the new proposed function of the building is found to be appropriate. The socio-cultural and economic sustainability is ensured. The success of this conservation acted as a catalyzer for historic buildings around the close surrounding neighborhood.

References

[1] Hill G F 1940. *A History of Cyprus*. 4 volumes. Cambridge: Cambridge University Press.

[2]UNDP-PFF. Restoration and re-use of the Bedestan (St. Nicholas Church). https://www.cy.undp.org/content/cyprus/en/home/operations/projects/partnershipforthefuture/upgrading-of-local-and-urban-infrastructure---phase-ii/restoration-and-re-use-of-the-bedestan--st--nicolas-church-.html [Accessed in September 2019].

[3] Enlart C 1987. *Gothic Art and the Renaissance in Cyprus*. London: Trigraph. (Translation of *L'Art Gothique en Chypre*, Paris: Ernest Leroux, 1899).

[4] Bağışkan T (2009). *Ottoman, Islamic and Islamised monuments in Cyprus*. Nicosia: Cyprus Turkish Education Foundation.

[5] Cavallaro A Cessari L and Gigliarelli E 2013. Site characterization by in situ and laboratory tests for the structural and architectural restoration of Saint Nicholas Church, Nicosia, Cyprus. In: E.

[6] Papacostas T 2010. Echoes of the Renaissance in the eastern confines of the stato da mar: Architectural evidence from Venetian Cyprus. *Acta Byzantina Fennica*. Vol. 3, No. 3, 136-172.

[7] Papacostas T 2012. Byzantine Nicosia. In: D. Michaelides (ed.). *Historic Nicosia*. Limassol: Rimal Publications, pp. 77-109.

[8] Hacışevki M A and Bozatlı N N 2013. Bedestan, Cyprus Vaqf Foundation

[9] Trélat P and Iliadou H 2011. Tracing the Market Place. Commercial and Artisan Activity in Nicosia Between the Latin and Ottoman Eras. *Cahiers du Centre d'Études Chypriotes*, Vol. 41, 299-328.

[10] Yıldız N 2009. The Vakf Institution in Ottoman Cyprus. In: M.N. Michael, M. Kappler, and E. Gavriel (eds.). *Ottoman Cyprus. A Collection of Studies on History and Culture*. Wiesbaden: Harrassowitz Verlag, pp. 117-159.

[11] Cobham C D 1908. *Excerpta Cypria. Materials for a History of Cyprus*. Cambridge: Cambridge University Press.

[12] Cobham C D 1909. *Travels in the Island of Cyprus. Translated from the Italian of Giovanni Mariti*. Cambridge: Cambridge University Press.

[13] Turkan Z and Özburak Ç 2018. Selimiye Square within the Historical City Texture of Nicosia. *Journal of History, Culture and Art Research*, Vol. 7, No. 2, 430-443.

[14] Yorucu V Mehmet O Alpar R and Ulucay P 2010. Cross-Border Trade Liberalization: The Case of Lokmaci/Ledra Gate in Divided Nicosia, Cyprus. *European Planning Studies*, Vol. 18, No. 10, 1749-1764.

[15] UNDP-PFF. Europa Nostra Event, Bedestan Presentation by UNDP Cyprus Programme Manager Tiziana Zennaro. https://www.slideshare.net/UNDP_PFF/europa-nostra-bedestan-nicosia-cyprus-18-april-2016. [Accessed in September 2019].

IOP Conf. Series: Materials Science and Engineering **800** (2020) 012039 doi:10.1088/1757-899X/800/1/012039

Effect of fly ash-limestone dust stabilization on volume change and strength of expansive soil

J Naji, H Bilsel and A P Balkis
Cyprus International University, Nicosia, N. Cyprus

Email:hbilsel@ciu.edu.tr

Abstract. The expansive soils in semi-arid and arid regions are affected by swelling-shrinking due to climatic effects, causing damage to lightweight structures. Various methods and materials have been employed to improve the engineering properties of these soils. This paper includes the findings of experimental work conducted studying the effectiveness of use of fly ash (FA) and limestone dust (LS) in the enhancement of the engineering properties of expansive soil sampled from Haspolat, Nicosia. Limestone quarry dust is a by-product produced in quarries as a result of the process of crushing and grinding. Laboratory experiments were conducted on the expansive soil with additive proportions of 5% LS, 10% FA, and 5% LS+ 10% FA by dry mass of the soil at curing periods of 1, 14 and 28 days. The 5% limestone dust and 10% fly ash mixed samples have shown significant increment of up to 220% in CBR value, 82% in volumetric shrinkage, and 100% in swell potential. The study has concluded that the introduction of admixtures to expansive soil generally enhanced the geotechnical properties, exhibiting greater durability and stability compared to the untreated soil. Hence limestone dust and fly ash can be regarded as effective expansive soil stabilizers, also providing cheap solutions to the disposal issue.

1. Introduction

The sensitivity of expansive soils to water content alteration has made them quite difficult for civil engineers to manage in the construction of roads. These soils expand when water content increases and shrink when desiccate. The repetitive cycles of wetting and drying due to climatic and environmental factors cause swelling-shrinkage, hence damaging the foundations or roads built on them [1]-[3]. Therefore, to alleviate the problems with expansive soils, so as to ensure safe construction, environmental-friendly and cost-effective solutions are used, which include employment of materials to improve the engineering properties of problematic soils. Chemical stabilization and physical stabilization are the two forms of soil stabilization. The process of enhancing soil properties by redistributing soil particles to increase the density through compaction with vibration and/or mechanical weights can be categorized as physical (mechanical) stabilization. Physical stabilization is characterized by rewetting, removal and replacement processes [4]. The employment of chemical admixtures to stabilize soil is considered when physical stabilization methods are ineffective, insufficient, too costly and/or impossible to use. Some of the chemical admixtures used are lime, silica fume, marble powder quarry dust or microorganisms [5].

The utilization of industrial by- products and wastes as secondary additives in soil stabilization has been widely implemented. One example is limestone quarry dust, which is a by- product produced in quarries as a result of the crushing and grinding of stones. Limestone is classified as a sedimentary rock which is constituted mainly of dolomite and calcium carbonate. Calcium carbonate rocks are formed by the re-crystallization of existing lime deposits. As a result of its application potential and resilience, limestone has a broad scope of use. Quarry crushers generate about 15% to 20% waste daily from crushed stones and the accumulation causes environmental degradation as well as public and environmental health risks [6]. The assessed engineering properties of clay soils treated with lime and stone powder show that the admixtures are efficient in stabilizing soil [7]. The improvement of engineering properties was clearly observed in the soil when the maximum dry density and CBR increased and the optimum moisture content decreased with the introduction of stone dust [8]. Satyanarayana et al. [9] observed that the free swell value of an expansive soil reduced from 100% to 83% and the soaked CBR value increased from 1.2 to 6.7 with 10% quarry dust. Whereas, Naman and Kumar [10] observed an increment in CBR value of 50% with 30% quarry dust content. Optimum moisture content and water affinity reduce as quarry dust content increases. Sonthwal and Guray [11] observed that a decrease in maximum dry density occurred with quarry dust content beyond 25%. The maximum dry density of an expansive soil rose from 1.58 g / cm3 to 1.68 g / cm3 and the optimum moisture content reduced from 32% to 26% with 20% stone dust content CBR value decreases beyond 25% quarry dust content and this percentage is effective for stabilizing and upgrade the engineering features of black cotton soil. Koranne and Ali [12] carried out a probe of the effect of a combination of stone dust and fly ash at equal proportion with varying admixture percentages from 10% to 50% at 10% interval on the properties of expansive clay. The research found that the swelling potential was almost completely controlled at 20% - 30% admixture percentage and the strength and stability properties increased with rise in admixture percentage. The combination of fly ash and stone dust brings about better improvement of expansive clay than a single homogenous admixture (fly ash or stone dust).

This study analyses the effectiveness of adopting soil-stabilization technique to positively modify the geotechnical features of expansive soil. In this study, fly ash and lime stone dust were used as additives to modify the geotechnical features of a locally obtained expansive soil. An experimental program was conducted on both expansive soil (control sample) and stabilized soil to prove the combined influence on geotechnical properties with reference to introducing limestone dust and fly ash in different percentages to expansive soil. This study is to be done for the aim of probing the effectiveness Fly ash and limestone dust in the enhancement of the engineering properties of expansive soil sampled from Haspolat at the outskirts of Nicosia. Secondly, to probe the influence of the treatment age on swelling and shrinkage behavior of the expansive soil. These aims will be achieved through the objectives highlighted below.

2. Materials

The soil sample was obtained from the outskirts of Taşkent region, Nicosia. The physical properties are presented in Table 1. It was established that the untreated soil possesses moderate to high swell potential and therefore requires stabilization. The study involves exploration of the influence of limestone dust and fly ash on Atterberg limits, compaction characteristics, and swelling-shrinking behavior. The experimental study was carried out in accordance to ASTM Standards. The expansive soil has dark grey color and is in the form of claystone chunks. The samples were oven dried for 5 days at 50±1°C and pulverized to be able to study geotechnical properties. Class C fly ash is mainly generated through the burning of sub-bituminous or lignite coal but it may also be manufactured by burning bituminous or anthracite coal (made entirely of carbon). Class C type of fly ash. Its calcium oxide (CaO) content can be greater than 10%. The tests done on the limestone dust (LS) were conducted with reference to all the particles of the grinded limestone passed through the 0.425mm sieve as found by the investigation done. The geotechnical features of the limestone fine dust are presented in Table 1.

Table 1. Physical properties of the materials used

Property	Soil	Fly Ash	Limestone dust
Colour	Dark gray	Light gray	White
Sand size (%)	14	10	15
Silt size (%)	50	90	73
Clay size (%)	36	--	12
Liquid Limit (%)	62	--	22
Plastic limit (%)	31	--	16
Plasticity Index (%)	31	NP	6
Specific Gravity	2.58	2.34	2.76

3. Results and Discussions

3.1. Standard Proctor Compaction Test

This test was conducted with reference to ASTM 698 (Method A) for all samples used and the compaction test parameters are presented in Table 2. The maximum dry density increased to 1.71 g/cm3 from 1.69 g/cm3 of the untreated soil when 5% fine dust of the limestone was added. Whereas with 10% fly ash addition, the maximum dry density was reduced to 1.57 gm/cm3and the optimum water content increased to 24% compared to the optimum water content of the natural soil which was 19%. When 5% LS and 10% FA were added to the soil, the maximum dry density was decreased to 1.66 g/cm3, whereas there were no significant changes in the maximum moisture content.

Table 2. Standard Proctor Compaction test parameters

Material	Maximum dry density (g/cm^3)	Optimum water content (%)
Natural soil	1.69	19
NS+10% FA	1.57	24
NS+5% LS	1.71	19.5
NS+10% FA+5%LS	1.66	17.5

3.2 One-dimensional swell test

This test is conducted to study the influence of admixtures and treatment age on swell potential according to ASTM D4546. Primary swell potential (S_p) and secondary swell rate (C_{ss}) were obtained from axial swell strain versus logarithm of time relationship, expressing the three stages of swell; initial, primary, and secondary. The swell potential is taken as the swell strain at the end of primary swell stage. All these parameters of untreated and treated soil with respect to treatment age are presented in Table 3.

Table 3. Swell parameters for untreated soil and its mixtures after different curing periods.

Curing time	Swell parameters	NS	5% LS	10% FA	10% FA+5% LS
1 day	S_P (%)	5.6	1.32	0.668	0.360
	C_{ss}	0.877	0.34	0.112	-
14 days	S_P (%)		2.846	0.0154	0.0508
	C_{ss}		0.772	-	-
28 days	S_P (%)		2.572	0.190	0.0231
	C_{ss}		0.91	-	-

The results indicate a decrease from 5.6% of the natural soil to 2.572% when 5% LS, 0.190% when 10% FA and 0.0231% when 10% FA+5%LS were added and aged for 28 days before testing. Therefore, the soil became a low swell potential soil after stabilization took place within 28-day curing period. The rate of secondary swell rate (C_{ss}) is observed to increase with 5% LS, whereas diminished completely when 10% FA and 10%FA+5%LS mixture are added.

3.3. Volumetric shrinkage results

In order to evaluate the shrinkage behaviour of untreated and stabilized soils, volumetric shrinkage tests were conducted by drying saturated specimens after treatment times of 1, 14 and 28 days. Table 4 depicts the void ratios at full saturation and completely dried conditions, as well as the percent reduction and time taken for shrinkage to be completed. The outcome of the volumetric change test for the untreated soil and its mixtures after different curing periods revealed that the best improvement in volume change is observed for 10%FA+5%LS treated for 28 days, while the largest volume change was observed for untreated soil. Moreover, there were no cracks observed on the specimens when completely dried except in the untreated and 5% LS added ones.

Table 4. Volumetric shrinkage test results.

Materials	Curing time (day)	Change in Void Ratio			Total Time of Shrinkage (min)
		Full Saturation Condition	Dried Condition	Reduction (%)	
NS	-	0.640	0.312	51.25	1570
5% LS Mixture	1	0.420	0.299	28.81	3182
	14	0.455	0.323	29.01	3094
	28	0.494	0.327	33.81	4340
10% FA Mixture	1	0.756	0.611	19.18	1420
	14	0.673	0.569	15.45	3087
	28	0.658	0.545	17.17	2848
10% FA+5% LS Mixture	1	0.581	0.452	22.20	4499
	14	0.519	0.468	9.83	2633
	28	0.613	0.556	9.30	1710

3.4. California bearing ratio test (CBR)

This test was done with reference to ASTM D 1883 through the employment of the standard Proctor method for unsoaked stabilized soil at 1, 14, and 28 days treatment periods. The 14.07% CBR value of the untreated soil increased to 16.26%, 15.27%, and 19.81% at 1, 14, 28 days treatments ages respectively when 5% fine limestone dust was introduced. Once 10% fly ash was introduced the CBR value rose to 15.69%, 26.92%, and 17.71% for the treatment ages 1, 14, and 28 days respectively. The variation in the rise of CBR value with 10% fly ash and 5% fine limestone dust was minimal. With the addition of a mixture of 5% LS + 10% FA, however, CBR increased with curing time up to 45.05% in 28-day treated specimens. Table 5 shows the CBR values of various mixtures at different treatment periods.

Table 5. California bearing ratio at various treatment periods.

Treatment period / Material	1 day	14 days	28 days
	CBR value %		
Natural soil	14.07		
Mix 5% LS	16.26	15.27	19.81
Mix 10% FA	15.69	26.92	17.71
Mix 5% LS + 1% FA	45.79	31.77	45.05

4. Conclusions

This experimental study has investigated the effectiveness of employing limestone dust and fly ash in stabilizing expansive soil, and the following conclusions were derived:

- Addition of 5% LS to expansive clay samples led to a 76.43% reduction in swell potential. While adding a mixture of 10% FA and 10%FA+5% LS caused considerable reductions of 88.07%, 93.57% respectively in swell potential. Based on these observations addition of 10% FA, 10% FA +5% LS have given the best results in improving the expansive soil.
- The largest reduction in volumetric shrinkage percentage is 81.27% in 10% FA +5% LS added specimens with respect to untreated condition. There were no cracks seen when complete dehydration of the samples was achieved except in 5% LS and NS specimens.
- With the addition of 5% LS, and 10% FA, the value of CBR for unsoaked samples was improved by 40.8% and 25.9% respectively, while giving excellent value of 220%, when a mixture of 10% FA+5% LS added and 28-day cured specimens were used.

Finally, it may be concluded that limestone dust and fly ash are relatively effective as expansive soil stabilizers, enhancing the geotechnical properties. The use of these materials provides a cheap stabilizing method, subsequently encouraging its use which eliminates the disposal issue. 10%FA+5%LS can be employed to stabilize soil for road base and lightweight structures, as well as for cutting down the accumulation of waste which may pose health and environmental hazards.

References

[1] Lucian C 2006 Geotechnical Aspects of Buildings on Expansive Soils in Kibaha, Tanzania, Preliminary Study. Division of Soil and Rock Mechanics. Department of Civil and Architectural Engineering Royal Institute of Technology Stockholm, Sweden, ISSN 1650-951X

[2] Avsar E Ulusay R and Sonmez H 2006 Assessments of swelling anisotropy of Ankara clay *Engineering Geology* 105 (1–2), pp 24–31

[3] Chen F H 2012 Foundations on expansive soils Vol. 12 *Elsevier*.

[4] Nelson J and Miller D J 1997 Expansive soils: problems and practice in foundation and pavement engineering John Wiley & Sons Kemp, J F 1896 Handbook of rocks. D. Van Nostrand; Princeton.

[5] Ouhadi V R Yong R N Amiri B Ouadi M H 2014Pozzolanic consolidation of stabilized soft clays *Applied Clay Science* 95.

[6] Soosan T G Sridharan A Jose B T Abraham B M 2005 Utilization of Quarry Dust to Improve the Geotechnical Properties of Soils in Highway Construction *Geotechnical Testing Journal* 28:4 ASTM International.

[7] Roohbakhshan A and Kalantari B 2013 Stabilization of Clayey Soil with Lime and Waste Stone Powder *International Journal of Scientific Research in Knowledge* (IJSRK) 1(12) pp 547-556.

[8] Lal B Alam S and Kiran S 2014 Use of Stone Dust from Construction Site As A Stabilizing Agent In Flexible Pavement Construction: A Case Study *Int. J. Struct. & Civil Eng. Research.*

[9] Satyanarayana P V V Raghu P Kumar R A and Pradeep N 2013 Performance of Crusher Dust in High Plastic Gravel soils as road construction material *IOSR Journal of mechanical and civil engineering*, vol10 Issue 3 pp01-05.

[10] Naman A and Kumar A 2015 Design of Stone Dust Stabilized Road Vol.6 IAEME, Issue 5, pp 28-33.

[11] Sonthwal, V and Gurav, S (2016), "Stabilization of soil reinforced with quarry dust", International Journal of Recent Trends in Engineering and Research, Volume 2, Issue 9, ISSN 2455-2457.

[12] Koranne Shudhad S and Ali Mir S 2011 Performance Analysis of expansive Soil Treated with Stone Dust and Fly Ash *Electronic Journal of Geotechnical Engineering*, Vol. 16 Bund. I, pp 973-

IOP Conf. Series: Materials Science and Engineering **800** (2020) 012040 doi:10.1088/1757-899X/800/1/012040

Some refurbishment particularities: An older building case study

R A F Oliveira[1*], J Lopes[2], M I Abreu[3]
Polytechnic Institute of Bragança - Civil Construction and Planning Department
Campus de Santa Apolónia - 5300-253 Bragança (Portugal)
*e-mail: roliveira@ipb.pt

Abstract. Portugal has many old buildings with historical and cultural values in advanced degradation. Many of them have an architectural interest, special values and some other characteristics with interest to preserve, which contribute to keeping the building originality and singularity. These building could be retrofitted for new facilities of uses increasing the levels of comfort similar to new buildings. However, all project management of that kind of buildings is more specific, complex and requires special knowledge in the management of the particularities of their work, even other factors related with misunderstandings, inheritance problems, and many others. This research describes a case study that involves the surveying of constraints and other problems concerning the retrofitting works of an old building located in Montesinho village (Bragança-Portugal). In their retrofitting works were applied some parameters of the "Retrofitting management system for buildings located in consolidated urban areas" to support the project design and the building existential particularities. Besides, the management system considers regulations obliged, sustainable practices and possible constraints particularities for that old building case study's project management. The results show constraints, problems, but also solutions application to solve and reduce them, being a contribution to many other old buildings with similar retrofitting needs.

1. Introduction

During the life cycle of existing building is of crucial importance to have conservation and maintenance practices which contributes to their assumed sustainable development, fulfilling of its everyday needs and keep comfort conditions. [1]. Original old buildings reflect the difference compared with recent buildings solutions, because sometimes that buildings revealing some rare and most often very ingenious technological solutions, meanwhile abandoned or forgotten [2]. In most cases, it is possible to understand the social context of that time, while they keep unique architectural and artistic features with interest to preserve, which is not the case in new buildings [3]. Monumental buildings are even more specific in these cases, possessing acknowledged details and cultural values with national, public or municipal interest [4].

Many of monumental buildings with private owners, such as manor houses, are degraded without habitability conditions and far away from the expectable investment needs. Nevertheless, it is possible to refurbishment and dignify the appearance of old buildings with or without architectural and cultural interest for several different purposes, such as touristic or even cultural aspects, attending to very specific commercial demands, develop historic centres or specific local areas [3].

However there are some problems linked to their refurbishment process, some of them related to localization aspects, cultural and patrimonial constraints, real estate pressures, demolitions of possible reusable elements, and difficulties in adaptation to realistic needs, mistakes in building planning, time overruns and increases in expected costs, among others [5].
The case study deals with the investigation of constraints and other specificities of an old building destined a house focusing on technical recommendations, sustainable practices and looking for adequate solutions that match the building existences. For that purpose, a toolkit "management system for refurbishment of consolidated urban centre localized buildings", here after called "management system", is used [6]. This toolkit considers the application of sustainable practices to help in the management of the whole building complex, besides contributing in the decision-making of the different parts involved in the refurbishment process.

2. Research methodology

This paper has adopted a case study to research the management of building refurbishment works in an old building located in the little village Montesinho, at 20 km distance from Bragança city (Portugal). The characterization of that building is attending in the research. The research methodology involves a qualitative study and uses available documentation (building project design) and direct observation ("in situ" real and direct observation) which may permit a better convergence of available information [7]. Table 1 contains strengths and weaknesses of data sources used in the case study.

Table 1. Strengths and weaknesses of data sources used in the case study (Adapted from Yin [7].

Data Resources	Strengths	Weaknesses
Documentation (Design project)	Stable (it could be revised sometimes); Discreet (not used as result); Exact (names, references and details); Wide coverage (time, events and distinct environments).	Recuperation low capacity; Tendency selection (if data collection is not complete); Pre-conceived ideas of the author; Information access could be deliberated.
Direct Observation	Events in real time; Contextual (take the event context).	Selectivity, much time and expensive; Reflexivity (The event could be differentiated).

The management system was developed in a research to support old buildings rehabilitation, retrofitting, refurbishment, reconstruction or other project types. Each parameter has 3 possible choice options, classified from less sustainable to more sustainable. The main goal consist in the reflection and option for viable solutions in old building project and report each parameters solutions recommendation to design projects. And whenever possible, it is chosen the option for solutions that have more benefits to sustainability and without forgetting costs aspects. Figure 1 show a framework scheme about the methodology used in this paper research which used about 39 parameters from a set of 50 parameters management system [6].

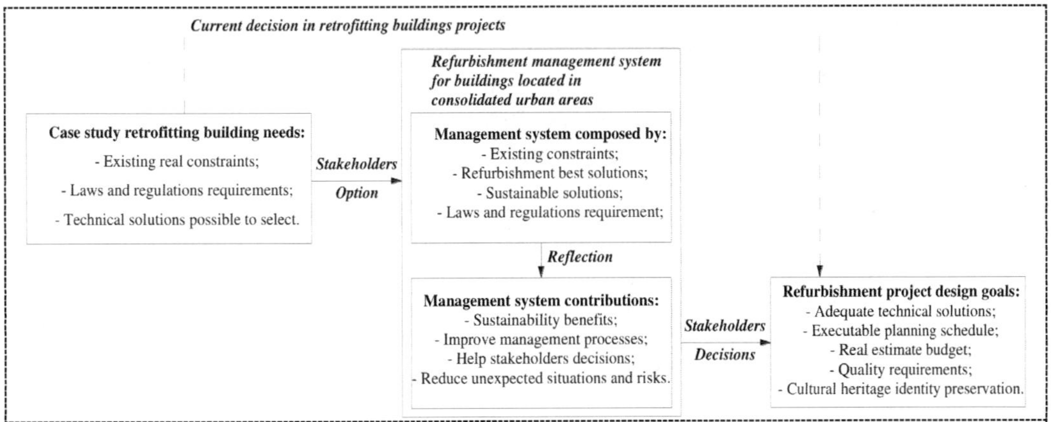

Figure 1. Framework scheme of management system uses for building refurbishment works

Each solution/option proposal or recommended has a careful reflection in the project and these selected solutions intend to give more benefits to sustainability when compared with conventional or similar solutions. Some management system solutions are proposed to improve building management process, such as reducing some unforeseen problems [8].

The management system could give a contribution in the project design phase about the adequacy of technical solutions and also in weighting some aspects about planning, budgeting, quality and preservation of existing heritage building. These aspects are frequently often neglected or are not managed in a correct way forgetting old building particularities [5].

3. Old buildings refurbishment particularities and proposals

Refurbishment is the "process to encompass rehabilitation, alteration, adaption, extension, improvement, modernization and repair work carried out to an existing building to permit it reuse for various specific reasons" [9]. However, this definition described the main reasons of refurbishment, which is to make sure a building is able to operate continuously, and offer better services that suit the demand of the tenants. And also, in technical point of view, refurbishment is the process of making as good as new, including essential modernization and renovations.

Old buildings refurbishment actions consider the re-use of pre-existing materials, fitting them to promotion of sustainable practices and keeping, as much as possible, the buildings' original features. Meanwhile some interventions on these buildings are built as new building constructions, with total ruin of the existing elements, using only the exterior walls, contradicting part the original ideal of refurbishment concept much as possible, Figures 2 and 3.

Figure 2. Steel beams to support and reinforcement existent walls.

Figure 3. Concrete slabs integrated in older stone walls structure.

3.1. Constraints connected to old building refurbishment works

Many old buildings refurbishment works need some specificity solutions attended in design projects to solve problems occurrences and also existent constraints and specificities or particularities [3]. Many factors could had contribution to them, especially according to old technologies used, villages or historic centres morphology which differ from recent regulations to attended new building construction [2]. Some constraints are connected to "Surroundings and location", such as [5]:

- Frequently, new buildings in recent areas are predicted than old buildings.
- Car parking spaces are limited and inexistence of gardens areas and common areas.
- Fire risk propagation is very difficult to control in fire cases.
- Narrow streets, opposite building proximity, low mobility people.
- Infrastructure networks maintenance or substitution needs.
- Main old buildings surroundings are degraded and uncharacterized.

3.2. Management system to support old building refurbishment works

The management system parameters are presented in Tables 2 and 3. Its uses are optional but it takes the design projects more consistent in technical point of view, with reflection and pondered solutions adjusted to refurbishment works. This tool is structured in 4 thematic groups, 15 indicators with 50 parameters. The main goal of this toolkit is to support old buildings works from viability to use, involving design project and construction works and also in other hand their stakeholders. It contains some solutions which help in reflection about technical aspects sometimes neglected in intervention works such as refurbishment works. The management system contains refurbishment constraints, frequent regulations to attend and better solutions to promote sustainable aspects.

Table 2. Management system thematic areas, indicators and parameters (part 1) [5].

Area	Indicators description	Parameters description
A1. Surroundings and location	I1. Mobility and amenities	P01. Public transport
		P02. Car parking
		P03. Local amenities
	I2. Local infrastructures	P04. Outward firefighting means
		P05. Technical networks in public space
		P06. Urban space quality
	I3. Land use occupation	P07. Land occupation
		P08. Total area and deployment area
		P09. Gardens and leisure places
	I4. Solar orientation and exposure	P10. Solar exposure
		P11. Solar orientation
A2. Project design	I5. Characterization of building conditions	P12. Request for technical studies
		P13. Characterization diagnoses of building conservation status
		P14. Project Design specificities
	I6. Architectonic organization and salubrity	P15. Conceptual architecture configuration and adaptability
		P16. Ratio useful floor area/Gross lettable area (GLA)
		P17. Acoustic insulation and indoor air quality
	I7. Infrastructures, foundations and	P18. Building technical networks
		P19. Peripheral retaining structures
		P20. Foundations

	structural elements conditions	P21. Structural elements
	I8. Materials	P22. Materials reuse P23. New materials P24. Fire safety
	I9. Sustainability promotion	P25. Water recovery and reuse P26. Solar collectors for hot water production P27. Electrical energy production P28. Energetic efficiency in thermal comfort P29. Other solutions for energetic efficiency P30. Bioclimatic solutions P31. Other sustainable solutions

Table 3. Management system thematic areas, indicators and parameters (part 2) [5]

Area	Indicators description	Parameters description
A3. Construction works and site works	I10. Initial works constraints	P32. Site works and surrounding space P33. Adjoining building conservation state P34. Stabilization and consolidation of building works and adjoining buildings P35. Adjoining buildings waterproofing
	I11. Industrialization/ execution of works	P36. Workforce P37. Specialized workforce and company's technical capacities P38. Specialized subcontracts P39. Technical requirements monitoring
	I12. Risk and constraints potential	P40. Propensity to project design changes P41. Propensity to the occurrence of unexpected works P42. Propensity to time overruns P43. Propensity to other work constraints
	I13. Other features resulting from works	P44. Archaeological Works prospection P45. Construction and demolition waste management P46. Needs of occupant's relocation
A4. Costs	I14. Work costs	P47. Costs of urban space works P48. Costs of general building works
	I15. Tax incentives and other costs	P49. Possibility to apply for benefits and tax incentives P50. Maintenance and conservation strategies

4. Case study

This case study approach the constraints study support related to an old building refurbishment works considering best solutions chosen during project management. This building attended village morphology and the façades characterization contribute to maintain the village preservation, identity and it is also an imposition requires in that place. Though this old building study in the case study is not a heritage building with cultural interest, their contribution to surrounding area is important in set of all building because these buildings define the streets and complement each other's in architectural and aspect characterization.

4.1. Old Building description and characterization

The old building studied in this case study research was in ruins and threatens collapse of neighbouring buildings (figure 4 and 5), predicting 120 years old from existence. It is located in Montesinho village, near from Bragança city (Portugal), and it is integrated in Montesinho Natural Park which is preserved by their lovely nature. The building has not cultural interest individually but it contributes to building set of village preservation. The original building was in reality 2 separated buildings very similar, each one with 2 floors, stone walls, floors in wood and also roof structure in wood with slate tiles. Only the stones colours aspect, positions integrated with black roof slate tiles differentiate all landscape near and away. The design project was made in 2009 while the refurbishment works started in 2010 and it was finished in 2014.

Figure 4. Principal façade with balcony.

Figure 5. Internal ruin collapse structures.

4.2. Old building constraints and particularities

The old building included in the case study had all the interior structure and part of the exterior walls in ruins without any technical or economical possibility of recuperation, figure 4 and 5. A direct observation put in evidence some constraints and particularities described in table 4.

Table 4. Constraints and particularities of the old building case study

Code Group	Description:
(A) Existent and design project aspects	- (A1) Principal façade has some elements to maintain and preserve exterior stairs and old balcony, figure 4;
	- (A2) The building had 2 storage floors. The first one had 2,2 m of finish ceiling height and the second one had 2,3m; which does not allow regulatory height;
	- (A3) Some internal walls in stone had no interest to preserve, figure 5;
	- (A4) Vacant state with internal flour and roof wooden structure and some exterior walls in ruins, though the floors and roof structure was supported by walls, figure 5;
	- (A5) Fragile and poorly consolidated exterior walls;
	- (A6) Lateral façades leaning against an adjoining building;
	- (A7) Principal façade define the street direction;
	- (A8) The posterior façade had not connection and continuity in wall, figure 6.
	- (A9) One of the first storage floor gate has 1.6m width and 1.8m height;
	- (A10) The adjoining building façades have more than 8meters of height and they were built in little stone pieces, figure 7;
	- (A11) Electrical cables networks crossing the building, figure 7;
	- (A12) Building foundations had structural problems in local;

	- (A13) The building had not water and electricity networks; - (A14) Some doors closure and their height need adaptation changes; - (A15) Thermal and acoustic insulation were not existent. - (A16) The refurbishment works has technical complexity and constraints;
(B) Cultural and historical aspects	- (B1) Important landscape views and village preservation history which give the name of the Montesinho Natural Park; - (B2) The building has lot of stone pieces to preserve and with possible use. - (B3) The owners have lot of interest in preservation old existing materials adapting them to increase internal comfort conditions.
(C) Economical, financial and sustainability	- (C1) The Priority intervention was to protect exterior and lateral walls from rain; - (C2) Maximal limit of economic resources were imposed without margins; - (C3) Costs of demolition, cleaning works and shoring façades are significant; - (C4) Water pipes were inexistent even solutions for consumption reduction; - (C5) Energy networks were inexistent even solution for consumption reduction; - (C6) Regulatory verification are above conventional minimal requirements; - (C7) Some materials with environmental concern were present in design project; - (C8) Procurement of local construction and materials supplied companies; - (C9) Construction supervision by engineer service's needs was imposed.
(D) Refurbishment works	- (D1) Inexistence of space for site works needs to put materials and tools deposit; - (D2) Site works was in internal building spaces and the access were narrow; - (D3) Cleaning needs were necessary and also demolition waste management, figure 5; - (D4) Support lateral façades with shoring structures - (D5) Narrow street which has low traffic cars movement.

Figure 6. Posterior façade without connection and continuity.

Figure 7. Adjoining building façades with little stone pieces. Electrical cables networks crossing the building

4.3. The contribution of the management system in the proposal studied

This building has the proposal to be used by a family and in possible cases seasonal rental for tourism. Besides architectural proposal by design project for the adaptation of that building, has a set of technical solutions [10] which made it possible to manage the refurbishment process, minimizing resources, costs and time [11]. These solutions were developed attending to the existing constraints and using the management system previously, mentioned in point 3.2. The main goal is assuring options for more sustainable solutions [5] [12] [13].

All management system parameters from (A1) "surroundings and location" area, parameters P1 to P11 were not considered in this study because the building was acquired yet with conditions accepted and assumed by the owners, which do not give any sustainable increment for the project in that thematises. In other words, it is not possible to improve suggested requests by the parameters P1 to P11 about surroundings and existing location.

Tables 5, 6 and 7 address the areas "Project design", "Construction works and site works" and "Costs" of the management system, which include from P12 to P50 parameters. In these areas were attended all constraints and other problems analysed in table 4, solved or lessened with recommendations of different parameters' attending sustainable solutions as possible [14, 15, 16].

Table 5. Parameters options from "Project Design" area.

Code and Parameter Solution selected	Constraints
P12 – Realization of geometrical drawings only requires traditional equipment's.	A4,A10,A16
P13 – Foundation solved by traditional engineering solutions without specific tests.	A12
P14 – Design project were revised internally by different engineers. The design project contains technical solutions to a construction sustainable.	A16
P15- Existing building constraints solved in design project, preserving some elements with historical and cultural values which contribute to maintain the authenticity and identity of the building such as the façades and also the village preservation. The design project study improve the existing building conditions, such as the constraint (A2) has 2.6m for ceiling height in each floor, readapting exterior and consolidate lateral walls.	A1, A2, A5, A6, A8, A10, A11, A14, A15, A16, B1, B2, B3
P16 – The refurbishment works promoted internal remodelling, create partial habitable area in roof which increasing construction area fulfilling legal regulations.	A2, A3, A4, A9, A14, A15
P17 – The design project considered legal regulations superior than minimal requirements in thermal, acoustic and air renovation requirements.	A15, A16
P18 – The building need all infrastructures and networks connections.	A13
P19 – The building has some peripheral containment needs, and it needs some simple solutions which promoted the existing elements reinforcement of all walls.	A6, A10, A12
P20 – Isolated foundations were built for pillars support and the exterior and lateral walls need reinforced concrete footing.	A5, A12, A16
P21 – The exterior walls are reutilized complying the regular requirements but lateral walls ante first floor level is reinforced with concrete walls (figure 8). The internal structure was built in steel profiles in pillars and beams. Some part of the beams transmits their loads to external walls with load levels very similar to original ones. The floors area was built in prefabricated concrete slabs. The roof structure was built in steel profiles and wood with slate tiles (figure 9 and 10)).	A15
P22 – Façades and up to 25% of other materials quantity (stone and wood) were reused.	A3, A4, A14
P23 – The design project had some concern in materials selection coming from manufacturing sites near from building localization which was not superior of 50% cost.	C7,C8
P24 – The building meets all fire legal regulations.	A4, A6, A16
P25 – The design project only contains water reduction in all taps and it has a deposit for rain water reuse for toilets and garage cleaning, which reduce 25% of consumption.	C4
P26 – Solar collectors will be used in a roof place not visible from the streets.	C5
P27 – The building contains installation for Photovoltaic panel's connection.	C5
P28 – The building had preview class energetic A+ with thermal insulation thickness superior than minimal requirements. The heating system is with biomass solution.	C6

P29 – The design project has some energetic efficiency concerns about natural light and LED artificial ones, renewable resources for building heating systems (biomass). It contains some monitoring equipment's for control and saving energy consumptions.	C5
P30 – The existing constraints do not only permit in the architecture design project the promotion of solutions for heating passive performance. Some glasses at South orientation is possible to put but it is not enough for needs.	C5, C7
P31 – The design project has preview an uncovered courtyard where is possible a garden	C7

Figure 8. Lateral walls at first floor in concrete.

Figure 9. Roof with metal steel beams and wood.

Figure 10. Roof with slate tiles.

Table 6. Parameters options from "Construction works and site works" area

Code and Parameter Solution selected	Constraints
P32 – Site works limitation space and access restrictions are presented in building equipment's assembly. Some examples: Deconstruction of principal façade and posterior façade for architectural and structural requirements. The design project preview ready-mixed concrete pumped as solution to eliminate aggregates in situ for concrete composition.	A9, A13, D1, D2, D4, D5
P33 – The adjoining building has good conservation state and stability but the walls from building studied need some stability works.	A6, A10
P34 – The lateral walls need some stabilization which is frequent in that works.	A10, A16
P35 – The adjoining building are in good maintenance performance but the roof need some waterproofing cares and drainage.	A6, A8
P36 – The refurbishment works requires smaller quantities and rhythms of labour than new construction. Ex: The steel structure was built in beams and pillars steel profiles.	A4
P37 – Building refurbishment works only require companies with some labour specialization in that works, namely with experience and technical knowledge in similar works. The companies might be regional and they had not requirements needs of management systems certification.	C8

P38 – Specialized subcontracting companies was necessary for reconstruction walls in stone and to put slate tiles in the roof which require complexity and preservation cares.	C8
P39 – This refurbishment works were quite similar to new construction where the dimension and technical works complexity were not required permanently an engineer's supervision.	C9
P40 – The design project are detailed without fails detection with unrequired changes.	C2, C9
P41 – The design project has a description of all the constraints described in that parameter management system and it also has some procedures to mitigate and control.	C2, C9
P42 – The design project has a description of a set of factors to develop a planning map adjusted to works complexity, without unforeseen works appearance but allowing with some difficulty the fulfilment of deadlines.	D2, D4
P43 – This refurbishment works has some constraints, such as: construction works signalization on the streets, some traffic problems, overhead electrical power lines over building, materials transport and supply, lack of parking, supplementary Safety and Health Plan measures are need.	A9, A11, D2, D3, D4, D5
P44 – The Montesinho village has no evidences in archaeological remains existence. In other hand the design project describe some notes to attend in occurrence cases.	A16
P45 – The design project preview some deconstruction works, façades reutilization, and it define also some proceedings for construction and demolition waste reutilization.	A1, A7, C1, D3
P46 – The building is vacant without occupants with relocation needs.	A4

Table 7. Parameters options from "Costs" area

Code and Parameter Solution selected	Constraints
P47 – The buildings surrounding are preserved.	B1
P48 – The building refurbishment costs was estimated in 600€/m² (plus VAT)	A16,C2,C6
P49 – These works were not covered by benefits, tax incentives, or other programs.	C2
P50 – The design project describes the forecast of reactive maintenance strategy with costs which could be larger in repairs of systems and equipment (repair after failure).	C2, C4, C5

The old building refurbishment works studied show a set of possible options [16] according to a selection from each management system parameters option possibilities present in table 2 and 3. Each parameter has 3 possible options, classified by levels from 1 (less sustainable) to 3 (more sustainable) [5]. The level 2 is considered with conventional options without increase sustainable benefits. Table 8 has a description of the management system thematic parameters by selection levels choices from each parameter.

Table 8. Levels of sustainability selection of the thematic parameters

Parameter level selection/option	Parameters
1 (less sustainable options)	P19, P34, P49
2 (conventional sustainable options)	P12, P13, P18, P20, P24, P30, P32, P33, P35, P36, P39, P43, P48, P50
3 (more sustainable options)	P14, P15, P16, P17, P21, P22, P23, P25, P26, P27, P28, P29, P31, P37, P38, P40, P41, P42, P44, P45, P46, P47

The case study only used 39 parameters from the management system predominating sustainability levels above conventional and better solutions to promote sustainability in 22 parameters. Conventional levels of sustainability and no surcharge for sustainability benefits were found in 14 parameters solution options. The solutions levels of sustainability used in parameters P19, P34 and P49 were less than conventional solutions.

Thus, according to the management system, the total options have been used to improve sustainability benefits in 56% of the parameters used. The detail and organization of the project design attending by designers and engineers team considered some sustainable solutions. In other hand, this old building project had so many constraints and a refurbishment operation with some particularities very specific to attend. The management system could be an important auxiliary by toolkit guide form to reducing failures, omissions and errors, deadlines control, improving quality level, costs control, managing risks and other unforeseen events [5, 17]. However, it requires some care and management specialized during old building refurbishment works because constraints are in considerable number with difficult resolution. But the result is a very beautiful building with external older aspect integrated in village architectural preservation and within internal comfort to their users, Figures 11 to 13.

Figure 11. Principal façade.

Figure 12. Internal building aspect.

Figure 13. Living room.

5. Conclusions

The old building in research study was no more than 120 years old without heritage classification which it is located in a Village of the Natural Park of Montesinho. All constraints and particularities were described in an exhaustive way, looking for solutions by the design project that enhances sustainability benefits and a more efficient management. However this refurbishment works shows a complex project to implement with solutions different than conventional ones such as the sidewall shoring, demolition works, comfort requirements, legal regulations imposed, rebuild stone walls and to keep a traditional context from exterior outside building. During design project develop the management system parameters were utilized from parameters P12 to P50, which including "project design", "construction works and site works" and "costs" areas.

In 56% (22 parameters) of used parameters were chosen more sustainable ideas than conventional ones, but in 14 parameters solutions (36%) the solutions were conventional without sustainable increasing. In different perspective some solutions were less than conventional ones and represent 8% (3 parameters). These solutions were required because specific building constraints which were impossible to change in a different way such as sidewall shoring, more structural reinforcement needs

IOP Conf. Series: Materials Science and Engineering **800** (2020) 012040 doi:10.1088/1757-899X/800/1/012040

and inexistence of benefits or tax incentives for refurbishment works in that time Portugal villages. Although many parameters were considered with conventional solutions for sustainability benefits and less that conventional ones, the majority parameters number were used in the design project with more sustainable solutions.

In other perspective, the study shows a balance between the realities of refurbishment works difficulties within pre-existences, technical requirements attending to legal regulations and also sustainable solution implementation in spite of the overall comprehensiveness of the whole project design. The management system was an important auxiliary during project phase, especially for design project phase, with a strong contribution for planning, budgeting and quality in building refurbishment works [17]. This toolkit is not obliged but their uses promote a reflection more consistent to take better support decisions based on particularities and unforeseen not only in the building existences but also in surroundings area, including lateral buildings, streets, traffic, site works, and many others forgotten frequently during design project phase by designers.

References

[1] Kibert C, 2012 *Sustainable Construction: Green Building Design and Delivery,* 3rd E John Wiley & Sons

[2] Appleton J 2011 Reabilitação de edifícios antigos - Patologias e tecnologias de intervenção, ed Orion (Amadora)

[3] Paiva J V, Aguiar J P, Pinho A, 2006 *Guia Técnico de Reabilitação habitacional*, ed Instituto Nacional da Habitação e Laboratório Nacional de Engenharia Civil (Lisboa)

[4] IPPAR 1995 *Informar para proteger-Critérios para classificação de imóveis*, ed IPPAR (Liboa)

[5] Oliveira R, Lopes J, Sousa H, Abreu I 2017 *A system for the management of old buildings retrofit projects in historical centres: The case of Portugal, International Journal of Strategic Property Management*, 21 (2), pp. 199-211, doi:10.3846/1648715X.2016.1251984

[6] Oliveira R, 2012 *Metodologia de gestão de obras de reabilitação em centros urbanos históricos, Unpublised PhD Thesis*, FEUP (Porto)

[7] Yin R, 2002 *Case Study Research, Design and Methods*, Sage Publications (Newbury Park)

[8] Fellows R and Liu A 2008 *Research methods for construction*, 3rd Ed, Wiley-Blackwell Publishing Lda (United Kingdom)

[9] Charles O E, Barbara A Y. and Victor B T 1996 *Refurbishment management practices in the shipping and construction industries - lessons to be learned*, Building Research & Information, 24:6, pp. 329-338, DOI:10.1080/09613219608727553.

[10] Sorace S, Terenzi G, 2013 *Structural assessment of a modern heritage building*, Engineering Structures 49 pp.743–755, https://doi.org/10.1016/j.engstruct.2012.12.012

[11] Kerzner H 2013 *Project management: A systems approach to planning scheduling, and controlling*, 11th Ed, Wiley (Canada)

[12] Ochoa C, Capeluto I 2015 *Decision methodology for the development of an expert system applied in an adaptable energy retrofit façade system for residential buildings*, Renewable Energy 78 pp.498–508, https://doi.org/10.1016/j.renene.2015.01.036.

[13] Jha K, Iyer K 2006 *Critical factors affecting quality performance in construction projects*, Total Quality Management 17(9): 1155–1170, https://doi.org/10.1080/14783360600750444

[14] Flick U, 2002 *Qualitative sozial forschung*, Rowohlt Taschenbuch Verlag GmbH, Reinbek bei (Hamburg)

[15] Mulliner E, Smallbone K, Maliene V 2013 *An assessment of sustainable housing affordability using a multiple criteria decision making method*, Omega 41(2) pp. 270–279, https://doi.org/10.1016/j.omega.2012.05.002

[16] Sdei A, Gloriant F, Tittelein P, Lassue S, Hanna P, Beslay C, Gournet R, McEvoy M 2015 *Social housing retrofit strategies in England and France: a parametric and behavioural analysis*, Energy Research & Social Science 10 pp. 62–71 https://doi.org/10.1016/j.erss.2015.07.001

[17] Murdock J and Hughes W 2008 *Construction Contracts – Law and Management*, 4th Ed, Taylor & Francis (New York)

IOP Conf. Series: Materials Science and Engineering **800** (2020) 012041 doi:10.1088/1757-899X/800/1/012041

The effect of stone column area replacement ratio on seismic behaviour of foundation

I Safkan[*], S Derogar, and J Anywar

Civil Eng. Department, European University of Lefke, Northern Cyprus, TR-10

**Email: isafkan@eul.edu.tr*

Abstract: Tall buildings require additional foundation solutions in order to resist the applied actions on soft soil sites. Nowadays the piled foundations are often altered by stone column solutions, as a cheap alternative. Structures built on soft soil or stiff soil, generally results in different behavior under earthquakes. However, the earthquake behavior of the stone column solution on different types of soil is still unclear. This paper aims to investigate the influence of different characteristics of soil structure interaction on the seismic response of stone column matt foundation. A reinforced concrete matt foundation was simulated with parametric soil stiffness and area replacement ratio of stone column. An analytical method was used to model the soil profile and appropriate ground motion data were incorporated for the analyses. The comparative results were presented and highlighted in terms of soil amplification and seismic response.

1. Introduction

Structures built on soft soil or rigid soil, generally interact in a different way under earthquakes. This was first observed in the earthquake disaster occurred in Mexico on the 19th and 21st of September 1985, two earthquakes with magnitude of 8.1 and 7.5 occurred at a part of the plate boundary besides the Mexican subduction zone prior known as the Michoacan gap [1]. A large city built on soft clay soil, approximately 360 kilometers away from the epicenter region had the earthquake effects, where it has left deep scares in the economy and the daily life in the city.

Champilo et al., [2] investigated the source characteristics and propagated waves from the epicenter to the Mexico City and the authors concluded that the earthquake catastrophe of Mexico City in 1985 was due to a large local amplification, caused by the extremely soft clay layer underlying the city. The importance of these events and of their consequences has brought an increased pressure to understand the behavior of Soil-Structure Interaction, in order to improve the seismic safety of multi-story structures.

Liquefaction during an earthquake is a major reason to causes hazardous damage to the buildings. Liquefaction damage prompts the soil to excessive settlement, loss of bearing capacity, sand boils, slope failure and landslides, lateral spread, etc. Structures built in seismic active regions are susceptible to experience liquefaction of the underlying soil caused by excessive pore pressure development. The two most commonly considered soil improvement techniques in practice are stone columns and piled foundations. Among the different improvement techniques, the most popular technique adopted for liquefiable soil improvement is the installation of stone columns. Yoshimi and Kuwabara [3] suggested the use of stone columns as an improvement for liquefaction hazards for the first time. Stone columns generally improve stiffness, permeability and strength properties of stone columns. A considerable number of research has been done regarding to the performance of stone columns to improve soil liquefaction. Seed and Booker [4] proposed a stone column model designed for liquefiable soil improvement by assuming the permeable structure of the stone column. Non dimensional charts have been introduced for the determination of the spacing of stone columns. Sasaki and Taniguchi [5] performed an experimental study for the effective area replacement ratio. Yang and KO [6] observed that the soil lacking drainage has more potential on liquefaction.

A study conducted by Asgari et al., [7], focused on stone column and pile-pinning methods, comparing the behavior of each method in retaining permanent seismic deformations. However, different stiffness properties of soil were not investigated.

This study, parametrically assess, different stiffness properties of clay on parametric stone column area replacement ratio and corresponding lateral behaviour. Relative difference in between the type or size of the foundation will be discussed in this paper.

2. Method

2.1. Modelling substructure Substructure was modeled by using the OpenseesPL software [8]. Parametric stone column foundation models were analyzed considering different soil conditions such as loose and dense clay. Table 1 below provides the soil parameters of clay utilized in the dynamic analyses.

Table 1. Soil parameters of clay

Parameters	Loose clay	Medium loose clay	Medium clay	Dense
Mass density (ρ) (kg/m^3)	1620	1870	2090	2130
Shear wave velocity (m/s)	150	300	500	700
Cohesion (c) (kPa)	22	32	55	75

*2.1.1 Stone column:*The main constitutive soil modeling parameters used in this study of stone column, for: loose clay, medium loose clay, medium clay, medium dense clay and dense clay, are the main constitutive soil modeling parameters used in the study of stone column which are represented in Table 1, with the soil layer having a total depth of 10 *m*.

Nowadays, the stone columns are preferred as an economical solution for the stabilization of underlying soil. Stone column case utilizes dense stabilization material property for but as column shape and with different area replacement ratio (ARR). Area replacement ratio of 5%, 10%, 25%, 50% and 100% were utilized for modelling in this study.

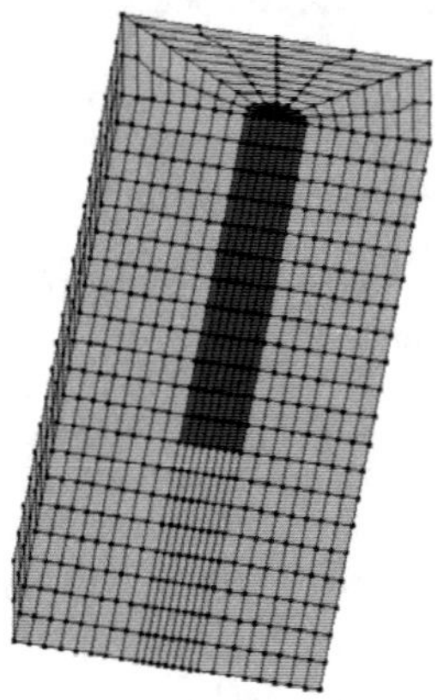

Figure 1. Stone Column Cell finite element model [8].

A pressure independent multi-yield elastoplastic material model [8] was used for modelling clay material. On the other hand, the characteristics of a pressure depended multi-yield material (See Fig. 2) require modelling the regular medium dense soil by replicating drained soil conditions. For the purpose of simulating a soil under fully undrained condition, the material should either be fitted in a fluid solid porous material, or with a solid-fluid coupled element with a low permeability. For the simulation of partially drained soil behavior, the pressure dependent multi-yield material must be coupled with a solid-fluid fully integrated element with the appropriate permeability characteristics [8].

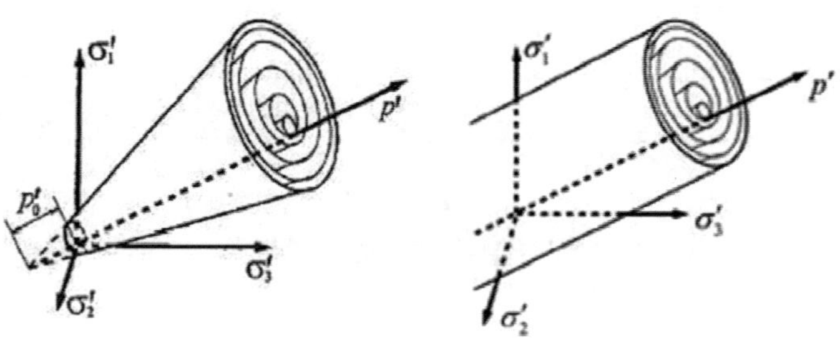

Figure 2. Pressure dependent multi-yield hysteretic model (Left) and pressure independent multi-yield hysteretic model (Rigth). (Opensees 2019)

Modal response difference was obtained for the system where relatively significant difference was obtained for different stone column area replacement ratios. In total 2 strong motion data were utilized for assessing the influence of foundation type on superstructure. Table 3 gives the statistics of selected ground motions. The relatively high peak ground acceleration characteristics of these two motions are considered to influence the amplification characteristics of soil in nonlinear behavior at broader spectrum range.

Table 2. Statistics of the utilized ground motions.

Earthquake	Year	Length of record (s)	Magnitude
El Centro	1940	31	7.1
Chile, Llolleo	1985	116	7.8

3. Results and discussions

The analyzed ground motions` frequency content were especially magnified at short period range. The longer period range for the Chile earthquake was monitored to not to change in a significant manner. On the other the El Centro data was observed to damp the energy on long period range when 4m deep embedment length was considered.

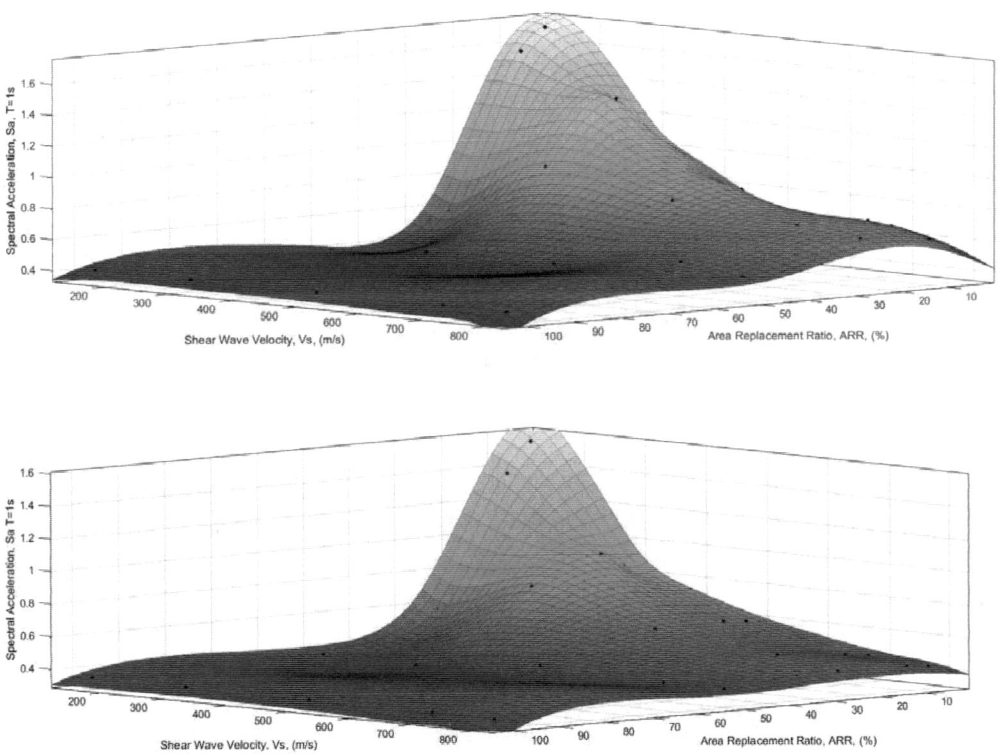

Figure 3. Long period spectral acceleration amplification respect to ARR of stone column and Vs conditions.

As result of analyses, a multidimensional relationship were developed in between ARR of stone column, Vs of surrounding soil and PGA (Fig. 5) When loose soil condition (Eg. Vs< 200) was considered, significant amplification of ground acceleration was observed for ARR<25%. Although the reduction was gradual for stiffer soil conditions, there is a jump at Vs: 500m/s and ARR: 20%. This may be argued as the secondary mode effects of stone column due to the specific stiffness-mass ratio. When compared with the code specified acceleration amplitudes at same period range, amplification safe region starts at 40-50% ARR range for the soft clay sites. A further study, assessing the structural behavior can be beneficial for this comparison.

Lateral maximum displacement of soil has considerably high influence on seismic demand especially on tall buildings with long period. A parametric study was conducted on clay profiles with different stiffness properties respect to different stone column area replacement ratios. The results (See Fig. 4) show insignificant difference between 50% and 100% ARR configurations. However, soil profile with 5% and 10% ARR resulted 37% more lateral displacement respect to 50% ARR profile. On the other hand insignificant difference on response was monitored for the medium and dense clay soil conditions. Furthermore, the difference in between the 50% and a100% ARR was also found to be negligible.

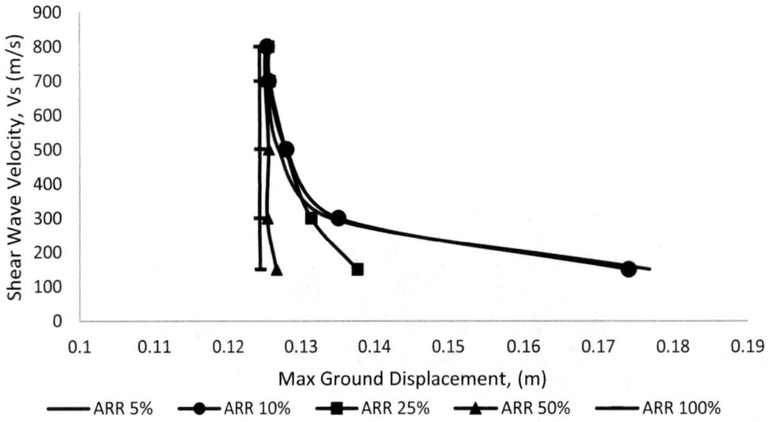

Figure 4. Comparison of ground displacement respect to ARR, Chile earthquake

On the other hand, no significant amplification difference was obtained for the firm soil conditions. This may be due to the insignificant difference at long period range of response spectra. Similar trend was also observed on firm soil conditions, where no amplification of acceleration characteristics were monitored.

4. Conclusion

In this study, the effect of stone column on seismic behavior of loose clay was investigated. Current practice on designing stone columns is usually limited to static bearing capacity estimates. However, the results indicate the importance of modelling stone columns based on seismic behavior. In addition, the effect of different soil types on the amplification or damping of spectral amplitudes were studied. Significantly high amplifications at longer periods were monitored for the ARR less than 25%. Finally, ARR equal to 50% was monitored to give the best response to loose soil under earthquake loading.

References

[1] Esteva, L. (1988). The Mexico earthquake of September 19, 1985 -- Consequences, lessons, and impact on research and practice. *Earthquake Spectra*, 4:413-426

[2] Campillo, M., Gariel, J. C., Aki, K. & Sfinchez-Sesma, F. J. (1989). Destructive strong ground motion in Mexico City: Source, path, and site effects during the great 1985 Michoacfin earthquake, *Bull. Seism. Soc.* Am., 79:1718 35.

[3] Yoshimi, Y., & Kuwabara, F. (1973). Effects of subsurface liquefaction on the strength of surface soil. *Soils Found.* 13, 67–81.

[4] Seed, H. B., & Booker, J. R. (1977). Stabilization of potentially liquefiable sand de- posits. *J. Geotech. Eng. Div.,* 103(7), 757–768.

[5] Sasaki, Y., & Taniguchi, E. (1982). Shaking table tests on gravel drains to prevent liquefaction sand deposits. *Soils Found.* 22(3), 1–14.

[6] Yang, T. F., & KO, H.-Y. (1998). Reduction of excess pore-water pressure by the gravel drainage method during earthquakes. *Centrifuge '98*, T. Kimura, O. Ku- sakabe, and J. Takemura, eds., A. A. Balkema, Rotterdam, Netherlands, 301–306.

[7] Asgari, A., Oliaei, M. and Bagheri, M. (2013). Numerical simulation of improvement of a liquefiable soil layer using stone column and pile-pinning techniques. *Soil Dynamics and Earthquake Engineering* (51): 77–79.

[8] Lu, J., Elgamal, A., & Yang, Z. (2011). OpenSeesPL: 3D lateral pile-ground interaction user manual (Beta 1.0). Department of Structural Engineering, University of California, San Diego. http://soilquake.net/openseespl/OpenSeesPL_UserManual.pdf

Pipe type and seismic performance in Christchurch, New Zealand

S Toprak[1], E Nacaroglu[2], M Ceylan[3], G Dundar[4], O Y Cirmiktili[5]

[1]Gebze Technical University, e-mail: stoprak@gtu.edu.tr
[2]Pamukkale University, e-mail: enacaroglu@pau.edu.tr
[3]Pamukkale University, e-mail: mceylan121@posta.pau.edu.tr
[4]Pamukkale University, e-mail: gdundar121@gmail.com
[5]Pamukkale University, e-mail: omercirmiktili@hotmail.com

Abstract. The pipeline damages observed in Christchurch, New Zealand during the 2010-2011 earthquakes provided new insights regarding the seismic performance of pipelines. The water pipeline damages were extensive and water service was interrupted for a long period of time. Meanwhile, unprecedented efforts have been underway following the period right after the first earthquake up to now to obtain and store data related to entire damaged structures, geology, soil conditions, seismicity etc. in the area for the recovery and future planning purposes. These data have been very crucial to study, understand and apply the lessons learnt from these earthquakes and get ready for future earthquakes. This paper focuses on seismic performance of pipelines regarding the pipe type during the M_W 6.2 22 February 2011 Christchurch earthquake. Majority of pipeline damages occurred in soil liquefaction areas where differential settlements and lateral strains affected the pipelines adversely. In addition, liquefaction severity parameters such as liquefaction potential index (LPI) and liquefaction severity number (LSN) have been considered in the evaluations of damages. Pipelines in Christchurch consisted of different pipe materials and their performance during the earthquake differed significantly. Geographical information systems (GIS), clustering analyses (fuzzy and k-means methods) and reliability aspects were used in their assessment.

1. Introduction

Water distribution pipelines are critical part of the infrastructure systems in urban areas. They transmit water for firefighting, drinking, cleaning, industrial and many other uses. Functionality of these water distribution systems or getting them back into the service in short times after the earthquakes are essential for the society. The only way to achieve this objective is to evaluate the systems before any earthquake occurs against seismic actions and take necessary steps to eliminate the deficiencies in the systems. The evaluations have to take into consideration whole aspects the condition of the system, the needs before and after the earthquake, the level of seismic loading, and soil conditions.

The Canterbury earthquake sequence which occurred in 2010-2011 and the respective damage and performance database of Christchurch City has been one of the best documented cases for the earthquakes up to today. The database includes but not limited to seismicity of the area [1], strong ground motions [2], pipeline systems [3], [4], soil investigations [5] and buildings [6]. Toprak presented ground displacement measurements from air photo, satellite, high resolution light detection and ranging (LiDAR), and ground surveys data with respect to their effects on pipeline damage assessments [7]. It is expected the results of the investigations and research related to these databases create new approaches, methodologies and applications to get ready for the future earthquakes. The research presented in this paper focus on buried pipeline systems and their performance during the 2011 Christchurch earthquake.

2. Christchurch pipeline system and earthquake effects

The water pipelines in Christchurch can be considered in two groups as main and submain pipelines. Main and submain pipelines are approximately 1700 km and 2000 km, respectively. The diameter range of main pipelines is between 75-600 mm and conveys the main portion of the water in the system. Because of some reliability problems on submain repairs, only main pipelines and repairs were considered in this study [8]. Figure 1 shows main pipelines in Christchurch. The pipelines are located almost exclusively on road corridors. As described by Cubrinovski [9], [10], the pipelines are laid in trenches 200-300 mm wider than the pipe diameter, at relatively shallow depths. The cover thickness depends on the pipe size, location and material, but is usually about 800mm (at least 750mm, but no more than 1.5m for the standard water mains diameters). Pipe trench characteristics have changed over time. From the first development of the network, trench backfill material was either the locally excavated soil material if soil particle size was appropriate (sands through to fine gravels), otherwise fine gravel material was imported from quarries situated on the city margins. From 1984 trench construction was standardized so that all pipes were emplaced in a sand layer covered with AP40 gravel mix. There was a program of backfill compaction testing to identify the optimum mix to prevent trench settling, which was impacting road surfaces. From 2005 an AP20 gravel mix has been used, as it was determined to be less abrasive and damaging to pipe materials than the AP40.

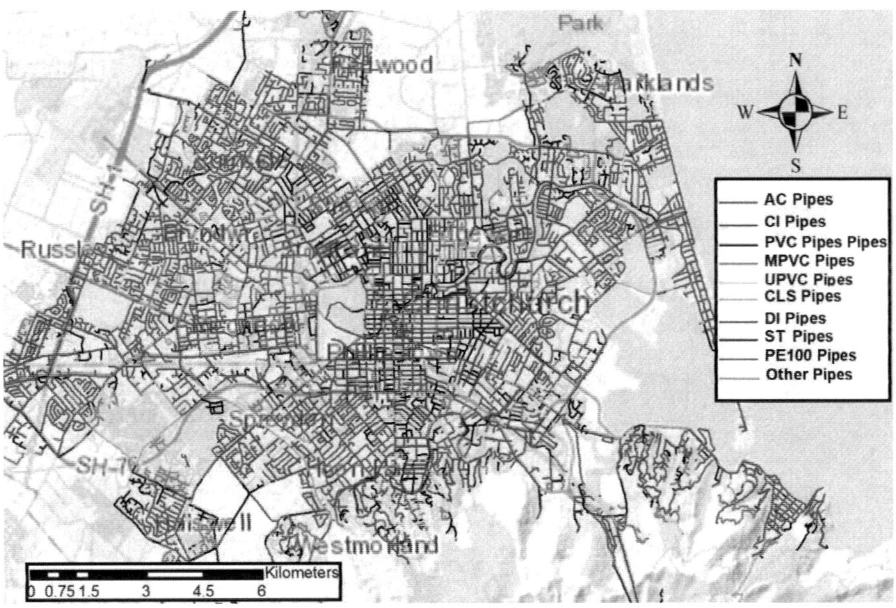

Figure 1. Pipelines in Christchurch with respect to pipe material

The water distribution system consists of various pipe materials. Figure 1 shows geographical distribution of Christchurch water main lines with respect to pipe material by the time the 22 February 2011 earthquake took place. Almost half of the pipelines are asbestos cement (AC) pipelines. Other major pipe types are cast iron (CI), polyvinyl chloride (PVC), modified polyvinyl chloride (MPVC), and unplasticised polyvinyl chloride (UPVC). About 9 % of the pipelines are concrete lined steel (CLS), ductile iron, concrete lined ductile iron (CLDI), steel (ST), and polyethylene 100. Less than one percent of the pipelines consists of polyethylene (PE), medium density polyethylene 80

(MDPE80), medium density polyethylene 100 (MDPE100), high density polyethylene (HDPE), mortar lined ductile iron (MLDI), reinforced concrete rubber ring (RCRR), and galvanized iron (GI).

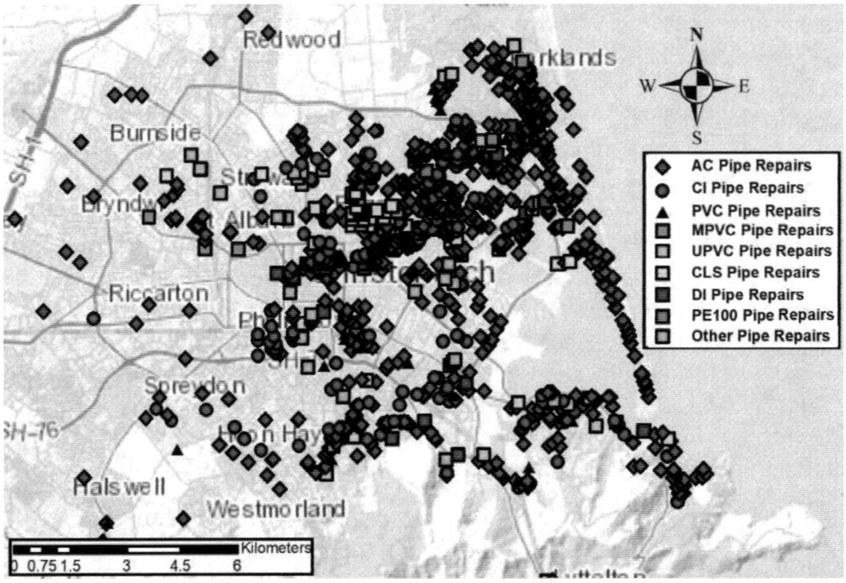

Figure 2. Pipeline repairs in Christchurch with respect to pipe material

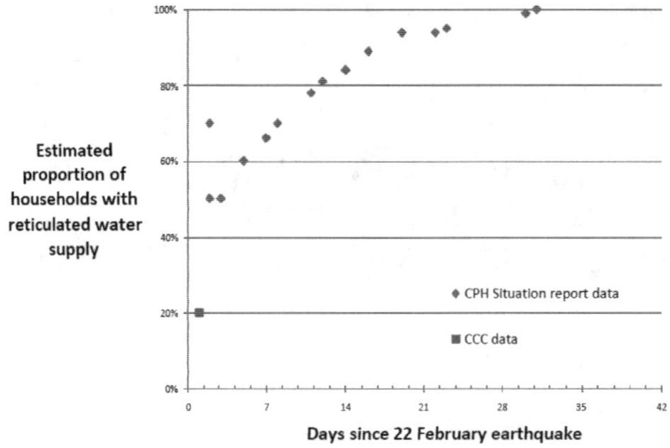

Figure 3. Water service recovery after the 22 February 2011 earthquake. [11]

The 22 February 2011 earthquake caused substantial damage to the pipelines (Figure 2). The total number of repairs to main pipelines was 1502. About two thirds of the damages occurred on AC pipelines. As shown in Figure 3, although service was provided to about half of the households a few days after the earthquake, the recovery took significant time. The City applied "boil water notice" to whole Christchurch between February 23rd and April 8th, 2011 to protect residents against disease. The "boil water notice" is about to treat tap or tanker water by boiling before any use for drinking, brushing teeth or preparing food.

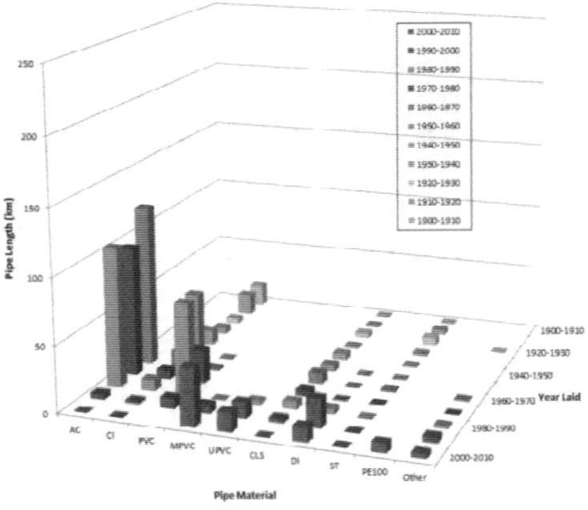

a) Length Distribution of Pipelines with respect to pipe material and year of construction

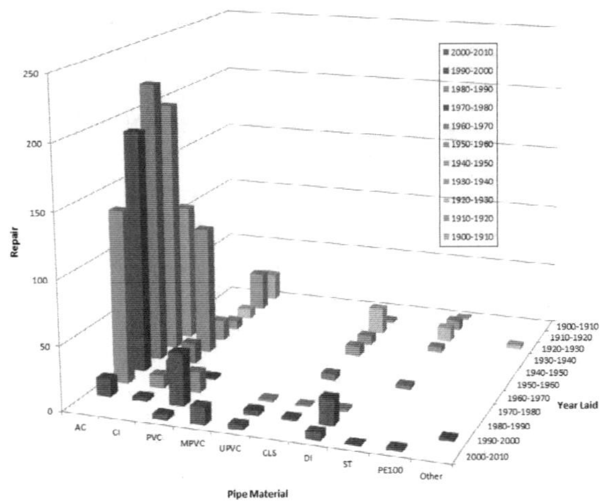

b) Number of repairs corresponding to pipe material and year of construction

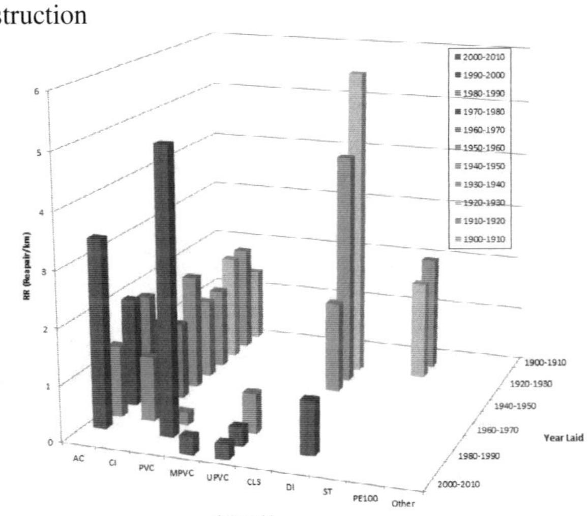

c) Repair rate corresponding to pipe material and year of construction

Figure 4. Pipelines in liquefaction zone with respect to pipe material and year of construction

3. Evaluation of Performance

One of the main characteristics of the 22 February 2011 earthquake was the extent of the soil liquefaction and its damage in Christchurch. Buried pipelines like buildings were affected by extensive liquefaction. Because the seismic actions are different inside and outside the liquefaction zones, their analyses performed separately. Pipelines in the liquefaction areas are affected by permanent ground deformations such as lateral or vertical displacement whereas the pipelines out of liquefaction zones are subject to ground shaking effects. Therefore, both cases should be considered separately for the assessment of pipeline damages.

Geographical information systems (GIS) were utilized during the preparation of the database and analyses stage. Figures 4a, 4b, and 4c show, respectively, the length in km, number of repairs and repair rate of pipelines in the liquefaction zone with respect to pipe material and year of construction.

Most of the pipelines were constructed after 1940s. Installments of AC and CI pipelines stopped even before the earthquake. Polyethylene pipelines were the preferred pipe type for the replacements after the earthquake. Repair rate (RR) is commonly used in pipeline damage assessment studies and shows the number of repairs divided by the length of the pipelines in the same area. In essence, the number of repairs are normalized by length of pipelines. It should be noted that not all RR values are shown in Figure 4c, because a screening criteria was applied. The fidelity of the RR statistics is sensitive to the pipeline length sampled and number of repairs observed within a given sampling length. Hence, in this study, the screening criteria developed by O'Rourke [3] were used during the RR analysis of pipeline damage to ensure fidelity of the RR comparisons.

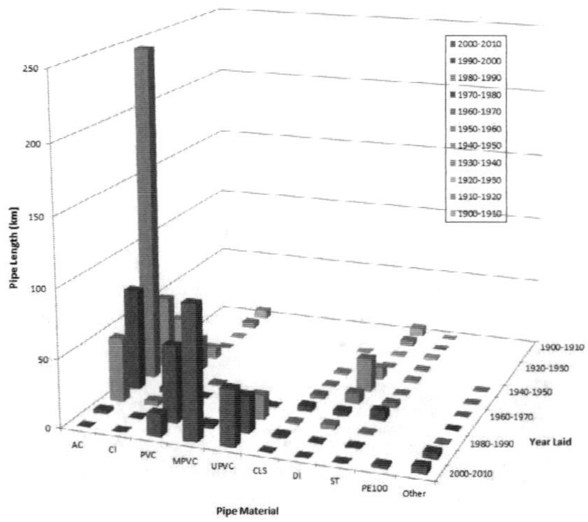

a) Length Distribution of Pipelines with respect to pipe material and year of construction

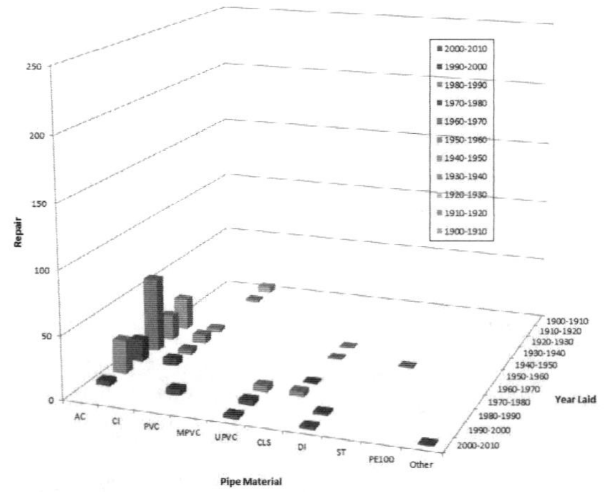

b) Number of repairs corresponding to pipe material and construction year of pipeline

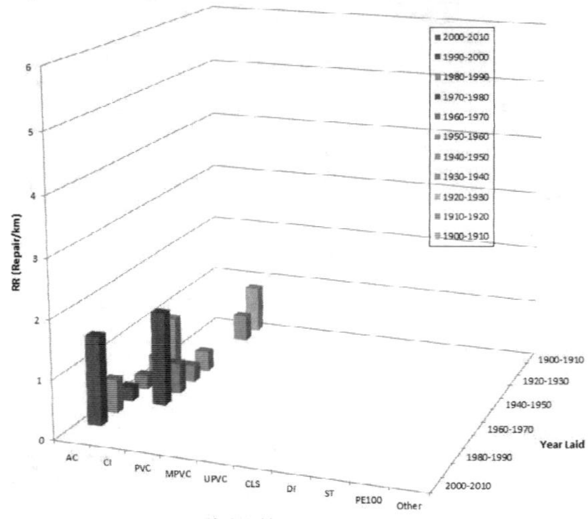

c) Repair rate corresponding to pipe material and construction year of pipeline

Figure 5. Pipelines in non-liquefaction zone with respect to pipe material and year of construction

Figures 5a, 5b, and 5c show, respectively the length in km, number of repairs and repair rate of pipelines in the non-liquefaction zone with respect to pipe material and year of construction. It should

IOP Conf. Series: Materials Science and Engineering **800** (2020) 012042 doi:10.1088/1757-899X/800/1/012042

be noted that Figure 5c shows the RR values only for the pipelines which pass the screening criteria explained above. Comparison of figures clearly shows that the repair rates are much higher in the liquefaction zone. Also AC pipelines performed worse compared to other pipelines. The spatial trends and pipe material effects in pipeline repairs can be analyzed by using clustering techniques. This method previously applied to City of Los Angeles water pipelines damaged by the 1994 Northridge earthquake [12]. An application of the k-means clustering analyses to Christchurch water pipeline damage data provided promising results [13]. Another method, fuzzy c-means clustering technique, is also one of the powerful methods which has been utilized in the analyses of pipeline damage in Christchurch (Figure 6).

Although pipe material type and pipeline instalment year are important for the seismic performance of pipelines, they are not the only factors that control the pipe behavior. The magnitude and the extent of seismic loading can be critical as well. The level of shaking in non-liquefaction zones and the level of ground deformations in liquefaction zones can force the pipeline to failure. For example, O'Rourke [3] and Toprak [7] utilized lateral strains and ground deformation parameters when developing pipeline damage correlations in liquefaction zones and ground velocity parameter (GMPGV) when developing pipeline damage correlations in non-liquefaction zones. Toprak [4] proposed the first time the use of liquefaction severity parameters in pipeline damage correlations and provided respective relationships. Those relationships are expected to contribute to loss estimation studies for urban areas. Figure 7 shows such a relationship only for one of the liquefaction vulnerability parameters, liquefaction severity number (LSN).

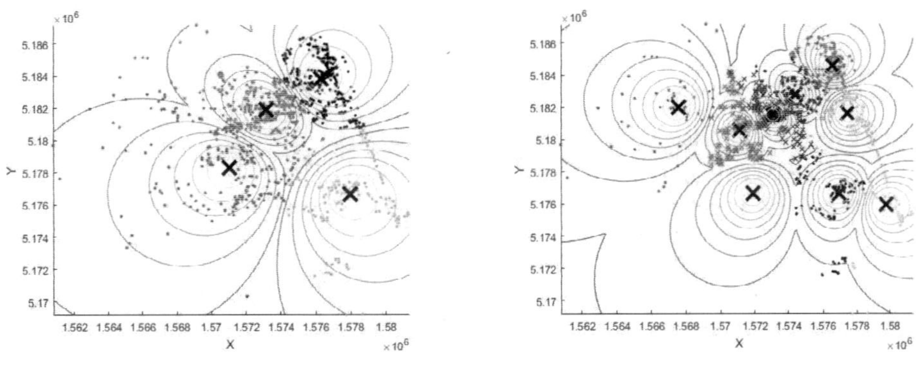

a) 4 clusters b) 9 clusters

Figure 6. Fuzzy c-means clustering application to Christchurch pipe repairs.

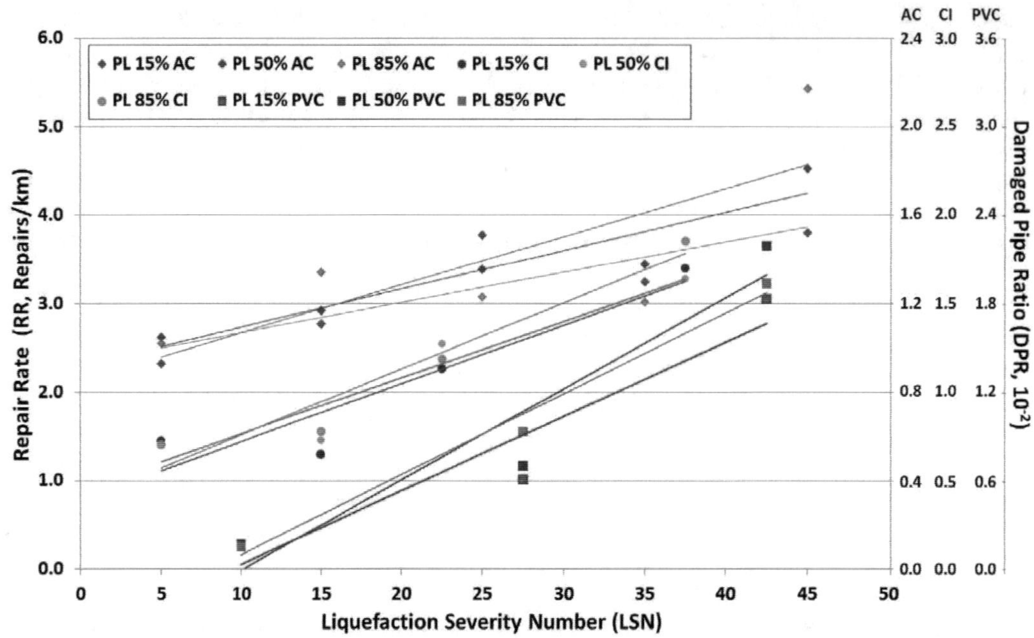

Figure 7. RR and DPR vs. LSN relationships for AC, CI, and PVC pipelines and probabilities of liquefaction of 15%, 50%, and 85%. [4]

4. Conclusions

The pipeline damages occurred during the Canterbury earthquake sequence provided unprecedented opportunity to evaluate seismic pipeline performance particularly in liquefaction zones and make recommendations for future applications. Within the scope of the research activities of the authors of this paper and collaborators, characteristics and performances of the damaged and undamaged pipelines have been investigated both in liquefied and non-liquefied areas in Christchurch and new pipeline damage correlations were proposed. The results show that AC pipelines have the highest damage rates, significantly higher than that for PVC pipelines. CI damages are about the average of AC and PVC pipelines. The ongoing studies focus on probabilistic and reliability assessment of pipelines along with clustering applications.

Acknowledgments

The Scientific and Technological Research Council of Turkey (TUBITAK) supported the research presented herein under Project No. 114M258. PAU BAP support was under Project No. 2019FBE013. The Christchurch Earthquake Recovery Authority (CERA), Stronger Christchurch Infrastructure Rebuild Team (SCIRT), Christchurch City Council (CCC), Earthquake Commission (EQC), Contact Energy are acknowledged for their assistance.

References

[1] GNS Science 2011 The Canterbury earthquake sequence and implications for seismic desing levels *GNS Science Consultancy Report 2011/183* commissioned by the Canterbury Earthquakes Royal Commission Wellington NZ

[2] Bradley B A, Quigley M C, Van Dissen R J, Litchfield N J 2014 Ground motion and seismic source aspects of the Canterbury earthquake sequence *Earthquake Spectra* **30** No 1 Pages 1-15
[3] O'Rourke T D, Jeon S S, Toprak S, Cubrinovski M, Hughes M, Ballegooy S and Dimitra B 2014 Earthquake response of underground pipeline networks in christchurch NZ *Earthquake Spectra* **30** No 1 Pages 183-204
[4] Toprak S, Nacaroglu E, Van Ballegooy S, Koc A C, Jacka M, Manav Y, Torvelainen E and O'Rourke T D 2019 Segmented pipeline damage predictions using liquefaction vulnerability parameters *Soil Dynamics and Earthquake Engineering* 125 105758 ISSN 0267-7261 https://doi.org/10.1016/j.soildyn.2019.105758
[5] Van Ballegooy S, Wentz F, Boulanger R W 2015 Evaluation of CPT-based liquefaction procedures at regional scale *Soil Dynamics and Earthquake Engineering* 79 B Pages 315-334 ISSN 0267-726 https://doi.org/10.1016/j.soildyn.2015.09.016
[6] Holmes W T, Luco N and Turner F 2014 Application of the recommendations of the canterbury earthquakes royal commission to the desing, construction and evaluation of buildings and seismic risk mitigation policies in the United States *Earthquake Spectra* **30** No 1 Pages 427-450
[7] Toprak S, Nacaroglu E, Koc A C, O'Rourke T D, Hamada M, Cubrinovski M, Ballegooy S 2018 Comparison of horizontal ground displacements in Avonside area, Christchurch from air photo, LİDAR and satellite measurements regarding pipeline damage assessment *Bull. Earthq. Eng.* **16** No 10 Pages 4497-4514 https://doi.org/10.1007/s10518-018-0317-9
[8] Bouziou D, Van Ballegooy S, Storie L, O'Rourke T D 2019 Spatial correlation of underground pipeline damage in Christchurch, Correlations with liquefaction-induced ground surface deformations and CPT-based liquefaction vulnerability index parameters Quake Center EQC
[9] Cubrinovski M, Hughes M, Bradley B, McCahon I, McDonald Y, Simpson H, Cameron R, Christison M, Henderson B, Orense R and O'Rourke T 2011 Liquefaction impacts on pipe networks, Recovery Project No 6 *Natural Hazards Research Platform Christchurch* (New Zealand: University of Canterbury)
[10] Cubrinovski M, Hughes M, Bradley B, Noonan J, Hopkins R, McNeil S, English G 2014 Performance of horizontal infrastructure in Christchurch city through the 2010-2011 Canterbury earthquake sequence *Research Report 2014-02 Christchurch* (New Zealand: Civil & Natural Resources Engineering University of Canterbury)
[11] Ministry of Health 2012 *Annual report on drinking-water quality 2010-2011* ISSN 1179-2604 (Online) https://www.health.govt.nz/system/files/documents/publications/annual-report-on-drinking-water-quality-2010-2011.pdf
[12] Toprak S, Nacaroglu E, Cetin O A, Koc A C 2009 Pipeline damage assessment using cluster analysis *In TCLEE 2009: Lifeline Earthquake Engineering in a Multihazard Environment* Pages 1-8
[13] Nacaroglu E, Toprak S, Dundar G, Ceylan M 2019 Investigation of pipeline damages by k-means clustering analysis method *ENGGEO2019: National Symposium on Engineering Geology and Geotechnics* (Turkey: Pamukkale University) Pages 631-638

AUTHOR INDEX

Abdeh, Amr 226
Abreu, M I 347
Abubaker, A 8
Akbas, S O 28
Akgün, M 208
Akpinar, P 1, 192
Al Attar, H M E 1
Ali, S I A 8, 18
Aljamal, J 234
Alshetwi, A B Z 8
Anafarta, D 28
Anywar, J 360
Atalar, C 202, 208
Ates, S 36
Aydin, E 47, 59
Ayva, B 248
Balkis, A P 341
Barros, A 69
Bekkouche, S Rehab 79
Benayoun, F 79
Bendada, L 79
Bengar, H A 143
Bilsel, H 324, 341
Bilsel, N 331
Bodnarova, L 186
Bok, M W 261
Bostanci, S C 89
Boumezerane, D 79
Celebi, E 36
Celikag, M 125, 143
Ceylan, M 366
Cirmiktili, O Y 366
Dawoud, Osama 98
Derogar, S 106, 360
Dindar, H 208
Doleželová, M 117
Dundar, G 366
Eissa, M 125
Eliášová, M 177
Elturki, F A A 18
Eren, O 153
Farah, R E 131
Gamadze, N 138
Gatabi, H R 143
Goktepe, F 36
Gökçekus, H 234
Gönenç, T 208
Habib, A 153, 170
Habib, M 160
Hela, R 186
Hofmanová, L 278
Houseen, Q Al Haj 192
Hussain, A 202
Hána, T 177
Ipek, Y 208
Iravanian, Anoosheh 226
Iravanian, A 215
Jibrel, S N 18
Kassem, Y 234
Kesten, S 248
Kirtel, O 36
Koppelhuber, D 271
Koppelhuber, J 261
Krejsová, J 117
Kucera, T 278
Kuda, D 290
Kunt, M M 170
Laama, I F G 297
Landy, M F B 310
Leto, A D 319
Lopes, J 347
Machalická, K V. 177
Magg, A 271
Matouk, A 160
Miftah, A 324
Misirlisoy, D 331
Nacaroglu, E 366
Naji, J 341
Nalbantoglu, Z 131
Nunes, E 69
Odabas, B 106
Oliveira, R A F 347
Onay, T 297
Ozdag, Ö C 208
Ozdal, D 297
Pamuk, E 208
Petrícková, M 290
Pirgalioglu, S 138
Romero, F 310
Safkan, I 106, 331, 360
Scheinherrová, L 117
Shlash, A 215
Sousa, S 69, 310
Tirkolaei, H Khodadadi 324
Toprak, S 366
Turk, E 28

Umut, Ö ... 248
Vimmrová, A ... 117
Vokác, M ... 177
Yildirim, U ... 153
Zulfikar, A C ... 36